Essential Stability Theory

Since their inception, the Perspectives in Logic and Lecture Notes in Logic series have published seminal works by leading logicians. Many of the original books in the series have been unavailable for years, but they are now in print once again.

Stability theory was introduced and matured in the 1960s and 1970s. Today stability theory influences and is influenced by number theory, algebraic group theory, Riemann surfaces, and representation theory of modules. There is little model theory today that does not involve the methods of stability theory.

In this volume, the 4th publication in the Perspectives in Logic series, Steven Buechler bridges the gap between a first-year graduate logic course and research papers in stability theory. The book prepares the student for research in any of today's branches of stability theory, and gives an introduction to classification theory with an exposition of Morley's Categoricity Theorem.

STEVEN BUECHLER works in the Department of Mathematics at the University of Notre Dame, Indiana.

PERSPECTIVES IN LOGIC

The *Perspectives in Logic* series publishes substantial, high-quality books whose central theme lies in any area or aspect of logic. Books that present new material not now available in book form are particularly welcome. The series ranges from introductory texts suitable for beginning graduate courses to specialized monographs at the frontiers of research. Each book offers an illuminating perspective for its intended audience.

The series has its origins in the old *Perspectives in Mathematical Logic* series edited by the Ω-Group for "Mathematische Logik" of the Heidelberger Akademie der Wissenschaften, whose beginnings date back to the 1960s. The Association for Symbolic Logic has assumed editorial responsibility for the series and changed its name to reflect its interest in books that span the full range of disciplines in which logic plays an important role.

For more information, see www.aslonline.org/books_perspectives.html

PERSPECTIVES IN LOGIC

Essential Stability Theory

STEVEN BUECHLER

University of Notre Dame, Indiana

ASSOCIATION FOR SYMBOLIC LOGIC

CAMBRIDGE
UNIVERSITY PRESS

CAMBRIDGE
UNIVERSITY PRESS

University Printing House, Cambridge CB2 8BS, United Kingdom

One Liberty Plaza, 20th Floor, New York, NY 10006, USA

477 Williamstown Road, Port Melbourne, VIC 3207, Australia

4843/24, 2nd Floor, Ansari Road, Daryaganj, Delhi – 110002, India

79 Anson Road, #06–04/06, Singapore 079906

Cambridge University Press is part of the University of Cambridge.

It furthers the University's mission by disseminating knowledge in the pursuit of education, learning, and research at the highest international levels of excellence.

www.cambridge.org
Information on this title: www.cambridge.org/9781107168398
10.1017/9781316717257

Association for Symbolic Logic
Richard A. Shore, Publisher
Department of Mathematics, Cornell University, Ithaca, NY 14853
http://www.aslonline.org

A catalogue record for this publication is available from the British Library.

ISBN 978-1-107-16839-8 Hardback

To my wife, Sally, and our children,
Ian, Jessica, Joel and Jeff

Table of Contents

Preface

This book grew out of lectures to graduate students in logic at the University of Notre Dame in the academic year 1992-93. The purpose of the course was to bridge the gap between the model theory in a first year graduate logic course (say, the first two chapters in Chang-Keisler) and research papers in stability theory. While the most basic definitions in model theory are repeated in Chapter 1, realistically, I expect the reader to have completed an introductory course in mathematical logic. My intention in writing this book was not to give a comprehensive treatment of elementary stability theory, but to get the student through the basics as quickly as possible. It was also written (hopefully) so that a well-prepared student can begin the book at a chapter appropriate to his or her needs.

Stability theory began in the early '60's with Morley's Categoricity Theorem. (See Section 3.1.) In the period 1965 to 1982 (or so) virtually all attention focused on Shelah's work on (and eventual proof of) Morley's Conjecture. The spectrum function of a complete first-order theory T is a map $I(-,T)$ such that for any cardinal λ, $I(\lambda,T)$ is the number of models of T of cardinality λ. In the late '60's Morley conjectured that the spectrum function of a complete countable first-order theory T is nondecreasing on uncountable cardinals; i.e., for all uncountable cardinals $\lambda < \kappa$, $I(\lambda,T) \leq I(\kappa,T)$. Shelah's proof of this conjecture spanned almost 15 years and is the main topic of [She90]. Part of the proof is the development of the forking dependence relation on a stable theory (see Section 5.1). (Examples of this relation are linear dependence in a vector space and algebraic dependence in an algebraically closed field.) The theme through much of Shelah's book is to find subsets of a model of a stable theory on which forking dependence is nice enough to admit a dimension theory (Sections 5.6 and 6.3). (Exactly what is meant by a well-behaved dimension theory is discussed in Section 6.3). Good summaries of this material can be found in [Har93], [She85] and [Hod87].

If I was writing in 1980 the analysis of dependence relations leading up to the proof of Morley's conjecture would be the sole theme of this book. However, in the past 15 years stability theory has grown dramatically in another direction. During the 1980-81 Jerusalem logic year researchers tried to understand Boris Zil'ber's work on the conjecture that a theory categorical in every infinite cardinal is not finitely axiomatizable. It was around this time that Zil'ber completed his proof of the conjecture and Cherlin, Harrington and Lachlan created an independent proof. This work is generally recognized as the birth of geometrical stability theory. (There are earlier results by Zil'ber and others that are properly placed in geometrical stability theory, however, they were not widely known, understood, or seen as closely related at the time.) From here the area took off through further research by the above four mathematicians, Poizat (on stable groups [Poi87]) and myself (see Section 6.2). The entry of Hrushovski around 1984 deepened the area at an tremendous rate. The underlying theme in much of geometrical stability theory is the characterization of certain critical subsets of a model as (essentially) modules or algebraically closed fields. This theme is found throughout Chapter 4 and Section 6.2.

I will not attempt to find a unifying thread in geometrical stability theory and Shelah-style classification theory for two reasons. First, stability theory is growing too rapidly for anyone to come forward and say "this is what it's all about". Secondly, I do not think an all-encompassing theme would be helpful to the reader at the level of this book.

Chapter 1 is simply a quick summary of the prerequisite definitions and theorems. Chapter 2 is a treatment of the classical first-order model theory relevant to stability theory. (Here classical model theory means, with a few exceptions, the model theory that existed prior to Morley's work on his categoricity theorem, circa, 1962.) Morley's Categoricity Theorem is proved in Section 3.1. The student with a good background in classical model theory (Chapters 1 and 2 in [CK73], for example) can begin the book here after absorbing the material on Cantor-Bendixson rank in Section 2.2. The dependence relation induced by Morley rank on a totally transcendental theory is developed in Section 3.3. This theory is applied in the remainder of Chapter 3 to prove the Baldwin-Lachlan Theorem and introduce $\omega-$stable groups.

Geometrical stability theory in an uncountably categorical theory is developed in Chapter 4. This is a good introduction to the area of geometrical stability theory; most of the key concepts are at least mentioned. The deepest results in the chapter are Zil'ber's Ladder Theorems (see Section 4.4) and the group existence results of Section 4.5.

In Chapter 5 we jump to stable theories in general. The forking dependence relation is developed "from scratch" in Section 5.1. The reader with a good understanding of model theory can begin the book in this section, provided she or he has mastered universal domains (Section 3.2) and T^{eq} (Section 4.1). There is very little in the first three sections of this chapter I would term nonessential. In a first reading a student should not feel guilty about skipping the proofs in the sections on prime and saturated models (although the statements of the theorems must be understood). The concepts in Section 5.6 (orthogonality, domination and weight) are at the heart of the dimension theory induced by forking dependence on the universal domain.

The study of superstable theories in Chapter 6 is a natural continuation of the material in Chapter 5. In particular, the dimension theory introduced in Section 5.6 is deepened in the third section on regular types. Geometrical stability theory in the context of a superstable theory of finite rank is introduced in Section 6.2. Section 7.1 contains an application of the dimension theory developed in Chapter 6 to the classification of certain ω−stable theories. Finally, the section on ranks (Section 7.2) contains important facts about Morley rank in an uncountably categorical theory and ∞−rank in a unidimensional theory.

Acknowledgments

I would like to thank the members of the model theory class in which I lectured on this material. They are: Tim Bahmer, Andras Benedek, Dan Gardner, Colleen Hoover, Byungham Kim, Grzegorz Michalski, John Thurber. Their feedback on my lectures has been invaluable. I also thank the colleagues who critiqued the early drafts I distributed. My secretaries, Ellen Victory and Tracy Mattix, have graciously helped me manage the necessary paper shuffling in the past 6 months.

Finally, I thank my wife and children for their patience. They each get a match when I throw the preliminary drafts in the fireplace.

1. The Basics

1.1 Preliminaries and Notation

We assume that the reader is familiar with the basic definitions and results normally found in a first course in mathematical logic. Specifically, we will freely use the concepts of a *first-order language*, a *structure* or *model* in that language, and the *satisfaction relation* between models and formulas. We also assume that the reader knows the Compactness and Omitting Types Theorems, and can carry out an elimination of quantifiers argument for a specific theory such as dense linear orders without endpoints or divisible abelian groups. In this first section we will review some of these results as a way of setting our notation and viewpoint and jogging the student's memory.

Notation. (Model Theory)

- A first-order language is denoted by L, L', L_0, etc. The cardinality of a language L, $|L|$, is simply the cardinality of the set of nonlogical symbols of L.
- Formulas are denoted by lower case Greek letters. Writing $\varphi(v_0, \ldots, v_n)$ indicates that the free variables in φ are in $\{v_0, \ldots, v_n\}$. If $t_0, \ldots, t_n$ are terms in the language, $\varphi(t_0, \ldots, t_n)$ is the formula obtained by substituting t_i for v_i. A sentence is a formula with no free variables.
- We use $\mathcal{M}$ or $\mathcal{N}$, decorated with various subscripts and superscripts, to denote a model or structure in a first-order language. The universe of, e.g., $\mathcal{M}_0$, is M_0. Elements of the universe are denoted by lower case letters such as a, b, c, etc. If X is an element of the language in which $\mathcal{M}$ is a model $X^{\mathcal{M}}$ denotes the interpretation of X in $\mathcal{M}$.
- Given models $\mathcal{M}$ and $\mathcal{N}$ in a language L, a function $f : M \longrightarrow N$ is an *isomorphism* if f is a bijection and for all symbols $X \in L$, $f(X^{\mathcal{M}}) = X^{\mathcal{N}}$. When there is an isomorphism of $\mathcal{M}$ onto $\mathcal{N}$ we write $\mathcal{M} \cong \mathcal{N}$ and say $\mathcal{M}$ and $\mathcal{N}$ are isomorphic. An *automorphism* of $\mathcal{M}$ is an isomorphism of $\mathcal{M}$ onto itself. For $\mathcal{M}$ a model, $\mathrm{Aut}(\mathcal{M})$ denotes the automorphism group of $\mathcal{M}$.
- A *theory* in the language L is a consistent set of sentences of L. A set of sentences need not be complete in order to be called a theory. (A set of sentences is *consistent* if it has a model. A theory is *complete* if all

models of the theory satisfy exactly the same sentences.) For T a theory, $Mod(T) = \{\mathcal{M} : \mathcal{M} \models T\}$. A class of models is *elementary* if it is $Mod(T)$ for some theory T. The complete theory of a model $\mathcal{M}$ is $Th(\mathcal{M}) = \{\sigma : \sigma$ is a sentence and $\mathcal{M} \models \sigma\}$. Models $\mathcal{M}$ and $\mathcal{N}$ are called *elementarily equivalent*, written $\mathcal{M} \equiv \mathcal{N}$, if $Th(\mathcal{M}) = Th(\mathcal{N})$. Given a theory T and $n < \omega$ define an equivalence relation $\sim_n$ on the formulas in n free variables by: $\varphi(\bar{v}) \sim_n \psi(\bar{v})$ if for all models $\mathcal{M}$ of T, $\mathcal{M} \models \forall\bar{v}(\varphi(\bar{v}) \leftrightarrow \psi(\bar{v}))$. The *cardinality of* T, denoted $|T|$, is the supremum over n of the number of $\sim_n$ –classes, which is always infinite when T has an infinite model.
- If $\mathcal{M}$ is a structure in the language L and L_0 is a sublanguage of L, then $\mathcal{M} \upharpoonright L_0$ is the restriction of $\mathcal{M}$ to L_0. This restriction is defined to be the model in the language L_0 with the same universe and the same interpretation for the elements of L_0.

Notation. (Set Theory) The set-theoretic notation used here is quite standard. Less basic concepts will be defined later when they are needed.

When discussing the "logical" properties of a model there is little difference between a finite sequence $(a_1, \ldots, a_n)$ and the set $\{a_1, \ldots, a_n\}$. We will muddy the difference by writing $\bar{a} \subset M$ when $\bar{a}$ is a finite subset of M or a finite sequence from M ($\mathcal{M}$ is a model). Given a finite set of elements $\{a_1, \ldots, a_n\}$ we may juxtapose the elements and write $a_1 \ldots a_n$ for the sequence $(a_1, \ldots, a_n)$.

If L is a language and X is a set, $L(X)$ denotes the language obtained from L by adding a new constant symbol for each element of X. We will usually use a to denote both the element $a \in X$ and the corresponding constant symbol. Given a structure $\mathcal{M}$ for L, if $X \subset M$, then $\mathcal{M}_X$ denotes the expansion of $\mathcal{M}$ to $L(X)$ which interprets the constant $a \in X \cap L(X)$ by the element a. If $X = \{a_0, \ldots, a_n\}$ we may also write $(\mathcal{M}, a_0, \ldots, a_n)$ for $\mathcal{M}_X$.

The satisfaction relation is defined in most books as a relation on triples $\langle \mathcal{M}, \varphi, s\rangle$, where $\mathcal{M}$ and φ are as usual, and s is an *assignment*; i.e., a function from variables (including those free in φ) into M. In this book we adopt an approach, developed by Shoenfield in [Sho67], which is more streamlined and reflects the way we view elements of a model as parameters which can be used in formulas of the language. Briefly, for $\mathcal{M}$ a model we define satisfaction in $\mathcal{M}_M$ of *sentences* in $L(M)$ by a standard inductive argument. For $\varphi(v_0, \ldots, v_n)$ a formula of L and elements $a_0, \ldots, a_n \in M$, we say that $(a_0, \ldots, a_n)$ *satisfies* $\varphi(v_0, \ldots, v_n)$ *in* $\mathcal{M}$, and write $\mathcal{M} \models \varphi(a_0, \ldots, a_n)$, if $\mathcal{M}_M \models \varphi(a_0, \ldots, a_n)$ (where $\varphi(a_0, \ldots, a_n)$ is treated as a sentence in $L(M)$).

When studying the relations defined on a model by the formulas of the language it is common to fix a certain set of elements A and study the relations between A and other elements of the universe. For example, given an algebraically closed field k and polynomials over $A \subset k$, the set A is fixed throughout the study of the sets defined on k by the polynomials.

Definition 1.1.1. *If $\mathcal{M}$ is a model and $A \subset M$, we call φ a* formula over A *if φ is a formula in $L(A)$. For φ a formula over A we let $\varphi(\mathcal{M})$ denote $\{\bar{a} \in M^n : \mathcal{M} \models \varphi(\bar{a})\}$, where φ has n free variables. We call $X \subset M^n$* definable over A in $\mathcal{M}$ *if $X = \psi(\mathcal{M})$ for some formula ψ over A having n free variables. We may say* A−definable in $\mathcal{M}$ *instead of definable over A in $\mathcal{M}$.*

Formulas with parameters from a model will be used frequently, and they will be introduced without formally changing the language.

All results in this book hold not just for a 1-sorted first-order language, but also for a many-sorted language (see, e.g., [End72]). To simplify the notation we will work in the context of a 1-sorted language (until Section 4.1 where we introduce a many-sorted expansion of a theory).

The starting point for model theory is the following result due to Gödel.

Theorem 1.1.1 (Compactness Theorem). *A theory T has a model if and only if every finite subset of T has a model.*

We assume that the reader has seen the Henkin construction of a model which proves the Compactness Theorem (see, e.g., [Hod93, 6.1.1]). The proof shows that when every finite subset of T has a model, T has a model of cardinality $\leq |T|$. As a first application of compactness we state

Corollary 1.1.1 (Löwenheim-Skolem Theorem). *A theory T in a language L which has an infinite model has a model in each infinite cardinality $\lambda \geq |T|$.*

Proof. Let $C = \{c_\alpha : \alpha < \lambda\}$ be a set of λ new distinct constant symbols, $L' = L \cup C$. Let $T' = T \cup \{c_\alpha \neq c_\beta : \alpha < \beta < \lambda\}$. By the Compactness Theorem T' has a model. As a corollary to the proof of the Completeness Theorem we know that T' has a model $\mathcal{M}'$ of cardinality $|L'| + \aleph_0 = \lambda$. The restriction of $\mathcal{M}'$ to L is the desired model of T.

A similar proof shows that a theory which has arbitrarily large finite models has an infinite model. Other elementary applications of the Compactness Theorem are stated in the exercises.

Recall that $\mathcal{M}$ is a *submodel* of a model $\mathcal{N}$, written $\mathcal{M} \subset \mathcal{N}$, if

- $M \subset N$,
- $c^{\mathcal{M}} = c^{\mathcal{N}}$, for any constant symbol c in L,
- for $F \in L$ an $n-$ary function symbol, $F^{\mathcal{M}} = F^{\mathcal{N}} \restriction M^n$, and
- for $R \in L$ an $n-$ary relation symbol, $R^{\mathcal{M}} = R^{\mathcal{N}} \cap M^n$.

Definition 1.1.2. *Let $\mathcal{M}$ and $\mathcal{N}$ be structures in the language L. We say that $\mathcal{M}$ is an* elementary submodel of $\mathcal{N}$, *and write $\mathcal{M} \prec \mathcal{N}$, if $M \subset N$ and for all formulas $\varphi(v_0, \ldots, v_n)$ of L, and $a_0, \ldots, a_n \in M$,*

$$\mathcal{M} \models \varphi(a_0, \ldots, a_n) \text{ if and only if } \mathcal{N} \models \varphi(a_0, \ldots, a_n)$$

(Notice that we could have stated the key condition in the definition as $\mathcal{M}_M \equiv \mathcal{N}_M$.) When $\mathcal{M} \prec \mathcal{N}$ we will also call $\mathcal{N}$ an *elementary extension of* $\mathcal{M}$. Our first question is: How can a submodel fail to be an elementary submodel? After doing Exercise 1.1.5 the reader will see that all failures are in the form: there are $a_0, \ldots, a_n \in M$ and $\varphi(v, v_0, \ldots, v_n)$ such that $\mathcal{N} \models \exists v \varphi(v, a_0, \ldots, a_n)$ and $\mathcal{M} \not\models \exists v \varphi(v, a_0, \ldots, a_n)$. This gives the elementary submodel relation the flavor of a closure condition; witnesses to existential quantifiers must be added to form an elementary submodel. This is exhibited in the next lemma, whose proof is left to the exercises.

Lemma 1.1.1 (Tarski-Vaught Test). *For models $\mathcal{M}$ and $\mathcal{N}$, $\mathcal{M} \prec \mathcal{N}$ if and only if*

- *$\mathcal{M} \subset \mathcal{N}$ and*
- *for all formulas $\varphi(v, v_0, \ldots, v_n)$ and $a_0, \ldots, a_n \in M$, if $\mathcal{N} \models \exists v \varphi(v, a_0, \ldots, a_n)$, there is a $b \in M$ such that $\mathcal{N} \models \varphi(b, a_0, \ldots, a_n)$.*

By the Compactness theorem any infinite model $\mathcal{M}$ has elementary extensions of any cardinality $\geq |M| + |T|$. (Just apply the Löwenheim-Skolem Theorem to $Th(\mathcal{M}_M)$.) A natural companion to this is

Theorem 1.1.2 (Downward Löwenheim-Skolem-Tarski Theorem). *Let T be a theory, λ, κ cardinals with $\lambda \geq \kappa \geq |T|$ and $\mathcal{M}$ a model of T of cardinality λ. Then for any $X \subset M$ with $|X| \leq \kappa$, $\mathcal{M}$ has an elementary submodel of cardinality κ containing X.*

Proof. Form a chain of sets $X = X_0 \subset X_1 \subset X_2 \subset \ldots$ such that if $\varphi(v, v_0, \ldots, v_n)$ is a formula of the language of T, $a_0, \ldots, a_n \in X_i$ and $\mathcal{M} \models \exists v \varphi(v, a_0, \ldots, a_n)$, then there is a $b \in X_{i+1}$ such that $\mathcal{M} \models \varphi(b, a_0, \ldots, a_n)$. Since there are $|T|$ many formulas to consider and $|X_i| + \aleph_0$ many tuples $(a_0, \ldots, a_n)$ from X_i, we may require each X_i to have cardinality κ. Let $N = \bigcup_{i<\omega} X_i$. Considering suitable choices for φ shows that N is the universe of a submodel $\mathcal{N}$ of $\mathcal{M}$. Furthermore, since any $n-$tuple from N is from some X_i, the Tarski-Vaught test implies that $\mathcal{N}$ is an elementary submodel of $\mathcal{M}$, as desired.

For models $\mathcal{M}$ and $\mathcal{N}$ in the same language, a function f is an *elementary embedding of* $\mathcal{M}$ *into* $\mathcal{N}$ if f is an isomorphism from $\mathcal{M}$ onto an elementary submodel of $\mathcal{N}$. We say $\mathcal{M}$ is *elementarily embeddable into* $\mathcal{N}$ if there is such an elementary embedding.

Let T be a theory in L. An *$n-$type in* $\bar{v} = (v_0, \ldots, v_{n-1})$ is a consistent set of formulas p in the variables $\bar{v}$. (Consistency is computed relative to the ambient theory T.) We may write $p(\bar{v})$ for p when it is helpful to exhibit the variables. Let p be an $n-$type in the theory T and $\mathcal{M} \models T$. Given $\bar{a} \in M^n$, we say that $\bar{a}$ *realizes* p *in* $\mathcal{M}$ (or *satisfies* p *in* $\mathcal{M}$) if $\mathcal{M} \models \varphi(\bar{a})$ for each $\varphi \in p$. Extending the notation used for formulas, $p(\mathcal{M})$ denotes $\{\bar{a} \in M^n : \bar{a} \text{ realizes } p \text{ in } \mathcal{M}\}$. The model $\mathcal{M}$ *realizes (or satisfies)* p if some tuple from $\mathcal{M}$ realizes p; otherwise, $\mathcal{M}$ *omits* p. As usual, this notion can be relativized over a set of parameters. Given $A \subset M$, an n*−type over* A *in* $T = Th(\mathcal{M})$ is a consistent set of formulas over A (where consistency is computed with respect to $Th(\mathcal{M}_A)$.) Given p a type over a set A, the *domain of* p, denoted $dom(p)$, is the minimal set $B \subset A$ such that p is a type over B.

The type $p(\bar{v})$ is called *complete* if it is a maximally consistent set of formulas in $\bar{v}$ (i.e., there is no formula $\varphi(\bar{v})$ such that both $p \cup \{\varphi\}$ and $p \cup \{\neg\varphi\}$ are consistent). We will deal extensively with complete types in complete theories.

Notation. Fix a model $\mathcal{M}$, $A \subset M$ and $\bar{v}$ an n−tuple of variables. Let $S_n(A)$ denote the set of complete n−types over A in $\bar{v}$. (If $p = p(\bar{v})$ and $\bar{w}$ is another sequence of n variables we equate p and $p(\bar{w})$ in almost all model-theoretic settings.) Let $S(A) = \bigcup_{n<\omega} S_n(A)$. Given an n−tuple $\bar{a}$ from M there is a unique $p \in S(A)$ realized by $\bar{a}$ in $\mathcal{M}$, called the *type of* $\bar{a}$ *over* A *in* $\mathcal{M}$ and denoted $tp_{\mathcal{M}}(\bar{a}/A)$. Explicitly,

$$tp_{\mathcal{M}}(\bar{a}/A) = \{\varphi : \varphi \text{ is a formula over } A \text{ and } \mathcal{M} \models \varphi(\bar{a})\}.$$

We may compress the notation and write, e.g.,

$$tp_{\mathcal{M}}(a_0, \ldots, a_n) \text{ for } tp_{\mathcal{M}}((a_0, \ldots, a_n)).$$

Definition 1.1.3. *Let* T *be a theory,* $\mathcal{M} \models T$, $A \subset M$ *and* p, q n*−types over* A. *We say that* p implies q in T, *written* $p \models q$, *if for any model* $\mathcal{N} \succ \mathcal{M}$, $p(\mathcal{N}) \subset q(\mathcal{N})$. *We say that* p *and* q *are* equivalent in T *if for every model* $\mathcal{N} \succ \mathcal{M}$, $p(\mathcal{M}) = q(\mathcal{M})$; *i.e.,* $p \models q$ *and* $q \models p$.

Let T be a theory and p, q n−types (over $\emptyset$, for simplicity). The Compactness Theorem gives the equivalent:

- $p \models q$ if and only if
- for all formulas $\varphi(\bar{v}) \in q$ there are $\psi_0(\bar{v}), \ldots, \psi_n(\bar{v}) \in p$ such that $T \models \forall\bar{v}(\bigwedge_{i \le n} \psi_i(\bar{v}) \to \varphi(\bar{v}))$.

(See the exercises). Additionally, $p \models q$ if and only if for every complete n−type r over A, $r \supset p \implies r \supset q$.

Definition 1.1.4. *Fix a theory* T. *A set of formulas* p *(in* n *variables) is said to be* isolated by φ *if* φ *is consistent with* T *and for all* $\psi \in p$, $T \models \forall\bar{v}(\varphi(\bar{v}) \to \psi(\bar{v}))$. p *is* isolated *if it is isolated by some formula. If* p *is not isolated it is called* nonisolated.

Notice that a nonisolated set of formulas need not be consistent. If p is isolated by φ and $\mathcal{M} \models T$ then $\varphi(\mathcal{M}) \subset p(\mathcal{M})$.

As the term "isolated" suggests there is a topology in the background. For $\varphi(\bar{v})$ a formula in n free variables, $\mathcal{O}_\varphi = \{p \in S_n(\emptyset) : \varphi \in p\}$. (Here, we equate the formulas $\varphi(\bar{v})$ and $\varphi(\bar{x})$, where $\bar{x}$ is another sequence of n variables: $\mathcal{O}_{\varphi(\bar{v})}$ is, by definition, the same as $\mathcal{O}_{\varphi(\bar{x})}$.) The sets of the form $\mathcal{O}_\varphi$ comprise the basic open sets of a topology on $S_n(\emptyset)$, called a *Stone space of T*. The topology is compact (by the Compactness Theorem) and Hausdorff. A type $p \in S_n(\emptyset)$ is isolated exactly when it is an isolated point in the Stone space topology on $S_n(\emptyset)$.

Given a model $\mathcal{M}$ and $A \subset M$, any $p \in S(A)$ is realized in *some* elementary extension $\mathcal{N}$ of $\mathcal{M}$. (This is a compactness argument, left to the reader in Exercise 1.1.8.) In general, though, there is no reason to think that p is realized in $\mathcal{M}$. For any formula $\varphi \in p$ (equivalently, any finite conjunction of formulas in p) the consistency of p requires that $\mathcal{M} \models \exists \bar{v}\varphi$; i.e., $\mathcal{M} \models \varphi(\bar{a})$ for some $\bar{a} \in M^n$. However, there may not be a single $\bar{a}$ which simultaneously satisfies all formulas in p. The obvious exception (which is immediate by the definition) is when p is isolated: If p is an isolated type over A then p is realized in any model of $Th(\mathcal{M}_A)$. That this is the single case when a type is realized in every model of a *countable* theory is proved in

Theorem 1.1.3 (Omitting Types Theorem). *If T is a countable theory and p is a nonisolated set of formulas in T, then T has a countable model which omits p.*

The proof is to use a Henkin construction to build a model which omits the nonisolated set of formulas. The restriction to a countable theory is necessary; there is a theory in which some nonisolated type is realized in every model. A related point is that there may not be an uncountable model omitting a nonisolated type even in a countable theory. (The reader is asked to find an example in the exercises.) The proof of the following slightly more complicated version of the Omitting Types Theorem is assigned as an exercise.

Corollary 1.1.2 (Extended Omitting Types Theorem). *Let T be a countable theory and for each i, let p_i be a nonisolated set of formulas in n_i variables. Then T has a countable model which omits each p_i.*

The Omitting Types Theorem is very useful when constructing nonisomorphic models of a theory. It enables us to show that a certain elementary class is "rich"; i.e., contains models with varying properties. The goal of the next chapter is to carry further this program of finding a wide variety of models of fixed theory.

1.1.1 Elimination of Quantifiers

The method of elimination of quantifiers provides model theorists with a powerful tool for understanding the definable subsets of a particular structure.

Definition 1.1.5. *Given a language L and a class $\mathcal{K}$ of structures in L, we say that a set Φ of formulas of L is an* elimination set for $\mathcal{K}$ *if for every formula $\varphi(\bar{v})$ of L there is a formula $\varphi'(\bar{v})$ such that*

– *φ' is a boolean combination of formulas in Φ and*
– *for every model $\mathcal{M} \in \mathcal{K}$, $\mathcal{M} \models \forall \bar{v}(\varphi(\bar{v}) \leftrightarrow \varphi'(\bar{v}))$.*

When T is a theory and $\mathcal{K}$ is the class of models of T we say elimination set for T.

The elimination of quantifiers program is: find an elimination set for a given class $\mathcal{K}$. Of course, the set of all formulas is always an elimination set for a class, but for many classes there is a simpler set which illuminates the model-theoretic properties of $\mathcal{K}$. Take, for example, the theory T in $L = \{<\}$ of dense linear orders without endpoints. (This is the theory axiomatized by the statements: $<$ defines a linear order, there is no least element or greatest element, and $\forall xy(x < y \rightarrow \exists z(x < z < y))$. $(\mathbb{Q}, <)$ is a model of T.) We leave it to the exercises to show that $\Phi = \{v_1 < v_2, v_1 = v_2\}$ is an elimination set for T. (Again, we do not distinguish between formulas which are obtained by a change of variables.)

In this example, the elimination set consists of atomic formulas — every formula is equivalent to a quantifier-free formula, so we really have eliminated the quantifiers. When the elimination set for T consists of atomic formulas we say that T *has elimination of quantifers*, T is *quantifier-eliminable*, or T is q.e. The term "elimination of quantifiers" has its origins, however, in the key step of a proof that Φ is an elimination set. The following is proved by induction on formulas (see the exercises). For Φ a set of formulas $\Phi^- = \{\neg\varphi : \varphi \in \Phi\}$.

Lemma 1.1.2. *Let $\mathcal{K}$ be a class of structures in L and Φ a set of formulas of L. Suppose that*

(1) *every atomic formula of L is in Φ, and*
(2) *for every formula $\theta(\bar{v})$ of the form $\exists w \bigwedge_{i<n} \psi_i(\bar{v}, w)$, where each $\psi_i \in \Phi \cup \Phi^-$, there is a formula $\theta'(\bar{v})$ which is a boolean combination of formulas from Φ and is equivalent to θ on every structure in $\mathcal{K}$.*

Then Φ is an elimination set for $\mathcal{K}$.

The following elimination of quantifier results will be used freely throughout the book. The reader is referred to [Hod93] for details.

– The class $\mathcal{K}$ of vector spaces over a division ring F has elimination of quantifiers. (The elements of $\mathcal{K}$ are structures in the language with a binary operation $+$, a unary function α for each $\alpha \in F$ and a constant symbol 0.)
– The class $\mathcal{K}$ of algebraically closed fields of a fixed characteristic has elimination of quantifiers.
– For $\mathcal{K}$ the class of real-closed fields $\{\exists y(y^2 = t(x)) : t(x)$ a term of L not containing $y\}$ is an elimination set for $\mathcal{K}$.
– The class of divisible abelian groups has elimination of quantifiers.

Historical Notes. The compactness theorem was proved for countable theories by Gödel in [Go30], for arbitrary propositional theories by Gödel in

[Go31] and in general by Mal'tsev [Mal36]. The Löwenheim-Skolem Theorem (as stated here) is found in [TV57], where the Tarski-Vaught test is also formulated. The Omitting Types Theorem can be attributed to several sources, Ehrenfeucht (in [Vau61]) and Grzegorczyk, Mostowski and Ryll-Nardzewski in [GMRN61]. The Downward Löwenheim-Skolem Theorem was proved first for some special languages in [Lo15] and in another form in [Sko20]. (They both consider the special case of showing that a countable theory with an infinite model has a countable one.)

Exercise 1.1.1. Prove that a theory which has arbitrarily large finite models has an infinite model.

Exercise 1.1.2. An ordered field $(F, +, \cdot, 0, 1, \leq)$ is called *Archimedean* if for any two positive elements a, $b \in F$ there is an n such that $na \geq b$. Show that $(\mathbb{R}, +, \cdot, 0, 1, \leq)$ has an elementary extension which is not Archimedean.

Exercise 1.1.3. Let σ be a sentence which holds in any algebraically closed field of characteristic 0. Then there is a $p \neq 0$ such that σ holds in all algebraically closed fields of characteristic $\geq p$.

Exercise 1.1.4. Let T be a theory. Prove that $p \models q$ if and only if for all formulas $\varphi(\bar{v}) \in q$ there are $\psi_0(\bar{v}), \ldots, \psi_n(\bar{v}) \in p$ such that $T \models \forall \bar{v}(\bigwedge_{i \leq n} \psi_i(\bar{v}) \rightarrow \varphi(\bar{v}))$.

Exercise 1.1.5. Which of the following submodel relations are elementary?
(a) $(\mathbb{Q}, \leq) \subset (\mathbb{R}, \leq)$.
(b) $(2\mathbb{Z}, +, 0) \subset (\mathbb{Z}, +, 0)$.
(c) $(\mathbb{Q}, +, \cdot, 0, 1) \subset (\mathbb{C}, +, \cdot, 0, 1)$.

Exercise 1.1.6. Use induction on formulas to prove the Tarski-Vaught Test.

Exercise 1.1.7. Prove that $\mathcal{M}$ is elementarily embeddable into $\mathcal{N}$ (both structures in the same language) if and only if $\mathcal{N}$ can be expanded to a model of $Th(\mathcal{M}_M)$.

Exercise 1.1.8. Show that if $\mathcal{M}$ is a model, $A \subset M$ and $p \in S(A)$, then p is realized in some elementary extension $\mathcal{N}$ of $\mathcal{M}$. Also show that if p is an isolated type over A then p is realized in any model of $Th(\mathcal{M}_A)$.

Exercise 1.1.9. Prove the Extended Omitting Types Theorem.

Exercise 1.1.10. Give an example of a countable theory T and a nonisolated type p which is realized in every uncountable model of T.

Exercise 1.1.11. Let $\mathcal{M}$ be a finite model in a language L. Show that

$$\mathcal{N} \equiv \mathcal{M} \implies \mathcal{N} \cong \mathcal{M}.$$

Exercise 1.1.12. Prove Lemma 1.1.2.

Exercise 1.1.13. Prove that the theory of dense linear orders without endpoints has elimination of quantifiers.

Exercise 1.1.14. Let $L = \{E\}$ where E is a binary relation and let T be the theory in L saying that E is an equivalence relation with infinitely many infinite classes and no finite classes. Prove that T has elimination of quantifiers.

2. Constructing Models with Special Properties

Much of the richness of model theory is attributable to the weakness of first-order logic. In the last chapter it was proved that a first-order theory cannot express, for example, that every model has cardinality $\aleph_0$. In fact, by the Löwenheim-Skolem Theorems an elementary class (in a countable language) containing an infinite model contains models in every infinite cardinality. Thus, a basic property expressible in first-order logic and true in an infinite model is true in a great variety of models. Besides the Löwenheim-Skolem Theorems, the Omitting Types Theorem is an example of such a result. Given a nonisolated type p in a countable theory there is a countable model $\mathcal{M}$ realizing p and a countable model $\mathcal{N}$ omitting p. The theme of this chapter is to continue this program of constructing models of a given theory with widely varying properties. In the first two sections we define several kinds of models (of complete theories having infinite models) distinguished by the elementary embeddings they admit and the types they realize or omit.

2.1 Prime and Atomic Models

The first special kind of model to be considered is one which is, intuitively, the "smallest" model of the theory.

Definition 2.1.1. *Given a theory T, we call $\mathcal{M} \models T$ a* prime *model of T if, for any $\mathcal{N} \models T$, $\mathcal{M}$ can be elementarily embedded into $\mathcal{N}$.*

For example, if T is the theory of algebraically closed fields of characteristic 0, the algebraic closure of the rationals, $\tilde{\mathbb{Q}}$, forms a prime model of T. (Since T has quantifier elimination, any model of T contains a copy of $\tilde{\mathbb{Q}}$ as an elementary submodel.) While the definition of a prime model makes sense for any theory only a complete theory can have a prime model (see the exercises). We will see, moreover, that the uniqueness of prime models and a useful condition sufficient for their existence can only be proved for countable complete theories. As a first observation, if $\mathcal{M}$ is a prime model of the countable theory T and $p \in S_n(\emptyset)$ is realized in $\mathcal{M}$ then p must be isolated. (Suppose to the contrary that p is nonisolated and realized in $\mathcal{M}$ by the n-tuple $\bar{a}$. By the Omitting Types Theorem, T has a countable model

$\mathcal{N}$ omitting p. Since $\mathcal{M}$ is prime there is an elementary embedding f of $\mathcal{M}$ into $\mathcal{N}$. The type of any tuple $\bar{b}$ in $\mathcal{M}$ is the same as the type of $f(\bar{b})$ in $\mathcal{N}$ (since f is elementary) hence, the type of $f(\bar{a})$ in $\mathcal{N}$ is p, contradicting the fact that $\mathcal{N}$ omits p.) This observation suggests the following definition.

Definition 2.1.2. *Let T be a complete theory. Given a model $\mathcal{M}$ of T, $A \subset M$ is called* atomic *if for each tuple $\bar{a}$ from A, $tp_{\mathcal{M}}(\bar{a})$ is isolated. We call $\mathcal{M}$ an* atomic model *if M is an atomic set.*

It was proved above that a prime model of a countable theory is necessarily atomic. That a countable atomic model is a prime model of its theory and elementarily equivalent countable atomic models are isomorphic are proved in the next proposition. If $\mathcal{M}$ and $\mathcal{N}$ are models and $A \subset M$, a function $f : A \longrightarrow N$ is called an *elementary map* if for all $a_0, \ldots, a_n \in A$, $tp_{\mathcal{M}}(a_0, \ldots, a_n) = tp_{\mathcal{N}}(f(a_0), \ldots, f(a_n))$. (An elementary embedding is simply an elementary map whose domain is a model.)

Proposition 2.1.1. *Let T be a countable complete theory.*

(i) A countable model $\mathcal{M}$ of T is prime if and only if $\mathcal{M}$ is atomic.

(ii) If $\mathcal{M}$ and $\mathcal{N}$ are both countable atomic models of T, then $\mathcal{M} \cong \mathcal{N}$.

Proof. If T has a finite model then all models of T are isomorphic (by Exercise 1.1.11) and every complete type is isolated, making both (i) and (ii) trivially true. Thus, we can assume T to have only infinite models.

(i) It only remains to show that a countable atomic model $\mathcal{M}$ is prime. Let $\mathcal{N}$ be an arbitrary model of T and $\{ a_i : i < \omega \}$ an enumeration of M.

Claim. There is a sequence $\{ b_i : i < \omega \} \subset N$ such that for each $i < \omega$, $tp_{\mathcal{M}}(a_0, \ldots, a_i) = tp_{\mathcal{N}}(b_0, \ldots, b_i)$.

The elements b_i are found by recursion on i. Suppose that $b_0, \ldots, b_{i-1}$ with the desired property have been selected and let $\varphi(v_0, \ldots, v_i)$ be a formula which isolates $p =$ the type of $(a_0, \ldots, a_i)$ in $\mathcal{M}$. When $i = 0$ the completeness of T yields a $b_0 \in N$ such that $\mathcal{N} \models \varphi(b_0)$. In general the existence of a $b_i \in N$ such that $\mathcal{N} \models \varphi(b_0, \ldots, b_i)$ is guaranteed by the inductive hypothesis: $tp_{\mathcal{M}}(a_0, \ldots, a_{i-1}) = tp_{\mathcal{N}}(b_0, \ldots, b_{i-1})$. Being complete, the theory T expresses the fact that φ isolates a complete type; i.e., for any formula $\psi \in p$, $T \models \forall \bar{v}(\varphi(\bar{v}) \longrightarrow \psi(\bar{v}))$. Thus, for all formulas $\psi(v_0, \ldots, v_i)$, $\mathcal{M} \models \psi(a_0, \ldots, a_i) \iff \mathcal{N} \models \psi(b_0, \ldots, b_i)$, proving the claim.

Let f be the map from M into N defined by: $f(a_i) = b_i$, for each $i < \omega$. The reader can verify that f is an elementary map of $\mathcal{M}$ onto $\{ b_i : i < \omega \}$ and $\{ b_i : i < \omega \}$ is the universe of an elementary submodel of $\mathcal{N}$. This proves that $\mathcal{M}$ is a prime model of T.

(ii) The manner in which the isomorphism between $\mathcal{M}$ and $\mathcal{N}$ is constructed is similar to the construction of the embedding in (i). Simply by quoting (i) we know that $\mathcal{M}$ can be elementarily embedded in $\mathcal{N}$, and vice-versa. The construction needs to be altered slightly to obtain an elementary embedding of $\mathcal{M}$ into $\mathcal{N}$ which is surjective. This is our first example of a *back and forth construction* of a map. This frequently used technique is a generalization of the standard uniqueness proof of countable dense linear orderings without endpoints (see the Historical Notes).

Well-order the universes M and N with order type ω. Let a_0 be the first element of M and $\varphi(v_0)$ the formula which isolates the type of a_0 in $\mathcal{M}$. As $\mathcal{N}$ is also a model of T there is a b_0 satisfying $\varphi(x)$ in $\mathcal{N}$. Now, let b_1 be the first element of $N \setminus \{b_0\}$ and find an $a_1 \in M$ such that $tp_{\mathcal{M}}(a_0, a_1) = tp_{\mathcal{N}}(b_0, b_1)$, as in the proof of (i). Let a_2 be the first element of $M \setminus \{a_0, a_1\}$ and find $b_2 \in N$ with $tp_{\mathcal{M}}(a_0, a_1, a_2) = tp_{\mathcal{N}}(b_0, b_1, b_2)$. Going back and forth ω times gives enumerations $\{ a_i : i < \omega \}$ and $\{ b_i : i < \omega \}$ of M and N, respectively, such that for each n, $tp_{\mathcal{M}}(a_0, \ldots, a_n) = tp_{\mathcal{N}}(b_0, \ldots, b_n)$. The map taking a_i to b_i (for $i < \omega$) then defines an isomorphism from $\mathcal{M}$ onto $\mathcal{N}$.

One question we need answered is: Does every complete theory have a prime model, or can we find a meaningful characterization of those which do? The previous proposition reduces the problem of finding a prime model of a countable complete theory to showing that an atomic model exists. The next example shows is not always possible.

Example 2.1.1. (A countable complete theory with no atomic model) Let $L = \{ P_i : i < \omega \}$, where each P_i is a unary relation symbol. Let X be the set of finite sequences of 0's and 1's. Each $s \in X$ is viewed as a function from $\{0, \ldots, m\}$ (for some m) into $\{0, 1\}$, and the *length of* $s = lh(s)$ is defined to be $m + 1$. The theory T will be defined so that for any model $\mathcal{M}$ of T and $s \in X$, the intersection of the family of sets $\{ P_i(\mathcal{M}) : s(i) = 0 \} \cup \{ M \setminus P_i(\mathcal{M}) : s(i) = 1 \}$ is nonempty. Let $P_i^0(v)$ denote the formula $P_i(v)$, and $P_i^1(v)$ the formula $\neg P_i(v)$. For $s \in X$ let $\varphi_s(v)$ be the formula $\bigwedge_{i<lh(s)} P_i^{s(i)}(v)$, $\sigma_s = \exists v \varphi_s(v)$ and $T = \{ \sigma_s : s \in X \}$. The reader can verify that T is a complete quantifier-eliminable theory. Thus, if $\mathcal{M} \models T$ and $a \in M$, the type of a in $\mathcal{M}$ is implied by $\{ P_i^j(v) : \mathcal{M} \models P_i^j(a),\ i < \omega,$ and $j = 0, 1 \}$. We claim that every complete 1–type in T is nonisolated. If, to the contrary, p is an isolated 1–type, then by the characterization of types just mentioned p is isolated by some $\varphi_s \in p$. However, if $j = lh(s)$, both $\exists v(\varphi_s(v) \wedge P_j(v))$ and $\exists v(\varphi_s(v) \wedge \neg P_j(v))$ are in T, proving that φ_s does not isolate a complete type in T. Since T has no isolated 1–types over $\emptyset$, no model of T can be atomic.

A more mathematically common example of a theory without a prime model is the theory of the model $(\mathbb{Z}, +)$, although this property is more difficult to verify for $Th(\mathbb{Z}, +)$ than the theory in the example. A complete description of the countable models of this theory is found in [BBGK73].

For a theory to have an atomic model it must satisfy the next condition, which we will also show is sufficient for countable theories in the subsequent proposition.

Definition 2.1.3. *For T a complete theory we say that the* isolated types of T are dense *if every formula φ in n variables consistent with T is contained in an isolated complete n−type over $\emptyset$.*

Such theories are called *atomic* in [CK73]. Our terminology is a literal description of the topological property which holds for such theories: The isolated types of T are dense when every basic open set in the topology on $S_n(\emptyset)$ contains an isolated point, for each n.

Proposition 2.1.2. *Let T be a countable complete theory. Then, T has a countable atomic model if and only if the isolated types of T are dense.*

Proof. For the left-to-right direction let $\mathcal{M} \models T$ be atomic, φ a formula consistent with T and $\bar{a}$ a tuple from M satisfying φ. Then $tp_{\mathcal{M}}(\bar{a})$ is an isolated type extending φ. (Note that this easy direction of the proposition does not require the theory or the atomic model to be countable.)

Now suppose that the isolated types of T are dense. For $n < \omega$ let $\Gamma_n = \{\neg\psi(v_0, \ldots, v_{n-1}) : \psi(v_0, \ldots, v_{n-1})$ isolates a complete n−types in $T\}$. If φ is a formula in n variables consistent with T, there is a formula ψ which isolates a complete type in T such that $\varphi \wedge \psi$ is consistent with T, hence Γ_n is nonisolated. By Corollary 1.1.2 (the Extended Omitting Types Theorem) T has a countable model $\mathcal{M}$ omitting each Γ_n. Then each tuple in M satisfies a formula isolating a complete type in T; i.e., $\mathcal{M}$ is atomic.

The next obvious question is: For which theories are the isolated types of T dense? The isolated types are dense for the theory of dense linear orders without endpoints and algebraically closed fields of a fixed characteristic, but not for the theory in Example 2.1.1. As these examples suggest, there is some connection between the density of the isolated types, and the size of $S(\emptyset)$. Specifically, we prove

Lemma 2.1.1. *If T is a complete theory with $|S(\emptyset)| < 2^{\aleph_0}$ then the isolated types of T are dense.*

Proof. This lemma is a special case of a stronger result (Proposition 2.2.6) proved in the next section using Cantor-Bendixson rank. A different proof is included here to improve our picture of atomic models.

Assume that the isolated types of T are not dense. Then there is some n for which $\{\varphi : \varphi$ is a formula in n variables which is consistent with T and not contained in an isolated type$\} = \Phi$ is nonempty. The fact that any formula in n variables which implies an element of Φ is also in Φ is used to construct continuum many complete types with the following recursion. Let X be the set of finite sequences of 0's and 1's and let $\varphi_\emptyset$ be any formula in Φ. Assuming that $s \in X$ and $\varphi_s \in \Phi$ has been defined (for $s \in X$) choose a formula ψ such that $\varphi_s \wedge \psi = \varphi_{s\frown 0}$ and $\varphi_s \wedge \neg\psi = \varphi_{s\frown 1}$ are consistent with T. Since $\varphi_{s\frown 0}$ and $\varphi_{s\frown 1}$ are both in Φ the recursion can continue. Let Y be the set of sequences of 0's and 1's of length ω, and for $f \in Y$ let $p_f = \{\varphi_s : s$ is an initial segment of f}. By construction, each p_f is consistent and $f \neq g \in Y \implies p_f \cup p_g$ is inconsistent. Extending each p_f to a complete type shows that $|S(\emptyset)| = 2^{\aleph_0}$, completing the proof of the lemma.

Thus, for a countable complete theory, having fewer than continuum many complete types is sufficient to guarantee the existence of a prime model. To see that this condition is not necessary consider, for example, the theory T of the model $(\mathbb{Z}, +, 1)$, where we have added to the cyclic group mentioned earlier a constant symbol for the generator. The reader will show in the exercises that $|S_1(\emptyset)| = 2^{\aleph_0}$. However, since every element of the model $(\mathbb{Z}, +, 1)$ interprets a term of the language, it is an elementary submodel of any model of T; i.e., $(\mathbb{Z}, +, 1)$ is the prime model. This example exhibits a cheap way to find theories with prime models: Given an arbitrary model $\mathcal{M}$ let $\mathcal{M}' = \mathcal{M}_M$. Then $T = Th(\mathcal{M}')$ has $\mathcal{M}'$ as its prime model. A slightly more general result can be obtained vis-a-vis the following notion.

Definition 2.1.4. *Let T be a theory.*

(i) A formula $\varphi(\bar{v})$ is called algebraic in T *if φ is consistent with T and for all $\mathcal{M} \models T$, $\varphi(\mathcal{M})$ is finite.*

(ii) A type is called algebraic in T *if it implies an algebraic formula.*

Remark 2.1.1. (i) For T a complete theory the algebraicity of φ can be tested with any model since if φ defines a finite set in one model then it defines a finite set in every model.

(ii) An algebraic formula is contained in only finitely many complete types in T, each of which is isolated (and algebraic).

(iii) The reader should be cautioned that a theory T may have a model $\mathcal{M}$ and a complete type p such that $p(\mathcal{M})$ is finite, but p is not algebraic (see the exercises).

(iv) A model realizing only algebraic types is atomic, in fact such a model is prime regardless of the cardinality of the language (see the exercises).

Following our conventions for working with parameters, if $\mathcal{M}$ is a model and $A \subset M$, $\mathcal{M}$ is called a *prime model over* A if $\mathcal{M}_A$ is a prime model of $Th(\mathcal{M}_A)$. Referring strictly to models of T, $\mathcal{M}$ is prime over A if whenever $\mathcal{N} \equiv \mathcal{M}$ and $f : A \longrightarrow N$ is an elementary map, there is an elementary embedding of $\mathcal{M}$ into $\mathcal{N}$ extending f. If $T = Th(\mathcal{M})$ is countable and has fewer than continuum many complete types and $A \subset M$ is finite then $Th(\mathcal{M}_A)$ also has fewer than continuum many complete types (see Exercise 2.2.3 and Lemma 2.2.4). Thus, a countable theory with fewer than continuum many complete types has a prime model over any finite subset of a model. There are few positive results concerning prime and atomic models which hold on the set of all countable theories. For instance, a countable theory may have a prime model over the empty set but not over some other finite set, and vice-versa (as we saw with $Th(\mathbb{Z}, +)$ which does not have a prime model, but does have a prime model over 1).

For uncountable theories the existence and uniqueness of prime and atomic models are all more complicated. The density of the isolated types does not, in general, guarantee the existence of an atomic model. (Remember that our proof in the countable case used the Omitting Types Theorem.) Harrington showed (unpublished) that a prime model of a theory need not be atomic (see Example 5.5.1), and there may be nonisomorphic prime models of a theory. Many of these issues are discussed in [Kni78]. We will see that within the stable theories sharper results are often possible (see Section 5.5).

A model $\mathcal{M}$ is *minimal* if $\mathcal{N} \prec \mathcal{M} \implies \mathcal{N} = \mathcal{M}$. In many of the examples given above the prime model of a theory is also minimal. For example, $\mathcal{M}$ is minimal and prime if it realizes only algebraic types (Exercise 2.1.8). To find a prime model which is not minimal we need look no further than $(\mathbb{Q}, <)$. It is easy to see that if a complete theory has a minimal model and a prime model then they must be isomorphic, however it is possible for a theory to have a minimal model which is not prime; in fact, $Th(\mathbb{Z}, +)$ has continuum many nonisomorphic minimal models (see [BBGK73]).

Historical Notes. The notions of prime and atomic models and the basic properties proved here are due to Vaught [Vau61]. Cantor proved the uniqueness of dense linear orders without endpoints in [Can95], but back-and-forth arguments were first isolated by Huntington in [Hun04] (as pointed out by Jack Plotkin).

Exercise 2.1.1. Show that a theory with a prime model is complete.

Exercise 2.1.2. Let T be a complete theory and φ a formula in n variables which is contained in only finitely many complete n−types of T. Show that every complete n−type containing φ is isolated.

Exercise 2.1.3. Suppose that $\bar{a}$ and $\bar{b}$ are sequences from a model $\mathcal{M}$ which have the same complete type in $\mathcal{M}$ and $\varphi(v, \bar{a})$ isolates a complete type over $\bar{a}$ (where $\varphi(v, \bar{x})$ is a formula over $\emptyset$). Show that $\varphi(v, \bar{b})$ isolates a complete type over $\bar{b}$.

Exercise 2.1.4. Let $\bar{a}$ and $\bar{b}$ be finite sequences from the universe of the model $\mathcal{M}$. Prove that $tp_{\mathcal{M}}(\bar{a}\bar{b})$ is isolated if and only if $tp_{\mathcal{M}}(\bar{a}/\bar{b})$ and $tp_{\mathcal{M}}(\bar{b})$ are both isolated. Using this fact show that when $\mathcal{M}$ is an atomic model and $\bar{a}$ is a finite sequence from M, then $\mathcal{M}$ is atomic over $\bar{a}$. Conversely, if $\mathcal{M}$ is atomic over $\bar{a}$ and $tp_{\mathcal{M}}(\bar{a})$ is isolated, then $\mathcal{M}$ is atomic.

Exercise 2.1.5. Show that the complete type realized by 1 in $(\mathbb{Z}, +)$ is nonisolated. (HINT: Use the preceding exercise).

Exercise 2.1.6. Show that $Th(\mathbb{Z}, +)$ has continuum many complete 1−types over $\emptyset$.

Exercise 2.1.7. The definition of an algebraic type is that it implies an algebraic formula, not that it has finitely many realizations in some model. Give an example of a model $\mathcal{M}$ containing an element a which is the only realization of $tp_{\mathcal{M}}(a)$ in $\mathcal{M}$, although this type is not even isolated. (HINT: There is an example in the language with infinitely many constant symbols.)

Exercise 2.1.8. Let $\mathcal{M}$ be a model such that the type in $\mathcal{M}$ of each tuple from M is algebraic. Prove that $\mathcal{M}$ is a prime and minimal model of its theory.

2.2 Saturated and Homogeneous Models

Prime and atomic models are intuitively "small" models of a theory; they are embedded in every model and realize a restricted set of types. In this section we consider "large" models; i.e., those which realize many types and have many models as elementary submodels. Before discussing these properties in full generality we consider their restrictions to the class of countable models, where the results are easier to understand and prove.

Definition 2.2.1. *Let T be a countable complete theory and $\mathcal{M}$ a countable model.*

(i) $\mathcal{M}$ is saturated *if for all finite $A \subset M$, $\mathcal{M}$ realizes every element of $S_1(A)$.*

(ii) $\mathcal{M}$ is homogeneous *if for all finite $A \subset M$, $a \in M$ and elementary maps $f : A \longrightarrow M$, there is an elementary map $g : A \cup \{a\} \longrightarrow M$ extending f.*

(iii) $\mathcal{M}$ is universal *if every countable model of T can be elementarily embedded into $\mathcal{M}$.*

The reader can verify that, in fact, if the countable model $\mathcal{M}$ is saturated then for all finite $A \subset M$, $\mathcal{M}$ realizes every type in $S(A)$ (see Exercise 2.2.7).

Example 2.2.1. (Theories having a saturated countable model)

(i) Let T be the theory in the empty language expressing that the universe is infinite. Then T has a unique countable model (up to isomorphism) and this model is saturated. Similarly, the theory T' of dense linear orders without endpoints has a unique countable model which is also saturated.

(ii) In the language having constant symbols c_i, $i < \omega$, let $T = \{c_i \neq c_j : i < j < \omega\}$. A saturated countable model (if one exists) must realize the n-type $p_n = \{v_i \neq v_j : i < j < n\} \cup \{v_i \neq c_j : i < n,\ j < \omega\}$, which expresses that $v_0, \ldots, v_{n-1}$ are distinct elements, none of which interprets a constant. Thus, a saturated countable model must contain infinitely many elements which do not interpret a constant. Using the quantifier-eliminability of the theory the reader can see that the countable model which contains infinitely many nonconstants is indeed saturated.

(iii) Let T be the theory in the language with unary relations P_0, $P_1, \ldots$ and constants c_{ij}, $i, j < \omega$, axiomatized by the statements: $\neg\exists x(P_i(x) \wedge P_j(x))$ for $i \neq j$; $P_i(c_{ij})$ for $i, j < \omega$; and $c_{ij} \neq c_{ik}$ for $j < k$. This theory is complete and has elimination of quantifiers. For each i there is an n−type saying that $v_0, \ldots, v_{n-1}$ are distinct elements satisfying P_i and none of the v_j's interpret a constant. There is also an n−type expressing that $v_0, \ldots, v_{n-1}$ are distinct elements not satisfying any of the P_i's. Thus, for $\mathcal{M}$ to be a countable saturated model it must at least have the properties that each $P_i(\mathcal{M})$ has infinitely many nonconstant elements and there are infinitely many elements of M not satisfying any P_i. Using the elimination of quantifiers the reader can verify that, in fact, any countable model satisfying these conditions is saturated.

(iv) Let T be the theory of algebraically closed fields of characteristic 0. For each n there is an n−type expressing that $v_0, \ldots, v_{n-1}$ are algebraically independent transcendental elements. So, for a model to have any hope of being saturated it must have infinite transcendence degree over the rationals. The quantifier-eliminability of the theory implies that every complete type is determined by the equations and inequalities it contains. Using this we see that a countable model with transcendence degree $\aleph_0$ over the rationals is saturated.

As a first result we offer

Lemma 2.2.1. *A saturated countable model $\mathcal{M}$ is homogeneous.*

Proof. Let $\bar{a} = (a_1, \ldots, a_n)$ be a tuple from M, $f : \bar{a} \longrightarrow M$ an elementary map, $(f(a_1), \ldots, f(a_n)) = (b_1, \ldots, b_n) = \bar{b}$ and c an element of M. Enumerate $q = tp_{\mathcal{M}}(c, a_1, \ldots, a_n)$ as $\{\varphi_j(v, v_1, \ldots, v_n) : j < \omega\}$ and let $p = \{\varphi_j(v, b_1, \ldots, b_n) : j < \omega\}$. Since $\bar{a}$ and $\bar{b}$ realize the same complete type over $\emptyset$, p is a complete 1−type over $\bar{b}$. There is a $d \in M$ realizing p since $\mathcal{M}$ is saturated. Then $(d, b_1, \ldots, b_n)$ realizes q, equivalently, the map from $(c, a_1, \ldots, a_n)$ to $(d, b_1, \ldots, b_n)$ which extends f and takes c to d is elementary. This proves that $\mathcal{M}$ is homogeneous.

(Later it is proved in the context of potentially uncountable models that a model is saturated if and only if it is homogeneous and universal.)

We turn now to question: When does a countable complete theory have a saturated countable model?

Definition 2.2.2. *A countable complete theory is called* small *if $S(\emptyset)$ is countable.*

Remark 2.2.1. A countable model can only realize countably many complete types, hence only a small theory can have a countable saturated model.

In fact, we will show that smallness is sufficient for the existence of a saturated countable model. The proof of this fact requires a new concept.

Following, is a natural way to construct the saturated countable algebraically closed field of characteristic 0. Let $\mathcal{M}_0 = \tilde{\mathbb{Q}}$, the algebraic numbers. Let a_0 be a transcendental element and $\mathcal{M}_1$ the algebraic closure of $M_0 \cup \{a_0\}$. Let a_1 be an element transcendental over M_1 and $\mathcal{M}_2$ the algebraic closure of $M_1 \cup \{a_1\}$. Continuing in this manner results in a chain $\mathcal{M}_0 \subset \ldots \subset \mathcal{M}_i \subset \ldots$, for $i < \omega$, of algebraically closed fields whose union has transcendence degree $\aleph_0$ (and hence is a saturated countable model). Our general construction of saturated countable models (and, subsequently, models with other properties) will use a generalization of this notion of the union of a chain of fields.

Definition 2.2.3. *A* chain of models *is an increasing sequence of models*

$$\mathcal{M}_0 \subset \mathcal{M}_1 \subset \ldots \subset \mathcal{M}_\beta \subset \ldots, \quad \beta < \alpha,$$

whose length is an ordinal α.

The union of the chain *is the model $\mathcal{M} = \bigcup_{\beta<\alpha} \mathcal{M}_\beta$ defined as follows:*

- *The universe of $\mathcal{M}$ is $M = \bigcup_{\beta<\alpha} M_\beta$.*
- *If R is a relation or function symbol then the interpretation of R on M is $\bigcup_{\beta<\alpha} R^{\mathcal{M}_\beta}$, and*
- *for c a constant symbol, $c^{\mathcal{M}} = c^{\mathcal{M}_\beta}$ for any (all) $\beta < \alpha$.*

Remark 2.2.2. The union of the chain $\{\mathcal{M}_\beta : \beta < \alpha\}$ is the minimal model $\mathcal{N}$ such that $\mathcal{M}_\beta \subset \mathcal{N}$, for all $\beta < \alpha$.

An important special case is obtained by replacing the submodel relation by the elementary submodel relation.

Definition 2.2.4. *An* elementary chain of models *is an increasing sequence of models*

$$\mathcal{M}_0 \prec \mathcal{M}_1 \prec \ldots \prec \mathcal{M}_\beta \prec \ldots, \quad \beta < \alpha,$$

whose length is an ordinal α.

Lemma 2.2.2. *Let $\mathcal{M}_\beta$, $\beta < \alpha$, be an elementary chain of models and $\mathcal{M} = \bigcup_{\beta<\alpha} \mathcal{M}_\beta$. Then $\mathcal{M}_\beta \prec \mathcal{M}$, for all $\beta < \alpha$, and $\mathcal{M}$ is the minimal such model in the sense that whenever $\mathcal{M}_\beta \prec \mathcal{N}$, for all $\beta < \alpha$, $\mathcal{M} \prec \mathcal{N}$.*

Proof. We prove by induction on formulas:

$$\text{for all } \varphi(v_1, \ldots, v_n), \text{ all } \beta < \alpha \text{ and all } a_1, \ldots, a_n \in M_\beta,$$
$$\mathcal{M}_\beta \models \varphi(a_1, \ldots, a_n) \iff \mathcal{M} \models \varphi(a_1, \ldots, a_n).$$

This is clear when φ is atomic since $\mathcal{M}_\beta \subset \mathcal{M}$. The only case in the induction requiring any work is $\varphi = \exists v \psi(v, v_1, \ldots, v_n)$. Suppose that $a_1, \ldots, a_n \in M_\beta$ and $\mathcal{M}_\beta \models \varphi(a_1, \ldots, a_n)$. Then there is a $b \in M_\beta$ such that $\mathcal{M}_\beta \models \psi(b, a_1, \ldots, a_n)$, hence $\mathcal{M} \models \psi(b, a_1, \ldots, a_n)$ and $\mathcal{M} \models \varphi(a_1, \ldots, a_n)$ (by induction). On the other hand, suppose that $\mathcal{M} \models \varphi(a_1, \ldots, a_n)$. Then, for some $\gamma \geq \beta$ and $b \in M_\gamma$, $\mathcal{M} \models \psi(b, a_1, \ldots, a_n)$. By induction, $\mathcal{M}_\gamma \models \psi(b, a_1, \ldots, a_n)$, hence $\mathcal{M}_\gamma \models \exists v \psi(v, a_1, \ldots, a_n)$. Because $\mathcal{M}_\beta \prec \mathcal{M}_\gamma$, $\mathcal{M}_\beta \models \varphi(a_1, \ldots, a_n)$. Thus, $\mathcal{M}_\beta \prec \mathcal{M}$ for all $\beta < \alpha$.

That $\mathcal{M}$ is the minimal such extension of the $\mathcal{M}_\beta$'s is left as an exercise.

Returning to the study of saturated countable models we prove:

Proposition 2.2.1. *A countable complete theory T has a saturated countable model if and only if it is small.*

Proof. If T has a saturated countable model $\mathcal{M}$ then $|S(\emptyset)| = \aleph_0$, since each complete type over the empty set is realized by one of the countably many finite tuples from M. Now assume T to be small. The converse will be established by constructing a saturated countable model as the union of an elementary chain of length ω defined as follows. Let $\mathcal{M}_0$ be any countable model of T, $i < \omega$, and suppose $\mathcal{M}_i$ has been defined. Since T is small, for any $\mathcal{M} \models T$ and finite $A \subset M$, $|S_1(A)| \leq \aleph_0$ (see Exercise 2.2.3). Thus, there is a countable elementary extension $\mathcal{M}_{i+1}$ of $\mathcal{M}_i$ such that for all finite $A \subset M_i$, every element of $S_1(A)$ is realized in $\mathcal{M}_{i+1}$. The union $\mathcal{M} = \bigcup_{i<\omega} \mathcal{M}_i$ is the desired countable saturated model.

The saturated countable model, when it exists, is unique:

Proposition 2.2.2. *Elementarily equivalent saturated countable models are isomorphic.*

Proof. The proof of this result is similar to the uniqueness proof for countable atomic models (Proposition 2.1.1(ii)). In fact, these two propositions are special cases of the same result about homogeneous models (Corollary 2.2.2).

Let $\mathcal{M}$ and $\mathcal{N}$ be saturated countable models of the same complete theory. Well-order the universes M and N with order type ω. Letting a_0 be the first element of M there is (by the saturation of $\mathcal{N}$) a $b_0 \in N$ such that $tp_{\mathcal{M}}(a_0) = tp_{\mathcal{N}}(b_0)$. Now let b_1 be the first element of $N \setminus \{b_0\}$ and find an $a_1 \in M$ such that $tp_{\mathcal{M}}(a_0, a_1) = tp_{\mathcal{N}}(b_0, b_1)$ (see the proof of Lemma 2.2.1 for a similar use of saturation). Let a_2 be the first element of $M \setminus \{a_0, a_1\}$ and find $b_2 \in N$ such that $tp_{\mathcal{M}}(a_0, a_1, a_2) = tp_{\mathcal{N}}(b_0, b_1, b_2)$. Going back and forth ω times we find enumerations $\{ a_i : i < \omega \}$ and $\{ b_i : i < \omega \}$ of M and N, respectively, such that for each n, $tp_{\mathcal{M}}(a_0, \ldots, a_n) = tp_{\mathcal{N}}(b_0, \ldots, b_n)$. The map taking a_i to b_i (for $i < \omega$) then defines an isomorphism from $\mathcal{M}$ onto $\mathcal{N}$.

Homogeneity can be viewed as trading some of the strengths of saturation for less stringent requirements for the existence of models with the property (as the next lemma illustrates).

Proposition 2.2.3. *A countable complete theory has a homogeneous countable model.*

Proof. A homogeneous countable model $\mathcal{M}$ is constructed as the union of an elementary chain of models $\mathcal{M}_i$, $i < \omega$, defined as follows. Let $\mathcal{M}_0$ be any countable model. Assuming $\mathcal{M}_i$ to be defined let $\mathcal{M}_{i+1}$ be a countable elementary extension of $\mathcal{M}_i$ such that

> if $\bar{a}$ and $\bar{b}$ are finite sequences from M_i realizing the same complete type in $\mathcal{M}_i$, then for all $c \in M_i$ there is a $d \in M_{i+1}$ such that $tp_{\mathcal{M}_i}(\bar{a}, c) = tp_{\mathcal{M}_{i+1}}(\bar{b}, d)$.

(A countable such $\mathcal{M}_{i+1}$ exists because there are countably many finite sequences from M_i.) The reader can verify that the model $\mathcal{M} = \bigcup_{i<\omega} \mathcal{M}_i$ is a homogeneous countable model of T.

Saturation is only achieved when a countable model realizes all types over finite subsets, however homogeneity can occur when enough types are omitted:

Proposition 2.2.4. *A countable atomic model $\mathcal{M}$ is homogeneous.*

Proof. Suppose that $\bar{a}$ and $\bar{b}$ are finite sequences realizing the same complete types in $\mathcal{M}$, and $c \in M$. Let $\varphi(\bar{y}x) \in tp_{\mathcal{M}}(\bar{a}c) = q$ be a formula isolating q. Since $\bar{b}$ and $\bar{a}$ realize the same complete type in $\mathcal{M}$, there is a $d \in M$ such that $\mathcal{M} \models \varphi(\bar{b}d)$. Since the formula φ isolates a complete type $tp_{\mathcal{M}}(\bar{a}c) = tp_{\mathcal{M}}(\bar{b}d)$, proving the homogeneity of $\mathcal{M}$.

Homogeneous countable models of the same complete theory are not necessarily isomorphic, indeed, many of the above examples have a countable saturated model and a prime model which are not isomorphic. There is, however, a relative uniqueness result:

Proposition 2.2.5. *If $\mathcal{M}$ and $\mathcal{N}$ are countable homogeneous models in the same language which realize the same elements of $S(\emptyset)$, then $\mathcal{M} \cong \mathcal{N}$.*

This proposition is a special case of Corollary 2.2.2, to be proved later, so we omit the proof for the sake of brevity. We recommend, however, that the reader construct an independent proof by adapting the argument used in Proposition 2.2.2.

Let T be a countable complete theory. We proved that T has a countable atomic model when $|S(\emptyset)| < 2^{\aleph_0}$ and T has a countable saturated model when $S(\emptyset)$ is countable. It is natural to ask if there is a countable complete theory with $|S(\emptyset)|$ strictly between $\aleph_0$ and $2^{\aleph_0}$ (in some model of set theory). The Cantor-Bendixson Theorem from point-set topology quickly gives a negative answer: $S_n(\emptyset)$ is countable or has cardinality $2^{\aleph_0}$. We will reproduce the proof of this theorem in the terminology of formulas and types, rather than open sets and points, not just for the sake of completeness but also to introduce Cantor-Bendixson rank which will have further applications. (Recall from Section 1.1 the definition of p implies φ, denoted $p \models \varphi$, where p is a type and φ is a formula.)

Definition 2.2.5. *Let T be a complete theory. The relation $CB(\varphi) = \alpha$, for φ a formula in n variables and α an ordinal or -1 is defined by the following recursion.*

(1) *$CB(\varphi) = -1$ if φ is inconsistent;*
(2) *Let $\Psi_\alpha = \{\psi : CB(\psi) = \beta$ for some $\beta < \alpha\}$.*
$CB(\varphi) = \alpha$ if $\{p \in S_n(\emptyset) : \varphi \in p$ and $\neg\psi \in p$ for all $\psi \in \Psi_\alpha\}$ is nonempty and finite.

For p any n-type, $CB(p)$ is defined to be

$$\inf\{CB(\varphi) : \varphi \text{ a formula implied by } p\}.$$

(For complete types $CB(p)$ is then $\inf\{CB(\varphi) : \varphi \in p\}$.) When $CB(p) = \alpha$ we say that the Cantor-Bendixson rank of p is α. *If there is no α with $CB(p) = \alpha$ we write $CB(p) = \infty$ and say that the Cantor-Bendixson rank of p does not exist.*

Extend the scope of $<$ so that $-1 < \alpha < \infty$ for all ordinals α. Then, $CB(p) \geq \alpha$ is a quick way to express that $CB(p) \neq \beta$ for all $\beta < \alpha$. Using these conventions (2) in the definition can be restated as: $CB(\varphi) = \alpha$ if $\{p \in S_n(\emptyset) : \varphi \in p$ and $CB(p) \geq \alpha\}$ is finite and nonempty. The term Cantor-Bendixson rank is usually shortened to *CB-rank*. It is clear from the definition that the CB relation defines a function.

The theme in the next basic lemma is the relationship between the CB-ranks of types and the CB-ranks of formulas implied by these types.

Lemma 2.2.3. *Let T be a complete theory, p an n-type and α an ordinal.*

(i) If p is complete then $CB(p) = 0$ if and only if p is isolated.

(ii) $CB(p) = \alpha$ if and only if there is a formula φ implied by p such that $\{q \in S_n(\emptyset) : \varphi \in q$ and $CB(q) = \alpha\}$ is finite and nonempty. Moreover, when $CB(p) = \alpha$ we can find a φ implied by φ such that $\{q \in S_n(\emptyset) : \varphi \in q$ and $CB(q) = \alpha\} = \{q \in S_n(\emptyset) : p \subset q$ and $CB(q) = \alpha\}$.

(iii) If $CB(p) = \alpha$ there is a $q \in S_n(\emptyset)$ such that $q \supset p$ and $CB(q) = \alpha$.

(iv) If p is complete and $CB(p) = \alpha$ there is a $\varphi \in p$ such that p is the only element of $\{q \in S_n(\emptyset) : \varphi \in q$ and $CB(q) \geq \alpha\}$.

(v) $CB(p) \geq \alpha$ if and only if, for all $\beta < \alpha$ and all φ implied by p, $\{q \in S_n(\emptyset) : \varphi \in q$ and $CB(q) \geq \beta\}$ is infinite.

(vi)

$$CB(\varphi) \text{ is the least ordinal } \alpha \text{ such that} \qquad (2.1)$$
$$\{p \in S_n(\emptyset) : \varphi \in p \text{ and } CB(p) \geq \alpha\} \text{ is finite.}$$

Proof. (i) If p is isolated by the formula φ then $CB(\varphi) = 0$, hence $CB(p) = 0$. Assuming, conversely, that p has CB-rank 0, there is a $\varphi \in p$ which is contained in only finitely many complete types, say $q_0, \ldots, q_k$, with $q_0 = p$. Let ψ be a formula in p implying φ and not in any of $q_1, \ldots, q_k$. Then ψ isolates p. (In fact, every completion of φ is isolated.)

(ii) Let Ψ be the set of formulas implied by p which have CB-rank α and $\Theta = \{\neg\theta : \theta$ is a formula in n variables with $CB(\theta) < \alpha\}$. Then $\psi \in \Psi \implies X_\psi = \{q \in S_n(\emptyset) : \psi \in q$ and $q \supset \Theta\}$ is finite and nonempty. Furthermore, if $\psi, \psi' \in \Psi$ and ψ implies ψ', $X_\psi \subset X_{\psi'}$. Thus, there is a $\varphi \in \Psi$ such that $X_\varphi = X_\psi$ for all $\psi \in \Psi$ which imply φ. Since p is implied by Ψ each element of X_φ contains p; i.e., $X_\varphi = \{q \in S_n(\emptyset) : p \subset q$ and $CB(q) = \alpha\}$. Since X_φ is finite and nonempty the proof of this part is complete.

(iii) This is just a repeat of part of (ii) but having an explicit reference is helpful.

(iv) This follows immediately from (ii). The details are left to the reader in Exercise 2.2.1.

(v) ($\Longleftarrow$) First notice that $CB(p) \geq \alpha$ if and only if

$$\text{for all } \varphi \text{ implied by } p,\ \{q \in S_n(\emptyset) : \varphi \in q \text{ and } CB(q) \geq \alpha\} \text{ is nonempty,}$$

which in turn is equivalent to

$$\text{for all } \varphi \text{ implied by } p,\ \{\varphi\} \cup \{\neg\psi : CB(\psi) < \alpha\} \text{ is consistent.}$$

Suppose that φ is implied by p and $\{\varphi\} \cup \{\neg\psi : CB(\psi) < \alpha\}$ is inconsistent. This inconsistency yields formulas $\psi_0, \ldots, \psi_n$ of CB-rank $< \alpha$ such that any complete type over $\emptyset$ containing φ also contains one of the ψ_i's. Since each of the ψ_i's has CB-rank $< \alpha$, $\beta = \max\{CB(\psi_0), \ldots, CB(\psi_n)\}$ is also $< \alpha$. For each $i \leq n$, $X_i = \{q \in S_n(\emptyset) : \psi_i \in q$ and $CB(q) \geq \beta\}$ is finite (simply because $CB(\psi_i) \leq \beta$), hence $X_0 \cup \ldots \cup X_n = \{q \in S_n(\emptyset) : \varphi \in q$ and $CB(q) \geq \beta\}$ is finite. Since $\beta < \alpha$ the right-hand-side of (v) fails, proving this direction.

($\Longrightarrow$) Suppose the right-hand-side of (v) to fail; i.e., there is a $\beta < \alpha$ and a φ implied by p such that $X = \{q \in S_n(\emptyset) : \varphi \in q$ and $CB(q) \geq \beta\}$ is finite. If X is empty then $CB(\varphi) < \beta$ (by (ii)), while if it is nonempty the definition of CB-rank yields $CB(\varphi) = \beta$. We conclude that $CB(\varphi)$, hence $CB(p)$ is not $\geq \alpha$, completing the proof.

(vi) This follows immediately from (v).

Occasionally, (2.1) is used as the definition of Cantor-Bendixson rank. Our definition is handy for proving properties of CB-rank, however (2.1) can be helpful in understanding CB-rank in particular examples (such as the ones given below).

When p is a complete type of CB-rank α and $\varphi \in p$ is such that p is the only completion of φ of CB-rank $\geq \alpha$ (see Lemma 2.2.3(iv)) we say that "φ isolates p relative to the types of CB-rank $\geq \alpha$."

Before applying Cantor-Bendixson rank to the problem of determining the possible cardinalities of $S(\emptyset)$ we give some illustrative examples.

Example 2.2.2. Let $L_1 = \{E\}$, where E is a binary relation. The theory T_1 in L_1 expressing that E is an equivalence relation with two classes, each infinite, is quantifier-eliminable. Let $\mathcal{M}$ be a model of T_1 and $T = Th(\mathcal{M}_M)$.

(So, by our conventions on parameters $S(\emptyset)$ in T is the same as $S(M)$ in T_1.) What does $S_1(\emptyset)$ in T look like? The isolated elements of $S_1(\emptyset)$, hence the elements of CB-rank 0, are exactly those containing $x = a$ for some $a \in M$. Let p be a complete 1–type containing $\{x \neq b : b \in M\}$. In any model of T the only E–classes are the two classes represented in M, so $E(x, a) \in p$ for some $a \in M$. If $b \in M$, then $E(x, b) \in p$ if and only if $\mathcal{M} \models E(a, b)$. By the quantifier-eliminability of T, p is the only element of $S_1(\emptyset)$ containing $\{E(x, a)\} \cup \{x \neq b : b \in M\} = \{E(x, a)\} \cup \{\neg\psi(x) : CB(\psi(x)) = 0\}$, hence $CB(E(x, a)) = CB(p) = 1$ (by (2.1)). Summarizing, $S_1(\emptyset)$ contains infinitely many isolated types and two types of CB-rank 1. (Notice that even though there is not a unique complete type of CB-rank ≥ 1, $CB(x = x) = 1$.)

Example 2.2.3. In the same language L_1 let T_2 be the theory saying that E is an equivalence relation with infinitely many infinite classes and no finite classes. Let $\mathcal{M}$ be a model of T_2, $T = Th(\mathcal{M}_M)$ and notice that T is also quantifier-eliminable. As in the previous example an element of $S_1(\emptyset)$ is isolated if and only if it contains $x = b$ for some $b \in M$. Also, for any $a \in M$, the formula $E(x, a)$ isolates a complete type q_a relative to the nonisolated types, hence $CB(E(x, a)) = CB(q_a) = 1$. Now consider any $p \in S_1(\emptyset)$ containing $\{x \neq b : b \in M\} \cup \{\neg E(x, a) : a \in M\}$, which exists by compactness. For any $\varphi \in p$ there are infinitely many $a \in M$ with $\varphi \in q_a$, so $CB(p) \geq 2$ by Lemma 2.2.3(v). Since p is the only complete 1–type which is nonisolated and not one of the q_a's (by quantifier elimination), $CB(x = x) = CB(p) = 2$ by (1).

Example 2.2.4. Here it is shown that for any ordinal α there is a theory with a type of CB-rank α. The theory is formulated as a chain of refining equivalence relations. Let α be an ordinal, and for $1 \leq \beta \leq \alpha$, let E_β be a binary relation. Let T_1 be the theory saying that each E_β is an equivalence relation with only infinite classes and for $1 \leq \beta < \gamma \leq \alpha$, E_β refines E_γ and each E_γ–class contains infinitely many E_β–classes. These two properties are guaranteed with the axioms:

$$\forall xy(E_\beta(x, y) \to E_\gamma(x, y)) \text{ and, for all } n < \omega,$$

$$\forall x \exists y_0 \cdots y_n(\bigwedge_{i \neq j \leq n} E_\gamma(x, y_i) \wedge \neg E_\beta(y_i, y_j)).$$

For $\mathcal{M} \models T_1$ let $T = Th(\mathcal{M}_M)$. As usual, T has elimination of quantifiers. Let $E_0(x, y)$ denote $x = y$.

Claim. For $\beta \leq \alpha$ and $a \in M$, $CB(E_\beta(x, a)) = \beta$.

This is proved by induction on β, with the case $\beta = 0$ being trivial. Let $\beta > 0$ and fix $\gamma < \beta$. There is an infinite set $B \subset M$ such that $E_\gamma(x, b)$ implies $E_\beta(x, a)$ and $b \neq b' \implies \neg E_\gamma(b, b')$ for all $b, b' \in B$. By induction, $E_\gamma(x, b)$ has CB-rank γ, for each $b \in B$, and extends to a complete 1–type of CB-rank γ by Lemma 2.2.3(iii). Since B is infinite this proves that $\{q \in S_1(\emptyset) :$

$E_\beta(x,a) \in q$ and $CB(q) \geq \gamma\}$ is infinite. By Lemma 2.2.3(v), $CB(E_\beta(x,a)) \geq \beta$. Furthermore, $X = \{q \in S_1(\emptyset) : E_\beta(x,a) \in q$ and $CB(q) \geq \beta\} \subset \{q \in S_1(\emptyset) : E_\beta(x,a) \in q$ and $\neg E_\gamma(x,b) \in q$, for $\gamma < \beta$ and $b \in M\} = Y$, X is nonempty (by Lemma 2.2.3(iii)) and Y contains a unique type by quantifier elimination. We conclude from Lemma 2.2.3(vi) that $CB(E_\beta(x,a)) = \beta$, as claimed.

From the axioms for the theory and the claim we conclude that T contains a complete type of CB-rank β for each $\beta \leq \alpha$.

While we restricted our attention to 1−types in these examples, similar arguments show that every element of $S(\emptyset)$ has CB-rank in each of these examples.

Example 2.2.5. In this final example a theory is specified (also involving equivalence relations) in which no nonalgebraic element of $S_1(\emptyset)$ has CB-rank. For $i < \omega$, let E_i be a binary relation. Let T_1 say that each E_i is an equivalence relation with infinitely many infinite classes and no finite classes and E_{i+1} refines E_i. Furthermore, each E_i−class is partitioned into infinitely many E_{i+1}−classes. Let $\mathcal{M} \models T_1$ and $T = Th(\mathcal{M}_M)$. An easy induction shows that for all nonalgebraic $p \in S_1(\emptyset)$ and ordinals α, $CB(p) \geq \alpha$.

In this section Cantor-Bendixson rank is applied to countable theories, however no assumption of countability is made in the definition or in the above examples.

The connection with the number of types in a countable theory is found in

Lemma 2.2.4. *For T a countable complete theory, the following are equivalent for each n.*

(1) $|S_n(\emptyset)| = \aleph_0$.
(2) $|S_n(\emptyset)| < 2^{\aleph_0}$.
(3) *Every $p \in S_n(\emptyset)$ has CB-rank equal to some $\alpha < \omega_1$.*

Proof. Trivially, (1) implies (2). To prove that (3) implies (1), let $\Phi = \{\varphi :$ φ is a formula in n variables having a unique completion p with $CB(p) = CB(\varphi)\}$. For each $\varphi \in \Phi$ let p_φ denote the completion of φ with the same CB-rank. By Lemma 2.2.3(iv) and the assumption that (3) holds, each element of $S_n(\emptyset)$ is p_φ for some $\varphi \in \Phi$. The countability of the theory then forces $S_n(\emptyset)$ to be countable; i.e., (1) holds.

To complete the proof we assume (3) to fail and prove that $|S_n(\emptyset)| = 2^{\aleph_0}$. Let Φ be the set of formulas in n variables which have CB-rank $< \omega_1$, and let $\alpha_{CB} = \sup\{CB(\varphi) : \varphi \in \Phi\}$. Since we have assumed (3) to fail there is a formula not in Φ.

Claim. For any formula φ in n variables not in Φ there is a formula ψ in n variables such that $\varphi \wedge \psi$ and $\varphi \wedge \neg\psi$ are both not in Φ.

Since $CB(\varphi) \geq \omega_1$, for each $\alpha < \omega_1$ there is a formula ψ_α such that $CB(\varphi \wedge \psi_\alpha) \geq \alpha$ and $CB(\varphi \wedge \neg\psi_\alpha) \geq \alpha$ (by Lemma 2.2.3(v)). Since there are countably many formulas there is a single formula ψ which is ψ_α for arbitrarily large $\alpha < \omega_1$. Thus, $\varphi \wedge \psi$ and $\varphi \wedge \neg\psi$ are both not in Φ.

Let X be the set of finite sequences of 0's and 1's.

Claim. There is a family of formulas φ_s, for $s \in X$, with the properties: (a) $\varphi_s \notin \Phi$, (b) if t is an initial segment of s then φ_s implies φ_t and (c) if t is not an initial segment of s and s is not an initial segment of t, then $\varphi_s \wedge \varphi_t$ is inconsistent with T.

The collection of φ_s's is defined by recursion. Given φ_s there is a ψ such that $\varphi_s \wedge \psi$ and $\varphi_s \wedge \neg\psi$ are both not in Φ (by the first claim). Let $\varphi_{s^\frown\langle 0\rangle} = \varphi_s \wedge \psi$ and $\varphi_{s^\frown\langle 1\rangle} = \varphi_s \wedge \neg\psi$. Properties (a), (b) and (c) are easy to verify.

Let Y denote the set of sequences of 0's and 1's of length ω and for $f \in Y$ let $p_f = \{\varphi_s : s \text{ is an initial segment of } f\}$. Each p_f is consistent (by (b)) and for distinct f and g in Y, $p_f \cup p_g$ is inconsistent (by (c)). Consistent completions of the p_f's form $2^{\aleph_0}$ many elements of $S_n(\emptyset)$, completing the proof.

In Lemma 2.1.1 it was proved that a small theory has a countable atomic model. The next proposition generalizes this result to potentially uncountable theories.

Proposition 2.2.6. *Let T be a complete theory in which each $p \in S(\emptyset)$ has CB-rank. Then the isolated types are dense in T.*

Proof. Let $\varphi(\bar{v})$ be a formula consistent with T and $\mathcal{O}_\varphi$ the set of complete types in $\bar{v}$ containing φ. Let $p \in \mathcal{O}_\varphi$ have minimal CB-rank among the elements of $\mathcal{O}_\varphi$ and let $\psi \in p$ be a formula isolating p relative to the types of CB-rank $\geq CB(p) = \alpha$. Without loss of generality, ψ implies φ, equivalently, $\mathcal{O}_\psi \subset \mathcal{O}_\varphi$. The following implications show that ψ isolates p: $q \in \mathcal{O}_\psi \implies q \in \mathcal{O}_\varphi \implies CB(q) \geq \alpha \implies q = p$.

A subset X of a topological space is *perfect* if it is closed and contains no isolated points, while X is *scattered* if every nonempty subset of X contains an isolated point. In topological terms the proof of Proposition 2.2.6 can be extended slightly to show that $S_n(\emptyset)$ is scattered when each complete n-type has CB-rank. Thus, by Lemma 2.2.4, a countable complete theory is small if and only if $S_n(\emptyset)$ is scattered, for each $n < \omega$. More generally, the set of complete n-types in a countable complete theory is the union of two disjoint sets, one scattered and one which is empty or perfect. The scattered set, which is countable, is the set of complete types having CB-rank, while the set of complete types without CB-rank, if nonempty, is perfect and of cardinality $2^{\aleph_0}$. (The reader is asked to prove this fact in Exercise 2.2.6.)

The remainder of this section is devoted to a complete treatment of saturated, homogeneous and universal models of potentially uncountable cardinalities.

Definition 2.2.6. *Let κ be an infinite cardinal and $\mathcal{M}$ a model.*

*(i) We call $\mathcal{M}$ $\kappa-$*saturated *if for all $A \subset M$ of cardinality $< \kappa$, $\mathcal{M}$ realizes every type in $S_1(A)$.*

*(ii) $\mathcal{M}$ is $\kappa-$*homogeneous *if for all $A \subset M$ with $|A| < \kappa$, $a \in M$ and elementary maps $f : A \longrightarrow M$, there is an elementary map $g : A \cup \{a\} \longrightarrow M$ extending f.*

*(iii) $\mathcal{M}$ is $\kappa-$*universal *if every model $\mathcal{N} \equiv \mathcal{M}$ of cardinality $< \kappa$ can be elementarily embedded into $\mathcal{M}$.*

If $\mathcal{M}$ is $|M|-$saturated, $|M|-$homogeneous or $|M|^+-$universal we call $\mathcal{M}$ saturated, homogeneous *or* universal, *respectively.*

As with countable saturated models, if $\mathcal{M}$ is $\kappa-$saturated then for all $A \subset M$ of cardinality $< \kappa$, $\mathcal{M}$ realizes every type in $S(A)$ (see Exercise 2.2.7).

Example 2.2.6. (Uncountable saturated models)

Let F be a countable field and T the theory of infinite dimensional vector spaces over F (which is complete and quantifier-eliminable). We will show that every uncountable model of T is saturated using the following.

Claim. If $\mathcal{M} \models T$ and $A \subset M$ there is a unique nonalgebraic type in $S_1(A)$.

Let p, $q \in S_1(A)$ be nonalgebraic. Taking an elementary extension of $\mathcal{M}$ if necessary we can assume that p and q are realized by elements a, $b \in M$, respectively. Since a and b are not in the subspace generated by A there is an automorphism of $\mathcal{M}$ fixing A and mapping a to b. Since automorphisms preserve types $p = q$.

Now let $\mathcal{M}$ be a model of T of cardinality $\kappa > \aleph_0$ and let $A \subset M$ have cardinality $< \kappa$. Any algebraic type over A is realized in $\mathcal{M}$, so it suffices to consider the unique nonalgebraic $p \in S_1(A)$. Since $|A| < \kappa$ and the field is countable the subspace generated by A has cardinality $< \kappa$. (The quantifier-eliminability of T implies that $tp(a/A)$ is algebraic if and only if a is in the subspace generated by A.) Thus, there is an $a \in M$ such that $tp(a/A)$ is p. Thus, $\mathcal{M}$ is saturated.

Not every countable complete theory has an $\aleph_0-$saturated model in every infinite power. For example, if the theory has continuum many complete types over $\emptyset$ then every $\aleph_0-$saturated model has cardinality $\geq 2^{\aleph_0}$ (since continuum many tuples from a model are needed to realize all of the types). There are similar (and often more complicated) restrictions on the cardinalities of $\lambda-$saturated models of arbitrary theories when λ is an uncountable cardinal. Consider, for instance, the following theory.

Example 2.2.7. The language L consist of two unary relations, P and Q, and a binary relation R. Let X be a set of cardinality $\aleph_0$, Y the power set of X and $E \subset X \times Y$ the relation which is satisfied by (x, y) exactly when x is an element of y. Let $\mathcal{M}$ be the model in this language with universe $X \cup Y$ where X is the interpretation of P, Y interprets Q and E interprets R. Let $T = Th(\mathcal{M})$. Then, for κ an infinite cardinal, any κ^+–saturated model $\mathcal{N}$ of T must have cardinality $\geq 2^\kappa$. (Let A be a subset of $P(\mathcal{N})$ of cardinality κ. For any $B \subset A$ the set of formulas $p_B = \{ R(b, v) : b \in B \} \cup \{ \neg R(a, v) : a \in A \setminus B \}$ is consistent and realized in $\mathcal{N}$, since $\mathcal{N}$ is κ^+–saturated. The realizations of the types p_B, as B ranges over the subsets of A, form a subset of $Q(\mathcal{N})$ of cardinality 2^κ.) Thus, given λ an infinite cardinal, if a saturated model of cardinality λ exists, then $\lambda \geq \sum_{\mu<\lambda} 2^\mu$.

In conclusion, the existence of a saturated model of cardinality κ may require κ to satisfy a relation of cardinal arithmetic which could fail in some model of set theory.

Recall that a cardinal λ is *regular* if λ is infinite and has cofinality λ. For all infinite cardinals κ, κ^+ is regular.

For an arbitrary theory the most general statement that can be made about the existence of models with some amount of saturation is

Lemma 2.2.5. *Let T be a complete theory and κ an infinite cardinal $\geq |T|$. Then T has a κ^+–saturated model of cardinality 2^κ.*

Proof. The targeted model will be constructed as the union of an elementary chain of models. At successor stages in the recursive definition of the elementary chain we will use

Claim. Let $\mathcal{M}$ be a model of T of cardinality 2^κ. Then there is an elementary extension $\mathcal{N}$ of $\mathcal{M}$ of cardinality 2^κ such that for every $A \subset M$ of cardinality κ, $\mathcal{N}$ realizes every element of $S_1(A)$.

First, let's count the number of types involved. If $|A| = \kappa$, then the number of formulas over A is $|A| + |T| = \kappa$, hence $|S_1(A)| \leq 2^\kappa$. The number of subsets of cardinality κ of a set X is $|X|^\kappa$ (or 0), so if $|X| = 2^\kappa$ there are $(2^\kappa)^\kappa = 2^\kappa$ many such subsets. Thus, $P = \bigcup\{ S_1(A) : A \subset M \text{ with } |A| = \kappa \}$ has cardinality 2^κ. Enumerate P as $\{ p_i : i < 2^\kappa \}$ and add to the language new constants c_i, $i < 2^\kappa$. Consider the theory $T' = Th(\mathcal{M}_M) \cup \bigcup\{ p_i(c_i) : i < 2^\kappa \}$. (By $p_i(c_i)$ we mean $\{ \varphi(c_i) : \varphi \in p_i \}$.) Compactness implies the consistency of T', hence it has a model of cardinality 2^κ. The restriction of this model to the original language is the model $\mathcal{N}$ required to prove the claim.

An elementary chain $\mathcal{M}_\beta$ for $\beta < \kappa^+$ is constructed as follows by recursion. Let $\mathcal{M}_0$ be any model of T of cardinality 2^κ. Assuming that $\mathcal{M}_\gamma$ has been defined let $\mathcal{M}_{\gamma+1}$ be an elementary extension of $\mathcal{M}_\gamma$ of cardinality 2^κ such that for every subset $A \subset M_\gamma$ of cardinality κ, $\mathcal{M}_{\gamma+1}$ realizes every element of $S_1(A)$ (the claim guarantees the existence of such a model). If δ is a limit ordinal let $\mathcal{M}_\delta = \bigcup_{\gamma<\delta} \mathcal{M}_\gamma$. Now let $\mathcal{M}$ be the union of the elementary chain, $\mathcal{M}_\beta$, $\beta < \kappa^+$. To verify that $\mathcal{M}$ is κ^+–saturated, let A be a subset of M of cardinality κ and p an element of $S_1(A)$. The elementary chain of $\mathcal{M}_\gamma$'s has length κ^+. Since κ^+ is regular there is a $\beta < \kappa^+$ such that $A \subset M_\beta$, hence p is realized in $\mathcal{M}_{\beta+1}$ (by construction). Since $\mathcal{M}_{\beta+1} \prec \mathcal{M}$, the same element realizes p in $\mathcal{M}$, proving the lemma.

As we will see later, saturated models are more useful than κ–saturated models where κ is less than the cardinality of the model (since they are homogeneous). The last lemma only guarantees the existence of a saturated model when $\kappa^+ = 2^\kappa$, a condition which is independent of the axioms of set theory, for many κ. A natural question is: What properties must a theory T and a cardinal λ possess in order for an adaptation of the above argument to yield a saturated model of T of cardinality λ? The special properties of κ^+ and 2^κ that were used to produce a κ^+–saturated model are: $|S_1(A)| \leq 2^\kappa$ when $|A| = \kappa$, and the cofinality of κ^+ is $> \kappa$. Thus, for the same argument to yield a saturated model of cardinality λ, λ must satisfy the conditions: $|A| < \lambda \implies |S_1(A)| < \lambda$, and the cofinality of λ is λ. (This condition on cofinality is used to guarantee that when $\mathcal{M}_\beta$, $\beta < \lambda$, is an elementary chain (λ a cardinal) and $A \subset \bigcup_{\beta<\lambda} M_\beta$ has cardinality $< \lambda$ there is a single M_β containing A.)

The reasoning in the previous paragraph leads to

Lemma 2.2.6. *Let T be a complete theory and $\kappa \geq |T|$ a regular cardinal such that for all models $\mathcal{M}$ of T and $A \subset M$, $|A| < \kappa \implies |S_1(A)| < \kappa$. Then T has a saturated model of cardinality κ.*

Proof. The proof is a rather straight-forward adaptation of the proof of the previous lemma. Arguing as in the claim a model $\mathcal{M}$ of cardinality $< \kappa$ has an elementary extension $\mathcal{N}$ of cardinality $< \kappa$ which realizes every element of $S_1(M)$. Now define an elementary chain $\mathcal{M}_\beta$, $\beta < \kappa$, such that

- every element of $S_1(M_\beta)$ is realized in $\mathcal{M}_{\beta+1}$, and $|M_{\beta+1}| < \kappa$, for all $\beta < \kappa$, and
- $\mathcal{M}_\delta = \bigcup_{\gamma<\delta} \mathcal{M}_\gamma$, when $\delta < \kappa$ is a limit ordinal.

Let $\mathcal{M}$ be the union of the chain, $\mathcal{M}_\beta$, $\beta < \kappa$ and notice that $|M| = \kappa$. To verify that $\mathcal{M}$ is saturated let A be a subset of M of cardinality $< \kappa$. Since κ is regular there is some $\beta < \kappa$ with $A \subset M_\beta$. Every element of $S_1(A)$ extends to an element of $S_1(M_\beta)$, which is realized in $\mathcal{M}_{\beta+1}$, hence in $\mathcal{M}$. (In the exercises the reader is asked to write out a complete proof of this lemma without referring to Lemma 2.2.5.)

A cardinal κ is called *strongly inaccessible* if κ is regular, uncountable and $2^\lambda < \kappa$ whenever $\lambda < \kappa$. It cannot be proved in ZFC that strongly inaccessible cardinals exist, however if λ is strongly inaccessible the last lemma shows that every theory of cardinality $\leq \lambda$ (with an infinite model) has a saturated model of cardinality λ.

We proved previously that every countable complete theory has a countable homogeneous model. Unfortunately, if we try to construct a homogeneous model of uncountable cardinality, say $\aleph_1$, set-theoretic problems arise similar to those which limit the existence of saturated models. For example, let's try generalizing the elementary chain argument used in the proof of Proposition 2.2.3. For $\mathcal{M}_\alpha$ a model of cardinality $\aleph_1$ and $A \subset M_\alpha$ countable, there are at most $\aleph_1$ many 1−types over A realized in $\mathcal{M}_\alpha$, however there are $2^{\aleph_0}$ many countable subsets of M_α to consider as sets of parameters. Thus, for a model $\mathcal{M}_\alpha$ of cardinality $\aleph_1$ there may not be an elementary extension $\mathcal{M}_{\alpha+1}$ of cardinality $\aleph_1$ (without assuming the continuum hypothesis) such that for all elementary maps $g : A \longrightarrow M_\alpha$, where $A \subset M_\alpha$ is countable, and $a \in M_\alpha$, there is an elementary $f : A \cup \{a\} \longrightarrow M_{\alpha+1}$ extending g.

Definition 2.2.7. *For $\mathcal{M}$ a model, the* type diagram of $\mathcal{M}$, *denoted $D(\mathcal{M})$, is* $\{p \in S(\emptyset) : p$ is realized in $\mathcal{M}\}$.

Notice that for $\mathcal{M}$ $\aleph_0$−saturated $D(\mathcal{M}) = S(\emptyset)$. The relative uniqueness of homogeneous models, and other results, will follow quickly from the next two lemmas.

Lemma 2.2.7. *Let $\mathcal{M}$ to be a κ−homogeneous model, $\mathcal{N}$ a model elementarily equivalent to $\mathcal{M}$ containing sets $A \subset B$ such that $|A| < |B| \le \kappa$ and $\{ tp_{\mathcal{N}}(\bar{b}) : \bar{b}$ is a finite sequence from $B \} \subset D(\mathcal{M})$. Then for any elementary map $f : A \longrightarrow M$ there is an elementary map $g : B \longrightarrow M$ which extends f.*

Proof. This is proved by induction on $|B|$. Enumerate $B \setminus A$ as $\{ b_\alpha : \alpha < \lambda \}$, where $\lambda = |B \setminus A|$. The desired extension g of f will be constructed by defining it (recursively) on b_α for successively larger $\alpha < \lambda$. Given $\delta < \lambda$, assume that the elementary map g extending f has been defined on b_α for all $\alpha < \delta$. Because $B_\delta = A \cup \{ b_\alpha : \alpha \le \delta \}$ has cardinality $< |B|$ there is an elementary map $h : B_\delta \longrightarrow M$ extending f. Since gh^{-1} is an elementary map on $\mathcal{M}$ taking $A \cup \{ h(b_\alpha) : \alpha < \delta \}$ to $A \cup \{ g(b_\alpha) : \alpha < \delta \}$, the κ−homogeneity of $\mathcal{M}$ yields a b such that $gh^{-1} \cup \{(h(b_\delta), b)\}$ is an elementary map. Define $g(b_\delta)$ to be b.

Lemma 2.2.8. *Suppose that $\mathcal{M}$ and $\mathcal{N}$ are homogeneous models of the same cardinality, $D(\mathcal{M}) = D(\mathcal{N})$, $A \subset M$ with $|A| < |M|$, and f is an elementary map of A into N. Then, f can be extended to an isomorphism from $\mathcal{M}$ onto $\mathcal{N}$.*

Proof. The isomorphism is constructed with a back and forth argument using the previous lemma. More precisely, a chain of elementary maps, f_α, $\alpha < |M|$, is constructed such that $f_0 = f$, every element of M is in the domain of some f_α, and every element of N is in the range of some f_α. Well-order the sets $M \setminus A$ and $N \setminus f(A)$ and suppose that f_γ has been defined for each $\gamma < \delta$. If δ is a limit ordinal let $f_\delta = \bigcup_{\gamma<\delta} f_\gamma$. Suppose that $\delta = \gamma + 2n$, where γ is a limit ordinal and $n \in \omega$ is > 0. Let a be the least element of

$M \setminus (dom f_{\gamma+2n-1})$ in the well-ordering of $M \setminus A$. By Lemma 2.2.7 there is an elementary map f_δ extending $f_{\gamma+2n-1}$ whose domain contains a. Finally, suppose $\delta = \gamma + 2n + 1$, for γ a limit ordinal and $n \in \omega$, and let b be the least element of $N \setminus (range f_{\gamma+2n})$. By the previous lemma there is an elementary map g extending the inverse of $f_{\gamma+2n}$ whose domain contains b. Let $f_\delta = g^{-1}$. It is clear that $\bigcup_{\alpha<|M|} f_\alpha$ is the desired isomorphism from $\mathcal{M}$ onto $\mathcal{N}$.

Corollary 2.2.1 (Extendibility of elementary maps). *If $\mathcal{M}$ is a homogeneous model, $A \subset M$ with $|A| < |M|$ and f is an elementary map from A into M, then f can be extended to an automorphism of $\mathcal{M}$.*

Corollary 2.2.2 (Relative uniqueness of homogeneous models). *Let $\mathcal{M}$ and $\mathcal{N}$ be homogeneous models of the same cardinality with $D(\mathcal{M}) = D(\mathcal{N})$. Then $\mathcal{M} \cong \mathcal{N}$.*

Since $D(\mathcal{M}) = S(\emptyset)$ for any saturated model $\mathcal{M}$,

Corollary 2.2.3 (Uniqueness of saturated models). *If $\mathcal{M}$ and $\mathcal{N}$ are saturated models of the same cardinality and the same complete theory, then $\mathcal{M} \cong \mathcal{N}$.*

Corollary 2.2.4 (Relative universality of homogeneous models). *Let $\mathcal{M}$ be κ–homogeneous and $\mathcal{N}$ a model of cardinality $\leq \kappa$ such that $D(\mathcal{N}) \subset D(\mathcal{M})$. Then $\mathcal{N}$ is elementarily embeddable into $\mathcal{M}$.*

Corollary 2.2.5 (Saturated=homogeneous+universal). *A model $\mathcal{M}$ which is κ–saturated is κ–homogeneous and κ^+–universal. As a partial converse, if $\mathcal{M}$ is κ–homogeneous and $D(\mathcal{M}) = S(\emptyset)$, then $\mathcal{M}$ is κ–saturated.*

Proof. The κ^+–universality of a κ–saturated model is by the previous corollary while its κ–homogeneity is clear from the definition. Now let $\mathcal{M}$ be κ–homogeneous with $D(\mathcal{M}) = S(\emptyset)$, $A \subset M$ of cardinality $< \kappa$ and $p \in S_1(A)$. By the consistency of p there is a model $\mathcal{N}$ containing A such that $\mathcal{M}_A \equiv \mathcal{N}_A$ and p is realized in $\mathcal{N}$ by some element a. By Lemma 2.2.7 there is an elementary map of $\{a\} \cup A$ into M which is the identity on A. The image of a under this map is the desired realization of p.

Corollary 2.2.6. *A homogeneous model which realizes every element of $S(\emptyset)$ is saturated.*

Example 2.2.8. (A countable universal model which is not saturated) Let T be the theory of the order $(\omega, \leq)$. The reader can show that T is quantifier-eliminable. Every model of T can be obtained in the following manner: Let $\mathcal{M} = (A, \leq)$ be any linear order. Form the model $\mathcal{N}$ of T by adding one copy of the ordering $(\mathbb{Z}, \leq)$ to the end of $(\omega, \leq)$ for each element of A, respecting the order induced by $\mathcal{M}$. More precisely,

(a) $N = \omega \cup (A \times \mathbb{Z})$,

(b) $\leq$ is the standard ordering on ω,
(c) $n \leq (a,m)$ for all $n \in \omega$ and $(a,m) \in (A \times \mathbb{Z})$,
(d) $(a,m) \leq (b,n)$ if and only if $a < b$ or $a = b$ and $m \leq n$, for all (a,m), $(b,n) \in (A \times \mathbb{Z})$.

It is not difficult to see that the model $\mathcal{M}$ obtained by letting $(A, \leq) = (\mathbb{Q}, \leq)$ in this algorithm, is a countable saturated model (hence a universal model). Regardless of the theory, any countable elementary extension of a countable saturated model is also universal. This observation quickly leads to a countable universal model which is not saturated: Let $\mathcal{N}$ be the model obtained by adding one copy of $(\mathbb{Z}, \leq)$ to the end of $\mathcal{M}$. ($\mathcal{N}$ cannot be saturated since it is not homogeneous.)

One of the conditions holding in a κ–saturated model $\mathcal{M}$ of cardinality κ, and perhaps failing in a κ–saturated model of cardinality $> \kappa$ is: any elementary map $f : A \to M$, where $A \subset M$ and $|A| < \kappa$, extends to an automorphism of $\mathcal{M}$. In some uses of saturated models they can be replaced by models which only satisfy this extendibility condition, which is given a name in the following definition.

Definition 2.2.8. *For κ an infinite cardinal, the model $\mathcal{M}$ is called* strongly κ–homogeneous *if for all elementary maps $f : A \to M$, where $A \subset M$ and $|A| < \kappa$, f extends to an automorphism of $\mathcal{M}$.*

Widespread existence of these models is proved through

Lemma 2.2.9. *Let $\mathcal{M}$ be an infinite model, $A \subset M$, and $f : A \to M$ an elementary map. Then there is an $\mathcal{N} \succ \mathcal{M}$ having an automorphism g extending f with $|N| = |M|$.*

Proof. The model $\mathcal{N}$ will be constructed in two stages. First we prove

Claim. Let $\mathcal{N}_0$ be an infinite model, $B \subset N_0$, and $f : B \longrightarrow N_0$ an elementary map. Then there is an $\mathcal{N}_1 \succ \mathcal{N}_0$ with $|N_1| = |N_0|$ and an elementary map $g : N_1 \longrightarrow N_1$ which extends f.

Let $\mathcal{M}_0 = \mathcal{N}_0$. We define, by recursion, an elementary chain of models $\mathcal{M}_i$, $i < \omega$, and elementary maps $g_i : M_i \to M_{i+1}$ such that $f \subset g_0 \subset \ldots \subset g_i \subset \ldots$. To begin, let $\kappa = |M_0|$ and let $\mathcal{M}'_1$ be a κ^+–saturated elementary extension. By Lemma 2.2.7 there is an elementary map $g_0 \supset f$ taking M_0 into M'_1. Let $\mathcal{M}_1$ be an elementary submodel of $\mathcal{M}'_1$ of cardinality κ containing both M_0 and $g_0(M_0)$. In general, let $\mathcal{M}'_{i+1} \succ \mathcal{M}_i$ be a κ^+–saturated model, $g_{i+1} : M_i \to M'_{i+1}$ an elementary map extending g_i, and $\mathcal{M}_{i+1}$ an elementary submodel of $\mathcal{M}'_{i+1}$ containing both M_i and $g_i(M_i)$. It is easily verified that $\mathcal{N}_1 = \bigcup_{i<\omega} \mathcal{M}_i$ and $g = \bigcup g_i$ satisfy the requirements of the claim.

The model $\mathcal{N}$ and automorphism g will be obtained as the limit of ω applications of the claim. Let $\kappa = |M|$, $\mathcal{N}_0 = \mathcal{M}$ and $g_0 = f$. To begin, the claim yields a model $\mathcal{N}_1 \succ \mathcal{N}_0$ of cardinality κ and an elementary map

$g_1 : N_1 \longrightarrow N_1$ extending g_0. Now apply the claim to $\mathcal{N}_1$ and g_1^{-1} to obtain an $\mathcal{N}_2 \succ \mathcal{N}_1$ of cardinality κ with an elementary map $g_2 : N_2 \longrightarrow N_2$ which extends g_1^{-1}. Continuing in this manner results in an elementary chain:

$$\mathcal{N}_0 \prec \mathcal{N}_1 \prec \mathcal{N}_2 \prec \ldots \prec \mathcal{N}_i \prec \ldots$$

and a chain of elementary maps $g_0 \subset g_1 \subset g_2^{-1} \subset g_3 \subset \ldots$ such that if i is odd the domain of g_i is $\mathcal{N}_i$ and if i is even (and > 0) the range of g_i^{-1} is $\mathcal{N}_i$. Then $\mathcal{N} = \bigcup_{i<\omega} \mathcal{N}_i$ is an elementary extension of $\mathcal{M}$ of cardinality κ and $g = \bigcup_{i<\omega} g_{2i+1}$ is an automorphism of $\mathcal{N}$ which extends f.

Proposition 2.2.7. *For κ an infinite cardinal and T a theory of cardinality $\leq \kappa$ there is a strongly κ−homogeneous model of cardinality $\leq 2^\kappa$.*

Proof. Given a model $\mathcal{N}$ of cardinality $\leq 2^\kappa$ let $\Phi = \{ f :$ for some $A \subset N$ of cardinality $\leq \kappa$, $f : A \longrightarrow N$ is elementary $\}$. Generalizing the proof of the previous lemma shows that there is an elementary extension $\mathcal{N}'$ of $\mathcal{N}$ of cardinality $\leq 2^\kappa$ such that every element of Φ extends to an automorphism of $\mathcal{N}'$. Using this fact the reader can construct an elementary chain whose union is the desired model.

Corollary 2.2.7. *Every model $\mathcal{M}$ has a strongly $\aleph_0$−homogeneous elementary extension of the same cardinality.*

Proof. Left to the reader.

Historical Notes. Cantor-Bendixson rank was introduced into model theory by Morley in [Mor65]. The notions of κ−saturated and saturated models go back to the η_α−sets of Hausdorff. (These are basically saturated models of the theory of dense linear orders without endpoints.) Their importance in model theory was not exploited until the late '50's. Universal and homogeneous models were developed by Fraïsé and Jónsson. Most of the results relating universal, saturated and homogeneous models are proved by Morley and Vaught in [MV62]. However, the relative uniqueness of homogeneous models, Corollary 2.2.2, was proved by Keisler and Morley in [KM67].

Exercise 2.2.1. Write out a proof of Lemma 2.2.3(iv).

Exercise 2.2.2. Write out the details in Example 2.2.5.

Exercise 2.2.3. Suppose that T is a complete theory in a countable language L which is not small and $T' \supset T$ is a complete theory in a language $L' \supset L$. Show that T' is also not small. On the other hand, if T is small, $\mathcal{M} \models T$ and $A \subset M$ is finite, then $T' = Th(\mathcal{M}_A)$ is small.

Exercise 2.2.4. Prove that the union of an elementary chain of atomic models is atomic. Use this fact to show that a countable complete theory having a prime model which is not minimal has an uncountable atomic model. (HINT: A prime model which is not minimal is isomorphic to a proper elementary extension of itself.)

Exercise 2.2.5. Show that a countable complete theory which is not small has $2^{\aleph_0}$ many nonisomorphic countable homogeneous models.

Exercise 2.2.6. Let T be a countable complete theory. Prove that $S_n(\emptyset)$ is the union of a scattered set and a set which is perfect or empty.

Exercise 2.2.7. Prove that if $\mathcal{M}$ is $\kappa-$saturated and A is a subset of M of cardinality $< \kappa$, then $\mathcal{M}$ realizes every complete $n-$type over A, for all n. (HINT: Use induction on n.)

Exercise 2.2.8. Let κ be an infinite cardinal, $\mathcal{M}$ a $\kappa-$saturated model and $A \subset M$ a set of cardinality $< \kappa$. Show that $\mathcal{M}_A$ is also $\kappa-$saturated.

Exercise 2.2.9. Let $\mathcal{M}$ be an $\aleph_0-$saturated model and p a complete $1-$type in $Th(\mathcal{M})$ such that $p(\mathcal{M})$ is finite. Prove that p is algebraic.

Exercise 2.2.10. Prove that a countable model $\mathcal{M}$ is saturated if and only if it is universal over every finite subset of M.

Exercise 2.2.11. Prove that the union of an elementary chain of $\aleph_0-$homogeneous models is $\aleph_0-$homogeneous.

Exercise 2.2.12. Show that every reduct of a saturated model to a sublanguage is saturated. (The corresponding result about homogeneous models is false. What do you think goes wrong in the proof?)

Exercise 2.2.13. Prove that if $\mathcal{M}$ is $\kappa-$saturated, A is a subset of M of cardinality $< \kappa$ and $p \in S(A)$, then the cardinality of the set of realizations of p in $\mathcal{M}$ is finite or $\geq \kappa$.

Exercise 2.2.14. Write out a complete proof of Lemma 2.2.6 which does not refer to the proof of Lemma 2.2.5.

Exercise 2.2.15. Suppose that κ and λ are infinite cardinals and the cofinality of λ is $\geq \kappa$. Prove that the union of an elementary chain $\mathcal{M}_\alpha$, $\alpha < \lambda$, of $\kappa-$saturated models is $\kappa-$saturated.

Exercise 2.2.16. Find a union of an elementary chain of $\aleph_1-$saturated models which is not $\aleph_1-$saturated. (HINT: See the last example concerning CB-rank.)

Exercise 2.2.17. Write out the details in the proof of Proposition 2.2.7.

Exercise 2.2.18. Prove Corollary 2.2.7.

Exercise 2.2.19. Prove that if $\mathcal{M}$ is strongly $\aleph_0$−homogeneous and $\bar{a}$ and $\bar{b}$ are sequences from $\mathcal{M}$ realizing the same type, then for all formulas $\varphi(\bar{x}, \bar{y})$, $|\varphi(\mathcal{M}, \bar{a})| = |\varphi(\mathcal{M}, \bar{b})|$.

2.3 Countable Models of Complete Theories

In the preceding section we studied models of complete theories which have special properties with respect to the types they realize and automorphisms they admit. The sharpest existence and uniqueness theorems were obtained for the countable models of complete theories. In this section we apply these results to study the following two problems about countable complete theories:

1. Characterize the countable complete theories which have a unique countable model, up to isomorphism.
2. For T a countable complete theory let $n(T)$ denote the number of countable models of T up to isomorphism. Determine the possible values of $n(T)$, as T ranges over the countable complete theories.

Definition 2.3.1. *For κ an infinite cardinal, a theory T is κ*−categorical *(or* categorical in *κ) if T has a unique model of cardinality κ, up to isomorphism.*

Later extensive attention will be given to countable theories which are categorical in some uncountable cardinality. Problem 1 asks for a characterization of $\aleph_0$−categorical theories. Here are some examples of such theories:

- Let T be the theory in a language with no nonlogical symbols expressing that there are infinitely many elements. The countable models of this theory are sets of cardinality $\aleph_0$, hence are all isomorphic.
- The theory of dense linear orders without endpoints is $\aleph_0$−categorical.
- All vector spaces of dimension $\aleph_0$ over a fixed field F are isomorphic. In the natural language of vector spaces over F there is no set of sentences which express that the dimension is $\aleph_0$ for an arbitrary field. However, assuming that F is a finite field any countable model of the theory T of *infinite* vector spaces over F must have dimension $\aleph_0$. Thus, T is categorical in $\aleph_0$.
- The theory of an equivalence relation with infinitely many infinite classes and no finite classes is $\aleph_0$−categorical.

Remark 2.3.1. Let T be a countable theory with an infinite model which is categorical in some infinite cardinal. Then T is complete. This is a classical result called the Łoś-Vaught Test (for completeness of a theory). (The proof is left to the reader in Exercise 2.3.1.)

The class of countable theories which are categorical in $\aleph_0$ admits the following characterization.

Theorem 2.3.1 (Ryll-Nardzewski, Engeler, Svenonius). *For a countable complete theory T the following are equivalent:*

(1) *T is $\aleph_0$-categorical.*
(2) *For each n, $S_n(\emptyset)$ is finite.*
(3) *For each n, there are finitely many formulas in n free variables, up to equivalence in T.*

Proof. First we establish a couple of preliminary claims about countable complete theories which are of interest in their own right.

Claim. Let T' be a countable complete theory with an infinite model such that every complete n-type in T is isolated. Then there are finitely many complete n-types.

Assuming every complete n-type to be isolated let $\Gamma = \{\neg\varphi : \varphi$ is a formula in n variables which isolates a complete type in $T'\}$. Since Γ is not contained in a complete type it is inconsistent. By compactness there are formulas $\varphi_0, \dots, \varphi_m$ which isolate distinct complete types in T' such that $T' \models \forall\bar{v}(\varphi_0(\bar{v}) \vee \dots \vee \varphi_m(\bar{v}))$. Thus, each element p of $S_n(\emptyset)$ contains one of the φ_i's, and for each i only one element of $S_n(\emptyset)$ contains φ_i. We conclude that $|S_n(\emptyset)| = m + 1$, proving the claim.

Claim. Let T' be a countable complete theory such that $S_n(\emptyset)$ is finite. Then there are finitely many formulas in n free variables up to equivalence with respect to T'.

For φ a formula in n variables, $\mathcal{O}_\varphi$ denotes the set of complete n-types in T' which contain φ. There are only finitely many subsets of $S_n(\emptyset)$ which can be $\mathcal{O}_\varphi$ for some formula φ in n variables. Since $\mathcal{O}_\varphi = \mathcal{O}_\psi$ if and only if $T' \models \forall\bar{v}(\varphi \leftrightarrow \psi)$, there are finitely many formulas in n variables in T'.

Turning to the main body of the proof, let T be $\aleph_0$-categorical. A countable theory which is not small has continuum many countable models, hence T is small and has a countable atomic model and a countable saturated model. Thus, the unique countable model $\mathcal{M}$ of T is both atomic and saturated. Since every element of $S_n(\emptyset)$ is realized in $\mathcal{M}$, every complete n-type is isolated. By the first claim, $S_n(\emptyset)$ is finite, proving that (1) $\Longrightarrow$ (2).

That (2) $\Longrightarrow$ (3) is proved in the second claim, so assume (3) to hold. The $\aleph_0$- categoricity of T will follow from the uniqueness of prime models (Proposition 2.1.1) once we show that every model of T is atomic. Let $\mathcal{M}$ be a model of T and $\bar{a}$ a finite tuple from M, and let $\varphi_1, \dots, \varphi_m$ be a list of the finitely many formulas, up to equivalence in T, which are satisfied by $\bar{a}$ in $\mathcal{M}$. Then $\varphi_1 \wedge \dots \wedge \varphi_m$ isolates $tp_{\mathcal{M}}(\bar{a})$. Thus, $\mathcal{M}$ is atomic, completing the proof of the theorem.

In the remainder of the section we discuss the progress to date on the second question stated above, namely:

For T a countable complete theory what are the possibilities for $n(T)$, the number of countable models of T up to isomorphism.

Simply for set-theoretic reasons we know that $n(T) \leq 2^{\aleph_0}$. A theory with continuum many complete types has $2^{\aleph_0}$ many countable models up to isomorphism (since every complete type is realized in some countable model). Thus, $n(T) = 2^{\aleph_0}$ is one possibility. (In the previous section we gave an example of a theory which is not small.) We have also seen examples of countable complete theories in which $n(T) = 1$; i.e., $\aleph_0$−categorical theories. We continue by giving examples which delineate less trivial possibilities.

For the remainder of this section an arbitrary theory is assumed to be countable, complete and have infinite models.

Example 2.3.1. (Of a theory with $\aleph_0$ many countable models) Let T be the theory of algebraically closed fields of characteristic 0. By quantifier elimination the isomorphism type of a model is determined by the transcendence degree of the model over the prime field. There are $\aleph_0$ many possibilities for countable transcendence degree.

Example 2.3.2. (Of a simpler theory with $\aleph_0$ many countable models) Consider the theory in a language with constant symbols c_i, $i < \omega$, which says that the constants are distinct. This theory has elimination of quantifiers and is complete. For each $n < \omega$ there is a model of T with exactly n elements which are not the interpretation of some constant. Furthermore, the isomorphism type of any model is determined by the number of such nonconstant elements. Thus, there are $\aleph_0$ many countable models.

Example 2.3.3. (Of a small theory with $2^{\aleph_0}$ many countable models) Example 2.2.1(iii) is an example of such a theory. The fact that the theory has a countable saturated model is equivalent to it being small. For any $X \subset \omega$ there is a countable model $\mathcal{M}_X$ of T such that $P_i(\mathcal{M}_X)$ contains a nonconstant if and only if $i \in X$. Thus, T has continuum many nonisomorphic countable models.

Example 2.3.4. (Of a theory with 3 countable models) This is a classical example due to Ehrenfeucht. Let T be the theory in the language with a binary relation $<$ and constants, c_i, $i < \omega$, saying that $<$ is a dense linear order without endpoints and $c_n < c_{n+1}$, for $n < \omega$. An elimination of quantifiers argument shows that T is complete and has the following 3 countable models. $\mathcal{M}_1$ is the model in which every element is $<$ some c_n. $\mathcal{M}_2$ is the model which contains elements $>$ every c_n and there is a least such. Lastly, $\mathcal{M}_3$ is the model which contains elements greater than every c_n and $\sup_{n<\omega} c_n$ does not exist in the model. Notice that $\mathcal{M}_1$ is the prime model of T and $\mathcal{M}_3$ is the countable saturated model of the theory.

Closer examination of the last example reveals some interesting interplay between saturation, universality and atomicity over a finite set. Since this theory T contains the axioms for dense linear orders without endpoints, if there is an element in a model which is greater than all of the c_i's, then there is a copy of the rationals greater than all of the c_i's in this model. In fact, there is an isomorphic embedding of $\mathcal{M}_3$ into $\mathcal{M}_2$ (viewing $\mathcal{M}_1$ as a submodel of each of these, embed the remaining elements of $\mathcal{M}_3$ into the elements of $M_2 \setminus (M_1 \cup \{\sup_{n<\omega} c_n\})$ in an order-preserving way). Since T has elimination of quantifiers, if $\mathcal{M}$ and $\mathcal{N}$ are models of T and $\mathcal{M} \subset \mathcal{N}$, then $\mathcal{M} \prec \mathcal{N}$. Thus, $\mathcal{M}_3$ is elementarily embedded into $\mathcal{M}_2$. Since $\mathcal{M}_3$ is saturated, we conclude that $\mathcal{M}_2$ is universal. Since $\mathcal{M}_2$ is not saturated, $\mathcal{M}_2$ must fail to be homogeneous (by Corollary 2.2.5) which is verified as follows. Let a be the element of M_2 which is the supremum of the c_i's. All elements of M_2 which are greater than all of the c_i's realize the same complete type in $\mathcal{M}_2$. However, any automorphism of $\mathcal{M}_2$ leaves a fixed.

As a final remark on the properties of the $\mathcal{M}_i$'s notice that $\mathcal{M}_2$ is prime over a. (A model $\mathcal{N}$ is prime over a if whenever $\mathcal{M} \models T$ and f is an elementary map of a into $\mathcal{M}$, f can be extended to an elementary embedding of $\mathcal{N}$ into $\mathcal{M}$. Since $\mathcal{M}_1$ does not realize $tp(a)$ there is no elementary map taking a into $\mathcal{M}_1$. Since $\mathcal{M}_3$ is saturated any elementary $f : \{a\} \longrightarrow \mathcal{M}_3$ extends to an elementary embedding of $\mathcal{M}_2$ into $\mathcal{M}_3$ (by Lemma 2.2.7). Thus, $\mathcal{M}_2$ is prime over a.)

By adding $n-2$ unary relations to the language it is possible to adapt the previous example to obtain a theory having exactly n countable models for each $n \geq 3$ (see the exercises). All such theories have a model behaving much like the model $\mathcal{M}_2$ in the Ehrenfeucht example, with the only difference being that the universality must be weakened.

Lemma 2.3.1. *Let T be a countable complete theory having finitely many but more than one countable model. Then T has a countable model which is prime over some finite set, is not saturated, and realizes every element of $S(\emptyset)$.*

Proof. Let $\mathcal{M}$ be a countable saturated model of T (which exists since the theory is small). Let $p_0, p_1, \ldots$ be an enumeration of $S(\emptyset)$. For $i < \omega$, let $\bar{a}_i$ be a finite sequence from M such that $j \leq i \implies p_j$ is realized by some subsequence of $\bar{a}_i$. Since T is small it has a model $\mathcal{N}_i$ which is prime over $\bar{a}_i$. We claim that $\mathcal{N}_i$ is not saturated, for all $i < \omega$. Assuming to the contrary that $\mathcal{N}_i \cong \mathcal{M}$, $\mathcal{M}$ is prime over some finite sequence $\bar{c}$ in M. Since $(\mathcal{M}, \bar{c})$ is also a saturated model (see Exercise 2.2.8), we conclude that $Th(\mathcal{M}, \bar{c})$ is $\aleph_0$-categorical. However, since T is not $\aleph_0$-categorical, there are infinitely many distinct complete n-types for some n. Distinct complete n-types in T extend to distinct complete n-types in $Th(\mathcal{M}, \bar{c})$, hence there are infinitely many complete n-types in $Th(\mathcal{M}, \bar{c})$, in contradiction to Theorem 2.3.1. Thus, none of the $\mathcal{N}_i$'s are saturated. Since T has finitely many countable models there is an infinite $X \subset \omega$ such that $\mathcal{N}_i \cong \mathcal{N}_j$ for all $i, j \in X$. By the conditions on the $\bar{a}_i$'s, each $\mathcal{N}_j$, for $j \in X$, must realize every complete type of T. Since $\mathcal{N}_j$ is not saturated we have proved the lemma.

The above examples raise the question: Is it possible for a countable complete theory to have exactly two countable models? Vaught, in the seminal paper [Vau61] proved that this is impossible. The bulk of his argument appears in the proof of the preceding lemma.

Proposition 2.3.1. *No complete countable theory has exactly two countable models, up to isomorphism.*

Proof. Suppose T to be a counterexample to the theorem. Since T is not $\aleph_0$-categorical it has a nonisolated complete type p. Since T is small it has a countable saturated model $\mathcal{M}$ and a prime model $\mathcal{N}$, which is not isomorphic to $\mathcal{M}$ (since there is a nonisolated complete type). By the preceding lemma T also has a model $\mathcal{N}'$ which is not saturated but realizes every complete type of T. Since p is realized in $\mathcal{N}'$, $\mathcal{N}'$ cannot be atomic, hence is not isomorphic to $\mathcal{N}$. Thus, the existence of $\mathcal{N}'$ contradicts that T has exactly two countable models, proving the proposition.

Turning to the other end of the spectrum, in each of our examples of a theory having uncountably many countable models we show without appealing to the continuum hypothesis that T has $2^{\aleph_0}$ many countable models. Vaught hypothesized in [Vau61] that this is always the case:

Conjecture 2.3.1 (Vaught's Conjecture). If T is a countable complete theory with fewer than $2^{\aleph_0}$ many countable models, up to isomorphism, then T has countably many countable models.

To date, this conjecture is open. Morley [Mor70] succeeded in showing that $n(T)$ (for T a countable complete theory) is always countable, $\aleph_1$ or $2^{\aleph_0}$. Researchers have approached Vaught's Conjecture by trying to prove it for classes of theories satisfying additional hypotheses. With respect to the stability heirarchy (see later chapters for the relevant definitions), Bouscaren and Lascar proved Vaught's Conjecture for ω-stable theories of finite Morley rank ([Bou83] and [BL83]), with Shelah handling all ω-stable theories in [SHM84]. Several years later a proof was obtained (by Newelski [New90] and Buechler [Bue87]) for properly superstable theories of ∞-rank 1 using significant results from geometrical stability theory. Buechler proved Vaught's conjecture for superstable theories of finite ∞-rank in [Bue93]. An informative discussion of Vaught's conjecture and isomorphism invariants can be found in [Las85].

While calculating $n(T)$ (or the number of models in an uncountable cardinal) is not today a central concern of stability theory, it was through work on these so-called spectrum problems that much of model theory (especially stability theory) was developed. Furthermore, progress on a problem like

Vaught's Conjecture often requires significant new tools which could see use elsewhere.

Historical Notes. Theorem 2.3.1 is due independently to the three researchers mentioned in the statement. See [RN59], [Eng59] and [Sve59]. Lemma 2.3.1 was extracted from Vaught's proof of Proposition 2.3.1 by J. Rosenstein.

Exercise 2.3.1. Prove Remark 2.3.1.

Exercise 2.3.2. Modify Ehrenfeucht's example to find a theory with exactly n countable models for each $n \geq 3$.

Exercise 2.3.3. Suppose that T is not $\aleph_0$−categorical and every countable model of T is homogeneous. Show that T has infinitely many countable models.

Exercise 2.3.4. Suppose that every countable model of T is homogeneous and T has uncountably many countable models. Show that T has $2^{\aleph_0}$ many countable models.

Exercise 2.3.5. Suppose that $\mathcal{M}$ and $\mathcal{N}$ are countable models, $\mathcal{M}$ can be elementarily embedded into $\mathcal{N}$ and $\mathcal{N}$ can be elementarily embedded into $\mathcal{M}$. Does it follow that $\mathcal{M}$ and $\mathcal{N}$ are isomorphic?

2.4 Indiscernible Sequences

In the next section we confront the problem of constructing potentially uncountable models with special properties using Skolem functions. Indiscernibles are introduced here because they play a part in most applications of Skolem functions. However, indiscernibles have applications in the context of stable theories (developed later) which far out distance their uses in conjunction with Skolem functions. In this section we only touch on the most basic properties.

Definition 2.4.1. *Let $\mathcal{M}$ be a model in the language L, and $X \subset M^m$ (for some m) a subset on which there is a linear order $<$. (This order need not be in L.) We call $(X, <)$ an* indiscernible sequence in $\mathcal{M}$ *if for all n and all sequences $x_1 < \ldots < x_n$ and $y_1 < \ldots < y_n$ from X, $tp_{\mathcal{M}}(x_1, \ldots, x_n) = tp_{\mathcal{M}}(y_1, \ldots, y_n)$. X is called an* indiscernible set in $\mathcal{M}$ *if $tp_{\mathcal{M}}(x_1, \ldots, x_n) = tp_{\mathcal{M}}(y_1, \ldots, y_n)$ for any $(x_1, \ldots, x_n)$, $(y_1, \ldots, y_n) \in X^n$ such that $x_i \neq x_j$ and $y_i \neq y_j$, for all $1 \leq i < j \leq n$. (In other words, X is an indiscernible set in $\mathcal{M}$ if $(X, <)$ is an indiscernible sequence in $\mathcal{M}$ for any linear ordering $<$ of X.)*

It is important to bear in mind that the ordering of X in the definition may or may not be in the language. When working with specific examples the following lemma (whose proof is assigned in the exercises) is a useful sufficient condition for indiscernibility. Note: If $\mathcal{M}$ is a model, $\bar{x}$ is the $n-$tuple $(x_1, \ldots, x_n)$ from M and f is an elementary map whose domain contains the x_i's, then $f(\bar{x})$ denotes $(f(x_1), \ldots, f(x_n))$.

Lemma 2.4.1. *Let $\mathcal{M}$ be a model and $(X, <)$ a linearly ordered set with $X \subset M^m$ such that for each pair of sequences $x_1 < \ldots < x_n$ and $y_1 < \ldots < y_n$ there is an automorphism f of $\mathcal{M}$ with $f(x_i) = y_i$ for $1 \le i \le n$. Then $(X, <)$ is an indiscernible sequence in $\mathcal{M}$.*

Example 2.4.1. Let $\mathcal{M} = (\mathbb{Q}, <)$ be the ordering on the rationals. We claim that $(\mathbb{Q}, <)$ is an indiscernible sequence in $\mathcal{M}$. Let $x_1 < \ldots < x_n$ and $y_1 < \ldots < y_n$ be sequences of rationals. Since the theory of dense linear orders without endpoints has elimination of quantifiers the type of any sequence is determined by the order relations within the sequence, hence $(x_1, \ldots, x_n)$ and $(y_1, \ldots, y_n)$ have the same type in $\mathcal{M}$.

Example 2.4.2. Let $\mathcal{M}$ be any model of the theory of algebraically closed fields of characteristic 0. Let X be an algebraically independent set in $\mathcal{M}$; i.e., a set of elements in $\mathcal{M}$ such that $x \in X \implies x$ is transcendental over $X \setminus \{x\}$. For $(x_1, \ldots, x_n)$ and $(y_1, \ldots, y_n)$ two $n-$tuples of distinct elements from X there is an automorphism of $\mathcal{M}$ taking $(x_1, \ldots, x_n)$ to $(y_1, \ldots, y_n)$. Thus, by the last lemma, X is an indiscernible set.

Example 2.4.3. Let V be a vector space over a field F and B a linearly independent set in V. Arguing as in the last example shows that B is a set of indiscernibles in V.

In our last example we specify two indiscernible sets A and B in the same model in which the type of an $n-$tuple from A is different from the type of an $n-$tuple from B.

Example 2.4.4. Let T be the theory in a language with a binary relation E saying that E is an equivalence relation with infinitely many infinite classes and no finite classes. Let $\mathcal{M}$ be any model of T. Let A be the set of elements of M comprising a single $E-$class, and let $B = \{b_i : i < \omega\}$ be a set of representatives from different classes; i.e., for $i \neq j$, $\mathcal{M} \models \neg E(b_i, b_j)$. Since T is quantifier eliminable it is easy to verify that both A and B are indiscernible sets.

The last three examples suggest a connection between the indiscernibility of a set and independence with respect to a dependence relation. This is the manner in which indiscernibles usually arise in the context of stable theories. In general, though, the following theorem is required to get an indiscernible sequence.

Theorem 2.4.1. *Let T be a theory with infinite models and $(X, <)$ any linearly ordered set. Then there is a model $\mathcal{M}$ of T with $X \subset M$ such that $(X, <)$ is an indiscernible sequence in $\mathcal{M}$.*

To prove the theorem a result from combinatorics is needed, whose proof can be found elsewhere. (See, for example [Hod93, 11.1.3].) Given a set A let $[A]^n$ denote the set of subsets of A with exactly n elements. Notice that if A is linearly ordered by $<$, there is a one-to-one correspondence between the elements of $[A]^n$ and increasing sequences of n elements from A.

Lemma 2.4.2 (Ramsey's Theorem). *Let I be an infinite set and $n < \omega$. If $\sim$ is an equivalence relation on $[I]^n$ with finitely many classes, then there is an infinite subset $J \subset I$ such that $[J]^n$ is contained in a single $\sim$ −class.*

Proof of Theorem 2.4.1: First expand the language L of T to $L(X)$, where there is a constant symbol for each element of X. Let T' be the result of adding to T all sentences of $L(X)$ of the form $\varphi(x_1, \ldots, x_n) \longleftrightarrow \varphi(y_1, \ldots, y_n)$, where $\varphi(v_1, \ldots, v_n)$ is a formula of L and $x_1 < \ldots < x_n$, $y_1 < \ldots < y_n$ are from X. The restriction of any model of T' to the original language will be the desired model having $(X, <)$ as an indiscernible sequence. The consistency of T' will follow (by compactness) from the consistency of $T \cup \{\varphi(x_1, \ldots, x_n) \longleftrightarrow \varphi(y_1, \ldots, y_n) : \varphi \in \Phi$, and $x_1 < \ldots < x_n$, $y_1 < \ldots < y_n$ from $X_0\}$, where Φ is an arbitrary finite set of formulas of L and X_0 a finite subset of X. We may assume that all formulas in Φ have exactly n free variables. Let $\mathcal{N}$ be any model of T, A an infinite subset of N and $<$ some linear ordering of A. For increasing n−tuples $a_1 < \ldots < a_n$, $b_1 < \ldots < b_n$ from A we define $\{a_1, \ldots, a_n\} \sim \{b_1, \ldots, b_n\}$ to hold if $\mathcal{N} \models (\varphi(a_1, \ldots, a_n) \longleftrightarrow \varphi(b_1, \ldots, b_n))$ for every $\varphi \in \Phi$. Since Φ is a finite set $\sim$ is an equivalence relation on $[A]^n$ with finitely many classes. By Ramsey's Theorem there is an infinite subset A_0 of A such that all increasing n−tuples from A_0 belong to the same $\sim$ −class. Thus, if we interpret the elements of X_0 by an appropriate number of elements of A_0, listed in increasing order, we get a model of $T \cup \{\varphi(x_1, \ldots, x_n) \longleftrightarrow \varphi(y_1, \ldots, y_n) : \varphi \in \Phi$, and $x_1 < \ldots < x_n$, $y_1 < \ldots < y_n$ from $X_0\}$. This proves the consistency of T', hence the theorem.

Notice that for an indiscernible sequence $(X, <)$ in $\mathcal{M}$, there is a unique complete type which is the type in $\mathcal{M}$ of an increasing n−tuple from X. The *type diagram of* X, denoted $D(X)$, is the set $\{p_n : n < \omega$ and p_n is the complete type realized by increasing n−tuples from $X\}$. Making a specific choice for the set A in the proof yields the following. Notice that while the notation in the proof fixed A as a subset of N it works equally well when A is a subset of N^k for some k. Thus, X need not be a subset of M in this corollary.

Corollary 2.4.1. *Let $(X, <)$ be an infinite sequence of indiscernibles in a model $\mathcal{M}$ and $(Y, <)$ any infinite linear ordering. Then there is some model*

$\mathcal{N} \equiv \mathcal{M}$ such that $(Y, <)$ is an indiscernible sequence in $\mathcal{N}$ and $D(Y) = D(X)$.

Historical Notes. Except for Ramsey's theorem (which was proved in [Ram30]) the results in this section are due to Ehrenfeucht and Mostowski [EM56]. The main results in that paper are covered in the next section.

Exercise 2.4.1. Give a proof of Lemma 2.4.1.

Exercise 2.4.2. Let X be an infinite indiscernible set in a model $\mathcal{M}$ and $n < \omega$. Let X' be a collection of n−tuples from X which are pairwise disjoint. Show that X' is also a set of indiscernibles.

Exercise 2.4.3. Let κ be an infinite cardinal and $\mathcal{M}$ a κ−saturated model. Show that for any nonalgebraic formula φ over M there is a countably infinite indiscernible sequence $(X, <)$ contained in $\varphi(\mathcal{M})$. Furthermore, for any such $(X, <)$ there is an indiscernible sequence $(Y, <')$ of size κ such that $Y \supset X$ and $<'$ extends $<$.

Exercise 2.4.4. Give an example of a model $\mathcal{M}$ containing an infinite set X such that any pair of distinct elements from X realize the same complete 2−type in $\mathcal{M}$, but there are two triples from X with different 3−types. (HINT: A projective plane)

2.5 Skolem Functions

Skolem functions, used in conjunction with indiscernibles, provide a way to construct uncountable models with various special properties. These theorems differ from those which yield, e.g., a κ−saturated model (for some κ) in that they may result in uncountable models which omit specified types.

In algebra it is common to speak of the object (e.g., the group or vector space) generated by a subset. Following is our formal definition of the notion of a "submodel generated by a set".

Definition 2.5.1. *Let $\mathcal{M}$ be a model in the language L and $X \subset M$.*

(i) The hull of X, *denoted $H(X)$, is the subset of M obtained by closing $X \cup \{a \in M : a$ interprets a constant of $L\}$ under $F^{\mathcal{M}}$, for every function F of L.*

(ii) $\mathcal{H}(X)$ is the submodel of $\mathcal{M}$ with universe $H(X)$ (when $H(X) \neq \emptyset$). We also call $\mathcal{H}(X)$ the hull of X.

For any model $\mathcal{M}$ and $X \subset M$, if $X \neq \emptyset$ or the language contains a constant symbol, then $\mathcal{H}(X)$ is the submodel generated by X in the sense that, $X \subset H(X)$, $\mathcal{H}(X) \subset \mathcal{M}$ and $\mathcal{H}(X) \subset \mathcal{N}'$ for every $\mathcal{N}' \subset \mathcal{M}$ containing X.

In general, however, there is no well-defined notion of "elementary submodel generated by a set". Consider, for example, $\mathcal{M} = (\mathbb{Q}, <)$ and X a finite nonempty subset of $\mathbb{Q}$. Then $(X, < \upharpoonright X)$ is the submodel generated by X but it is certainly not an elementary submodel (for one thing, it is finite). In fact, if $\mathcal{N}$ is any elementary submodel of $\mathcal{M}$ containing X there is $\mathcal{N}'$ containing X which is a proper elementary submodel of $\mathcal{N}$. Here, we will show how to expand a theory T to a theory T^* in a larger language so that whenever $\mathcal{M} \models T^*$ and $\mathcal{N} \subset \mathcal{M}$, $\mathcal{N} \prec \mathcal{M}$. The specific goal is

Theorem 2.5.1 (Skolem). *Let T be a theory in a language L. Then there is a theory T^* in a language L^* such that:*

(1) $L \subset L^*$, $T \subset T^*$ *and* $|T^*| = |T|$;
(2) *every model of T can be expanded to a model of T^*, and*
(3) *if $\mathcal{M}^* \models T^*$ and $\mathcal{N}^* \subset \mathcal{M}^*$ then $\mathcal{N}^* \prec \mathcal{M}^*$.*

Proof. The language L^* and theory T^* will be the unions of chains which approximate (3) with increasing precision. The inductive step in the construction of the chain is handled by

Claim. Let T be a theory in a language L. Then there is a theory T' in a language L' such that:

(a) $L \subset L'$, $T \subset T'$ and $|T'| = |T|$;
(b) every model of T can be expanded to a model of T', and
(c) if $\mathcal{M}' \models T'$ and $\mathcal{N}' \subset \mathcal{M}'$ then $\mathcal{N}' \upharpoonright L \prec \mathcal{M}' \upharpoonright L$.

We define $L' \supset L$ to be the language which adds to L a new constant symbol $c_{\exists x\varphi}$ for every sentence $\exists x\varphi$ of L, and a new n–ary function symbol $F_{\exists x\varphi}$ for every formula $\exists x\varphi(x, v_1, \ldots, v_n)$. Let T' be the union of T and the set of all sentences of L' of the form: $\exists x\varphi(x) \longrightarrow \varphi(c_{\exists x\varphi})$ or $\forall v_1 \ldots v_n[\exists x\varphi \longrightarrow \varphi(F_{\exists x\varphi}(v_1, \ldots, v_n), v_1, \ldots, v_n)]$, for all appropriate formulas $\exists x\varphi$ of L. That (a) holds is clear, and it is easy to interpret the new functions and constants on a model of T to ensure (b). The Tarski-Vaught Test guarantees that (c) holds.

Turning to the proof of the theorem define: $L_0 = L$, $T_0 = T$, $L_{n+1} = (L_n)'$, $T_{n+1} = (T_n)'$, $L^* = \bigcup_{n<\omega} L_n$ and $T^* = \bigcup_{n<\omega} T_n$. The verification of (1)-(3) is now easy using the claim and the fact that any formula of L^* is a formula of some L_n.

The functions and constants added to L to obtain L^* are called the *Skolem functions for* L and $\Gamma = T^* \setminus T$ is the set of *Skolem axioms for* L. Notice that Γ depends only on the language; i.e., if T and T_0 are two theories in L, then $T^* = T \cup \Gamma$ and $T_0^* = T_0 \cup \Gamma$ satisfy (1)–(3) of the theorem. A theory T having the property that whenever $\mathcal{M} \models T$ and $\mathcal{N} \subset \mathcal{M}$ then $\mathcal{N} \prec \mathcal{M}$ is said to *have Skolem functions*. This terminology is used even when T is not constructed from some other theory using the theorem.

If $\mathcal{M}$ is a model of a theory having Skolem functions then for any $X \subset M$, $\mathcal{H}(X) \prec \mathcal{M}$, and we call $\mathcal{H}(X)$ the *Skolem hull of* X. As the elements of $H(X)$ interpret terms of the language applied to the elements of X, $X \subset Y \subset M \implies \mathcal{H}(X) \prec \mathcal{H}(Y)$. A model which is the Skolem hull of a sequence of indiscernibles is often called an *Ehrenfeucht-Mostowski model*, after the researchers who developed indiscernibles.

In theories with Skolem functions the Skolem hulls of indiscernible sequences have properties which are rigidly tied to the indiscernibles. This is made explicit in

Lemma 2.5.1. *Let T be a complete theory with Skolem functions. Let $\mathcal{M}$ and $\mathcal{N}$ be models of T and $(X, <)$, $(Y, <)$ infinite indiscernible sequences in $\mathcal{M}$, $\mathcal{N}$, respectively, with $D(X) = D(Y)$.*

(i) $\mathcal{H}(X)$ and $\mathcal{H}(Y)$ have the same type diagrams.

(ii) If h is an order-preserving map of X into Y then h extends uniquely to an elementary embedding h^* of $\mathcal{H}(X)$ into $\mathcal{H}(Y)$. In fact, $h^*(\mathcal{H}(X)) = \mathcal{H}(h(X))$.

(iii) If h is an order-preserving map of X onto itself then h extends uniquely to an automorphism of $\mathcal{H}(X)$.

Proof. Recalling the definition of the hull of X an arbitrary element of $H(X)$ is of the form $t^{\mathcal{M}}(x_1, \ldots, x_n)$ where t is some term of L and $x_1 < \ldots < x_n$ in X. Thus, if $\mathcal{H}(X)$ realizes a $1-$type p there is some such $t^{\mathcal{M}}(x_1, \ldots, x_n)$ such that $\mathcal{M} \models \varphi(t(x_1, \ldots, x_n))$, for every $\varphi \in p$. This is to say that $\varphi(t(v_1, \ldots, v_n))$ belongs to the type satisfied by increasing $n-$tuples from X. Since $D(Y) = D(X)$, it follows that $t^{\mathcal{N}}(y_1, \ldots, y_n)$ realizes p in $\mathcal{H}(Y)$ for any $y_1 < \ldots < y_n$ in Y. The proof for types in n variables is just notationally more complicated, so we have proved (i).

Continuing the notation of the previous paragraph, the extension h^* needed to obtain (ii) is seen to be the map defined by:

$$h^*(t^{\mathcal{M}}(x_1, \ldots, x_n)) = t^{\mathcal{N}}(h(x_1), \ldots, h(x_n)),$$

for $x_1 < \ldots < x_n$ in X and terms t. Part (iii) follows immediately from (ii).

Corollary 2.5.1 (Ehrenfeucht-Mostowski). *Let T be a complete theory with an infinite model and $(X, <)$ some linearly ordered set. Then there is a model $\mathcal{M}$ of T in which $(X, <)$ is an indiscernible sequence and every automorphism of this linear order extends to an automorphism of $\mathcal{M}$.*

Proof. Let T' be an expansion of T to a complete theory with Skolem functions. Let $\mathcal{M}'$ be a model of T' in which $(X, <)$ is an indiscernible sequence. We may choose $\mathcal{M}'$ to be $\mathcal{H}(X)$. By Lemma 2.5.1(iii) every automorphism of $(X, <)$ extends to an automorphism of $\mathcal{M}'$. Such an automorphism is, a fortiori, an automorphism of $\mathcal{M} = \mathcal{M}' \restriction L$, proving the corollary.

Our next application of the lemma illustrates how Skolem functions and indiscernibles can be used to find very special uncountable models.

Theorem 2.5.2. *Let T be a countable complete theory in L having an infinite model. Then T has a countable model $\mathcal{M}$ such that for all cardinals $\kappa \geq \aleph_0$ there is a model $\mathcal{N}$ of cardinality κ with $D(\mathcal{N}) = D(\mathcal{M})$.*

Proof. First, let T' be an expansion of T to a complete theory with Skolem functions. By Theorem 2.4.1 T' has a countable model $\mathcal{M}_0$ containing an infinite indiscernible sequence X. Let $\mathcal{M}^* = \mathcal{H}(X)$ and $\mathcal{M} = \mathcal{M}^* \restriction L$. Given an infinite cardinal κ, let $(Y, <)$ be any linear order of cardinality κ. By Corollary 2.4.1 T' has a model $\mathcal{N}_0$ in which $(Y, <)$ is an indiscernible sequence with $D(X) = D(Y)$. Let $\mathcal{N}^* = \mathcal{H}(Y)$, the hull of Y in $\mathcal{N}_0$, and $\mathcal{N} = \mathcal{N}^* \restriction L$. By Lemma 2.5.1(i) $D(\mathcal{M}^*) = D(\mathcal{N}^*)$. Restricting to the original language, $D(\mathcal{M}) = D(\mathcal{N})$, as desired.

Further results can be obtained by varying the properties of the linear order $(Y, <)$ used in the proof of the theorem. For example, if $\kappa \geq \aleph_0$ there is a dense linear order without endpoints $(Y, <)$ with 2^κ many automorphisms. By (iii) of the lemma each of these extends to an automorphism of the Skolem hull of the indiscernible sequence $(Y, <)$. A result which will see important duty in the next chapter is

Lemma 2.5.2 (Morley). *Let T be a countable complete theory with infinite models. Then for every infinite cardinal κ, T has a model $\mathcal{M}$ of cardinality κ such that for every $A \subset M$, $\mathcal{M}$ realizes at most $|A| + \aleph_0$ many complete types over A.*

Proof. Let T^* be an expansion of T to a theory with Skolem functions, L the language of T and L^* the language of T^*. Let $(X, <)$ be a well-ordering of order type κ, considered as a sequence of indiscernibles in some model of T^*. Let $\mathcal{M}^*$ be the Skolem hull of X and $\mathcal{M} = \mathcal{M}^* \restriction L$. To verify that $\mathcal{M}$ satisfies the requirements, let A be a subset of M. For the purposes of this lemma we may as well require that $A = H(Y)$, for some $Y \subset X$. (For any A there is a Y such that $A \subset H(Y)$ and $|Y| \leq |A| + \aleph_0$.) We call two sequences $x_1 < \ldots < x_n$, $y_1 < \ldots < y_n$ from X *equivalent over* Y if for $1 \leq k \leq n$ and all $z \in Y$, $x_k = z$ if and only if $y_k = z$ and $x_k < z$ if and only if $y_k < z$. (That is, the x_i's and y_i's satisfy the same order relations with the elements of Y.) Because X is an indiscernible sequence, whenever $x_1 < \ldots < x_n$ and $y_1 < \ldots < y_n$ are equivalent over Y they have the same complete type over Y in $\mathcal{M}$. In fact, for any term $t(v_1, \ldots, v_n)$ of $L^*(Y)$ the two elements $t(x_1, \ldots, x_n)$ and $t(y_1, \ldots, y_n)$ realize the same complete type over A. Similarly for n-tuples from M. Thus, to complete the proof it suffices to show that the equivalence relation of being equivalent over Y has at most $|Y| + \aleph_0$ many classes. To see this we define x', for $x \in X \setminus Y$, by

- $x' = \infty$ if there is no $z \in Y$ with $x < z$, and
- $x' =$ the least $z \in Y$ such that $x < z$, if there is such a $z \in Y$.

Then, $x_1 < \ldots < x_n$ and $y_1 < \ldots < y_n$ are equivalent over Y if and only if $x'_1 = y'_1, \ldots, x'_n = y'_n$. Hence there are $\leq |Y| + \aleph_0$ many equivalence classes of $n-$tuples, as required.

Historical Notes. Skolem functions date back to Skolem's 1920 paper [Sko20]. Morley proved Lemma 2.5.2 in [Mor65]. The other results in the section were proved by Ehrenfeucht and Mostowski in [EM56].

Exercise 2.5.1. Let $\mathcal{M}$ and $\mathcal{N}$ be models, X and Y nonempty subsets of M and N, respectively, and f an elementary bijection from X onto Y. Show that f extends to an isomorphism between the submodels generated by X and Y.

Exercise 2.5.2. Let $\mathcal{M} = (\omega, \leq), T = Th(\mathcal{M})$ and T^* the Skolem expansion of T. Show that $\mathcal{M}$ has two expansions to a model of T^* which are not elementarily equivalent. (Thus, T^* is not complete.)

Exercise 2.5.3. Let T be a complete theory with Skolem functions, $(X, <)$ an indiscernible sequence in $\mathcal{H}(X) \models T$. Show that there is an embedding of the automorphism group of $(X, <)$ into the automorphism group of $\mathcal{H}(X)$. Also, if both X and $<$ are definable this embedding is an isomorphism.

Exercise 2.5.4. Elements x and y of a model $\mathcal{M}$ are said to have the *same automorphism type* if there exists an automorphism f of $\mathcal{M}$ such that $f(x) = y$. Show that if T is a countable complete theory with infinite models, then T has a model in each cardinality which has only countably many automorphism types.

Exercise 2.5.5. Suppose that X is an infinite set of indiscernibles, $\mathcal{M}$ is a Skolem hull of X and the language of $\mathcal{M}$ is countable. How many automorphisms are there of $\mathcal{M}$?

3. Uncountably Categorical and $\aleph_0$–stable Theories

In this chapter we will study the results which laid the foundation for stability theory, namely Morley's Categoricity Theorem and the Baldwin-Lachlan Theorem. Some of the concepts arising in their proofs will be redeveloped later for stable theories. We feel, however, that these proofs present an excellent introduction to the key concepts encountered later, and are historically important enough to warrant individual treatment.

In Section 1 a proof of Morley's Categoricity Theorem is given. In the third section of the chapter totally transcendental theories, which arose in Morley's original proof of Morley's Categoricity Theorem, will be studied more deeply. Again, ideas will be introduced which are seen throughout stability theory. In the fourth section these new concepts are applied to prove the Baldwin-Lachlan Theorem. Groups definable in totally transcendental theories are studied in the fifth section.

3.1 Morley's Categoricity Theorem

Throughout this section an arbitrary theory is assumed to be countable and have infinite models. For emphasis this assumption may be repeated within the statements of theorems. Recall that a theory T is said to be *categorical in* κ (or κ*–categorical*), where κ is an infinite cardinal, if T has a unique model of cardinality κ, up to isomorphism. A theory is called *uncountably categorical* if it is categorical in every uncountable cardinality. In the previous chapter theories were exhibited which are:

- categorical in every infinite cardinal;
- categorical in $\aleph_0$ but not in any uncountable cardinal;
- categorical in every uncountable cardinal, but not in $\aleph_0$;
- not categorical in any infinite cardinal.

It was conjectured by Łoś that every countable complete theory satisfies one of these four possibilities. Morley proved this conjecture with

Theorem 3.1.1 (Morley's Categoricity Theorem). *If a countable complete theory T is categorical in some uncountable cardinality then it is categorical in every uncountable cardinality.*

This section is devoted to a proof of this theorem. Some examples of theories categorical in some uncountable cardinality are:

1. the theory in the empty language with only infinite models;
2. infinite Abelian groups in which all elements have order p, for p some prime;
3. divisible torsion-free Abelian groups;
4. algebraically closed fields of a fixed characteristic;
5. the theory of a model (A, σ), where A is an infinite set and σ is a permutation of A with no finite cycles;
6. the theory of the model (ω, S), where S is the successor function.

The uncountable categoricity of the theories in 2, 3 and 4 above follow from well-known classical results. For instance, it is known that any divisible torsion-free Abelian group is a direct sum of copies of $(\mathbb{Q}, +)$. Thus, the isomorphism type of a divisible torsion-free Abelian group G is determined by the number of copies of the rationals used in such a decomposition. If G has cardinality $\kappa > \aleph_0$, then κ copies of the rationals must appear in a decomposition. Hence, any two uncountable divisible torsion-free Abelian groups are isomorphic. Steinitz's Theorem says that the isomorphism type of an algebraically closed field is determined by its characteristic and transcendence degree. For uncountable algebraically closed fields the transcendence degree is the same as the cardinality, hence the theory in 4 is uncountably categorical. The uncountable categoricity of the theories in 1, 5 and 6 follow quickly from quantifier-elimination. Close examination shows that in each of the examples above the isomorphism type of a model is determined by some cardinal invariant. Furthermore, this invariant is the dimension of some subset of the model with respect to a dependence relation. We will see, in fact, that whenever a theory is categorical in an uncountable cardinal the models are determined by the dimension on a definable subset of the model with respect to a particular dependence relation.

Remark 3.1.1. The assumption that T is complete in Morley's Categoricity Theorem was only made to avoid distracting the reader from the main issues. A classical result known as the Łoś-Vaught Test implies that a first-order theory categorical in some $\kappa \geq |T|$ is complete. See Remark 2.3.1.

Definition 3.1.1. *Let λ be an infinite cardinal and T a complete theory (of any cardinality) with an infinite model. T is said to be $\lambda-$stable if for all $\mathcal{M} \models T$ and $A \subset M$ of cardinality $\leq \lambda$, $|S_1(A)| \leq \lambda$. The term $\omega-$stable may be used in place of $\aleph_0-$stable. A model $\mathcal{M}$ is called $\lambda-$stable if $Th(\mathcal{M})$ is $\lambda-$stable.*

A straight-forward induction on n shows that if T is $\lambda-$stable, A is a subset of a model of T and $|A| \leq \lambda$, then $|S_n(A)| \leq \lambda$. Observe that an $\aleph_0-$stable theory must be countable and small. While the definition of a

$\lambda-$stable theory could easily be rewritten for possibly incomplete theories, the benefits of the added generality are negligible.

Lemma 3.1.1. *If the countable theory T is categorical in some uncountable cardinal, then T is $\aleph_0-$stable.*

Proof. Here is where Skolem functions come into play. Let T be categorical in $\lambda \geq \aleph_1$. By Lemma 2.5.2, T has a model $\mathcal{M}$ of cardinality λ such that for any countable $A \subset M$, $\mathcal{M}$ realizes only countably many complete types over A. Assuming that T is not $\aleph_0-$stable there is a model $\mathcal{N}$ of T containing a countable set B such that $|S(B)| > \aleph_0$. Without loss of generality, $\mathcal{N}$ is countable. By a simple compactness argument $\mathcal{N}$ has an elementary extension $\mathcal{N}'$ of cardinality λ realizing uncountably many complete types over B. Since T is $\lambda-$categorical $\mathcal{N}'$ must be isomorphic to $\mathcal{M}$. This contradiction proves the lemma.

As stated above, on a model of an uncountably categorical theory there is a dependence relation and a corresponding notion of dimension, which gives rise to an isomorphism invariant. It is the $\aleph_0-$stability of the uncountably categorical theory which gives rise to this dependence relation. These dependence relations are developed in the next few pages.

Definition 3.1.2. *Let $\mathcal{M}$ be a model and φ a nonalgebraic formula (in n variables) over M. We call φ* strongly minimal *if for every $\mathcal{N} \succ \mathcal{M}$ and every formula ψ (in n variables) over N, $\varphi(\mathcal{N}) \cap \psi(\mathcal{N})$ or $\varphi(\mathcal{N}) \cap \neg\psi(\mathcal{N})$ is finite. We call a complete theory T* strongly minimal *if the formula $x = x$ is strongly minimal.*

Slightly rewording the definition in terms of definable sets, φ is strongly minimal if for all $\mathcal{N} \succ \mathcal{M}$ every subset of $\varphi(\mathcal{N})$ definable in $\mathcal{N}$ (over N) is finite or cofinite.

Remark 3.1.2. Let $\bar{a}$ be a sequence from a model $\mathcal{M}$, $\varphi(v, \bar{a})$ a formula over $\bar{a}$, and let $\bar{b}$ be a sequence from a model $\mathcal{N}$ such that $tp_{\mathcal{M}}(\bar{a}) = tp_{\mathcal{N}}(\bar{b})$. Then $\varphi(v, \bar{a})$ is strongly minimal if and only if $\varphi(v, \bar{b})$ is strongly minimal. (The proof of this is left to the reader in Exercise 3.1.12.)

Example 3.1.1. (Strongly minimal theories)

(i) (The theory of infinite sets in the empty language) For L the empty language, the theory in L saying that there are infinitely many elements is quantifier-eliminable. Let $\mathcal{M}$ be an arbitrary model of T. A formula in the single variable v over M is equivalent to a boolean combination of formulas $v = a_i$, for some $a_0, \ldots, a_n \in M$. Thus, any subset of M definable over M is finite or cofinite. That is, T is strongly minimal.

(ii) (The theory of vector spaces over a field F) For a field F the theory T of infinite vector spaces over F is quantifier-eliminable (in the natural language). Let $\mathcal{M}$ be an arbitrary model of T. A subset of M defined by a linear equation over F with coefficients from M consists of a single element or is all of M. Since any subset of M definable over M is a boolean combination of sets defined by linear equations, the formula $x = x$ is strongly minimal.

(iii) (The theory of algebraically closed fields of a fixed characteristic) In this case the theory T is also quantifier-eliminable. Given a model $\mathcal{M}$ of T, a subset of M definable by some equation over M is either finite or the entire field. As in the previous example, it follows that T is strongly minimal.

The properties of strongly minimal formulas are best described using the algebraic closure relation.

Definition 3.1.3. *Let A be a subset of a model $\mathcal{M}$. A finite tuple $\bar{a}$ from M is said to be* algebraic over A *if $tp_{\mathcal{M}}(\bar{a}/A)$ is algebraic. The* algebraic closure of A *(in $\mathcal{M}$), denoted $acl(A)$, is $\{ a \in M : a$ is algebraic over $A \}$. Sequences $\bar{a}$ and $\bar{b}$ are* interalgebraic over A *if $\bar{a} \in acl(A \cup \{\bar{b}\})$ and $\bar{b} \in acl(A \cup \{\bar{a}\})$.*

Remark 3.1.3. Let A be a subset of a model $\mathcal{M}$.

(i) Notice that the model plays no active role in the definition of $acl(A)$; if $\mathcal{N} \succ \mathcal{M}$ then $a \in N$ is algebraic over A only when $a \in M$ and $tp_{\mathcal{M}}(a/A)$ is algebraic. See Exercise 3.1.4.

(ii) In the exercises the reader is asked to verify that $|acl(A)| \leq |A| + |T|$, where T is the theory of $\mathcal{M}$.

Algebraic closure is most naturally studied in the context of closure operators.

Definition 3.1.4. *Let S be some set and $c\ell$ a unary operator on the set of subsets of S.*

(i) $c\ell$ is a closure operator *if for all $X, Y \subset S$:*

(a) $X \subset c\ell(X)$,
(b) $c\ell^2(X) = c\ell(X)$, *and*
(c) $X \subset Y \implies c\ell(X) \subset c\ell(Y)$.

A closure operator $c\ell$ is called finitary *(or* of finite character*) if $c\ell(X) = \bigcup\{ c\ell(Y) : Y \subset X$ and Y is finite$\}$. (Standard terminology uses "algebraic" where we use "finitary", but we feel this leaves too much room for confusion with other uses of the word algebraic.) A subset X of S is called* closed *if $X = c\ell(X)$.*

(ii) If $c\ell$ is a finitary closure operator on S, then $\mathbf{S} = (S, c\ell)$ is a pregeometry *if it satisfies the* exchange property*: for all $a, b \in S$ and $A \subset S$, if $a \in c\ell(A \cup \{b\}) \setminus c\ell(A)$ then $b \in c\ell(A \cup \{a\})$. A pregeometry is a* geometry *if $c\ell(\emptyset) = \emptyset$ and for all singletons $a \in S$, $c\ell(\{a\}) = \{a\}$.*

(iii) Let $\mathbf{S} = (S, c\ell)$ be a pregeometry and $A, B \subset S$. We say that A is $c\ell$−independent over B *if for all $a \in A$, $a \notin c\ell(B \cup (A \setminus \{a\}))$. For $X \subset S$ we call A a* basis of X over B *if A is a maximal subset of $c\ell(X \cup B)$ which is $c\ell$−independent over B. A standard argument using the exchange property*

and the transitivity of closure shows that all bases of X *over* B *have the same cardinality, which is called the* dimension of X over B *and denoted* $\dim(X/B)$.

(iv) Let $\mathbf{S} = (S, cl)$ *be a pregeometry,* X, Y *and* $Z \subset Y$ *subsets of* S. *We say* X *is* dim-independent from Y over Z *or simply* independent from Y over Z *if for all finite* $X_0 \subset X$, $\dim(X_0/Y) = \dim(X_0/Z)$.

Remark 3.1.4. Let S be a pregeometry.

(i) Notice that dimension on S is additive:

$$\text{For all } A, B \subset S, \ \dim(A \cup B) = \dim(A/B) + \dim(B).$$

(See Exercise 3.1.8.)

(ii) When $S \neq \emptyset$, cl-independent is different from dim-independent. (Given a nonempty $X \subset S$, X is dim-independent from X over X, but X is not cl-independent from X over X.)

Lemma 3.1.2. *Algebraic closure forms a finitary closure operator on the universe of a model.*

Proof. Left to the reader in Exercise 3.1.7.

Elimination of quantifiers can be used to verify that algebraic closure on an algebraically closed field or a vector space is a pregeometry. Actually, this is typical of sets defined by strongly minimal formulas:

Lemma 3.1.3. *Let* $\mathcal{M}$ *be a model of a theory of cardinality* κ, φ *a strongly minimal formula over* $A \subset M$ *and* $D = \varphi(\mathcal{M})$. *Let* B *be a subset of* M *containing* A *and let* cl *be the restriction to* D *of algebraic closure over* B. *(That is, for* $X \subset D$ *and* $a \in D$, $a \in cl(X)$ *if* $a \in acl(X \cup B) \cap D$.*).*

(i) There is a unique nonalgebraic $p \in S(B)$ *containing* φ.

(ii) (D, cl) *is a pregeometry.*

(iii) If $\{a_0, \ldots, a_n\}$, $\{b_0, \ldots, b_n\} \subset D$ *are* cl*-independent over* B, *then*

$$tp_{\mathcal{M}}(a_0, \ldots, a_n/B) = tp_{\mathcal{M}}(b_0, \ldots, b_n/B).$$

Thus, any subset of D *which is* cl*-independent over* B *is an indiscernible set over* B. *Furthermore, if* I, $J \subset D$ *are infinite and* cl*-independent over* B, *then* $D(I) = D(J)$.

Proof. (i) For ψ an arbitrary formula over B only one of $\varphi \wedge \psi$ and $\varphi \wedge \neg\psi$ is nonalgebraic (by the strong minimality of φ). Thus, φ has a unique nonalgebraic completion over B.

(ii) By Lemma 3.1.2 it only remains to verify that cl satisfies the exchange property. (To simplify the notation we assume φ to be a formula in one variable; i.e., $D \subset M$ instead of M^n, for some n. The proof is almost identical in general.) First we prove:

Claim. Suppose that, for $i = 0, 1$, a_i, $b_i \in D$, $a_i \notin acl(B)$ and $b_i \notin acl(B \cup \{a_i\})$. Then $tp_{\mathcal{M}}(a_0 b_0/B) = tp_{\mathcal{M}}(a_1 b_1/B)$.

By (i), $tp_{\mathcal{M}}(a_0/B) = tp_{\mathcal{M}}(a_1/B)$. Since the types of elements do not change when passing between a model and an elementary extension, we are free to replace $\mathcal{M}$ by an elementary extension. By Lemma 2.2.9 there is a model $\mathcal{N} \succ \mathcal{M}$ having an automorphism f such that f is the identity on B and $f(a_0) = a_1$. Then $f(b_0)$ is an element of $\varphi(\mathcal{N})$ which is not in the algebraic closure of $B \cup \{a_1\}$. Again by (i), $tp_{\mathcal{N}}(f(b_0)/B \cup \{a_1\}) = tp_{\mathcal{N}}(b_1/B \cup \{a_1\})$. Since $f(a_0) = a_1$ and automorphisms preserve types, $tp_{\mathcal{N}}(a_0 b_0/B) = tp_{\mathcal{N}}(f(a_0)f(b_0)/B) = tp_{\mathcal{N}}(a_1 f(b_0)/B) = tp_{\mathcal{N}}(a_1 b_1/B)$.

Now suppose the assertion in (ii) to fail; i.e., for some $C \subset D$ and $a, b \in D$, $a \in c\ell(C \cup \{b\}) \setminus c\ell(C)$ and $b \notin c\ell(C \cup \{a\})$. Let $B' = B \cup C$. Let $\mathcal{N}$ be an elementary extension of $\mathcal{M}$ in which $D' = \varphi(\mathcal{N})$ has cardinality $> (|B'| + \kappa)^+ = \lambda$. Choose Y, $D \subset Y \subset D'$ of cardinality λ and let $X = acl(B' \cup Y) \cap D'$, which has cardinality $\lambda + \kappa = \lambda$ (by Exercise 3.1.5). Since $|D'| > \lambda$ there is a $d \in D' \setminus X$. By (i), d and b realize the same complete type over $B' \cup \{a\}$ in $\mathcal{N}$. In fact, by the claim, if c is any element of $X \setminus acl(B')$, $tp_{\mathcal{N}}(dc/B') = tp_{\mathcal{N}}(ba/B')$. Thus, $X \subset acl(B' \cup \{d\})$. This contradicts that $|X| = \lambda > |B'| + \kappa = |B' \cup \{d\}| + \kappa = |acl(B' \cup \{d\})|$, to prove (ii).

(iii) This follows directly from the proof of the above claim and an induction on n.

Remark 3.1.5. Let $\mathcal{M}$ be a model of the theory of algebraically closed fields of a fixed characteristic. A subset of M is *acl*−independent if and only if it is algebraically independent. Other notions, like basis and dimension also agree with their standard algebraic interpretations.

If $\mathcal{M}$ is a model of the theory of vector spaces over a field F, then $I \subset M$ is *acl*−independent if and only if I is linearly independent.

Corollary 3.1.1. *If T is a countable strongly minimal theory, then T is categorical in every uncountable cardinal.*

Proof. Let $\mathcal{M}$ and $\mathcal{N}$ be models of T of cardinality $\kappa > \aleph_0$. Let I and J be bases for the closed sets M and N, respectively. Thus, $M = acl(I)$ and $N = acl(J)$. Since $|acl(I)| \leq |I| + \aleph_0$ and κ is uncountable, $|I|$ and $|J|$ must both be κ. Since I and J are indiscernible sets and $D(I) = D(J)$ (by Lemma 3.1.3(iii)) any bijection f from I onto J is an elementary map. An elementary map between two sets can be extended to an elementary map between their algebraic closures (see Exercise 3.1.10). Thus, f extends to an isomorphism of $\mathcal{M}$ onto $\mathcal{N}$.

Strongly minimal formulas enter the proof of Morley's Categoricity Theorem through

Lemma 3.1.4. *Let T be an $\aleph_0$−stable theory and let $\mathcal{M}$ be a countable saturated model of T. Then there is a strongly minimal formula over $\mathcal{M}$.*

Proof. Since T is $\aleph_0$−stable $|S(A)|$ is countable for any countable subset A of a model of T. Hence, T does have a countable saturated model $\mathcal{M}$ and, for each n, every element of $S_n(M)$ has Cantor-Bendixson rank. First notice that any isolated $p \in S(M)$ is algebraic. (Let $\varphi(\bar{v}) \in p$ isolate p. There is an $\bar{a}$ from M satisfying φ; i.e., $\{\varphi(\bar{v}), \bar{v} = \bar{a}\}$ is consistent. Since φ isolates p, $\bar{v} = \bar{a}$ must be in p.) Let p be a type in $S_1(M)$ of least Cantor-Bendixson rank among the nonisolated complete 1−types, and let $\varphi \in p$ isolate p relative to the nonisolated types. Hence,

$$p \text{ is the unique nonalgebraic element of } S_1(M) \text{ containing } \varphi. \tag{3.1}$$

Claim. φ is strongly minimal.

For ψ any formula over M (in one variable) one of $\varphi \wedge \psi$ or $\varphi \wedge \neg\psi$ is algebraic (by (3.1)). To prove that φ is strongly minimal, however, this condition must be true when ψ is a formula over an arbitrary elementary extension of $\mathcal{M}$. Let $\mathcal{N} \succ \mathcal{M}$ and let ψ be a formula over N in one variable. Let $\bar{a}$ be the parameters in φ and $\psi = \psi(x, \bar{b})$, where $\psi(x, \bar{y})$ is over $\emptyset$. Suppose, towards a contradiction, that $\varphi \wedge \psi$ and $\varphi \wedge \neg\psi$ are both nonalgebraic. Then for each n, $\mathcal{N}$ is a model of the sentences saying that there are $\geq n$ elements satisfying $\varphi \wedge \psi$ and there are $\geq n$ elements satisfying $\varphi \wedge \neg\psi$. Since $\mathcal{M}$ is saturated there is a $\bar{c}$ in M such that $tp_{\mathcal{M}}(\bar{c}/\bar{a}) = tp_{\mathcal{N}}(\bar{b}/\bar{a})$, hence $\mathcal{M} \models \exists^{\geq n} x(\varphi(x, \bar{a}) \wedge \psi(x, \bar{c}))$ and $\mathcal{M} \models \exists^{\geq n} x(\varphi(x, \bar{a}) \wedge \neg\psi(x, \bar{c}))$, for each n. Thus, both $\varphi(x, \bar{a}) \wedge \psi(x, \bar{c})$ and $\varphi(x, \bar{a}) \wedge \neg\psi(x, \bar{c})$ are nonalgebraic. Since $\bar{c}$ is from M this contradicts (3.1) to prove the lemma.

Examining the proof of the lemma yields:

Corollary 3.1.2. *Let T be $\aleph_0$−stable and $\psi(\bar{v})$ any nonalgebraic formula over a countable saturated model $\mathcal{M}$. Then, over M, there is a strongly minimal formula φ which implies ψ.*

Another property of $\aleph_0$−stable theories used critically below is the existence of prime models, even over uncountable sets.

Lemma 3.1.5. *Let T be $\aleph_0$−stable, $\mathcal{M}$ a model of T and $A \subset M$. Then there is a model $\mathcal{N}$ which is prime over A and atomic over A.*

Proof. The bulk of the work is contained in

Claim. For every $B \subset M$ the isolated types are dense in $S(B)$.

(By our conventions for dealing with parameters, "the isolated types are dense in $S(B)$" simply means that the isolated types are dense in $Th(\mathcal{M}_B)$.) The proof of this claim is quite similar to the proof of Lemma 2.1.1, where it was shown that the isolated types are dense in a small countable complete theory. Here, B is potentially uncountable. However, we will show that if the isolated types are not dense in $S(B)$ there is a countable $B' \subset B$ with $S(B')$ uncountable, contradicting the $\aleph_0$−stability of T.

Suppose to the contrary that there is a formula φ over B not contained in an isolated element of $S(B)$. Let X be the set of finite sequences of 0's and

1's and Y the set of sequences of length ω from $\{0,1\}$. Define by recursion a family of formulas φ_s, for $s \in X$, with the properties: (a) $\varphi_\emptyset = \varphi$, (b) φ_s is not contained in an isolated complete type over B, (c) if t is an initial segment of s then $\mathcal{M} \models \forall \bar{v}(\varphi_s \longrightarrow \varphi_t)$ and (d) if t is not an initial segment of s and s is not an initial segment of t, then $\varphi_s \wedge \varphi_t$ is inconsistent with $Th(\mathcal{M}_B)$. (This is possible since given φ_s there is a ψ such that $\varphi_s \wedge \psi$ and $\varphi_s \wedge \neg\psi$ are consistent, neither of which can be contained in an isolated complete type over B.) For $f \in Y$ let $p_f = \{\varphi_s : s$ is an initial segment of $f\}$. Then each p_f is consistent and for distinct f and g in Y, $p_f \cup p_g$ is inconsistent. Since there are countably many φ_s's there is a countable set $B' \subset B$ such that each p_f is over B'. Consistent completions of the p_f's form $2^{\aleph_0}$ many elements of $S(B')$, contradicting the $\aleph_0$−stability of the theory to prove the claim.

A prime model over A is constructed as follows. Let $\lambda = |A| + \aleph_0$. Define by recursion a set $\{c_\alpha : \alpha < \lambda\} \subset M$ such that, letting $C_\alpha = \{c_\beta : \beta < \alpha\}$,

- for $\alpha < \lambda$, $tp_{\mathcal{M}}(c_\alpha/A \cup C_\alpha)$ is isolated;
- if $\varphi(x)$ is a consistent formula over $A \cup C_\alpha$, then $\varphi \in tp_{\mathcal{M}}(c_\beta/A \cup C_\beta)$ for some $\beta \geq \alpha$.

In detail, the construction proceeds as follows. Without loss of generality, $\mathcal{M}$ has cardinality λ. Let ψ_α, $\alpha < \lambda$, be a list of all consistent formulas in one variable over M. Assume c_β to be defined for each $\beta < \alpha$, and let $\Gamma = \{\gamma < \lambda : \psi_\gamma$ is over $A \cup C_\alpha$ and is not satisfied by some element of $C_\alpha\}$. If $\Gamma \neq \emptyset$ let φ be ψ_γ for γ the least element of Γ; if $\Gamma = \emptyset$, let φ be $v = v$. Since the isolated types are dense in $S_1(A \cup C_\alpha)$ there is some element d in M such that $\mathcal{M} \models \varphi(d)$ and $tp_{\mathcal{M}}(d/A \cup C_\alpha)$ is isolated. Let $c_\alpha = d$. This completes the definition of the set C_λ. It follows quickly from the Tarski-Vaught Test that C_λ is the universe of an elementary submodel $\mathcal{N}$ of $\mathcal{M}$.

It is easy to see that $\mathcal{N}$ is an atomic model over A. Let $\mathcal{M}'$ be an arbitrary model containing A such that $\mathcal{M}'_A \equiv \mathcal{M}_A$. To show that $\mathcal{N}$ is a prime model over A we need to prove

Claim. There are elements d_α in M', for $\alpha < \lambda$, such that the map f which is the identity on A and takes c_α to d_α is an elementary.

Assume that d_β has been defined, for $\beta < \alpha$, so that the mapping g which fixes A pointwise and takes c_β to d_β, for $\beta < \alpha$, is elementary. Let $\varphi(x, \bar{c})$ be a formula which isolates $tp_{\mathcal{M}}(c_\alpha/A \cup C_\alpha)$. Let $\bar{d} = g(\bar{c})$ and choose d_α to be any element of M' which satisfies $\varphi(x, \bar{d})$. Since $\varphi(x, \bar{c})$ isolates a complete type over $A \cup C_\alpha$, $\varphi(x, \bar{d})$ isolates a complete type over $A \cup D_\alpha$, and the extension of g which takes c_α to d_α is elementary. This proves the claim.

Clearly, $N' = \{d_\alpha : \alpha < \lambda\}$ is the universe of a model $\mathcal{N}' \prec \mathcal{M}'$ and f is an isomorphism of $\mathcal{N}$ onto $\mathcal{N}'$ fixing A. This proves that $\mathcal{N}$ is a prime model over A.

Remark 3.1.6. As with countable theories, it follows quickly from the previous lemma that *any* prime model over a set in an $\aleph_0$−stable theory is atomic.

It is also the case that there is a *unique* (up to isomorphism) prime model over a set in an $\aleph_0$−stable theory. Unlike the corresponding result for countable theories, however, this is rather difficult to prove (see Corollary 5.5.1).

Most readers of this book will have heard the term "prime field". The notion of a prime element of a class of structures is found throughout mathematics, especially algebra. Likewise in model theory we will occasionally speak of the prime model relative to a nonelementary class of models. Next we develop the notion of a prime model over a set, relative to the $\aleph_0$−saturated models.

Notation. If p is a type over a set A and $B \subset A$ we let $p \upharpoonright B$ denote $\{\varphi \in p : \varphi \text{ is over } B\}$, called the *restriction of p to B*.

Definition 3.1.5. *Let $\mathcal{M}$ be a model and $A \subset M$.*

(i) We say that $p \in S_n(A)$ is $\aleph_0$−isolated over B *if $B \subset A$ is finite and p is the only extension of $p \upharpoonright B$ in $S_n(A)$, in which case we say that $p \upharpoonright B$* $\aleph_0$−isolates p. *p is* $\aleph_0$−isolated *if it is $\aleph_0$−isolated over some finite subset of A.*

(ii) A set $B \subset M$ is said to be $\aleph_0$−atomic over A *if for each finite sequence $\bar{a}$ from B, $tp_{\mathcal{M}}(\bar{a}/A)$ is $\aleph_0$−isolated.*

(iii) $\mathcal{M}$ is $\aleph_0$−prime over A *if*

- *$\mathcal{M}$ is $\aleph_0$−saturated, and*
- *for any $\aleph_0$−saturated model $\mathcal{N}$ containing A with $\mathcal{N}_A \equiv \mathcal{M}_A$, there is an elementary embedding of $\mathcal{M}$ into $\mathcal{N}$ which is the identity on A.*

In a sense, the definition of an $\aleph_0$−prime model over a set can be obtained from the definition of a prime model over a set by uniformly replacing formulas by complete types over finite sets. Continuing the parallel development:

Lemma 3.1.6. *Let T be $\aleph_0$−stable, $\mathcal{M}$ a model of T and $A \subset M$. Then there is a model $\mathcal{N}$ which is $\aleph_0$−prime over A and $\aleph_0$−atomic over A.*

Proof. Only an outline of the proof is given, leaving the details (which are very similar to those found in Lemma 3.1.5) to the reader. The first claim in that previous proof is replaced by the following.

Claim. Let B be a subset of M and q a complete 1−type over a finite subset of B. Then there is $p \in S_1(B)$ which extends q and is $\aleph_0$−isolated.

In the verification that the constructed model $\mathcal{N}$ is $\aleph_0$−prime we use the fact that if $\mathcal{M}'$ is $\aleph_0$−saturated, $B \subset M'$ and $p \in S(B)$ is $\aleph_0$−isolated, then p is realized in $\mathcal{N}$. (p is $\aleph_0$−isolated by a complete type q over a finite set, which is realized by some $a \in N$. Then $tp_{\mathcal{N}}(a/B) = p$.)

We now turn towards properties more specific to theories categorical in some uncountable cardinal.

Definition 3.1.6. *Let $\mathcal{M}$ and $\mathcal{N}$ be models and T a countable complete theory.*

(i) We say that $(\mathcal{M}, \mathcal{N}, \varphi)$ is a Vaughtian triple *if $\mathcal{N}$ is a proper elementary submodel of $\mathcal{M}$, φ is a nonalgebraic formula in one free variable over N and $\varphi(\mathcal{M}) = \varphi(\mathcal{N})$.*

(ii) The pair $(\mathcal{M}, \mathcal{N})$ is called a Vaughtian pair *if for some φ, $(\mathcal{M}, \mathcal{N}, \varphi)$ is a Vaughtian triple.*

(iii) T is said to admit a Vaughtian pair *(or* triple*) if there is a Vaughtian triple $(\mathcal{M}, \mathcal{N}, \varphi)$, where $\mathcal{M}$ and $\mathcal{N}$ are models of T, in which case we call $(\mathcal{M}, \mathcal{N}, \varphi)$ a Vaughtian triple for T.*

Proposition 3.1.1. *Let T be categorical in some uncountable power. Then there is no Vaughtian pair consisting of models of T.*

Remark 3.1.7. This proposition, which is proved in the next few lemmas, comprises a major portion of the work in the proof of Morley's Categoricity Theorem. Let κ be an uncountable cardinal in which T is categorical. To prove the proposition we assume, to the contrary, that T has a Vaughtian pair. We then prove (making heavy use of the $\aleph_0$−stability of T) that there is a Vaughtian pair $(\mathcal{M}, \mathcal{N})$ with $|M| = \kappa$ and $|N| = \aleph_0$. Thus some N−definable subset of M is countable. However a relatively straightforward elementary chain argument shows that T (in fact any countable theory) has a model $\mathcal{M}'$ of cardinality κ such that every M'−definable relation in $\mathcal{M}'$ is uncountable. This contradiction proves the proposition.

Lemma 3.1.7. *If T is a small theory and admits a Vaughtian pair, then T admits a Vaughtian pair $(\mathcal{M}, \mathcal{N})$ where $\mathcal{N}$ and $\mathcal{M}$ are countable saturated models.*

Proof. Let $(\mathcal{M}_0, \mathcal{N}_0, \varphi)$ be a Vaughtian triple for T. Let L be the language of T, let $\bar{a}$ be the parameters in φ and $L(\bar{a})$ the expansion of L obtained by adding constants for the elements of $\bar{a}$. The proof centers on finding a theory (in a larger language) expressing that a triple is a Vaughtian triple for T. Specifically, we show

Claim. There is a theory $T' \supset Th(\mathcal{N}_0, \bar{a})$ in a language containing $L(\bar{a})$ and a new unary relation P such that whenever $\mathcal{M}' \models T'$,

(a) $P(\mathcal{M}')$ is the universe of an elementary submodel $\mathcal{N}$ of $\mathcal{M}' \restriction L = \mathcal{M}$,
(b) $\bar{a}$ is from N, and
(c) $(\mathcal{M}, \mathcal{N}, \varphi)$ is a Vaughtian triple.

Let P be a new unary relation symbol and $L' = L(\bar{a}) \cup \{P\}$. Let T' be the set of sentences in L' expressing the following:

– $T' \supset T$;

– for $\psi(v_0, \ldots, v_n)$ a formula of L,

$$\forall v_0 \ldots v_{n-1}(\bigwedge_{i<n} P(v_i) \wedge \exists v \psi(v_0, \ldots, v_{n-1}, v) \longrightarrow \exists v(\psi(v_0, \ldots, v_{n-1}, v) \wedge P(v)));$$

– every element of $\bar{a}$ satisfies P;
– $\forall v(\varphi(v) \longrightarrow P(v))$ and
– there is an element not satisfying P.

Interpreting P by N_0 gives an expansion of $(\mathcal{M}_0, \bar{a})$ which is a model of T', so the theory is consistent. To verify that T' satisfies the requirements of the claim let $\mathcal{M}'$ be any model of T'. The second item in the definition implies that $N = P(\mathcal{M}')$ is the universe of a model $\mathcal{N}$ (in L) which is an elementary submodel of $\mathcal{M} = \mathcal{M}' \upharpoonright L$. The last three items verify (c) of the claim.

Claim. There is a countable model $\mathcal{M}'$ of T' in which $\mathcal{M} = \mathcal{M}' \upharpoonright L$ and $P(\mathcal{M}')$ are each a countable saturated model of T.

The targeted model $\mathcal{M}'$ is constructed using a standard elementary chain argument. Let $\mathcal{M}'_0$ be any countable model of T. Assuming $\mathcal{M}'_n$ to be defined, let $\mathcal{M}'_{n+1}$ be a countable elementary extension of $\mathcal{M}'_n$ with the property:

> If p is a complete type in L over a finite subset A of M'_n then p is realized in $\mathcal{M}'_{n+1}$. Furthermore, if $A \subset P(\mathcal{M}'_n)$, then there is a realization of p in $P(\mathcal{M}'_{n+1})$.

Then $\mathcal{M}' = \bigcup_{i<\omega} \mathcal{M}'_i$ satisfies the requirements of the claim.

Letting $\mathcal{M}$ be the restriction of $\mathcal{M}'$ to L and $\mathcal{N}$ the elementary submodel of $\mathcal{M}$ with universe $P(\mathcal{M}')$ gives the required Vaughtian pair.

The next step in the proof of Proposition 3.1.1 is to show that we can "stretch" the larger model in a Vaughtian pair of $\aleph_0$–saturated models while fixing the smaller one. Continuing this through κ steps results in a Vaughtian pair which contradicts the κ–categoricity of T, as described immediately after the statement of the proposition. This stretching of a Vaughtian pair is accomplished using the nonsplitting relation on types. The importance of the nonsplitting relation to the study of arbitrary $\aleph_0$–stable theories justifies this rather lengthy diversion.

Definition 3.1.7. *Let $\mathcal{M}$ be a model, $A \subset B \subset M$, and $p \in S(B)$.*

(i) We say that p does not split over *A if for all tuples $\bar{a}$, $\bar{b}$ from B and formulas $\varphi(\bar{x}, \bar{v})$ over $\emptyset$,*

$$\text{if } tp_{\mathcal{M}}(\bar{a}/A) = tp_{\mathcal{M}}(\bar{b}/A), \text{ then } (\varphi(\bar{x}, \bar{a}) \in p \Longleftrightarrow \varphi(\bar{x}, \bar{b}) \in p).$$

The negation of "p does not split over A" is p splits over *A.*

(ii) Suppose that p does not split over A, $B \subset C \subset M$, and $q \in S(C)$. Then q is called a strong heir *of p if $q \supset p$ and q does not split over A.*

Remark 3.1.8. The reader should supply proofs for the following elementary facts about the splitting relation. Let $\mathcal{M}$ be a model, $A \subset B \subset M$, and $p \in S(B)$.

(i) If p does not split over A and $A \subset A' \subset B$, then p does not split over A'.

(ii) If p does not split over A and $\varphi(\bar{x}, \bar{v})$ is a formula over A, then for all tuples $\bar{a}$, $\bar{b}$ from B realizing the same complete type over A,

$$\varphi(\bar{x}, \bar{a}) \in p \Longleftrightarrow \varphi(\bar{x}, \bar{b}) \in p.$$

(In other words, the clause defining the nonsplitting relation holds for formulas over A as well as formulas over $\emptyset$.)

(iii) p does not split over B.

Example 3.1.2. (i) Let T be the theory of a single equivalence relation E having infinitely many infinite classes and no finite classes.

Claim. Let $\mathcal{M}$ be a model of T. Each complete 1−type over $\mathcal{M}$ does not spilt over some element of M. Furthermore, each element of $S_1(M)$ has a strong heir in $S_1(A)$ for any $A \supset M$.

The theory T is complete and has elimination of quantifiers. If $p \in S_1(M)$ is algebraic there is a $b \in M$ such that $x = b \in p$. In this case, p does not split over b, and for any $A \supset M$, a subset of an elementary extension of $\mathcal{M}$, the unique extension of p in $S_1(A)$ is a strong heir of p. Now let $\mathcal{N}$ be a proper elementary extension of $\mathcal{M}$, $a \in N \setminus M$ and $p = tp_{\mathcal{N}}(a/M)$. Suppose that $E(a, b)$ holds for some b in M. We claim that p does not split over b. Every formula is equivalent in T to a boolean combination of instances of E and equality. Since $a \notin M$, $x \neq c \in p$, for all $c \in M$. Let c, $d \in M$ realize the same complete type over b. Then, $E(x, c) \in p \Longleftrightarrow \mathcal{M} \models E(c, b) \Longleftrightarrow \mathcal{M} \models E(d, b) \Longleftrightarrow E(x, d) \in p$. Using the elimination of quantifiers we conclude that p does not split over b. (Notice that p does split over $\emptyset$: for c an element not E−equivalent to b, $tp_{\mathcal{N}}(c) = tp_{\mathcal{N}}(b)$ while $E(x, b) \in p$ and $\neg E(x, c) \in p$.) Now let $A \supset M$ be a subset of the model $\mathcal{M}'$. There is a $q \in S_1(A)$ containing p such that for all $c \in A$, $x \neq c \in q$, and $E(x, c) \in q \Longleftrightarrow \mathcal{M}' \models E(c, b)$. The type q is a strong heir of p.

Supposing that $a \in N \setminus M$ is not E−equivalent to any element of M, a simple argument shows that $p = tp_{\mathcal{N}}(a/M)$ does not split over $\emptyset$. Let $\mathcal{M}' \prec \mathcal{M}$, $M' \supset A \supset M$ and $q \in S_1(A)$ a type such that for all $c \in A$, $x \neq c \in q$ and $\neg E(x, c) \in q$. Then, q is a strong heir of p, proving the claim.

(ii) Let $\mathcal{M}$ be a dense linear order without endpoints; i.e., a model of $Th(\mathbb{Q}, <)$. A *cut* in $\mathcal{M}$ is a subset J of M such that, whenever $a \in J$ and $b < a$, $b \in J$. For $B \cup \{a\} \subset M$, $\sup B = a$ if every $b \in J$ is $\leq a$ and, for any $a' \in M$ such that $b \in J \Longrightarrow b \leq a'$, $a \leq a$. For subsets B and C of M we say that $\sup(B) = \sup(C)$ if for all $b \in B$ there is a $c \in C$ such that $b > c$ and for all $c \in C$ there is a $b \in B$ greater than c. The relation $\inf(B) = \inf(C)$ is defined similarly.

Let A be a cut in $\mathcal{M}$ such that $\sup(A)$ does not exist in M and $M \setminus A$ is nonempty. Let p be the unique element of $S_1(M)$ containing $\{x > a : a \in A\} \cup \{x < b : b \in M \setminus A\}$. We show that there are very strict limitations on the subsets of M over which p does not split in

Claim. Given $B \subset M$, p does not split over B if and only if $\sup(B \cap A) = \sup(A)$ or $\inf(B \setminus A) = \inf(M \setminus A)$.

First suppose that $\sup(B \cap A) = \sup(A)$ and $c, d \in M$ have the same type over $B \cap A$. If $b \in B \cap A$ and $c < b$, then $d < b$, so $x > c \in p$ and $x > d \in p$. If $c > b$ for all $b \in B \cap A$, then $x < c$ and $x < d$ are both in p. It follows quickly from elimination of quantifiers that p does not split over $B \cap A$. Similarly, p does not split over B if $\inf(B \setminus A) = \inf(M \setminus A)$. Suppose, on the other hand, that there are $a \in A$ greater than every element of $B \cap A$ and $c \in M \setminus A$ less than each element of $B \setminus A$. Then, $tp_{\mathcal{M}}(b/B) = tp_{\mathcal{M}}(c/B)$, $x > b \in p$ and $x < c \in p$, proving that p splits over B.

Among other things, we conclude that p splits over any finite subset of M.

These definitions reflect the following view of types. Let $\mathcal{M}$ be a model, $B \subset M$, $p \in S_1(B)$ and a a realization of p in $\mathcal{M}$. The formulas in p define the relations holding on $(a, \bar{b})$ for sequences $\bar{b}$ from B. If p does not split over $A \subset B$, a definable relation holding on a and sequences from B is determined by A in the following sense. For any formula $\varphi(x, \bar{y})$ there is $P_\varphi \subset S(A)$ such that $\varphi(x, \bar{b})$ belongs to p if and only if $tp_{\mathcal{M}}(\bar{b}/A) \in P_\varphi$. We think of the family $\mathcal{P} = \{P_\varphi : \varphi \text{ a formula}\}$ as being a kind of oracle which tells us which formulas go into p. If $q \in S(C)$ is a strong heir of p no essentially new relations are being defined using the elements of C; the family $\mathcal{P}$ still determines which formulas enter the type. A strong heir is a "freest" possible extension in that only the traits which are inherited from p are found in q. (We use the term strong heir as "heir" has been reserved for a different but closely related concept defined in [LP79] (see Definition 5.1.13). Such notions of "free" extensions of types are the foundation of stability theory.)

Definition 3.1.8. *Let $\mathcal{M}$ be a model, $A \subset M$.*

(i) Let p be a type over A and f an elementary map whose domain contains A. Then $f(p)$ denotes $\{\varphi(\bar{v}, f(\bar{a})) : \varphi(\bar{v}, \bar{a}) \in p\}$, a set of formulas over $f(A)$.

(ii) If $\bar{b}$ and $\bar{c}$ are sequences and there is an elementary map f such that f is the identity on A and $f(\bar{b}) = \bar{c}$, then we say $\bar{b}$ is conjugate to *$\bar{c}$* over *A. This terminology is applied to infinite sequences $\bar{b}$ and $\bar{c}$ as well as finite sequences. Occasionally we will say, e.g., "B and C are conjugate over A", leaving the relevant ordering of B and C to be understood.*

(iii) When p and q are types over a model $\mathcal{M}$ and there is an elementary map f whose domain contains A such that $f(p) = q$, we say that p and q are conjugate over *A.*

Let p, A, and f be as in (i) of the definition. Because f is elementary, $f(p)$ is itself a type (i.e., it is consistent) and $f(p)$ is complete whenever p is complete. It is easily verified that if, e.g., f is an automorphism of a model $\mathcal{M}$ containing $B \cup \{\bar{a}\}$ and $p = tp(\bar{a}/B)$, then $f(p) = tp(f(\bar{a})/f(B))$.

If $\mathcal{M}$ is a model and $p \in S_1(M)$, the automorphisms of $\mathcal{M}$ act on p to produce other elements of $S_1(M)$. This action is effected by nonsplitting in the following way.

Lemma 3.1.8. *Let $B \subset M$, where $\mathcal{M}$ is a model, p an element of $S(B)$ which does not split over $A \subset B$, $A \subset B_0 \subset B$, and f an elementary map fixing A pointwise and taking B_0, to $B_1 \subset B$. Then, $f(p \restriction B_0) = p \restriction B_1$. In particular, if f maps B onto B, then $f(p) = p$.*

Proof. Let $\bar{b}$ be a sequence from B_0 and $\bar{b}' = f(\bar{b})$. Since f is the identity on A, $tp(\bar{b}/A) = tp(\bar{b}'/A)$. Since p does not split over A, $\varphi(\bar{x}, \bar{b}) \in p \iff \varphi(\bar{x}, \bar{b}') \in p$ for any formula $\varphi(\bar{x}, \bar{y})$. However,

$$\varphi(\bar{x}, \bar{b}) \in p \restriction B_0 \iff \varphi(\bar{x}, \bar{b}') \in f(p \restriction B_0),$$

from which we conclude that $p \restriction B_1 = f(p \restriction B_0)$.

In other words, for $\mathcal{M}$ a model and $p \in S_1(M)$ which does not split over $A \subset M$, if $q \in S_1(M)$ is conjugate to p over A, then $q = p$.

To set the stage for the next lemma, let $\mathcal{M}$ be a countable saturated model of an $\aleph_0$–stable theory and $p \in S_1(M)$. There are continuum many automorphisms of $\mathcal{M}$, each generating a conjugate of p in $S_1(M)$. Assuming that p splits over every finite subset of M we prove (in the next lemma) that continuum many of these conjugates of p are distinct (a contradiction).

Lemma 3.1.9. *Let T be $\aleph_0$–stable, A a subset of a model of T and $p \in S(A)$. Then there is a finite set $B \subset A$ such that p does not split over B.*

Proof. Assume, to the contrary, that there is no finite $B \subset A$ over which p does not split. The $\aleph_0$–stability of T will be contradicted by constructing continuum many types over some countable set. Let $\mathcal{M}$ be an $\aleph_0$–saturated model of T containing A, and let X be the set of all finite sequences of 0's and 1's.

Claim. For $s, t \in X$ there are: $A_s \subset A$ finite, $B_s \subset M$, $q_s \in S(B_s)$, and elementary maps f_s from A_s onto B_s such that

(a) $f_s(p \restriction A_s) = q_s$,
(b) if t is an initial segment of s then $f_t \subset f_s$,
(c) if t is not an initial segment of s and s is not an initial segment of t, then $q_s \cup q_t$ is inconsistent.

To begin let $A_\emptyset = B_\emptyset = f_\emptyset = \emptyset$. Assume that A_t, B_t and f_t have been defined for all $t \in X$ of length k. For an arbitrary $s \in X$ of length k we show how to define A_r, B_r and f_r for $r = s\hat{}\langle i\rangle = si$, $i = 0, 1$. By assumption, p splits over the finite set A_s; i.e., there are $\bar{a}$ and $\bar{b}$ from A such that $tp(\bar{a}/A_s) = tp(\bar{b}/A_s)$ and a formula φ, such that $\varphi(\bar{x}, \bar{a}) \in p$ and $\neg\varphi(\bar{x}, \bar{b}) \in p$. Let $A_{s0} = A_s \cup \bar{a}$ and $A_{s1} = A_s \cup \bar{b}$. Since $\mathcal{M}$ is $\aleph_0$-saturated there is a $\bar{c}$ in M and elementary maps f_{s0}, f_{s1} extending f_s such that $f_{s0}(\bar{a}) = \bar{c}$ and $f_{s1}(\bar{b}) = \bar{c}$. Let $B_{si} = B_s \cup \bar{c}$ and $q_{si} = f_{si}(p \restriction A_{si})$, for $i = 0, 1$. Since $\varphi(\bar{x}, \bar{c}) \in q_{s0}$ and $\neg\varphi(\bar{x}, \bar{c}) \in q_{s1}$ all of the required conditions are satisfied, completing the proof of the claim.

Let $B = \bigcup_s B_s$, Y the set of all sequences of 0's and 1's of length ω, and for $s \in Y$ let $q_s = \bigcup\{q_t : t = s \restriction k, \text{ for some } k\}$. By conditions (a) and (b) in the claim, when t is an initial segment of $r \in X$, $q_t = f_t(p \restriction A_t) = f_r(p \restriction A_t) \subset f_r(p \restriction A_r) \subset q_r$, hence q_s is consistent for each $s \in Y$. For $s \in Y$ let q'_s be a completion of q_s in $S(B)$. There are continuum many such q'_s's by (c). Since B is countable this contradicts the $\aleph_0$-stability of T to prove the lemma.

Corollary 3.1.3. *Let T be an $\aleph_0$-stable theory.*

(i) T is κ-stable for all $\kappa \geq \aleph_0$.

(ii) For every regular cardinal κ, T has a saturated model of cardinality κ.

Proof. (i) Let A_0 be a subset of a model with $|A_0| = \kappa$, and let $\mathcal{M}$ be an $\aleph_0$-saturated model of cardinality κ which contains A_0. Since distinct elements of $S(A_0)$ extend to distinct elements of $S(M)$, it suffices to show that $|S(M)| = \kappa$. Every element of $S(M)$ does not split over some finite subset of M. Since there are κ many finite subsets of M it suffices to prove

Claim. For all finite $A \subset M$ there are only countably many elements of $S(M)$ which do not split over A.

Suppose to the contrary that there are distinct $p_i \in S_n(M)$, for $i < \omega_1$, such that each p_i does not split over A. Let $\mathcal{N}$ be a countable saturated elementary submodel of $\mathcal{M}$ containing A and let $q_i = p_i \restriction N$, for $i < \omega_1$. We claim that $q_i \neq q_j$, for $i \neq j < \omega_1$. Let i, j be distinct ordinals $< \omega_1$, and $\varphi(\bar{x}, \bar{a})$ a formula such that $\varphi(\bar{x}, \bar{a}) \in p_i$ and $\neg\varphi(\bar{x}, \bar{a}) \in p_j$. Let $q = tp(\bar{a}/A)$ and $\bar{c}$ a realization of q in $\mathcal{N}$. Since p_i and p_j both do not split over A, $\varphi(\bar{x}, \bar{c}) \in p_i$ and $\neg\varphi(\bar{x}, \bar{c}) \in p_j$. Thus, the q_i's form an uncountable set of complete types over the countable set N, contradicting the $\aleph_0$-stability of the theory. This completes the proof of the claim and this part of the lemma.

(ii) This follows from (i) and Lemma 2.2.6.

Lemma 3.1.10. *Let T be an $\aleph_0$-stable theory and $\mathcal{M} \prec \mathcal{N}$ models of T with $\mathcal{M}$ $\aleph_0$-saturated. Then every element of $S(M)$ has a strong heir in $S(N)$.*

Proof. Let $p \in S(M)$ and let A be a finite subset of M over which p does not split. Define a set of formulas Γ over N by the scheme: Given a formula $\varphi(x, \bar{v})$ over $\emptyset$ and tuple $\bar{a}$ from N, $\varphi(x, \bar{a})$ is in Γ if there is a sequence $\bar{b}$ from M such that $tp_{\mathcal{N}}(\bar{a}/A) = tp_{\mathcal{N}}(\bar{b}/A)$ and $\varphi(x, \bar{b}) \in p$. This set Γ is well-defined since p does not spilt over A. (Given $\varphi(x, \bar{a})$, if there is a sequence $\bar{b}$ from M such that $q = tp_{\mathcal{N}}(\bar{a}/A) = tp_{\mathcal{N}}(\bar{b}/A)$ and $\varphi(x, \bar{b}) \in p$, then for any sequence $\bar{c}$ from M realizing q, $\varphi(x, \bar{c}) \in p$.) Also, $\Gamma \supset p$. To see that Γ is consistent, let $\{\varphi_0(x, \bar{a}_0), \ldots, \varphi_n(x, \bar{a}_n)\}$ be a finite subset of Γ. Since $\mathcal{M}$ is $\aleph_0$-saturated there are $\bar{b}_i \in M$ with $tp_{\mathcal{N}}(\bar{a}_0 \ldots \bar{a}_n/A) = tp_{\mathcal{N}}(\bar{b}_0 \ldots \bar{b}_n/A)$. The definition of Γ implies that $\{\varphi_0(x, \bar{b}_0), \ldots, \varphi_n(x, \bar{b}_n)\} \subset p$, hence Γ is consistent. It also follows quickly from its definition that Γ is a complete type which does not split over A.

Lemma 3.1.11 (Stretching a Vaughtian pair). *Let T be an $\aleph_0$-stable theory with a Vaughtian triple $(\mathcal{M}, \mathcal{N}, \varphi)$, where $\mathcal{M}$ and $\mathcal{N}$ are $\aleph_0$-saturated models of T and $\mathcal{N}$ is countable. Then there is a proper elementary extension $\mathcal{M}'$ of $\mathcal{M}$ which is $\aleph_0$-saturated such that $(\mathcal{M}', \mathcal{N}, \varphi)$ is a Vaughtian triple.*

Proof. Let a be any element of $M \setminus N$ and $p = tp_{\mathcal{M}}(a/N)$. There is a finite $A \subset N$ over which p does not split (by Lemma 3.1.9) and there is a $q \in S(M)$ which is a strong heir of p (by Lemma 3.1.10). Without loss of generality, A contains the parameters in φ. Let b be a realization of q in some elementary extension of $\mathcal{M}$ and let $\mathcal{M}'$ be an $\aleph_0$-prime model over $M \cup \{b\}$ (which exists by Lemma 3.1.6). It remains to verify that $(\mathcal{M}', \mathcal{N}, \varphi)$ is a Vaughtian triple. Assume to the contrary that there is a $c \in M' \setminus N$ satisfying φ. Since $(\mathcal{M}, \mathcal{N}, \varphi)$ is a Vaughtian triple, $c \in M' \setminus M$, hence $r = tp_{\mathcal{M}'}(c/M \cup \{b\})$ is $\aleph_0$-isolated. We will contradict that $(\mathcal{M}, \mathcal{N}, \varphi)$ is a Vaughtian triple by finding a $c_0 \in M \setminus N$ satisfying φ.

Let B, $M \supset B \supset A$ be a finite set such that r is $\aleph_0$-isolated over $B \cup \{b\}$. Since B is finite there is a countable saturated model $\mathcal{N}_0 \prec \mathcal{M}$ containing B. Since $tp_{\mathcal{M}'}(c/N_0 \cup \{b\})$ is also $\aleph_0$-isolated (over $B \cup \{b\}$) there is an $\mathcal{M}'_0 \prec \mathcal{M}$ which is $\aleph_0$-prime over $N_0 \cup \{b\}$ and contains c. Since $\mathcal{N}$ and $\mathcal{N}_0$ are both countable saturated models there is an isomorphism f_0 of $\mathcal{N}_0$ onto $\mathcal{N}$ which is the identity on A. Since $q = tp(b/M)$ does not split over A, $f_0(q \restriction N_0) = q \restriction N = p = tp(a/N)$ (by Lemma 3.1.8). Thus, f_0 extends to an elementary map f_1 of $N_0 \cup \{b\}$ onto $N \cup \{a\}$. Since $\mathcal{M}'_0$ is $\aleph_0$-prime over $N_0 \cup \{b\}$ and $\mathcal{M}$ is an $\aleph_0$-saturated model containing $N \cup \{a\}$ there is an elementary embedding f of $\mathcal{M}'_0$ into $\mathcal{M}$ which extends f_1. Then, $f(c)$ is an element of $M \setminus N$ which satisfies φ. This contradicts that $(\mathcal{M}, \mathcal{N}, \varphi)$ is a Vaughtian triple, completing the proof of the lemma.

Proof of Proposition 3.1.1. Let κ be an uncountable cardinal in which T is categorical and assume the proposition to fail. By Lemma 3.1.7 there is a Vaughtian triple $(\mathcal{M}, \mathcal{N}, \varphi)$ with $\mathcal{M}$ and $\mathcal{N}$ countable saturated models.

Claim. There is an elementary chain $\mathcal{M}_\alpha$, $\alpha < \kappa$, such that

(1) $\mathcal{M}_\alpha$ is an $\aleph_0$–saturated model of cardinality $< \kappa$,
(2) $(\mathcal{M}_\alpha, \mathcal{N}, \varphi)$ is a Vaughtian triple,
(3) if $\beta < \alpha$, $\mathcal{M}_\beta \neq \mathcal{M}_\alpha$.

This chain is defined by recursion using the previous lemma. Let $\mathcal{M}_0 = \mathcal{M}$. Suppose that $\mathcal{M}_\beta$ has been defined for all $\beta < \alpha$ and (1)-(3) hold relative to these models. If α is a limit ordinal let $\mathcal{M}_\alpha = \bigcup_{\beta<\alpha} \mathcal{M}_\beta$. If $\alpha = \beta+1$ apply Lemma 3.1.11 to the Vaughtian triple $(\mathcal{M}_\beta, \mathcal{N}, \varphi)$ to obtain an $\aleph_0$–saturated proper elementary extension $\mathcal{M}_\alpha$ such that $(\mathcal{M}_\alpha, \mathcal{N}, \varphi)$ is a Vaughtian triple. Of course, we can assume that $|M_\alpha| = |M_\beta|$. Notice that (1)-(3) hold for all $\alpha < \kappa$, proving the claim.

Let $\mathcal{M}' = \bigcup_{\alpha<\kappa} \mathcal{M}_\alpha$. Then $\mathcal{M}'$ is a model of cardinality κ (by (1) and (3)) and $(\mathcal{M}', \mathcal{N}, \varphi)$ is a Vaughtian triple. Since $\mathcal{N}$ is countable, $\varphi(\mathcal{M}')$ is countable. However, as noted in Remark 3.1.7 the κ–categoricity of T implies that every relation definable over M' is uncountable. This contradiction proves the proposition.

The proof of Morley's Categoricity Theorem is completed by showing

Theorem 3.1.2. *Let T be a countable complete $\aleph_0$–stable theory with no Vaughtian pair. Then, T is categorical in every uncountable cardinal.*

The proof of the theorem is separated into the following three steps.

(a) First it is shown that there is a strongly minimal formula φ (in one variable) over a prime model $\mathcal{M}_0$ of T.
(b) Next, given a model $\mathcal{M} \succ \mathcal{M}_0$ of T we prove that $\mathcal{M}$ is prime over $\varphi(\mathcal{M})\cup$ (the parameters in φ).
(c) Finally, we show that the theorem follows from (b).

Presently we only know that there is a strongly minimal formula over a countable saturated model (by Lemma 3.1.4). This is improved to obtain (a) in:

Proposition 3.1.2. *Let T be a countable complete $\aleph_0$–stable theory with no Vaughtian pair. Then there is a strongly minimal formula over a prime model.*

The proposition will be proved using

Lemma 3.1.12. *Suppose that T is $\aleph_0$–stable, has no Vaughtian pair and $\varphi(\bar{x}, \bar{v})$ is a formula. Then there is a natural number n such that for all models $\mathcal{M}$ of T and all $\bar{a}$ from M, $\varphi(\mathcal{M}, \bar{a})$ is infinite or of cardinality at most n.*

Proof. For notational simplicity we assume $\varphi(\bar{x}, \bar{v}) = \varphi(x, \bar{v})$; i.e., $\bar{x}$ has length 1. Suppose, to the contrary, that there is no such n for $\varphi(x, \bar{v})$. If ψ is an algebraic formula over a model $\mathcal{N}$ and $\mathcal{N}' \succ \mathcal{N}$, then $\psi(\mathcal{N}') \subset N$. Thus,

(∗) For all $n < \omega$ there is a model $\mathcal{N}$ and a sequence $\bar{a}$ from N such that $|\varphi(\mathcal{N}, \bar{a})| \geq n$ and for all $\mathcal{N}' \succ \mathcal{N}$, $\varphi(\mathcal{N}', \bar{a}) \subset N$.

We will define a theory T' such that a model of T' represents a pair of models which is a Vaughtian pair for T. The consistency of T' will follow from (∗). This contradiction to Proposition 3.1.1 will prove the lemma.

Let L be the language of T and $L' = L \cup \{P, c_1, \ldots, c_k\}$, where P is a unary predicate symbol and $c_1, \ldots, c_k$ are new constant symbols (k = the length of $\bar{v}$). Let $T' \supset T$ be a theory in L' such that for any $\mathcal{M}' \models T'$,

- $P(\mathcal{M}')$ is the universe of a proper elementary submodel (with respect to L) of $\mathcal{M}' \restriction L$, and $\bar{c} \subset P(\mathcal{M}')$,
- for each n, $|\varphi(\mathcal{M}', \bar{c})| \geq n$ and
- $\varphi(\mathcal{M}', \bar{c}) \subset P(\mathcal{M}')$.

(The formulation of the actual sentences in L' comprising T' is left to the reader.) By (∗) and compactness, T' is consistent. Let $\mathcal{M}'$ be a model of T', $\mathcal{M} = \mathcal{M}' \restriction L$, $\mathcal{N}$ = the proper elementary submodel of $\mathcal{M}$ with universe $P(\mathcal{M}')$ and $\bar{b}$ the interpretation of $\bar{c}$. The second and third items say that $\varphi(\mathcal{M}, \bar{b})$ is infinite and contained in $N \neq M$. This contradicts that T does not have a Vaughtian pair, completing the proof of the lemma.

Proof of Proposition 3.1.2. The argument is like the proof of Lemma 3.1.4 but we work over a prime model instead of an $\aleph_0$-saturated model. Let $\mathcal{M}$ be a prime model of T. Let $\varphi(v)$ be a formula which has a unique nonalgebraic extension in $S(M)$ and let $A \subset M$ be the set of parameters in φ. The strong minimality of φ is proved as follows. Let $\mathcal{N}$ be an elementary extension of $\mathcal{M}$ and $\psi(v, \bar{b})$ a formula over N, where $\psi(v, \bar{x})$ is over $\emptyset$. Assume, towards a contradiction, that both $\sigma(v, \bar{b}) = \varphi(v) \wedge \psi(v, \bar{b})$ and $\tau(v, \bar{b}) = \varphi(v) \wedge \neg\psi(v, \bar{b})$ are nonalgebraic. By Lemma 3.1.12 there is an integer n such that for all elementary extensions $\mathcal{M}'$ of $\mathcal{M}$ and $\bar{c}$ from M', if $\sigma(\mathcal{M}', \bar{c})$ and $\tau(\mathcal{M}', \bar{c})$ both have cardinality $\geq n$, then $\sigma(x, \bar{c})$ and $\tau(x, \bar{c})$ are nonalgebraic. Since $\mathcal{N} \models \exists^{\geq n} v\sigma(v, \bar{b}) \wedge \exists^{\geq n} v\tau(v, \bar{b})$, there is a $\bar{d}$ from M such that $\mathcal{M} \models \exists^{\geq n} v\sigma(v, \bar{d}) \wedge \exists^{\geq n} v\tau(v, \bar{d})$. By the choice of n, both $\sigma(v, \bar{d}) = \varphi(v) \wedge \psi(v, \bar{d})$ and $\tau(v, \bar{d}) = \varphi(v) \wedge \neg\psi(v, \bar{d})$ are nonalgebraic. Since φ has a unique nonalgebraic completion over M we have obtained a contradiction which proves the proposition.

The following example shows that in Proposition 3.1.2 we cannot eliminate the hypothesis that the theory does not have a Vaughtian pair.

Example 3.1.3. Let E be a binary relation and T the theory expressing that (a) E is an equivalence relation and (b) for all $n < \omega$, there is an E-class containing exactly n elements. T is quantifier-eliminable, complete, and $\aleph_0$-stable. By the Omitting Types Theorem there is a model $\mathcal{M}$ of T (namely the prime model) such that each E-class in $\mathcal{M}$ is finite. By compactness there is an $\mathcal{N} \succ \mathcal{M}$ containing infinitely many infinite classes. It is easy to verify from the elimination of quantifiers that when the class of $a \in N$ is infinite, $E(x, a)$ is a strongly minimal formula. Of course, when $a \in M$, $E(x, a)$ is not strongly minimal, and neither is $\neg E(x, a)$ (since $E(x, b)$, for $b \in N \setminus M$, defines an infinite and coinfinite subset of $\neg E(\mathcal{N}, a)$). There are other possibilities for strongly minimal formulas over $\mathcal{M}$, however they all basically reduce to $E(x, a)$ or $\neg E(x, a)$ by the elimination of quantifiers. That is, there is no strongly minimal formula over the prime model of T.

Note: There are models $\mathcal{N}$ and $\mathcal{N}'$, with $\mathcal{N}'$ a proper elementary extension of $\mathcal{N}$, such that for some $a \in N$, $E(\mathcal{N}', a)$ is an infinite subset of N. Thus, T has a Vaughtian pair.

Item (b) in the outline of the proof of Theorem 3.1.2 is handled in

Corollary 3.1.4. *Let T be a countable complete $\aleph_0$-stable theory with no Vaughtian pair, $\mathcal{M}$ a model of T and φ a strongly minimal formula over some finite set $A \subset M$. Then,*

(i) $\mathcal{M}$ is prime over $\varphi(\mathcal{M}) \cup A$ and minimal over $\varphi(\mathcal{M}) \cup A$, and

(ii) if $\mathcal{M}$ is uncountable, $\dim(\varphi(\mathcal{M})/A) = |M|$.

Proof. (i) Simply by the $\aleph_0$-stability of the theory there is $\mathcal{N} \prec \mathcal{M}$ which is a prime model over $\varphi(\mathcal{M}) \cup A$ (see Lemma 3.1.5). Since T does not have a Vaughtian pair and $\varphi(\mathcal{N}) = \varphi(\mathcal{M})$, $\mathcal{M}$ cannot be a proper elementary extension of $\mathcal{N}$. Thus $\mathcal{M}$ is a prime model over $\varphi(\mathcal{M}) \cup A$, which is also minimal over this set by the same reasoning.

(ii) Let $\kappa = |M|$ be uncountable and I a basis for $\varphi(\mathcal{M})$ over A. Then $|\varphi(\mathcal{M})| \leq |acl(I \cup A)| \leq |I| + \aleph_0$ (since $\varphi(\mathcal{M}) \subset acl(I \cup A)$ and A is finite). Suppose, towards a contradiction, that $|I| < \kappa$. Then $|\varphi(\mathcal{M})| < \kappa$, in which case there is a proper elementary submodel $\mathcal{N}$ of $\mathcal{M}$ containing $\varphi(\mathcal{M}) \cup A$. This is impossible because T does not have a Vaughtian pair, proving the corollary.

Proof of Theorem 3.1.2. Let $\mathcal{M}$ and $\mathcal{N}$ be two models of T of the same uncountable cardinality κ. By Lemma 3.1.2 there is a formula $\varphi(v, \bar{x})$ over $\emptyset$ and an $\bar{a}$ from M such that $tp_{\mathcal{M}}(\bar{a})$ is isolated and $\varphi(v, \bar{a})$ is strongly minimal. Since $tp_{\mathcal{M}}(\bar{a})$ is isolated there is a sequence $\bar{b}$ from N such that $tp_{\mathcal{N}}(\bar{b}) = tp_{\mathcal{M}}(\bar{a})$. Then, $\varphi(v, \bar{b})$ is also strongly minimal (by Remark 3.1.2). Let I be a basis for $\varphi(\mathcal{M}, \bar{a})$ over $\bar{a}$ and J a basis for $\varphi(\mathcal{N}, \bar{b})$ over $\bar{b}$. By (ii) of the previous corollary, I and J both have cardinality κ. Let f be a mapping which takes $\bar{a}$ to $\bar{b}$ and is a bijection from I onto J. If we could assume that both $\bar{a}$ and $\bar{b}$ are empty, so that I and J are from the same strongly minimal set, then Lemma 3.1.3 would imply that f is an elementary map (see the proof of Corollary 3.1.1). In general, a slight adaptation of Lemma 3.1.3, whose proof is left to the reader in Exercise 3.1.13, shows that f is elementary. Since $\varphi(\mathcal{M}, \bar{a})$ and $\varphi(\mathcal{N}, \bar{b})$ are contained in the algebraic closures of $I \cup \bar{a}$ and $J \cup \bar{b}$, respectively, f extends to an elementary map from $\varphi(\mathcal{M}, \bar{a})$ onto $\varphi(\mathcal{N}, \bar{b})$ (see Exercise 3.1.10). By Corollary 3.1.4(i), $\mathcal{M}$ is prime over $\varphi(\mathcal{M}, \bar{a}) \cup \bar{a}$, so f extends to an elementary embedding g of $\mathcal{M}$ into $\mathcal{N}$ which maps $\bar{a}$ to $\bar{b}$ and $\varphi(\mathcal{M}, \bar{a})$ onto $\varphi(\mathcal{N}, \bar{b})$. Since $\mathcal{N}$ is a minimal model over $\varphi(\mathcal{N}, \bar{b}) \cup \bar{b}$, g maps $\mathcal{M}$ onto $\mathcal{N}$. Thus, g is an isomorphism from $\mathcal{M}$ onto $\mathcal{N}$ proving the theorem.

Proof of Theorem 3.1.1(**Morley's Categoricity Theorem**). Combining Theorems 3.1.1 and 3.1.2 shows that a countable theory T categorical in some uncountable power is categorical in every uncountable cardinality.

Corollary 3.1.5. *For T a countable complete theory the following are equivalent.*

(i) T is uncountably categorical.

(ii) T is $\aleph_0$−stable and does not have a Vaughtian pair.

(iii) For every regular $\kappa > \aleph_0$, every model of T of cardinality κ is saturated.

(The restriction to regular κ in (iii) will be removed in Lemma 3.4.10.)

The proof of Morley's Categoricity Theorem yields the following information about an uncountably categorical theory T. Over any model $\mathcal{M}$ of T there is a strongly minimal formula φ. Remember that $\varphi(\mathcal{M})$ is a pregeometry under algebraic closure. Letting A be the set of parameters in φ, $\mathcal{M}$ is prime over $\varphi(\mathcal{M}) \cup A$. In this way $\mathcal{M}$ is represented by a pregeometry. When $\mathcal{M}$ is uncountable assign to $\mathcal{M}$ a cardinal number $\mathcal{I}(\mathcal{M})$ = the dimension of $\varphi(\mathcal{M})$, for φ any strongly minimal formula over M. Given any two uncountable models $\mathcal{M}$ and $\mathcal{N}$ of T, $\mathcal{M} \cong \mathcal{N}$ if and only if $\mathcal{I}(\mathcal{M}) = \mathcal{I}(\mathcal{N})$. The number $\mathcal{I}(\mathcal{M})$ is called a *cardinal isomorphism invariant*, or simply a *cardinal invariant*, for $\mathcal{M}$. (When T is the theory of infinite vector spaces over a field F the dimension of $\mathcal{M} \models T$ is a cardinal invariant for $\mathcal{M}$. When T is the theory of algebraically closed fields of a fixed characteristic and $\mathcal{M} \models T$ the transcendence degree of $\mathcal{M}$ is a cardinal invariant of $\mathcal{M}$.) Such an assignment of cardinal invariants is known as a *structure theorem.* (See page 326 for further discussion.) An important feature of the proof of Theorem 3.1.2 is that this cardinal invariant is independent of the choice of strongly minimal formula: if $\varphi(v)$ and $\varphi'(v)$ are strongly minimal formulas (over the finite sets $A \subset M$ and $A' \subset M$, respectively), then $\dim(\varphi(\mathcal{M})/A) = |M| = \dim(\varphi'(\mathcal{M})/A')$. What we have not yet proved is that there are also cardinal invariants for the countable models of T. This is basically the Baldwin-Lachlan Theorem which is proved in Section 3.4 using the more powerful machinery developed in Section 3.3.

Historical Notes. Morley's Categoricity Theorem was proved by Morley in [Mor65]. His proof involved Morley rank and other tools developed in Section 3.3. The term "λ−stable" was introduced by Rowbottom in [Row64]. Strongly minimal formulas were defined by Marsh [Mar66], and developed further by Baldwin and Lachlan [BL71]. It was in this later paper where

Theorem 3.1.2 and its key component, Proposition 3.1.2, were proved. The concept of splitting was developed in the early papers of Shelah (see, e.g., [She69]).

Exercise 3.1.1. Let T be a theory which is categorical in some $\kappa \geq |T|$. Prove that T is complete.

Exercise 3.1.2. Show that if T is λ–stable and $|A| \leq \lambda$, then $|S_n(A)| \leq \lambda$.

Exercise 3.1.3. Let T be a countable uncountably categorical theory and M an uncountable model of T. Show that M is $\aleph_0$–saturated.

Exercise 3.1.4. Prove: If $\mathcal{M}$ is a model, $\mathcal{N} \prec \mathcal{M}$ and $a \in N$ is algebraic over $A \subset M$, then $a \in M$ and $tp_{\mathcal{M}}(a/A)$ is algebraic.

Exercise 3.1.5. Let A be a subset of a model $\mathcal{M}$ in a theory of cardinality κ. Prove that $|acl(A)| \leq |A| + \kappa$.

Exercise 3.1.6. Prove: If the type of $\bar{a} = (a_1, \ldots, a_n)$ over A (in some model) is algebraic, then $a_1, \ldots, a_n \in acl(A)$.

Exercise 3.1.7. Prove Lemma 3.1.2.

Exercise 3.1.8. Let S be a pregeometry. Show that for all A, $B \subset S$,

$$\dim(A \cup B) = \dim(A/B) + \dim(B).$$

Exercise 3.1.9. Let T be an $\aleph_0$–stable theory in the language L and let T_0 be the restriction of T to a sublanguage of L. Show that T_0 is $\aleph_0$–stable.

Exercise 3.1.10. Let $\mathcal{M}$ and $\mathcal{N}$ be models, $A \subset M$, $B \subset N$ and f an elementary mapping from A onto B. Prove that f can be extended to an elementary mapping from $acl(A)$ onto $acl(B)$.

Exercise 3.1.11. Find an example of a model $\mathcal{M}$ and formula $\varphi(x)$ such that every subset of $\varphi(\mathcal{M})$ definable over M is finite or cofinite, but φ is not strongly minimal. (Thus, in the definition of strongly minimal we cannot avoid checking elementary extensions of $\mathcal{M}$. HINT: There is an example in this section.)

Exercise 3.1.12. Prove Remark 3.1.2.

Exercise 3.1.13. Let $\mathcal{M}$ be a model, $\bar{a}$ and $\bar{b}$ sequences from M realizing the same complete type and $\varphi(x, \bar{a})$ strongly minimal. By Remark 3.1.2, $\varphi(x, \bar{b})$ is also strongly minimal. Let I and J be bases for $\varphi(\mathcal{M}, \bar{a})$ over $\bar{a}$ and $\varphi(\mathcal{M}, \bar{b})$ over $\bar{b}$, respectively. Show that if $|I| = |J|$ and f is an elementary mapping from $\bar{a}$ onto $\bar{b}$ then f can be extended to an elementary mapping from $I \cup \bar{a}$ onto $J \cup \bar{b}$.

Exercise 3.1.14. Suppose that T is a countable theory which is categorical in $\aleph_1$, but not categorical in $\aleph_0$. Prove that the prime model of T is minimal.

Exercise 3.1.15. Let $\mathcal{M}$ be a model, $A \subset M$, $\varphi(x)$ a strongly minimal formula over A and $p \in S(M)$ the unique nonalgebraic extension of φ. Prove that p does not split over A.

Exercise 3.1.16. Suppose that $\mathcal{M}$ is a model, $\mathcal{N} \succ \mathcal{M}$, $a \in N\backslash M$ and A is a subset of M over which $tp_{\mathcal{N}}(a/M)$ does not split. Show that for all sequences $\bar{b}$, $\bar{c}$ from M, $tp_{\mathcal{N}}(\bar{b}/A) = tp_{\mathcal{N}}(\bar{c}/A) \implies tp_{\mathcal{N}}(\bar{b}/A \cup \{a\}) = tp_{\mathcal{N}}(\bar{c}/A \cup \{a\})$.

Exercise 3.1.17. Let T be $\aleph_0$−stable, $\mathcal{M}$ a countable saturated model of T and let $CB(-)$ denote Cantor-Bendixson rank as computed in $S_1(M)$. Prove that for all sequences $\bar{a}$, $\bar{b}$ from M with $tp(\bar{a}) = tp(\bar{b})$ and formulas $\varphi(x, \bar{y})$, $CB(\varphi(x, \bar{a})) = CB(\varphi(x, \bar{b}))$.

Exercise 3.1.18. Give a detailed proof of Lemma 3.1.3(iii).

Exercise 3.1.19. Give a detailed proof of Lemma 3.1.6.

Exercise 3.1.20. Let T be an uncountably categorical theory and $\mathcal{M}$ a countable model of T. Show that $Th(\mathcal{M}_M)$, the theory of $\mathcal{M}$ with constants added for the elements of $\mathcal{M}$, is also uncountably categorical.

Exercise 3.1.21. Let T be any complete countable theory and κ an uncountable cardinal. Prove that there is a model $\mathcal{M}$ of T of cardinality κ such that for any formula φ over $\mathcal{M}$, $\varphi(\mathcal{M})$ is finite or of cardinality κ.

3.2 A Universal Domain

Before proceeding with our study of totally transcendental theories we introduce some conventions to simplify the notation when dealing with the models of a fixed complete theory.

In Section 2.2 we introduced saturated models and proved some of their basic properties. One of the more useful properties of a saturated model $\mathcal{M}$ is its universality; i.e., if $\mathcal{N}$ is a model elementarily equivalent to $\mathcal{M}$ and $|N| \leq |M| = \kappa$ then $\mathcal{N}$ is isomorphic to an elementary submodel of $\mathcal{M}$. Thus, any property invariant under isomorphism and possessed by a model of the theory of cardinality $\leq \kappa$ holds in some elementary submodel of $\mathcal{M}$.

Suppose that a theory T has saturated models of arbitrarily large cardinality. Many theorems in model theory assert that a property holds for all models of a theory. If we fix a saturated model $\mathcal{M} \models T$ of cardinality κ and prove the theorem relative to the elementary submodels of $\mathcal{M}$, we seem to have limited ourselves to the models of cardinality $\leq \kappa$. However, with few exceptions it is possible to give a proof which does not depend on a particular κ, in the following sense. Let τ denote a theorem concerning the models of T and for $\mathcal{M}$ a model of T, let $\tau \upharpoonright \mathcal{M}$ denote the relativization of τ to the elementary submodels of $\mathcal{M}$. For $\mathcal{M}'$ another saturated model of T, the proof of $\tau \upharpoonright \mathcal{M}'$ can be obtained from a proof of $\tau \upharpoonright \mathcal{M}$ simply by replacing a reference to $\mathcal{M}$ to $\mathcal{M}'$. Thus, to prove τ (which is equivalent to "$\tau \upharpoonright \mathcal{M}'$ holds for all saturated models of T") it suffices to prove $\tau \upharpoonright \mathcal{M}$. For these reasons we adopt the following conventions without changing the validity of any theorems.

We assume that any theory under discussion has saturated models of arbitrarily large cardinality. (As noted in Section 2.1 this will be true assuming that there are arbitrarily large strongly inaccessible cardinals. The reader who is uncomfortable with such an assumption can reword any of the subsequent proofs to see that they do not depend on any additional set-theoretic assumptions.)

Definition 3.2.1. *Given a complete theory T we let $\mathfrak{C}$ denote a saturated model of T of arbitrarily large cardinality. $\mathfrak{C}$ is called the* universal domain of T *or simply* the universe of T. *(Such a model is called the "monster model" of T in some sources.)*

From hereon, we will say "$\mathcal{M}$ is a model of T" only when $\mathcal{M}$ is an elementary submodel of T and $|M| < |\mathfrak{C}|$. (The restriction on the cardinality of $\mathcal{M}$ is discussed below.) Narrowing our attention to elementary submodels of $\mathfrak{C}$ eliminates the need to specify an ambient model whenever we speak of the type of an element. If a is an element (of $\mathfrak{C}$) and $\mathcal{M}$ is a model (which is, by fiat, an elementary submodel of $\mathfrak{C}$) containing a, then $\mathcal{M} \models \varphi(a)$ if and only if $\mathfrak{C} \models \varphi(a)$. Thus, we may as well drop the reference to the model altogether and write $\models \varphi(a)$ instead of $\mathfrak{C} \models \varphi(a)$. Also, simply being told that the set M is the universe of a model completely determines the model. Since the term "model" is only applied to an elementary submodel of $\mathfrak{C}$, the interpretation on M of the elements of the language is uniquely determined. The full list of our conventions follows.

- $\mathfrak{C}$ denotes a model of the relevant complete theory which is saturated and of arbitrarily large cardinality.
- The term "A is a set" is interpreted as: $A \subset \mathfrak{C}$ and $|A| < |\mathfrak{C}|$. Similarly, "a is an element" means $a \in \mathfrak{C}$.
- By the term "formula" we always mean a formula over $\mathfrak{C}$. Similarly, a "type" is a type over $\mathfrak{C}$.
- For $\bar{a}$ a tuple and $\varphi(\bar{v})$ a formula we write $\models \varphi(\bar{a})$ for $\mathfrak{C} \models \varphi(\bar{a})$, and say "$\bar{a}$ satisfies $\varphi(\bar{v})$". If A is a set $tp(\bar{a}/A)$ denotes $tp_{\mathfrak{C}}(\bar{a}/A)$.
- We say "M is a model" if M is a set (hence of cardinality $< |\mathfrak{C}|$) and M is the universe of an elementary submodel of $\mathfrak{C}$. (Note: for M a model and φ a formula $M \models \varphi(\bar{a})$ if and only if $\bar{a}$ a tuple from M and $\models \varphi(\bar{a})$.) Models are denoted by M, N, M', etc.

The main reason for restricting to sets of cardinality $< |\mathfrak{C}|$ is so that types over sets can be realized. (If $A \subset \mathfrak{C}$, $|A| < |\mathfrak{C}|$ and $p \in S(A)$, then p is realized in $\mathfrak{C}$, which may not be true if $|A| = |\mathfrak{C}|$.) A closely related benefit if the restriction is:

– Every elementary map extends to an element of $\mathrm{Aut}(\mathfrak{C})$.

(Let f be an elementary map from A onto B. Then, by assumption, A and B are subsets of $\mathfrak{C}$ with $|A| = |B| < |\mathfrak{C}|$. By the homogeneity of $\mathfrak{C}$, f is the restriction to A of an automorphism of $\mathfrak{C}$.)

Definition 3.2.2. *Given a complete theory with universal domain $\mathfrak{C}$, X is a* definable set *if $X = \varphi(\mathfrak{C})$ for some formula φ.*

Remark 3.2.1. (i) Comparing Definition 3.2.2 with the definition of "definable in $\mathcal{M}$" in Definition 1.1.1 we see that X is a definable set exactly when it is a definable set in $\mathfrak{C}$.

In most sources the term "definable set" means any set of the form $\varphi(M)$, where M is a model and φ is a formula over M. For the subject matter of this book it is most natural to reserve the unqualified term for sets of the form $\varphi(\mathfrak{C})$, and to say explicitly "definable set in M" when working in a particular model M.

(ii) For φ a nonalgebraic formula the saturation of $\mathfrak{C}$ forces $\varphi(\mathfrak{C})$ to have the same cardinality as $\mathfrak{C}$, thus strict adherence to our terminology prohibits $\varphi(\mathfrak{C})$ from being called a set. However, "definable set" is the established term for an object of the form $\varphi(\mathfrak{C})$ and since its usage should cause no confusion we will stick with it.

Notation. Given a universal domain $\mathfrak{C}$, formula φ and $X = \varphi(\mathfrak{C})$, we call X a *strongly minimal set* if φ is strongly minimal.

By convention an element of $S(\mathfrak{C})$ is not called a type since its domain is not a set. However, these "ideal types" or "limit types" do provide us with a convenient way to discuss all possible extensions of a type. For example, a formula $\varphi(x)$ over A is strongly minimal if for all sets $B \supset A$ there is a unique nonalgebraic $p \in S(B)$ which contains φ. A quicker way to say this is: φ is strongly minimal if there is a unique nonalgebraic $p \in S(\mathfrak{C})$ containing φ. Elements of $S(\mathfrak{C})$ will only be used to streamline notation in this manner.

As an exercise to familiarize the reader with this notation we recommend rewriting several definitions, lemmas and proofs from the previous section under the assumption that we are working in a universal domain.

3.3 Totally Transcendental Theories

The totally transcendental theories were defined and used by Morley in his proof of Morley's Categoricity Theorem (Theorem 3.1.1). Instead of total transcendentality our proof used $\aleph_0$–stability, which we will show is equivalent to total transcendentality for countable theories. Here, totally transcendental theories are investigated without any categoricity or cardinality assumptions and results are proved which go far beyond those obtained in Section 3.1 for $\aleph_0$–stable theories. Besides shedding light on a rich class of theories this section introduces many of the central concepts of stability theory. These results will be applied in the next two sections to prove the Baldwin-Lachlan Theorem for uncountably categorical theories (mentioned at the end of Section 3.1), and to begin the study of t.t. groups (i.e., those totally transcendental theories which are the theory of a group).

Our proof of Morley's Categoricity Theorem depended heavily on the following properties of a strongly minimal set (i.e., a set defined by a strongly minimal formula) in an uncountably categorical theory. Let T be uncountably categorical and let $\varphi(v)$ be a strongly minimal formula which, for simplicity, we assume to be over $\emptyset$. Let M be a model of T and $D = \varphi(M)$.

(1) $c\ell(-)$ = the restriction of algebraic closure to D, defines a pregeometry on D.
(2) Let I be a basis for $\varphi(M)$, $N \models T$ and J a basis of $\varphi(N)$. Then any elementary map from J into I extends to an elementary map of $\varphi(N)$ into $\varphi(M)$.
(3) M is prime over $\varphi(M)$.

In this way the pregeometry on a strongly minimal set represents a model of T. In an arbitrary $\aleph_0$–stable theory T there is a strongly minimal formula φ (over $\mathfrak{C}$), but

- there may not be a strongly minimal formula over the prime model, and
- even when φ is over M property (3) may fail for M.

Thus, we must look beyond strongly minimal formulas to find a dependence relation which influences the whole model. We will find a relation on all subsets of the universe (of an $\aleph_0$–stable theory) which meets the following conditions. The term "free" is used instead of "independent" to avoid confusion with later specific interpretations of the term.

Definition 3.3.1. *A ternary relation $\mathcal{F}$ on subsets of the universe is called a* freeness relation *if it satisfies the following conditions. ($\mathcal{F}(A, B; C)$ is read* A is free from B over C.*)*

(1) *(finite character and monotonicity) A is free from B over C if and only if for all finite $A_0 \subset A$ and $B_0 \subset B$, A_0 is free from B_0 over C.*
(2) *For any $\bar{a}$ and B there is a $B_0 \subset B$ of cardinality $\leq |T|$ such that $\bar{a}$ is free from B over B_0.*
(3) *(transitivity of independence) If $A \subset B \subset C$ then $\bar{a}$ is free from C over A if and only if $\bar{a}$ is free from C over B and $\bar{a}$ is free from B over A.*

(4) *(symmetry) If A is free from B over C, then B is free from A over C.*

(5) *If A is free from B over C and $f \in Aut(\mathfrak{C})$ then $f(A)$ is free from $f(B)$ over $f(C)$.*

(6) *There is a cardinal λ such that for $B \supset A$ and $p \in S(A)$, the set of types $\{\, tp(\bar{a}/B) : \bar{a}$ is free from B over A and $tp(\bar{a}/B) \supset p\}$ is nonempty and of cardinality $\leq \lambda$.*

(7) *(reflexivity) If $b \notin acl(A)$ then b is not free from $A \cup \{b\}$ over A.*

No formal abstract properties of freeness relations will be proved. The conditions (1)–(7) will only be used as a basic minimal list of properties that any usable notion of freeness must satisfy. As a first example of a freeness relation, let D be a strongly minimal set (in the universe of some theory) and define, for sets A, B, C,

$$\mathcal{F}_0(A, B; C) \iff \text{for all finite } \ A' \subset A, \ \dim(A'/B \cup C) = \dim(A'/C).$$

Previously proved properties of strongly minimal sets show that $\mathcal{F}_0$ is a freeness relation. The axioms for an abstract dependence relation given by van der Waerden in [VdW49] include the transitivity of dependence: If a depends on X and each $x \in X$ depends on Y, then a depends on Y. However, van der Waerden's notion is formulated as: "the point a depends on the set $X \ldots$". Virtually any dependence relation applying to all subsets of the universe fails to satisfy transitivity. Indeed transitivity fails for the freeness relation $\mathcal{F}_0$ defined above. (Let a, b, c and d be four independent nonzero elements of a vector space. Let $A = \{a, b\}$, $B = \{b, c\}$ and $C = \{c, d\}$. Then A depends on B over $\emptyset$ and B depends on C over $\emptyset$, but A is independent from C over $\emptyset$.)

Item (5) in the definition says that freeness is determined by types; i.e., if $tp(\bar{a}/B \cup C) = tp(\bar{b}/B \cup C)$, then $\bar{a}$ is free from B over C if and only if $\bar{b}$ is free from B over C. Thus, it makes sense to define a freeness relation as a relation on types, which is done below with Morley rank.

Some of the properties of $\aleph_0$−stable theories proved in Section 3.1 (for example, the existence of prime models) used the fact that for A any countable set every element of $S(A)$ has Cantor-Bendixson rank. Morley rank is basically Cantor-Bendixson rank computed over universe, instead of over a fixed set. Remember that the term "formula" means "formula over $\mathfrak{C}$".

Definition 3.3.2. *Let T be a complete theory. The relation $MR(\varphi) = \alpha$, for φ a formula in n variables and α an ordinal or -1, is defined by the following recursion.*

(1) *$MR(\varphi) = -1$ if φ is inconsistent;*

(2) *$MR(\varphi) = \alpha$ if*

$$\{\, p \in S_n(\mathfrak{C}) : \ \varphi \in p \text{ and } \neg\psi \in p \text{ for all formulas } \psi \text{ with } \ MR(\psi) < \alpha \,\}$$

is nonempty and finite.

For p any $n-$type, $MR(p)$ is defined to be

$$\inf\{MR(\varphi) : \varphi \text{ a formula implied by } p\}.$$

(Thus, for $p \in S(\mathfrak{C})$, $MR(p)$ is $\inf\{MR(\varphi) : \varphi \in p\}$.*) When $MR(p) = \alpha$ we say that the* Morley rank of p is α. *If there is no α with $MR(p) = \alpha$ we write $MR(p) = \infty$ and say that the Morley rank of p does not exist.*

Extend the scope of $<$ so that $-1 < \alpha < \infty$ for all ordinals α. Then, $MR(p) \geq \alpha$ is a quick way to express that $MR(p) \neq \beta$ for all $\beta < \alpha$. Using these conventions (2) in the definition can be restated as: $MR(\varphi) = \alpha$ if $\{p \in S_n(\mathfrak{C}) : \varphi \in p \text{ and } MR(p) \geq \alpha\}$ is finite and nonempty.

The only reason Morley rank is not simply Cantor-Bendixson rank in $S(\mathfrak{C})$ is because CB-rank is computed for types over a fixed *set* A, which by convention must have cardinality $< |\mathfrak{C}|$. This difference is strictly due to a notational convention. All properties of CB-rank proved in Section 2.2 hold for Morley rank with virtually identical proofs. The statements of the most critical properties will be repeated for ease of reference.

It follows immediately from the definition that for p and q types with $q \models p$, $MR(p) \geq MR(q)$. We leave it as an exercise to the reader to show that conjugate types have the same Morley rank.

Remark 3.3.1. Let $\mathfrak{C}$ be the universal domain of a complete theory. Given a type p and formula φ such that $p \models \varphi$ there is a finite $p_0 \subset p$ such that $\bigwedge p_0 \models \varphi$, hence $MR(\varphi) \geq MR(\bigwedge p_0)$. Thus, if p is closed under finite conjunctions there is a formula $\varphi \in p$ such that $MR(\varphi) = MR(p)$.

The following is an almost literal restatement of Lemma 2.2.3.

Lemma 3.3.1. *Let T be a complete theory, p an $n-$type and α an ordinal.*

(i) If $p \in S(\mathfrak{C})$ then $MR(p) = 0$ if and only if p is algebraic.

(ii) $MR(p) = \alpha$ if and only if there is a formula φ implied by p such that $\{q \in S_n(\mathfrak{C}) : \varphi \in q \text{ and } MR(q) = \alpha\}$ *is finite and nonempty, and this set is equal to* $\{q \in S_n(\mathfrak{C}) : p \subset q \text{ and } MR(q) = \alpha\}$.

(iii) If $MR(p) = \alpha$ there is a $q \in S_n(\mathfrak{C})$ such that $q \supset p$ and $MR(q) = \alpha$.

(iv) If $p \in S(\mathfrak{C})$ and $MR(p) = \alpha$ there is a $\varphi \in p$ such that p is the only element of $\{q \in S_n(\mathfrak{C}) : \varphi \in q \text{ and } MR(q) \geq \alpha\}$.

(v) $MR(p) \geq \alpha$ if and only if, for all $\beta < \alpha$ and all φ implied by p, $\{q \in S_n(\mathfrak{C}) : \varphi \in q \text{ and } MR(q) \geq \beta\}$ *is infinite.*

(vi)

$$\begin{gathered} MR(\varphi) \text{ is the least ordinal } \alpha \text{ such that} \\ \{q \in S_n(\mathfrak{C}) : \varphi \in q \text{ and } MR(q) \geq \alpha\} \text{ is finite.} \end{gathered} \tag{3.2}$$

Proof. A complete type over a model is isolated if and only if it is algebraic. Thus, (i) is really a restatement of Lemma 2.2.3(i). A proof of each remaining part can be obtained from the proof of the corresponding part in Lemma 2.2.3 simply by changing the notation.

Definition 3.3.3. *Let p be an n-type in a complete theory and suppose $MR(p) = \alpha < \infty$. By (ii) of the previous lemma $\{q \in S_n(\mathfrak{C}) : p \subset q$ and $MR(q) = \alpha\}$ is finite. The* Morley degree of p, *denoted* $\deg(p)$, *is defined to be* $|\{q \in S_n(\mathfrak{C}) : p \subset q \text{ and } MR(q) = \alpha\}|$. *If* $\deg(p) = 1$ *we say that p is* stationary.

When $p = tp(\bar{a}/B)$ we may write $MR(\bar{a}/B)$ for $MR(p)$.

Notation. Given a universal domain $\mathfrak{C}$ and definable set $X = \varphi(\mathfrak{C})$, we may write $MR(X)$ for $MR(\varphi)$ and $\deg(X)$ for $\deg(\varphi)$.

Definition 3.3.4. *A complete theory T is called* totally transcendental *(t.t.) if every $p \in S(\mathfrak{C})$ has Morley rank.*

Notice that a complete theory T is t.t. if and only if for every n and $\bar{v} = (v_1, \ldots, v_n)$, $MR(\bar{v} = \bar{v}) < \infty$. (See Remark 3.3.2.)

Definition 3.3.5. *Let $\mathfrak{C}$ be the universal domain of a t.t. theory. For A, B, C subsets of $\mathfrak{C}$ we say A is* Morley rank independent from B over C *and write $A \underset{C}{\perp} B$ if for all finite tuples $\bar{a}$ from A, $MR(\bar{a}/B \cup C) = MR(\bar{a}/C)$; writing $A \underset{C}{\not\perp} B$ for the negation of Morley rank independence (called* Morley rank dependence*). We write $A \perp B$ and $A \not\perp B$ for $A \underset{\emptyset}{\perp} B$ and $A \underset{\emptyset}{\not\perp} B$, respectively. If p is a complete type over $B \supset C$ and $q = p \upharpoonright C$ we call p a* free extension of q *if $MR(p) = MR(q)$. When p is a free extension of $p \upharpoonright C$ we may say p is* free over C.

Let T be t.t. Most of the properties of a freeness relation are easily verified for Morley rank independence. That the relation has finite character, is transitive, and is preserved under automorphisms is clear from the basic properties of Morley rank. For $p \in S(A)$ there is a finite $B \subset A$ such that p is a free extension of $p \upharpoonright B$. If $p \in S(A)$ and $B \supset A$ then there is at least one and only finitely many $q \in S(B)$ which are free extensions of p. Reflexivity follows from the fact that $MR(\bar{a}/B) = 0$ if and only if $\bar{a} \in acl(B)$. The symmetry property (Definition 3.3.1(4)) however, will take a significant amount of work to prove.

In this section the term "Morley rank independent" (Morley rank dependent) is shortened to "independent" (dependent).

In Section 2.2 we gave examples of theories formulated using equivalence relations such that, for M any model of the theory, every element of $S(M)$ has CB-rank. Letting $M = \mathfrak{C}$ in each example and rewording the justifications shows that each of the theories is t.t.

Here are some basic facts about Morley rank and independence in a t.t. theory.

Lemma 3.3.2. *Let T be totally transcendental.*

(i) For $\bar{a}'$ a subsequence of the sequence $\bar{a}$, $MR(\bar{a}'/B) \leq MR(\bar{a}/B)$.

(ii) For all $\bar{a}$, $\bar{b}$ and A, $\bar{a} \in acl(A \cup \{\bar{b}\}) \implies MR(\bar{a}\bar{b}/A) = MR(\bar{b}/A)$.

(iii) For all $\bar{a}$ and A, $\bar{a} \mathop{\smile\!\!\!\!\!\!\!|}_{A} acl(A)$.

(iv) For all $\bar{a}$ and sets B there is a finite $B_0 \subset B$ such that $\bar{a}$ is independent from B over B_0.

(v) **(Pairs Lemma)** *For all $\bar{a}$, $\bar{b}$ and $B \supset A$,*

$$B \mathop{\smile\!\!\!\!\!\!\!|}_{A} \bar{a}\bar{b} \iff B \mathop{\smile\!\!\!\!\!\!\!|}_{A\cup\bar{b}} \bar{a} \text{ and } B \mathop{\smile\!\!\!\!\!\!\!|}_{A} \bar{b}.$$

Proof. Part (i) is left to the exercises, and (iv) follows immediately from the definition of Morley rank independence.

(ii) Without loss of generality, $A = \emptyset$. By (i) it suffices to show that $MR(\bar{a}\bar{b}) \leq MR(\bar{b}) = \beta$. Below, $tp(\bar{b})$ is viewed as a type in the variables $\bar{y}$ and $tp(\bar{a}\bar{b})$ as a type in $\bar{x}\bar{y}$. For $\bar{v}$ a sequence of variables, let $\Psi_{\bar{v}} = \{\psi(\bar{v}) : MR(\psi(\bar{v})) < \beta\}$. Let $\theta(\bar{x}\bar{y})$ be any formula in $tp(\bar{a}\bar{b})$ such that $MR(\exists\bar{x}\theta(\bar{x},\bar{y})) = \beta$ and, for any $\bar{d}$, $\theta(\bar{x},\bar{d})$ is algebraic. Then, $\{q(\bar{x}\bar{y}) \in S(\mathfrak{C}) : \theta(\bar{x}\bar{y}) \in q \text{ and } \neg\psi(\bar{x}\bar{y}) \in q \text{ for all } \psi \in \Psi_{\bar{x}\bar{y}}\} \subset \{q(\bar{x}\bar{y}) \in S(\mathfrak{C}) : \theta(\bar{x}\bar{y}) \in q \text{ and } \neg\psi(\bar{y}) \in q \text{ for all } \psi \in \Psi_{\bar{y}}\} = Q$. Since $MR(\exists\bar{x}\theta(\bar{x},\bar{y})) = \beta$, $R = \{r(\bar{y}) \in S(\mathfrak{C}) : r$ is the restriction to $\bar{y}$ of some element of $Q\}$ is finite. Since $\theta(\bar{x},\bar{d})$ is algebraic for any $\bar{d}$, $r(\bar{y}) \cup \{\theta(\bar{x}\bar{y})\}$ has finitely many completions in $S(\mathfrak{C})$, for any $r \in R$. Hence, Q is finite, from which we conclude that $MR(\bar{a}\bar{b}) \leq MR(\theta(\bar{x}\bar{y})) \leq \beta$.

(iii) Suppose, to the contrary, that $\varphi(\bar{v},\bar{y})$ is a formula over A and $\varphi(\bar{v},\bar{e}) \in tp(\bar{a}/acl(A))$ has Morley rank $\beta < MR(\bar{a}/A)$. We may assume that $\exists\bar{v}\varphi(\bar{v},\bar{y})$ isolates $tp(\bar{e}/A)$. Let p be a free extension of $tp(\bar{a}/A)$ in $S(\mathfrak{C})$. Since $\exists\bar{v}\varphi(\bar{v},\bar{y}) \in p$ isolates an algebraic type over A, $\varphi(\bar{v},\bar{e}') \in p$ for some $\bar{e}'$. Since $\varphi(\bar{v},\bar{e}')$ is conjugate to $\varphi(\bar{v},\bar{e})$ it also has Morley rank $\beta < MR(\bar{a}/A) = MR(p)$. This contradiction proves (iii).

(v) ($\Longrightarrow$) Suppose that B is independent from $\bar{a}\bar{b}$ over A, and let $\bar{c} \subset B$. Since $MR(\bar{c}/A \cup \bar{a}\bar{b}) = MR(\bar{c}/A)$, $MR(\bar{c}/A \cup \bar{b}) = MR(\bar{c}/A)$ and $MR(\bar{c}/A \cup \bar{a}\bar{b}) = MR(\bar{c}/A \cup \bar{b})$.

($\Longleftarrow$) This direction follows immediately from the transitivity of independence.

Remark 3.3.2. Let $\mathfrak{C}$ be the universal domain of a complete theory and let $\varphi(x,\bar{y})$ be a formula over $\emptyset$. If $MR(\exists x\varphi(x,\bar{y})) < \infty$ and for each $\bar{a}$, $MR(\varphi(x,\bar{a})) < \infty$, then $MR(\varphi(x,\bar{y})) < \infty$. In fact a bound on $MR(\varphi(x,\bar{y}))$ can be computed in terms of $MR(\exists x\varphi(x,\bar{y}))$ and a bound on $MR(\varphi(x,\bar{a}))$, as $\bar{a}$ ranges over tuples from $\mathfrak{C}$.

This fact is due to Erimbetov [Eri75]. Proofs can also be found in [Lac80] and [She90, V.7.8]. This lemma is occasionally helpful in showing that a particular theory is t.t. To show that a complete theory T is t.t., instead of showing that every $p \in S(\mathfrak{C})$ has Morley rank, it is enough to show that each

$p \in S_1(\mathfrak{C})$ has Morley rank. Equivalently, T is t.t. if and only if $MR(v = v) < \infty$.

Morley rank and Cantor-Bendixson rank are further connected by

Lemma 3.3.3. *Let T be a complete theory, M an $\aleph_0$–saturated model of T and let $CB(-)$ denote Cantor-Bendixson rank computed in $S_n(M)$. Then, for all n–types p over M,*

(i) $MR(p) = CB(p)$;

(ii) if $MR(p) < \infty$ and p is complete, then p is stationary.

(iii) Hence, if every element of $S(M)$ has CB-rank, T is totally transcendental.

Proof. (i) It suffices to consider the case when p is a formula φ. That $CB(\varphi) \geq MR(\varphi)$ is a straight-forward exercise left to the reader. To show the reverse inequality we prove (by induction on α) that $CB(\varphi) \geq \alpha \implies MR(\varphi) \geq \alpha$. Assuming that $MR(\varphi) \not\geq \alpha$ yields the inconsistency of $\{\varphi\} \cup \{\neg\psi : MR(\psi) < \alpha\}$. Thus, there are formulas $\psi_1, \ldots, \psi_n$ such that

$$\models \forall\bar{v}(\varphi(\bar{v}) \to \bigvee_{1 \leq i \leq n} \psi_i(\bar{v})) \text{ and } MR(\psi_i) < \alpha, \text{ for } i = 1, \ldots, n.$$

By Exercise 3.3.5, $\psi = \bigvee_{1 \leq i \leq n} \psi_i$ has Morley rank $< \alpha$. Letting A be a finite set containing the parameters in φ, the $\aleph_0$–saturation of M yields a formula ψ' over M conjugate over A to ψ. By induction, $CB(\psi') = MR(\psi') < \alpha$. Since φ implies ψ', $CB(\varphi) < \alpha$; i.e., $CB(\varphi) \not\geq \alpha$.

(ii) Since $MR(p) = CB(p) = \alpha < \infty$ there is a formula $\varphi \in p$ such that $\{ q \in S(M) : \varphi \in q \text{ and } MR(q) \geq \alpha \} = \{p\}$ (by Lemma 2.2.3(iv)). Supposing that $\deg(p) > 1$ produces two contradictory formulas ψ_1 and ψ_2 over $\mathfrak{C}$ such that $MR(\varphi \wedge \psi_i) = \alpha$, for $i = 1, 2$. Arguing as in (i) yields contradictory formulas ψ_1' and ψ_2' over M such that $MR(\varphi \wedge \psi_i') = \alpha$, for $i = 1, 2$. This contradicts that p is the unique completion of φ over M having Morley rank α.

(iii) T is t.t. since, letting $\bar{v} = (v_1, \ldots, v_n)$, $MR(\bar{v} = \bar{v}) = CB(\bar{v} = \bar{v}) = \sup\{ CB(p) : p \in S_n(M) \}$, for each $n < \omega$.

Let φ be a strongly minimal formula (over $\emptyset$) in a t.t. theory and $D = \varphi(\mathfrak{C})$. Remember that D and $c\ell(-)$, the restriction of $acl(-)$ to D, form a pregeometry and the resulting dependence relation defines a freeness relation (called dim –independence above). Since $\mathfrak{C}$ is t.t., Morley rank defines another freeness relation on D.

The next lemma connects dim –independence and Morley rank independence in a strongly minimal D. (Note: In the lemma the ambient theory is not required to be totally transcendental.)

Lemma 3.3.4. *Let φ be a strongly minimal formula over A^* (in some complete theory) and $D = \varphi(\mathfrak{C})$. If $\bar{a}$ is a tuple from D and A is any set containing A^*, then $MR(\bar{a}/A) = \dim(\bar{a}/A)$. Thus, for all subsets X, Y and*

$Z \subset Y$ of D, X is Morley rank independent from Y over Z if and only if X is dim *−independent from Y over Z.*

Furthermore, if $\bar{a} \cap acl(A^*) = \emptyset$, $\bar{a}$ is acl−independent over A if and only if $\bar{a}$ is Morley rank independent from A and A^* and $\bar{a}$ is Morley rank independent over A.

Proof. Without loss of generality, $A^* = \emptyset$. The lemma is proved by induction on $n = \dim(\bar{a}/A)$. Let $p = tp(\bar{a}/A)$. By Lemma 3.3.1(i), $MR(p) = 0$ if and only if $\bar{a} \in acl(A)$, hence the result is true when $n = 0$. Without loss, $\bar{a} = (a_1, \ldots, a_n, a_{n+1}, \ldots, a_k)$ where, $\dim(\{a_1, \ldots, a_n\}/A) = n > 0$. We first show that $MR(p) \geq n$. Let $B = \{ b_1^i : i < \omega \} \subset D$ be a set which is acl−independent over A. Since $tp(b_1^i/A) = tp(a_1/A)$, for all i, there is a tuple $\bar{b}^i$ realizing p such that b_1^i is the first entry in $\bar{b}^i$. We can furthermore assume that $\bar{b}^i$ is acl−independent from $A \cup B$ over $A \cup \{b_1^i\}$, hence, $\dim(\bar{b}^i/A \cup B) = \dim(\bar{b}^i/A \cup \{b_1^i\}) = n - 1$ (by additivity, Remark 3.1.4) and $tp(\bar{b}^i/A \cup B) \neq tp(\bar{b}^j/A \cup B)$ when $i \neq j$. By induction $MR(\bar{b}^i/A \cup B) = n - 1$, hence $\{ tp(\bar{b}^i/A \cup B) : i < \omega \}$ forms an infinite set of extensions of p, each having Morley rank $n - 1$. Thus $MR(p) \geq n$.

To prove that $MR(p) \leq n$ it suffices to show that p does not have infinitely many contradictory extensions of Morley rank $\geq n$. Assume, to the contrary, that there is a set $B \supset A$ and, for each $i < \omega$, there is a $\bar{b}^i$ realizing p such that for all $i < j < \omega$, $MR(\bar{b}^i/B) \geq n$ and $tp(\bar{b}^i/B) \neq tp(\bar{b}^j/B)$. By the first paragraph, $n \leq \dim(\bar{b}^i/B) \leq \dim(\bar{a}/A) = n$, hence $\dim(\bar{b}^i/B) = n$ (for $i < \omega$). Then, letting $\bar{d}^i = (b_1^i, \ldots, b_n^i)$, the fact that $tp(\bar{b}^i/A \cup \bar{d}^i)$ is algebraic forces $\{b_1^i, \ldots, b_n^i\}$ to be acl−independent over B. Since an acl−independent subset of a strongly minimal set is indiscernible (by Lemma 3.1.3(iii)), $tp(\bar{d}^i/B) = tp(\bar{d}^j/B)$ for all i and j. Hence, for each $i < \omega$ there is an $f_i \in \mathrm{Aut}(\mathfrak{C})$ fixing B pointwise and taking $\bar{d}^i$ to $\bar{d}^0$. Let $\bar{a}' = (a_1, \ldots, a_n)$ and let $\varphi(\bar{x}, \bar{a}')$ be an algebraic formula satisfied by $\bar{a}$. Then, $\models \varphi(\bar{b}^i, \bar{d}^i)$, hence $\models \varphi(f_i(\bar{b}^i), \bar{d}^0)$, for all i. Since $\varphi(\bar{x}, \bar{d}^0)$ is an algebraic formula there are distinct i and j such that $f_i(\bar{b}^i) = f_j(\bar{b}^j)$. This implies (since the f_i's are automorphisms fixing B) that $tp(\bar{b}^i/B) = tp(\bar{b}^j/B)$. This contradiction completes the first part of the proof.

The furthermore clause in the lemma follows from the definitions and the first part of the lemma.

Notation. For A, B, C subsets of a strongly minimal set D we can use *A is independent from B over C* for "A is Morley rank independent from B over C" or "A is dim −independent from B over C". The acl−independence relation on D will be termed "algebraically independent".

In Lemma 2.2.4 the existence of Cantor-Bendixson rank was connected to the number of complete types in the theory. A similar connection exists for Morley rank.

Proposition 3.3.1. *Let T be a complete theory. If T is t.t. then for all sets A, $|S(A)| \leq |A| + |T|$; i.e., T is λ–stable for all infinite cardinals $\lambda \geq |T|$. If T is countable, then T is $\aleph_0$–stable if and only if T is totally transcendental.*

Proof. Let A be an arbitrary subset of $\mathfrak{C}$. For any formula φ over A let $U_\varphi = \{p \in S(A) : \varphi \in p \text{ and } MR(p) = MR(\varphi)\}$. Every element of $S(A)$ is in some U_φ and each U_φ is finite, hence $|S(A)|$ is equal to the number of formulas over A, which is $|A| + |T|$.

If T is countable and $\aleph_0$–stable then T has a countable saturated model M and every element of $S(M)$ has CB-rank by Lemma 2.2.4. By Lemma 3.3.3, T is t.t., completing the proof.

Warning: There is an uncountable theory T which is $|T|$–stable but not t.t.

We saw in previous sections the usefulness of indiscernible sequences in the construction of uncountable models with special properties. Indiscernible *sets* also played an important role in our proof of Morley's Categoricity Theorem, e.g., if M is a model of an uncountably categorical theory and φ is a strongly minimal formula (over say $\emptyset$) then a basis for $\varphi(M)$ is an indiscernible set over which the model is prime. Indiscernible sets will also play a vital role in the analysis of Morley rank independence in a t.t. theory. Preliminarily, we show that every infinite indiscernible sequence is actually an indiscernible set. This is an instance of a basic theme in stability theory: the presence of an order gives rise to many types over sets.

Lemma 3.3.5. *Let T be a complete theory which is λ–stable for some $\lambda \geq |T|$, A a set and $(I, <)$ an infinite indiscernible sequence over A. Then*

(i) I is an indiscernible set over A.

(ii) For any formula $\varphi(\bar{x}, \bar{y})$ over A there is an $n < \omega$ such that for all $\bar{a}$,

$$|\{\bar{b} \in I : \models \varphi(\bar{a}, \bar{b})\}| \leq n \text{ or } |\{\bar{b} \in I : \models \neg\varphi(\bar{a}, \bar{b})\}| \leq n.$$

Proof. (i) Suppose, to the contrary, that there is a formula $\varphi(\bar{v}_1, \ldots, \bar{v}_n)$ over A satisfied by increasing n–tuples from I, but not satisfied by some n–tuple of distinct elements from I. Let P_n denote the group of permutations of $\{1, \ldots, n\}$. Let P_n^+ be the set of elements σ of P_n such that for $\bar{a}_1 < \ldots < \bar{a}_n$ from I, $\models \varphi(\bar{a}_{\sigma 1}, \ldots, \bar{a}_{\sigma n})$. By assumption, $P_n^+ \neq P_n$. Every element of P_n can be written as a product of transpositions of the form $(k, k+1)$ (which denotes the permutation fixing every $i \notin \{k, k+1\}$ and switching k and $k+1$). Thus, there are $\sigma \in P_n^+$, $\tau \in P_n \setminus P_n^+$ and k such that $\tau = (k, k+1) \cdot \sigma$. Letting $\psi(\bar{v}_1, \ldots, \bar{v}_n)$ be the formula $\varphi(\bar{v}_{\sigma 1}, \ldots, \bar{v}_{\sigma n})$ we have

$$\models \psi(\bar{a}_1, \ldots, \bar{a}_n) \text{ and } \models \neg\psi(\bar{a}_1, \ldots, \bar{a}_{k-1}, \bar{a}_{k+1}, \bar{a}_k, \bar{a}_{k+2}, \ldots, \bar{a}_n),$$
$$\text{for } \bar{a}_1 < \ldots < \bar{a}_n \text{ from } I.$$

Let $(Y, <)$ be a dense linear order without endpoints of cardinality $> \lambda$ containing a dense subset X of cardinality λ. (First let μ be the least cardinal such that $2^\mu > \lambda$. Let X_0 be the set of sequences of 0's and 1's of length $< \mu$, ordered lexicographically by $<$. Then $(X_0, <)$ is a dense linear order with 2^μ cuts. Let $(Y, <)$ be an order of cardinality $> \lambda$ in which X_0 is dense and let X be any subset of Y of cardinality λ which contains X_0.) By Corollary 2.4.1 we may assume $(Y, <)$ to be an indiscernible sequence of tuples with $D(Y) = D(I)$. Let $y, y' \in Y$ with $y < y'$. Then, there are $x_i \in X$, for $1 \leq i \leq n$ and $i \neq k$, such that

$$x_1 < \ldots < x_{k-1} < y < x_{k+1} < y' < x_{k+2} < \ldots < x_n.$$

Consequently,

$$\models \psi(x_1, \ldots, x_{k-1}, y, x_{k+1}, \ldots, x_n) \text{ and } \models \neg\psi(x_1, \ldots, x_{k-1}, y', x_{k+1}, \ldots, x_n).$$

Thus, distinct elements of Y realize distinct types over X. Since $|X| = \lambda$ and $|Y| > \lambda$ this contradicts the λ–stability of T, proving (i).

Turning to (ii), by (i) we may assume I to be an indiscernible set. Suppose (ii) to fail for $\varphi(\bar{x}, \bar{y})$. Then for all n there are $I_0, I_1 \subset I$, each of cardinality $> n$, such that $\{ \varphi(\bar{x}, \bar{b}) : \bar{b} \in I_0 \} \cup \{ \neg\varphi(\bar{x}, \bar{b}) : \bar{b} \in I_1 \}$ is consistent. By compactness and the indiscernibility of I, for I_0, I_1 any two disjoint subsets of I, $\{ \varphi(\bar{x}, \bar{b}) : \bar{b} \in I_0 \} \cup \{ \neg\varphi(\bar{x}, \bar{b}) : \bar{b} \in I_1 \}$ is consistent. Again, by Corollary 2.4.1, there is an indiscernible set J with $D(J) = D(I)$ and $|J| = \lambda$. Thus, for any disjoint pair of subsets J_0, J_1 of J, $\{ \varphi(\bar{x}, \bar{b}) : \bar{b} \in J_0 \} \cup \{ \neg\varphi(\bar{x}, \bar{b}) : \bar{b} \in J_1 \}$ is consistent. Thus, $|S(J)| = 2^{|J|} = 2^\lambda$. This contradiction of the λ–stability of T, proves the lemma.

Let φ be a strongly minimal formula (over some set A). Assuming that a_β has been defined for $\beta < \alpha$ let a_α be a realization of the unique nonalgebraic extension of φ over $\{ a_\beta : \beta < \alpha \} \cup A$. Then, for any α, $\{ a_\beta : \beta < \alpha \}$ is an indiscernible set over A. The following concept is a generalization of this construction to an arbitrary stationary type. If I is an ordered set and $C = \{ c_i : i \in I \}$ is a set indexed by I, then for $i \in I$, C_i denotes $\{ c_j : j < i \}$.

Definition 3.3.6. *Let T be a totally transcendental theory, $p \in S(C)$ a stationary type and $B \supset C$. We call $A = \{ \bar{a}_\beta : \beta < \alpha \}$ a* Morley sequence over B in p *if for all $\beta < \alpha$, $tp(\bar{a}_\beta / B \cup A_\beta)$ is the unique free extension of p in $S(B \cup A_\beta)$.*

As with bases of strongly minimal sets such sequences are indiscernible:

Lemma 3.3.6. *Let T be a totally transcendental theory, $p \in S(C)$ a stationary type, $B \supset C$ and $A = \{ \bar{a}_\beta : \beta < \alpha \}$ a Morley sequence over B in p. Then A is an indiscernible set over B.*

Proof. The following argument is a slight generalization of the proof of Lemma 3.1.3(iii). Define an order $<$ on A by: $\bar{a}_\beta < \bar{a}_\gamma$ if $\beta < \gamma$. By

Lemma 3.3.5 it suffices to show that $(A,<)$ is an indiscernible sequence. To accomplish this we prove by induction on k,

$$\text{for all } \bar{a}_{\beta_1} < \ldots < \bar{a}_{\beta_k} \text{ and } \bar{a}_{\gamma_1} < \ldots < \bar{a}_{\gamma_k} \text{ from } A, \quad tp(\bar{a}_{\beta_1}\ldots\bar{a}_{\beta_k}/B) = tp(\bar{a}_{\gamma_1}\ldots\bar{a}_{\gamma_k}/B).$$

Assuming this is true for $k < n$ let $\bar{a}_{\beta_1} < \ldots < \bar{a}_{\beta_n}$ and $\bar{a}_{\gamma_1} < \ldots < \bar{a}_{\gamma_n}$ be from A. By induction there is an automorphism f of $\mathfrak{C}$ fixing B pointwise and mapping $\bar{a}_{\beta_i}$ to $\bar{a}_{\gamma_i}$, for $1 \le i < n$. Since f fixes B pointwise and preserves Morley rank, $MR(f(\bar{a}_{\beta_n})/B \cup \{\bar{a}_{\gamma_1},\ldots,\bar{a}_{\gamma_n}\}) = MR(p)$ and $f(\bar{a}_{\beta_n})$ realizes p, hence $f(\bar{a}_{\beta_n})$ and $\bar{a}_{\gamma_n}$ both realize the unique free extension of p over $B\cup\{\bar{a}_{\gamma_1},\ldots,\bar{a}_{\gamma_{n-1}}\}$. We conclude that $(\bar{a}_{\beta_1},\ldots,\bar{a}_{\beta_n})$ and $(\bar{a}_{\gamma_1},\ldots,\bar{a}_{\gamma_n})$ have the same type over B, proving the lemma.

As with independent subsets of a strongly minimal set, if I is a Morley sequence over $B \supset A$ in the stationary $p \in S(A)$, then (by Lemma 3.3.5) for all $\bar{a} \in I$, $\bar{a}$ is independent from $B \cup (I \setminus \{\bar{a}\})$ over A.

Example 3.3.1. (Two examples of Morley sequences.) Let E be a binary relation and T the theory saying that E is an equivalence relation with infinitely many infinite classes and no finite classes. Let $M \models T$. The unique $p \in S_1(\emptyset)$ is stationary. To obtain a Morley sequence in p simply take any $I \subset M$ such that $a \neq b \in I \implies \models \neg E(a,b)$. Now let a be any element of M and $q \in S(a)$ the unique nonalgebraic completion of $E(x,a)$ (which also happens to be stationary). Then, any $J \subset E(M,a) \setminus \{a\}$ is a Morley sequence in q.

With these tools in hand we can complete the proof that Morley rank independence satisfies all of the properties of a freeness relation:

Proposition 3.3.2 (Symmetry Lemma). *For all sets A, B and C,*

$$A \underset{C}{\perp} B \implies B \underset{C}{\perp} A.$$

Proof. Assuming the symmetry property to fail, the finite character of dependence yields a set C and (finite sequences) $\bar{a}$ and $\bar{b}$ such that $\bar{b}$ depends on $\bar{a}$ over C and $\bar{a}$ is independent from $\bar{b}$ over C; i.e.,

$$MR(\bar{b}/C \cup \bar{a}) < MR(\bar{b}/C) = \alpha \text{ and } MR(\bar{a}/C \cup \bar{b}) = MR(\bar{a}/C) = \beta. \quad (3.3)$$

Claim. There are $\bar{a}$, $\bar{b}$ and C satisfying (3.3) with C the universe of an $\aleph_0$-saturated model M.

Let $M \supset C$ be an $\aleph_0$-saturated model. Let $\bar{b}'$ be a realization of $tp(\bar{b}/C)$ which is independent from M over C, and choose $\bar{a}'$ such that $tp(\bar{b}'\bar{a}'/C) = tp(\bar{b}\bar{a}/C)$ and $\bar{a}'$ is independent from $M \cup \bar{b}'$ over $C \cup \bar{b}'$. Then, $MR(\bar{b}'/M) = MR(\bar{b}'/C) = \alpha$, $MR(\bar{b}'/M \cup \bar{a}') \le MR(\bar{b}'/C \cup \bar{a}') < \alpha$ and $MR(\bar{a}'/M \cup \bar{b}') =$

$MR(\bar{a}'/C \cup \bar{b}') = MR(\bar{a}'/C) = \beta$. Replacing $\bar{b}$ by $\bar{b}'$ and $\bar{a}$ by $\bar{a}'$ proves the claim.

By Lemma 3.3.3(ii), both $p = tp(\bar{b}/M)$ and $q = tp(\bar{a}/M)$ are stationary. Let $\varphi(\bar{x}, \bar{y})$ be a formula in $tp(\bar{b}\bar{a}/M)$ such that $MR(\varphi(\bar{x}, \bar{a})) < \alpha = MR(p)$. The formula φ will be used to contradict Lemma 3.3.5(ii). First, let I be an infinite Morley sequence over M in q and let $\bar{b}'$ be a realization of p which is independent from I over M. For any $\bar{a}' \in I$, $\varphi(\bar{x}, \bar{a}')$ has Morley rank $< \alpha$ (since it is conjugate to $\varphi(\bar{x}, \bar{a})$) hence $\models \neg\varphi(\bar{b}', \bar{a}')$. Now let J be an infinite Morley sequence over $M \cup \bar{b}' \cup I$ in q. The unique free extension of q over $M \cup \bar{b}$ (considered as a type in $\bar{y}$) is $tp(\bar{a}/M \cup \bar{b})$, hence contains $\varphi(\bar{b}, \bar{y})$. Since $tp(\bar{b}/M) = tp(\bar{b}'/M)$, $q \cup \{\varphi(\bar{x}, \bar{b}')\}$ also has Morley rank β, hence the unique free extension of q over $M \cup \bar{b}'$ contains $\varphi(\bar{b}', \bar{y})$. Thus, for every $\bar{a}' \in J$, $\models \varphi(\bar{b}', \bar{a}')$. Checking the definitions of I and J, $I \cup J$ is a Morley sequence over M in q, hence an indiscernible set over M. Since $\{\,\bar{c} \in I \cup J : \models \varphi(\bar{b}', \bar{c})\,\} = J$ and $\{\,\bar{c} \in I \cup J : \models \neg\varphi(\bar{b}', \bar{c})\,\} = I$ are both infinite we have contradicted Lemma 3.3.5(ii), proving the lemma.

This lemma allows us to say, for instance, "A and B are independent over C," instead of A is independent from B over C or B is independent from A over C. A collection of sets (or elements) I is called *independent over* A or A*–independent* if for each $X \in I$, X is independent from $I \setminus X$ over A. Note: a Morley sequence over A is independent over A.

Symmetry is such a basic property of independence that the lemma will usually be applied tacitly.

Corollary 3.3.1. *Let T be t.t., $p \in S(A)$ a stationary type, I a Morley sequence over A in p, and $\bar{b}$ a finite sequence. Then there is a finite $J \subset I$ such that $I \setminus J$ is a Morley sequence over $A \cup J \cup \bar{b}$ in p.*

Proof. Let J be a finite subset of I such that $MR(\bar{b}/A \cup I) = MR(\bar{b}/A \cup J) = \beta$. Let $\bar{a} \in I \setminus J$ and $I' = I \setminus \{\bar{a}\}$. Then $MR(\bar{b}/A \cup I' \cup \bar{a}) = MR(\bar{b}/A \cup I')$ (since $I' \supset J$), so by the Symmetry Lemma, $MR(\bar{a}/A \cup I' \cup \bar{b}) = MR(\bar{a}/A \cup I') = MR(\bar{a}/A)$. Thus, $I \setminus J$ is a Morley sequence over $A \cup J \cup \bar{b}$ in p.

In Section 3.1 the relation "$p = tp(\bar{a}/B)$ does not split over $A \subset B$" was promoted as a way of expressing that $\bar{a}$ is (at least intuitively) free from B over A. We now introduce a strengthening of the nonsplitting relation called definability. The relation "$p \in S(B)$ does not split over $A \subset B$" is equivalent to: for all formulas $\varphi(\bar{x}, \bar{y})$ there is a set $P_\varphi \subset S(A)$ such that for any $\bar{a}$ from B, $\varphi(\bar{x}, \bar{a}) \in p$ if and only if $tp(\bar{a}/A) \in P_\varphi$. Then p will be definable over A if p does not split over A *and* for each φ, P_φ is a basic open set in the appropriate Stone Space. Explicitly,

Definition 3.3.7. *Let p be a complete type over B and A a set. We say that p is* definable over A *if for every formula $\varphi(\bar{x}, \bar{y})$ over $\emptyset$ there is a formula $\psi(\bar{y})$ over A such that for all sequences $\bar{a}$ from B, $\varphi(\bar{x}, \bar{a}) \in p \iff \models \psi(\bar{a})$.*

For p and φ as above, let $p \upharpoonright \varphi$ be the type $\{\varphi(\bar{x},\bar{b}) : \varphi(\bar{x},\bar{b}) \in p\} \cup \{\neg\varphi(\bar{x},\bar{b}) : \neg\varphi(\bar{x},\bar{b}) \in p\}$. When ψ has the property that for all sequences $\bar{a}$ from B, $\varphi(\bar{x},\bar{a}) \in p \iff \models \psi(\bar{a})$, we say that ψ defines *$p \upharpoonright \varphi$.*

If $p \in S(B)$ is definable there is a function d such that for each formula $\varphi(\bar{x},\bar{y})$ over $\emptyset$, $d\varphi$ is a formula $\psi(\bar{y})$ which defines $p \upharpoonright \varphi$. Both d and the collection of formulas $\{ d\varphi : \varphi$ a formula over *$\emptyset \}$ are called a* defining scheme for p.

Clearly, if p is definable over A then p does not split over A.

If $p \in S(\mathfrak{C})$ and d is a defining scheme for p consisting of formulas over A, then d is also a defining scheme for any complete $q \subset p$. For this reason, many of the results stated below for elements of $S(\mathfrak{C})$ can actually be applied to any complete type.

Definition 3.3.8. *For A a set and φ a formula, φ is* almost over A *if the set $\{ f(\varphi) : f \in \mathrm{Aut}(\mathfrak{C})$ and f fixes A pointwise $\}$ contains finitely many formulas, up to equivalence; i.e.,*

$$\{ f(\varphi(\mathfrak{C})) : f \in \mathrm{Aut}(\mathfrak{C}) \text{ and } f \text{ fixes } A \text{ pointwise} \}$$

contains finitely many sets.

Much of the remainder of the section is devoted to the proof of the following theorem which ties together freeness, definability and nonsplitting.

Theorem 3.3.1. *Let T be a t.t. theory, $p \in S(\mathfrak{C})$ and A a set.*

(i) The following are equivalent.

(1) *p is a free extension of $p \upharpoonright A$ and $p \upharpoonright A$ is stationary.*
(2) *p is definable over A.*
(3) *p does not split over A.*

(ii) The following are also equivalent.

(1) *p is a free extension of $p \upharpoonright A$.*
(2) *There is a defining scheme for p consisting of formulas which are almost over A.*
(3) *p is definable over any model containing A.*

This theorem is actually a compilation of many lemmas and propositions, which we have collected as a focal point for the remainder of the section. One part of the theorem, namely (3) $\Longrightarrow$ (2) of (i), will not be proved until Section 4.1.1. With the additional tools developed in Section 4.1 the proof of this implication is easier than it would be if we forced it into this section. It is stated as part of a theorem in this section to present a coherent picture of the relationship between freeness and definability. The most difficult part of the theorem is (1) $\Longrightarrow$ (2) of (i), which is handled in

Lemma 3.3.7 (Definability Lemma). *Suppose that T is t.t., A is a set, $A \subset B \subset \mathfrak{C}$, $p \in S(B)$ is a free extension of $p \upharpoonright A$ and $p \upharpoonright A$ is stationary. Then, p is definable over A.*

Part (i) of the following elementary result is used in the proof of the lemma.

Lemma 3.3.8. *(i) Let φ be a formula and A a set such that for all automorphisms f of $\mathfrak{C}$ fixing A pointwise, φ is equivalent to $f(\varphi)$. Then φ is equivalent to a formula over A.*

(ii) Let $\varphi(\bar{x}, \bar{y})$ be a formula over A. If $\varphi(\bar{x}, \bar{c})$ is almost over A there is a formula E over A which defines an equivalence relation with finitely many classes and satisfies: for all $\bar{d}$ realizing $tp(\bar{c}/A)$, $\models E(\bar{d}, \bar{d}') \iff \varphi(\mathfrak{C}, \bar{d}) = \varphi(\mathfrak{C}, \bar{d}')$.

Proof. (i) Let $p \in S(A)$ be the type over A of some tuple $\bar{a}$ satisfying φ. If $\bar{b}$ also realizes p there is an $f \in \mathrm{Aut}(\mathfrak{C})$ fixing A and taking $\bar{a}$ to $\bar{b}$. Since φ is equivalent to $f(\varphi)$, $\bar{b}$ also satisfies φ. Thus, by compactness there is a formula $\psi \in p$ such that $\models \forall \bar{x}(\psi \to \varphi)$. A further compactness argument, left to the reader, shows that there is a formula ψ' over A equivalent to φ.

(ii) Let $E_0(\bar{y}, \bar{y}')$ be the equivalence relation defined by:

$$E_0(\bar{y}, \bar{y}') \iff \forall \bar{x}(\varphi(\bar{x}, \bar{y}) \leftrightarrow \varphi(\bar{x}, \bar{y}')).$$

Since $\varphi(\bar{x}, \bar{y})$ is over A, so is E_0. Let $p = tp(\bar{c}/A)$. Since $\varphi(\bar{x}, \bar{c})$ is almost over A there are finitely many, say k, E_0-classes containing a realization of p. By compactness there is a formula $\psi \in p$ such that

$$\models \exists \bar{y}_1 \cdots \bar{y}_k \forall \bar{y}[\psi(\bar{y}) \longrightarrow \bigvee_{1 \le i \le k} E_0(\bar{y}, \bar{y}_i)].$$

The equivalence relation $E(\bar{y}, \bar{y}') = (E_0(\bar{y}, \bar{y}') \wedge \psi(\bar{y}) \wedge \psi(\bar{y}')) \vee (\neg\psi(\bar{y}) \wedge \neg\psi(\bar{y}'))$ has the desired properties.

Proof of the Definability Lemma. First suppose that B is a set; i.e., $|B| < |\mathfrak{C}|$. We may assume that $B = M$ is a κ-saturated model, where $\kappa > |A| + |T|$. (Let $M \supset B$ be such a κ-saturated model and p' the unique free extension of p over M. If p' is definable over A then so is p.) Let $q = p \upharpoonright A$. Since M is κ-saturated and q is stationary M contains an infinite Morley sequence I in q. Let $\bar{a}$ be a realization of p and note that $I \cup \{\bar{a}\}$ is also a Morley sequence in q (since $p \upharpoonright (I \cup A)$ is the unique free extension of q over $I \cup A$). Let $\varphi(\bar{x}, \bar{y})$ be a formula over $\emptyset$.

Claim. For any sequence $\bar{b}$ from M of length $\ell = lh(\bar{y})$, $\varphi(\bar{x}, \bar{b}) \in p$ if and only if $\{ \bar{c} \in I : \models \varphi(\bar{c}, \bar{b})$ is cofinite in $I \}$.

By Corollary 3.3.1 there is a cofinite set $I' \subset I$ which is a Morley sequence over $A \cup \bar{b}$ in q, in fact, $I' \cup \{\bar{a}\}$ is a Morley sequence over $A \cup \bar{b}$ in q. Thus, for $\bar{b}$ a sequence from M of length ℓ,

$$\varphi(\bar{x}, \bar{b}) \in p \iff \models \varphi(\bar{a}, \bar{b}) \iff (\models \varphi(\bar{a}', \bar{b}) \text{ for every } \bar{a}' \in I').$$

This proves the claim.

Bringing Lemma 3.3.5(ii) into play, there is an n such that for all $\bar{b} \in M^\ell$, $|\{\bar{a}' \in I : \models \varphi(\bar{a}', \bar{b})\}| < n$ or $|\{\bar{a}' \in I : \models \neg\varphi(\bar{a}', \bar{b})\}| < n$. Thus,

$$\begin{aligned} \varphi(\bar{x}, \bar{b}) \in p \iff & (\text{ there is } J \subset I \text{ of cardinality } \geq n \text{ such that} \\ & \quad \bar{a}' \in J \Longrightarrow \models \varphi(\bar{a}', \bar{b})\ ) \qquad (3.4) \\ \iff & (\text{ there is no } J \subset I \text{ of cardinality } \geq n \text{ such that} \\ & \quad \bar{a}' \in J \Longrightarrow \models \neg\varphi(\bar{a}', \bar{b})\ ). \qquad (3.5) \end{aligned}$$

A formula defining p relative to $\varphi(\bar{x}, \bar{y})$ is found through this equivalence. Let $J = \{\bar{a}_0, \ldots, \bar{a}_{2n}\}$ be a subset of I of cardinality $2n+1$ and $\bar{c}$ an enumeration of J. Let

$$\psi(\bar{y}, \bar{c}) = \bigvee \left\{ \bigwedge_{i \in w} \varphi(\bar{a}_i, \bar{y}) : w \subset 2n+1,\ |w| = n \right\}.$$

If $\bar{b}$ is any sequence from M which satisfies $\psi(\bar{y}, \bar{c})$, there are certainly n elements of I satisfying $\varphi(\bar{x}, \bar{b})$, hence $\varphi(\bar{x}, \bar{b}) \in p$ (by (3.4)). Conversely, if $\models \neg\psi(\bar{b}, \bar{c})$ there must be n elements of I satisfying $\neg\varphi(\bar{x}, \bar{b})$, hence $\neg\varphi(\bar{x}, \bar{b}) \in p$ (by (3.5)).

To prove the lemma we must obtain a defining formula for $p \restriction \varphi$ which is over A, while the defining formula $\psi(\bar{y}, \bar{c})$ is over $A \cup I$.

Claim. For any $\bar{d}$ realizing $tp(\bar{c}/A)$ in M, $\psi(\bar{y}, \bar{d})$ is equivalent to $\psi(\bar{y}, \bar{c})$.

Let $\bar{d}$ in M realize $tp(\bar{c}/A)$. Then $\bar{d}$ is also an enumeration of a Morley sequence $\{\bar{d}_0, \ldots, \bar{d}_{2n}\}$ in q, and there is an infinite Morley sequence $J \subset M$ in q containing $\{\bar{d}_0, \ldots, \bar{d}_{2n}\}$. In obtaining $\psi(\bar{y}, \bar{c})$ above we can take I to be any infinite Morley sequence in q which is contained in M and $\bar{c}$ any subset of I of cardinality $2n+1$. Thus, for all $\bar{b}$ from M, $\varphi(\bar{x}, \bar{b}) \in p \iff \models \psi(\bar{b}, \bar{d})$. Hence, $\models \forall \bar{y}(\psi(\bar{y}, \bar{c}) \leftrightarrow \psi(\bar{y}, \bar{d}))$, proving the claim.

Thus, all conjugates over A of $\psi(\bar{y}, \bar{c})$ which are over M are equivalent. Since M is κ−saturated, this implies that all conjugates over A of $\psi(\bar{y}, \bar{c})$ are equivalent. By Lemma 3.3.8, $\psi(\bar{y}, \bar{c})$ is equivalent to a formula $\psi'(\bar{y})$ over A, proving the lemma in the case when B is a set.

Now let $B \subset \mathfrak{C}$ be arbitrary and suppose, towards a contradiction, that there is some formula $\varphi(\bar{x}, \bar{y})$ such that $p \restriction \varphi$ is not defined by a formula over A. Then there is a $p' \subset p$ whose domain is a set $B' \supset A$ such that $p' \restriction \varphi$ is not defined by a formula over A, in contradiction to the first part of the proof. This completes the proof of the Definability Lemma.

Theorem 3.3.1(i), (2)$\iff$(3) is proved in

Lemma 3.3.9. *Let T be t.t. and $p \in S(\mathfrak{C})$. Then p does not split over A if and only if p is definable over A.*

Proof. If p is definable over A, then clearly p does not split over A. Now suppose that p does not split over A. By the Definability Lemma, p is definable over some set. Let $\varphi(\bar{x}, \bar{y})$ be an arbitrary formula and let $\psi(\bar{y})$ be a defining formula of $p \restriction \varphi$. If $f \in \mathrm{Aut}(\mathfrak{C})$ fixes A pointwise, then $f(p) = p$ since p does not split over A. Also,

$$\begin{aligned} \models \psi(\bar{a}) \quad & \text{if and only if} \quad \varphi(\bar{x}, \bar{a}) \in p \\ & \text{if and only if} \quad \varphi(\bar{x}, f(\bar{a})) \in f(p) = p \\ & \text{if and only if} \quad \models \psi(f(\bar{a})) \end{aligned}$$

That is , ψ is preserved by the automorphisms of $\mathfrak{C}$ which fix A. Thus ψ is equivalent to a formula over A, (by Lemma 3.3.8(i)), as required.

The remainder of the proof of part (i) of the main theorem is delayed until Lemma 4.1.4.

Proof of Theorem 3.3.1(ii). (1) $\Longrightarrow$ (2). First notice that if ψ and ψ' are formulas which define $p \restriction \varphi$ then ψ and ψ' are equivalent.

Since p is a free extension of $q = p \restriction A$,

$$\{p' \in S(\mathfrak{C}) : p' \text{ is conjugate to } p \text{ over } A\}$$

is a finite set of types, which we enumerate as $\{p_0, \ldots, p_k\}$. (Every conjugate of p is an extension of q in $S(\mathfrak{C})$ with the same Morley rank as q.) Suppose that ψ defines $p \restriction \varphi$ (where φ is some formula over $\emptyset$).

Claim. If $f, g \in \mathrm{Aut}(\mathfrak{C})$ fix A pointwise and $f(p) = g(p)$, then $f(\psi)$ is equivalent to $g(\psi)$.

For f and g as hypothesized and $r = f(p) = g(p)$, both $f(\psi)$ and $g(\psi)$ define $r \restriction \varphi$. Hence, $f(\psi)$ is equivalent to $g(\psi)$.

Since there are finitely many elements of $S(\mathfrak{C})$ conjugate over A to p, ψ has finitely many conjugates over A. This proves (1) $\Longrightarrow$ (2).

(2) $\Longrightarrow$ (3). It suffices to show that a formula $\psi(\bar{y}, \bar{c})$ which is almost over A is equivalent to a formula over any model $M \supset A$. By Lemma 3.3.8(ii) there is an A−definable equivalence relation E with finitely many classes such that, for all $\bar{c}'$, $\models E(\bar{c}, \bar{c}')$ if and only if $\psi(\mathfrak{C}, \bar{c}') = \psi(\mathfrak{C}, \bar{c})$. Since E has only finitely many classes every class has a representative in M, in particular, there is a sequence $\bar{c}'$ from M such that $\models E(\bar{c}, \bar{c}')$. Then $\psi(\bar{y}, \bar{c}')$ is the desired formula over M equivalent to $\psi(\bar{y}, \bar{c})$.

(3) $\Longrightarrow$ (1). Suppose p is definable over any model containing A and, to the contrary, there is a formula $\varphi(\bar{x}, \bar{a}) \in p$ (where $\varphi = \varphi(\bar{x}, \bar{y})$ is over $\emptyset$) such that

$$MR(\varphi(\bar{x}, \bar{a})) < MR(p \restriction A) = \alpha.$$

Let $M \supset A$ be an $\aleph_0$-saturated model independent from $\bar{a}$ over A. By hypothesis, there is a formula ψ over M defining $p \restriction \varphi$. Since $r = tp(\bar{a}/M)$ is stationary there is an infinite I which is a Morley sequence in r over M. Let $\bar{b}$ realize $p \restriction M \cup I$ and let $I' \subset I$ be a finite set such that $\bar{b}$ is independent from $M \cup I$ over $M \cup I'$. Let $\bar{a}' \in I \setminus I'$.

Claim. $\bar{a}'$ and $\bar{b}$ are independent over A.

Simply because $\bar{a}' \in I$, $\bar{a}'$ is independent from $\bar{b}$ over $M \cup I'$. Since I is a Morley sequence over M and $\bar{a}' \in I \setminus I'$, $\bar{a}'$ is independent from I' over M. Furthermore, since $tp(\bar{a}'/M) = tp(\bar{a}/M)$ and $\bar{a}$ is independent from M over A, $\bar{a}'$ is also independent from M over A. Applying the transitivity of independence we conclude that $\bar{a}'$ is independent from $M \cup I' \cup \bar{b}$ over A, hence $\bar{a}'$ and $\bar{b}$ are independent over A. This proves the claim.

Since $\models \psi(\bar{a}')$ and ψ defines $p \restriction \varphi$, $\varphi(\bar{x}, \bar{a}') \in p$. Hence, $\models \varphi(\bar{b}, \bar{a}')$. Thus, $MR(\bar{b}/\bar{a}') \leq MR(\varphi(\bar{x}, \bar{a}')) = MR(\varphi(\bar{x}, \bar{a})) < \alpha = MR(\bar{b}/A)$, contradicting the claim and completing the proof of Theorem 3.3.1(ii).

Most properties of Morley rank independence in a t.t. theory can be derived quickly from Theorem 3.3.1. We include a few corollaries for ease of reference.

The following corollary is left as an exercise to the reader.

Corollary 3.3.2. *Let T be t.t. and $p \in S(\mathfrak{C})$. If p is a free extension of $q = p \restriction A$, then for all models $M \supset A$ there is a finite $B \subset M$ such that p is definable over B and $p \restriction B$ is stationary.*

The following observation lets us pick the parameters in the formulas of a defining scheme to have a very particular form.

Corollary 3.3.3. *Let T be t.t. and $p \in S(\mathfrak{C})$. Let $q \in S(A)$ be any complete stationary type such that p is the unique free extension of q and let I be an infinite Morley sequence in q over A. Then, p is definable over I.*

Proof. This follows immediately from the proof of the Definability Lemma.

The following consequence of definability will play a minor role in the proof of the Baldwin-Lachlan Theorem. This is occasionally called "The Open Mapping Theorem", although it is actually a corollary of that result proved for stable theories in Lemma 5.1.11.

Lemma 3.3.10. *If T is t.t., $tp(\bar{a}/A \cup \{\bar{b}\})$ is isolated and $tp(\bar{a}/A)$ is non-isolated, then $\bar{a} \underset{A}{\not\perp} \bar{b}$.*

Proof. Suppose, to the contrary, that $\bar{a}$ is independent from $\bar{b}$ over A. Let $\varphi(\bar{x}, \bar{b})$ isolate $tp(\bar{a}/A \cup \bar{b})$, where $\varphi(\bar{x}, \bar{y})$ is a formula over A.

Claim. There is a formula $\psi(\bar{x})$ over A such that whenever $\models \psi(\bar{c})$, $q = tp(\bar{b}/A)$ has a free extension over $\bar{c}$ containing $\varphi(\bar{c}, \bar{y})$.

To see this, let $p(\bar{y}) \in S(\mathfrak{C})$ be a free extension of q and $\psi_0(\bar{x})$ a formula almost over A defining $p \restriction \varphi$. Let $\psi_0, \ldots, \psi_k$ be a list of the conjugates of ψ_0 over A (up to equivalence), and let $\psi = \psi_0 \vee \ldots \vee \psi_k$. Since ψ is invariant under any automorphism which fixes A pointwise, ψ is equivalent to a formula over A (by Lemma 3.3.8). Suppose $\models \psi(\bar{c})$. Then $\models \psi_i(\bar{c})$, for some i, and for p' a conjugate over A to p such that ψ_i defines $p' \restriction \varphi$, $\varphi(\bar{c}, \bar{y}) \in p'$. This proves the claim.

A contradiction will be reached by showing that ψ isolates $tp(\bar{a}/A)$. Suppose that $\bar{c}$ and $\bar{c}'$ satisfy ψ. By the claim there are $\bar{d}$ and $\bar{d}'$ realizing q such that $\models \varphi(\bar{c}, \bar{d})$ and $\models \varphi(\bar{c}', \bar{d}')$. Since $tp(\bar{c}/\bar{d})$ and $tp(\bar{c}'/\bar{d}')$ are isolated by $\varphi(\bar{c}, \bar{d})$ and $\varphi(\bar{c}', \bar{d}')$, respectively, there is an automorphism which is the identity on A and takes $\bar{c}\bar{d}$ to $\bar{c}'\bar{d}'$. Hence $tp(\bar{c}/A) = tp(\bar{c}'/A)$. This contradiction proves the lemma.

The proof of the following is left to the exercises.

Corollary 3.3.4. *Suppose that T is t.t., M is a model and N is a prime model over $M \cup A$. Then $b \in N \setminus M \implies b \not\downarrow_M A$.*

The final topic to be covered in this section could be called "relativization". Suppose that T is t.t., $\varphi(x)$ is a formula over A and $D = \varphi(\mathfrak{C})$. In some studies it is natural to "restrict the universe to D", defined formally as follows.

Definition 3.3.9. *Let $\mathfrak{C}$ be the universal domain of a complete theory and $D = \varphi(\mathfrak{C})$ an A–definable subset of $\mathfrak{C}^k$ for some k. The* relativization of $\mathfrak{C}$ to D *is the model N (in a language L^*) defined as follows.*

(1) *The universe of N is D.*
(2) *For each A–definable relation $X \subset D^n$ (for some n) there is a relation symbol R in L^* whose interpretation on N is X.*

We may alternatively call N the relativization of $Th(\mathfrak{C})$ to φ *or the* restriction of $\mathfrak{C}$ to D.

When N is a relativization of $\mathfrak{C}$ to some definable D it is natural to ask: What is the difference between N and the structure on D induced by all definable relations in $\mathfrak{C}$? The next proposition says that there is no difference when T is t.t.

Proposition 3.3.3. *Let T be t.t. and D a subset of $\mathfrak{C}^n$ (for some n) which is definable over A. Then, for any k and definable $H \subset D^k$ there is a $B \subset D$ such that H is definable over $A \cup B$.*

Preliminarily, we prove

Lemma 3.3.11. *(i) If T is complete and there is a defining scheme for $p \in S(A)$ consisting of formulas almost over A, then there is a defining scheme for p consisting of formulas over A.*

(ii) If T is t.t., then every $p \in S(A)$ is definable over A.

Proof. (i) Let $\psi(\bar{y}, \bar{a})$ be a formula almost over A which defines $p \restriction \varphi$ (where $\psi(\bar{y}, \bar{z})$ is over A). Since $\psi(\bar{y}, \bar{a})$ is almost over A there is a formula $\theta(\bar{z})$ over A (by compactness) such that $\models \theta(\bar{b})$ implies $\psi(\bar{y}, \bar{b})$ is equivalent to a conjugate over A of $\psi(\bar{y}, \bar{a})$. If $\psi(\bar{y}, \bar{a}')$ is conjugate over A to $\psi(\bar{y}, \bar{a})$, then for $\bar{c}$ any tuple from A, $\models \psi(\bar{c}, \bar{a}) \iff \models \psi(\bar{c}, \bar{a}')$. Thus, $\psi'(\bar{y}) = \exists \bar{z}(\theta(\bar{z}) \wedge \psi(\bar{y}, \bar{z}))$ defines $p \restriction \varphi$.

(ii) Let $q \in S(\mathfrak{C})$ be a free extension of p. By Theorem 3.3.1(ii) there is a defining scheme d for q consisting of formulas almost over A. Since d is also a defining scheme for p, (i) implies that p is definable over A.

Remark 3.3.3. A was not assumed to be a set in this last lemma; when $|A| = |\mathfrak{C}|$ the same proof works.

Proof of Proposition 3.3.3. Let $H \subset D^k$ be $\psi(\mathfrak{C}, \bar{a})$. By the previous lemma, there is a defining scheme for $tp(\bar{a}/D \cup A)$ consisting of formulas over $D \cup A$. Thus, there is a formula $\theta(\bar{x})$ over $D \cup A$ such that $\models \psi(\bar{b}, \bar{a})$ if and only if $\models \theta(\bar{b})$, for all $\bar{b}$ from $D \cup A$; i.e., θ defines H. This proves the proposition.

The proof of the following is left to the reader.

Corollary 3.3.5. *Suppose that T is t.t., D is a definable set in $\mathfrak{C}$ and T_D is the theory of the relativization of $\mathfrak{C}$ to D. Then, the Morley rank of D in T is the same as the Morley rank of the universe in T_D.*

Corollary 3.3.6. *Suppose that T is t.t., D is a definable set in $\mathfrak{C}$ and T_D is the theory of the relativization of $\mathfrak{C}$ to D. If T is uncountably categorical then so is T_D.*

Proof. See Exercise 3.3.22.

Corollary 3.3.7. *Let M be a model of a t.t. theory and $p \in S_n(M)$. Let φ be a formula over M such that any realization of p is a tuple from $\varphi(\mathfrak{C})$. Then, $p \restriction \varphi(M)$ implies p.*

Proof. Let A be a finite set such that φ is over A and let $\psi(\bar{v}, \bar{b}) \in p$ be such that $\psi(\mathfrak{C}, \bar{b})$ is a set of tuples from $D = \varphi(\mathfrak{C})$. By Proposition 3.3.3 there is a formula $\theta(\bar{v}, \bar{a})$ equivalent to $\psi(\bar{v}, \bar{b})$, where $\bar{a} \subset D$ and $\theta(\bar{v}, \bar{w})$ is over A. Since M is a model there is such an $\bar{a}$ from $\varphi(M)$. In other words there is a formula in $p \restriction \varphi(M)$ (namely $\theta(\bar{v}, \bar{a})$) which implies $\psi(\bar{v}, \bar{b})$. This proves the corollary.

Much more can be said about the relationships between definability, independence and stationarity (see Section 5.1) but this basic foundation is sufficient to prove the results in the next chapter and the remainder of this one.

Historical Notes. Morley rank and the concept of a totally transcendental theory were developed by Morley in [Mor65], where we also find the notion of a Morley sequence. Definability of types was introduced by Shelah [She71], although not in the setting of a t.t. theory. Overall, the treatment of t.t. theories given here was motivated by properties of the forking dependence relation on stable theories as proved by Shelah; [She90] is the definitive source.

Exercise 3.3.1. Let T be the theory in the language with two binary relations E_1, E_2 which says that the E_i's define equivalence relations with infinitely many infinite classes and no finite classes, E_2 refines E_1 and each E_1−class contains infinitely many E_2−classes. Compute $MR(x = x)$.

Exercise 3.3.2. Let T be the theory in the first exercise, M a model of T and b an element such that $E_1(b, a)$ for some $a \in M$, but b is not E_2−equivalent to any element of M. Show that there is a defining scheme d over a for $p = tp(b/M)$. What are dE_1 and dE_2?

Exercise 3.3.3. Prove (i) of Lemma 3.3.2.

Exercise 3.3.4. Write out the proof that Morley rank independence satisfies properties (1)-(3) and (5)-(7) in the definition of a freeness relation.

Exercise 3.3.5. Let $\psi_0(x)$ and $\psi_1(x)$ be formulas in some complete theory. Show that $MR(\psi_0 \vee \psi_1) = \max\{MR(\psi_0), MR(\psi_1)\}$.

Exercise 3.3.6. Let X be a definable set of Morley rank $\alpha < \infty$ in the universal domain of some theory. Show that $\deg(X) = 1$ if and only if for all definable $Y \subset X$, $MR(Y) < \alpha$ or $MR(X \setminus Y) < \alpha$.

Exercise 3.3.7. Show that for T a countable complete theory, T is $\aleph_0$−stable if and only if $MR(x = x) < \infty$. (See Remark 3.3.2.)

Exercise 3.3.8. Let $\mathfrak{C}$ be the universal domain of a t.t. theory and let A be a subset of $\mathfrak{C}$. Let $T = Th(\mathfrak{C}_A)$, the theory of $\mathfrak{C}$ in the language with a constant symbol for each element of A. Show that T is also t.t.

Exercise 3.3.9. Prove: If T is t.t., then for all sets A and B there is $C \subset B$ such that $A \underset{C}{\perp} B$ and $|C| \leq |A|$.

Exercise 3.3.10. Show that if T is t.t., I is independent and $I \perp A$, then I is A−independent.

Exercise 3.3.11. Let T be a complete theory, M an $\aleph_0$−saturated model of T and let $CB(-)$ denote Cantor-Bendixson rank computed in $S_n(M)$. Prove that for all n−types p over M, $MR(p) \geq CB(p)$.

Exercise 3.3.12. Let I be an infinite indiscernible set in a t.t. theory. Show that there is a type $p \in S(I)$ such that $\bar{b}$ realizes p if and only if $I \cup \{\bar{b}\}$ is indiscernible. Conclude that in a κ−saturated model M any infinite set of indiscernibles is contained in an indiscernible set $J \subset M$ of cardinality κ.

Exercise 3.3.13. Suppose that T is a t.t. theory, $tp(a/A)$ and $tp(b/A \cup \{a\})$ are both stationary. Then $tp(ba/A)$ is stationary.

Exercise 3.3.14. Give an example of a t.t. theory and a $p \in S_1(\mathfrak{C})$ which is a free extension of $p \restriction \emptyset$ but not definable over $\emptyset$.

Exercise 3.3.15. Let T be a t.t. theory, $\varphi(u, v)$ a formula and $A = \{a_i : i < \omega\}$ a set such that $\models \varphi(a_i, a_j)$ if and only if $i \leq j$. Then A is finite. (See Lemma 5.1.6 for a proof of this property in stable theories.)

Exercise 3.3.16. Suppose that T is t.t., $p \in S(A)$ is a stationary type and I and J are Morley sequences over A in p with $|I| = |J|$. Then there is an automorphism of $\mathfrak{C}$ fixing A and mapping I onto J. Also, if f is an elementary map from A onto A' and I' is a Morley sequence in $p' = f(p)$ with $|I'| = |I|$, then f can be extended to an elementary map g taking I onto I'.

Exercise 3.3.17. Let T be the theory of infinite vector spaces over a field (formulated in the usual language so that T is strongly minimal). Show that for all tuples $\bar{a}$ and sets A, $tp(\bar{a}/A)$ is stationary.

Exercise 3.3.18. Let T be an uncountably categorical theory, M a model of T, p, $q \in S(M)$ strongly minimal types and a a realization of p. Then there is a realization b of q which is interalgebraic with a over M.

Exercise 3.3.19. Complete the proof of Lemma 3.3.8(i).

Exercise 3.3.20. Prove Corollary 3.3.2.

Exercise 3.3.21. Prove Corollary 3.3.4.

Exercise 3.3.22. Prove Corollary 3.3.6.

3.4 The Baldwin-Lachlan Theorem

In this section we prove that an uncountably categorical theory has 1 or $\aleph_0$ many countable models. This is the Baldwin-Lachlan proof of a conjecture due to Vaught. We will actually prove the following stronger result.

Theorem 3.4.1 (Baldwin-Lachlan). *Let T be a countable theory which is uncountably categorical but not $\aleph_0$–categorical. Then, T has $\aleph_0$ many countable models. Moreover,*

(i) If M is a countable model and $\varphi(v, \bar{a})$ is a strongly minimal formula with $tp(\bar{a})$ isolated, and $\bar{b}$ is a sequence from M with $tp(\bar{b}) = tp(\bar{a})$, then

$$\dim(\varphi(M, \bar{a})/\bar{a}) = \dim(\varphi(M, \bar{b})/\bar{b}).$$

(ii) Every model of T is homogeneous.

Throughout this section

- T denotes a countable uncountably categorical theory;
- $\varphi(v, \bar{x})$ is a formula and $\bar{a}^*$ is a tuple with $\varphi(v, \bar{a}^*)$ strongly minimal and
- $q = tp(\bar{a}^*)$ isolated;
- when $N \models T$ and $\bar{b} \subset N$ realizes q, $\mathrm{Dim}_{\bar{b}}(N)$ denotes $\dim(\varphi(N, \bar{b})/\bar{b})$.

(We do not assume here that T is not $\aleph_0$-categorical.) By Corollary 3.1.4(i), any model M is prime over $\varphi(M, \bar{a}) \cup \bar{a}$, where $\bar{a}$ is any realization of q in M. In fact M is prime over $I \cup \bar{a}$, for I any basis for $\varphi(M, \bar{a})$ over $\bar{a}$.

The results in the proof of Morley's Categoricity Theorem are enough to prove the requisite upper bound:

Lemma 3.4.1. *T has countably many countable models.*

Proof. Let M and N be countable models. By Corollary 3.1.4(i), M and N are isomorphic if there are $\bar{a} \subset M$ and $\bar{b} \subset N$ realizing q such that $\mathrm{Dim}_{\bar{a}}(M) = \mathrm{Dim}_{\bar{b}}(N)$. Since $\{\, \mathrm{Dim}_{\bar{c}}(M') : M'$ is a countable model and $\bar{c}$ realizes q in $M' \,\}$ is countable, T has countably many countable models.

For convenience all parts of Theorem 3.4.1 are stated for theories which are not categorical in $\aleph_0$, however both (i) and (ii) are true for *all* uncountably categorical theories. (Theorem 3.4.1(i) is true for theories which are also $\aleph_0$-categorical since $\mathrm{Dim}_{\bar{a}}(M) = |M|$ for any model M and $\bar{a} \in q(M)$. Part (ii) is proved for *all* uncountably categorical theories in Lemma 3.4.10 below.)

The next lemma ties categoricity in $\aleph_0$ to properties of strongly minimal sets in uncountably categorical theories.

Lemma 3.4.2. *The following are equivalent:*

(1) *T is $\aleph_0$-categorical.*
(2) *For $\bar{a}$ a realization of q, $acl(A) \cap \varphi(\mathfrak{C}, \bar{a})$ is finite for all finite sets $A \supset \bar{a}$.*
(3) *For all $M \models T$ and all $\bar{a}$ realizing q in M, $\mathrm{Dim}_{\bar{a}}(M)$ is infinite.*

Proof. Assume (1), let $\bar{a}$ be a realization of q and $A \supset \bar{a}$ a finite set. Since there are only finitely many formulas over A in one free variable, $acl(A) \cap \varphi(\mathfrak{C}, \bar{a})$ must be finite, proving (2). Assume (2), let $M \models T$, $\bar{a} \subset M$ a realization of q and I be a basis for $\varphi(M, \bar{a})$ over $\bar{a}$. Since $\varphi(M, \bar{a})$ is infinite, (2) forces I to be infinite. Thus (3) holds.

The proof that (3) $\Longrightarrow$ (1) is virtually identical to the proof of Theorem 3.1.2. For M and N countable models and $\bar{a}$, $\bar{b}$ realizations of q in M, N, respectively, let I be a basis for $\varphi(M, \bar{a})$ and J a basis for $\varphi(N, \bar{b})$. By assumption, $|I| = |M| = |N| = |J|$, so there is an elementary map f taking $I \cup \bar{a}$ onto $J \cup \bar{b}$. Since M is prime over $I \cup \bar{a}$ and N is prime over $J \cup \bar{b}$, f extends to an isomorphism of M onto N. This proves the $\aleph_0$-categoricity of T.

– **From hereon we assume that T is not $\aleph_0$-categorical.**

By the previous lemma, the prime model M of T contains a realization $\bar{a}$ of q in M for which $\mathrm{Dim}_{\bar{a}}(M)$ is finite. (In fact, since M is homogeneous, $\mathrm{Dim}_{\bar{b}}(M)$ is finite for all $\bar{b}$ from M.) Let I be a (finite) basis for $\varphi(M, \bar{a})$ over $\bar{a}$, and $p \in S(I \cup \bar{a})$ the unique nonalgebraic completion of $\varphi(v, \bar{a})$ over this set. Since p is not realized in M it must be nonisolated. No previously set conditions are invalidated by incorporating I into $\bar{a}$, hence we can assume that the unique nonalgebraic completion of $\varphi(v, \bar{a})$ in $S(\bar{a})$ is nonisolated. Notice that the corresponding behavior carries over to other realizations of q, so we can require

- for any $\bar{b}$ realizing q the nonalgebraic completion of $\varphi(v, \bar{b})$ in $S(\bar{b})$ is nonisolated.

The assumption that T is not $\aleph_0$-categorical is first used to prove

Lemma 3.4.3. *If M is a countable model and there is an $\bar{a}$ from M such that $\mathrm{Dim}_{\bar{a}}(M)$ is infinite, then M is saturated and not prime over a finite set.*

Proof. Suppose that M and $\bar{a}$ are as hypothesized and let N be a countable saturated model. Repeating he proof of (3) $\Longrightarrow$ (1) in the last lemma produces an isomorphism from M onto N. Furthermore, since T is not $\aleph_0$-categorical, the countable saturated model cannot be prime over a finite set.

Remark 3.4.1. If M is a countable model and there is some $\bar{a} \subset M$ realizing q such that $\mathrm{Dim}_{\bar{a}}(M)$ is finite, then $\mathrm{Dim}_{\bar{b}}(M)$ is finite for all $\bar{b}$ realizing q in M.

It will be important in this section to understand the behavior of nonisolated complete strongly minimal types. The next lemma sheds much light on the situation. Notice that the theory is not required to be uncountably categorical here.

Lemma 3.4.4. *Let T_0 be a complete theory, B a set and θ a strongly minimal formula over B. Let $p \in S(B)$ be the unique nonalgebraic type in $S(B)$ containing θ.*

(i) Then, p is isolated if and only if $acl(B) \cap \theta(\mathfrak{C})$ is finite.

(ii) If p is nonisolated and $A \supset B$, then the unique nonalgebraic extension of p in $S(A)$ is nonisolated.

Proof. Since (ii) follows immediately from (i), we only need to prove the first part. Assuming that $acl(B) \cap \theta(\mathfrak{C}) = X$ is finite, there is a formula ψ over B such that X is $\psi(\mathfrak{C})$. Since there is a unique nonalgebraic element of $S(B)$ containing θ, p is isolated by $\theta \wedge \neg\psi$.

Conversely, suppose $acl(B) \cap \theta(\mathfrak{C}) = X$ is infinite. To prove that p is nonisolated it suffices to show that any $\psi \in p$ is satisfied by an element of X. Let

$$\Sigma = \{\sigma : \sigma \text{ is an algebraic formula over } B\}.$$

Since there is a unique nonalgebraic element of $S(B)$ containing θ the set of formulas $\{\theta \wedge \neg\psi\} \cup \{\neg\sigma : \sigma \in \Sigma\}$ is inconsistent. By compactness there are $\sigma_0, \ldots, \sigma_n \in \Sigma$ such that $\models \forall v(\theta(v) \wedge \bigwedge_{i\leq n} \neg\sigma_i(v) \longrightarrow \psi(v))$. Since X is infinite it contains an element satisfying $\bigwedge_{i\leq n} \neg\sigma_i$, hence ψ. This proves the lemma.

We proved in Lemma 2.3.1 that a theory with finitely many but more than one countable model has a countable model which is prime over a finite set, not saturated, and realizes every complete $n-$type over $\emptyset$. In the next lemma we show that such a model also exists under the (apparently) weaker assumption that (i) of the theorem fails.

Lemma 3.4.5. *Suppose that M is a countable model of T containing realizations $\bar{a}$ and $\bar{b}$ of q such that* $\mathrm{Dim}_{\bar{a}}(M) \neq \mathrm{Dim}_{\bar{b}}(M)$. *Then M realizes every complete $n-$type over $\emptyset$, for all n, and M is prime over some finite set.*

Proof. For some $\bar{a}$ from M realizing q, $\mathrm{Dim}_{\bar{a}}(M)$ is finite. Furthermore, $\mathrm{Dim}_{\bar{c}}(M)$ must be finite for any realization $\bar{c}$ of q in M by Remark 3.4.1. Fix $\bar{a}$ realizing q in M such that $i = \mathrm{Dim}_{\bar{a}}(M)$ is minimal and let I be a basis for $\varphi(M, \bar{a})$. By Corollary 3.1.4(i) M is prime over $I \cup \bar{a}$. The main step in the proof is

Claim. For every n there is a $\bar{c}$ with $\mathrm{Dim}_{\bar{c}}(M) \geq n$.

Assume, to the contrary, that there is a bound on these dimensions. Let $\bar{b}$ be a realization of q in M such that $j = \mathrm{Dim}_{\bar{b}}(M)$ is maximal and let J be a basis for $\varphi(M, \bar{b})$. By hypothesis, $i < j$. Let J' be a subset of J of cardinality i and $N \subset M$ a prime model over $J' \cup \bar{b}$. Since there is an elementary map from $I \cup \bar{a}$ onto $J' \cup \bar{b}$, M and N are isomorphic. Thus, there is a sequence $\bar{d}$ from N with $\mathrm{Dim}_{\bar{d}}(N) = j$. Since T does not have any Vaughtian pairs $\varphi(M, \bar{d}) \not\subset N$, in fact, $\mathrm{Dim}_{\bar{d}}(M) > Dim_{\bar{d}}(N)$. This contradicts the maximality assumption on j to prove the claim.

Now let p be an arbitrary element of $S(\emptyset)$, $\bar{c}$ a realization of p and N a prime model over $\bar{c}$. Let $\bar{d}$ be a realization of q in N, in which case N is prime over $\varphi(N, \bar{d}) \cup \bar{d}$. There is a finite $\bar{d}-$independent set $J \subset \varphi(N, \bar{d})$ such that $r = tp(\bar{c}/J \cup \bar{d})$ is isolated. By the claim there is a $\bar{b}$ in M realizing q and a $\bar{b}-$independent set $K \subset \varphi(M, \bar{b})$ of cardinality $|J|$. Hence there is an elementary map f taking $\bar{d}$ to $\bar{b}$ and J onto K. Since $f(r)$ is isolated it is realized by some $\bar{c}'$ in M. This sequence $\bar{c}'$ is a realization of p in M.

The combination of the previous lemma and the next proposition will lead quickly to a contradiction.

Proposition 3.4.1. *Let T' be any uncountably categorical theory and $\bar{a}$ any tuple. Then there is a k such that whenever $\{\bar{b}_0, \ldots, \bar{b}_k\}$ is independent, there is some $\bar{b}_i$ independent from $\bar{a}$.*

The bound k obtained in the proposition is given the formal name "pre-weight":

Definition 3.4.1. *Let T' be t.t., A a set and $p \in S(A)$. The* pre-weight *of p, denoted $pwt(p)$, is*

$$\sup\{\, \kappa : \text{ there is an } \bar{a} \text{ realizing } p \text{ and } A-\text{independent set } I \text{ of cardinality } \kappa \text{ such that } \bar{b} \in I \Longrightarrow \bar{a} \not\perp_A \bar{b} \,\}.$$

We may write $pwt(\bar{a}/A)$ for $pwt(tp(\bar{a}/A))$.

(When the supremum of a class X of cardinals does not exist we write $\sup X = \infty$ and extend the order on cardinals so that $\kappa < \infty$ for all cardinals κ.)

We will prove

Proposition 3.4.2. *If T' is uncountably categorical then every complete type has finite pre-weight.*

Remark 3.4.2. Let T' be t.t. and $p = tp(\bar{a}/A)$. If I is an A–independent set such that each $\bar{b} \in I$ depends on $\bar{a}$ over A, then I is finite. (Assuming there is an infinite such I let $J \subset I$ be a finite set such that $\bar{a}$ is independent from $I \cup A$ over $J \cup A$. Then any $\bar{b} \in I \setminus J$ is independent from $\bar{a}$ over A; contradiction.) That there is a finite bound to $|I|$, as I ranges over all such sets, is not immediately clear.

Proposition 3.4.1 follows immediately from Proposition 3.4.2 by letting $k = pwt(tp(\bar{a}))$. Pre-weight (and weight) will be studied extensively in stable theories (after replacing Morley rank independence by forking independence) in Sections 5.6 and 6.3. Proposition 3.4.2 is implicit in Theorem 5.6.1.

Lemma 3.4.6. *Let T be a t.t. theory and p, q complete types with q a free extension of p. Then $pwt(p) \leq pwt(q)$.*

Proof. Suppose that $p \in S(A)$, $q \in S(B)$, $\bar{a}$ realizes p and I is an A–independent set such that each $\bar{b} \in I$ depends on $\bar{a}$ over A. Let $\bar{a}'$ realize q. Let J be a set such that $\{\bar{a}\} \cup I$ is conjugate to $\{\bar{a}'\} \cup J$ over A and $\{\bar{a}'\} \cup J$ is independent from B over A. Using standard properties of independence the reader can show that J is B–independent and each $\bar{b} \in J$ depends on $\bar{a}'$ over B. Thus, $pwt(q) \geq |J| = |I|$. This inequality is true for any A–independent set I such that each element of I depends on $\bar{a}$ over A, hence $pwt(q) \geq pwt(p)$, to complete the proof.

The proof of Proposition 3.4.2 requires some preliminary lemmas involving the following notion.

Definition 3.4.2. *For A, B and C sets in a t.t. theory we say that A is* dominated by B over C *if for all sets D,*

$$D \perp_C B \Longrightarrow D \perp_C A.$$

Notice that for all sets A and B, $acl(A)$ is dominated by A over B.

Lemma 3.4.7. *Let T' be a t.t. theory, M an $\aleph_0$-saturated model of T' and A a set which is atomic over $M \cup B$. Then $A \cup B$ is dominated by B over M.*

Proof. Without loss of generality, $A = \bar{a}$ and $B = \bar{b}$ are finite. Suppose, towards a contradiction, that there is a $\bar{c}$ such that

$$\bar{c} \underset{M}{\perp} \bar{b} \text{ and } \bar{c} \underset{M}{\not\perp} \bar{a} \cup \bar{b}.$$

Then (by symmetry and the transitivity of independence) $\bar{a}$ depends on $\bar{c}$ over $M \cup \bar{b}$. Let $A' \subset M$ be a finite set such that $tp(\bar{a}/M \cup \bar{b})$ is isolated over $A' \cup \bar{b}$, and $\bar{a}$ depends on $\bar{c}$ over $A' \cup \bar{b}$. We can require, furthermore, that $\bar{b}$ is independent from M over A' and $tp(\bar{b}/A')$ is stationary. Let $\varphi(\bar{x})$ be a formula over $A' \cup \bar{b}$ which isolates $tp(\bar{a}/M \cup \bar{b})$ and observe that $MR(\varphi) = MR(\bar{a}/M \cup \bar{b}) = MR(\bar{a}/A' \cup \bar{b}) = \alpha$. Let $\psi(\bar{x}, \bar{c}, \bar{b})$ be a formula in $tp(\bar{a}/A' \cup \bar{b} \cup \bar{c})$ which implies φ and has Morley rank $< \alpha$.

Since $\bar{b}$ is independent from $\bar{c}$ over M and $tp(\bar{b}/A')$ is stationary, $tp(\bar{b}/M \cup \bar{c})$ does not split over A' (by Theorem 3.3.1(i)). Since M is $\aleph_0$-saturated it contains a realization $\bar{c}'$ of $tp(\bar{c}/A')$, in fact, $tp(\bar{c}'/A' \cup \bar{b}) = tp(\bar{c}/A' \cup \bar{b})$. By the conjugacy over A of $\psi(\bar{x}, \bar{c}, \bar{b})$ and $\psi(\bar{x}, \bar{c}', \bar{b})$, $\psi(\bar{x}, \bar{c}', \bar{b})$ is consistent, implies φ and has Morley rank $< \alpha$. Since $\bar{c}'$ is from M this contradicts that φ isolates a complete type over $M \cup \bar{b}$ of Morley rank α.

Corollary 3.4.1. *Let M be a countable $\aleph_0$-saturated model of an uncountably categorical theory, φ a strongly minimal formula over M and $\bar{a}$ a tuple of elements. Then there is a sequence $\bar{c} \subset \varphi(\mathfrak{C})$ such that $\bar{a} \cup \bar{c}$ is dominated by $\bar{c}$ over M and $\bar{a} \cup \bar{c}$ is dominated by $\bar{a}$ over M.*

Proof. By Exercise 3.1.20, $T' = Th(M_M)$, the theory of M in the language with constant symbols for the elements of M, is also uncountably categorical. By Corollary 3.1.4(i) there is a tuple $\bar{c}$ from $\varphi(\mathfrak{C})$ such that $tp(\bar{a}/\bar{c})$ is isolated in T'. In fact, $\bar{c}$ can be chosen in the prime model over $\bar{a}$ (in T'), hence $tp(\bar{c}/\bar{a})$ in T' is also isolated. Returning to the original theory $T = Th(M)$, $tp(\bar{a}/M \cup \bar{c})$ and $tp(\bar{c}/M \cup \bar{a})$ are isolated. By Lemma 3.4.7, $\bar{a} \cup \bar{c}$ is dominated by $\bar{c}$ over M and $\bar{a} \cup \bar{c}$ is dominated by $\bar{a}$ over M, proving the corollary.

Two more facts about pre-weight will be enough to prove the finiteness of pre-weight in uncountably categorical theories.

Lemma 3.4.8. *Let T be a t.t. theory.*

(i) Let D be a strongly minimal set over a set A and $\bar{c}$ a sequence from D. Then $pwt(\bar{c}/A) = \dim(\bar{c}/A)$.

(ii) If $\bar{a}$ is dominated by $\bar{b}$ over A, then $pwt(\bar{a}/A) \leq pwt(\bar{b}/A)$.

Proof. (i) It is easy to see that $pwt(\bar{c}/A) \geq \dim(\bar{c}/A)$. That $pwt(\bar{c}/A) \leq \dim(\bar{c}/A)$ is proved by induction on $\dim(\bar{c}/A)$ (uniformly for all A). Let I be a nonempty A–independent set such that any $\bar{b} \in I$ is A–dependent on $\bar{c}$. Let $\bar{b}_0 \in I$ and $I' = I \setminus \{\bar{b}_0\}$. By Lemma 3.3.4, $\dim(\bar{c}/A \cup \bar{b}_0) < \dim(\bar{c}/A)$. Since I' is independent over $A \cup \bar{b}_0$ and any $\bar{b} \in I'$ depends on $\bar{c}$ over $A \cup \bar{b}_0$, the inductive hypothesis implies $pwt(\bar{c}/A \cup \bar{b}_0) \leq \dim(\bar{c}/A \cup \bar{b}_0)$. Thus, $|I| = |I'| + 1 \leq pwt(\bar{c}/A \cup \bar{b}_0) + 1 \leq \dim(\bar{c}/A \cup \bar{b}_0) + 1 \leq \dim(\bar{c}/A)$. This proves (i).

(ii) Let I be an A–independent set such that each $\bar{d} \in I$ depends on $\bar{a}$ over A. Since $\bar{a}$ is dominated by $\bar{b}$ over A, each $\bar{d} \in I$ must depend on $\bar{b}$ over A. Thus, $|I| \leq pwt(\bar{b}/A)$.

Proof of Proposition 3.4.2. Let A be a set and $p \in S(A)$ a complete type. Let M be an $\aleph_0$–saturated model containing A and $q \in S(M)$ a free extension of p. By Lemma 3.4.6 $pwt(p) \leq pwt(q)$, hence it suffices to show that $pwt(q)$ is finite. Let $\bar{a}$ be a realization of q, φ a strongly minimal formula over M and $\bar{c}$ a (finite) tuple from $\varphi(\mathfrak{C})$ such that $\bar{a} \cup \bar{c}$ is dominated by $\bar{c}$ over M (see Lemma 3.4.6). By Lemma 3.4.8, $pwt(q) = pwt(\bar{a}/M) \leq pwt(\bar{c}/M) = \dim(\bar{c}/M)$, proving the proposition.

As noted above, this also proves Proposition 3.4.1.

The following lemma draws together several results of the section to prove Theorem 3.4.1(i).

Lemma 3.4.9. *Let M be a countable model of T. Then for all sequences $\bar{a}$, $\bar{b}$ from M realizing q,*

$$\mathrm{Dim}_{\bar{a}}(M) = \mathrm{Dim}_{\bar{b}}(M).$$

Proof. Suppose the lemma fails for M. By Lemma 3.4.5 M is prime over a finite sequence $\bar{c}$ and realizes every element of $S(\emptyset)$. Let k $(= pwt(\bar{c}/\emptyset))$ be as guaranteed by Proposition 3.4.1 for $\bar{c}$. Since T is not $\aleph_0$–categorical there is a nonisolated $p \in S(\emptyset)$. Let $\{\bar{b}_0, \ldots, \bar{b}_k\}$ be an independent set of realizations of p and $r = tp(\bar{b}_0, \ldots, \bar{d}_k)$. Since r is realized in M we may as well assume that $\{\bar{b}_0, \ldots, \bar{b}_k\} \subset M$. By our choice of k there is some i such that $\bar{c} \perp \bar{b}_i$. However, since M is prime over $\bar{c}$, $tp(\bar{b}_i/\bar{c})$ is isolated, contradicting Lemma 3.3.10 (since p is nonisolated). This proves the lemma.

Remark 3.4.3. This lemma is also true for an uncountably categorical theory T' which is $\aleph_0$–categorical. (If M is a countable model of T' and $\varphi'(x, \bar{a})$ is a strongly minimal formula over M then for all $\bar{b}$ from M realizing $tp(\bar{a})$, $\dim(\varphi'(M, \bar{b})/\bar{b}) = \aleph_0$.)

From Lemma 3.4.9 we quickly obtain the main assertion of the Baldwin-Lachlan Theorem:

Corollary 3.4.2. *T has $\aleph_0$ many countable models.*

Proof. Let $\bar{a}$ be a realization of q and for each $k < \omega$ let I_k be a subset of $\varphi(\mathfrak{C}, \bar{a})$ of dimension k. We have assumed that the unique nonalgebraic completion of $\varphi(\bar{x}, \bar{a})$ in $S(\bar{a})$ is nonisolated. For each $k < \omega$ let M_k be a prime model over $I_k \cup \bar{a}$.

Claim. For $l \neq k < \omega$, $M_l \not\cong M_k$.

By Lemma 3.4.4(ii), the unique nonalgebraic extension of $\varphi(\bar{x}, \bar{a})$ in $S(I_j \cup \bar{a})$ is nonisolated, hence not realized in M_j. That is, $Dim_{\bar{a}}(M_j) = j$ for each $j < \omega$. Suppose, towards a contradiction, that for some $k \neq l$ there is an isomorphism f from M_l onto M_k. Then, for $\bar{b} = f(\bar{a})$, $Dim_{\bar{b}}(M_k) = l$, contradicting Lemma 3.4.9 to prove the claim and the corollary.

It remains only to prove (ii) of the Baldwin-Lachlan Theorem. We showed in Corollary 3.1.5 that for T' an uncountably categorical countable theory and $M \models T'$, if $|M|$ is uncountable and regular then M is saturated. Included in the next lemma is the removal of the restriction to regular cardinals in that earlier corollary.

Lemma 3.4.10. *If T' is an uncountably categorical countable theory then every model of T' is homogeneous and every uncountable model of T' is saturated.*

Proof. The reader should first verify:

(∗) Let $M \supset N$ be models, φ' a strongly minimal formula defined over $A \subset N$, I a basis for $\varphi'(N)$ over A and J a basis for $\varphi'(M)$ over N. Then $I \cup J$ is a basis for $\varphi'(M)$ over A, and $\dim(\varphi'(M)/A) = \dim(\varphi'(N)/A) + \dim(\varphi'(M)/N)$.

Let M be a model of T'. Certain parts of the proof are handled differently depending on whether M is countable or uncountable, so we split the argument into two (very similar) cases.

First suppose M to be uncountable. Let $A \subset M$, with $|A| < |M| = \kappa$, and let f be an elementary map with $f(A) = B \subset M$. For $N_0 \subset M$ a prime model over A we can assume f to be an isomorphism from N_0 onto a model $N_1 \subset M$ which is prime over B. There is an isolated type $q' \in S(\emptyset)$ and a $\varphi'(\bar{x}, \bar{y})$ such that for any realization $\bar{c}$ of q', $\varphi'(\bar{x}, \bar{c})$ is strongly minimal. Let $\bar{a}'$ be a realization of q' in N_0 and $\bar{b}' = f(\bar{a}')$. Since f is an isomorphism $\dim(\varphi'(N_0, \bar{a}')) = \dim(\varphi'(N_1, \bar{b}')) = \lambda$. Also $\dim(\varphi'(M, \bar{a}')/\bar{a}') = \dim(\varphi'(M, \bar{b}')/\bar{b}') = \kappa$ by Corollary 3.1.4. By (∗), if I_0 is a basis for $\varphi'(M, \bar{a}')$ over N_0 and I_1 is a basis for $\varphi'(M, \bar{b}')$ over N_1, then $\kappa = \lambda + |I_0| = \lambda + |I_1|$. Since $\lambda < \kappa$, $|I_0| = |I_1|$. Thus, by extending f we can assume f to be an elementary map from $N_0 \cup I_0$ onto $N_1 \cup I_1$. There is a submodel M' of M which is prime over $N_0 \cup I_0$. Since I_0 is a basis for $\varphi'(M, \bar{a}')$ over N_0, $\varphi'(M', \bar{a}') = \varphi'(M, \bar{a}')$. Since T' has no Vaughtian pairs, $M' = M$; i.e., M is prime over $N_0 \cup I_0$. Hence f can be extended to an elementary embedding of M into itself so that $f(M) \supset \varphi'(M, \bar{b}')$. Again, since T' has no Vaughtian pairs, this embedding must be onto M, proving that M is homogeneous.

Now suppose that M is countable. The proof is virtually identical to the uncountable case, but individual steps may be justified differently. Certainly

M is homogeneous if T' is $\aleph_0$−categorical, so we can assume that T' is not $\aleph_0$−categorical. Let A be a finite subset of M and f an elementary map from A onto $B \subset M$. Choose N_0, N_1, φ', $\bar{a}'$ and $\bar{b}'$ exactly as above. Since N_0 and N_1 are isomorphic $\dim(\varphi'(N_0, \bar{a}')) = \dim(\varphi'(N_1, \bar{b}')) = k$, and $k < \aleph_0$ by Lemma 3.4.3. By Lemma 3.4.9 $\dim(\varphi'(M, \bar{a}')/\bar{a}') = \dim(\varphi'(M, \bar{b}')/\bar{b}') = \mu$. Again by (∗), if I_0 is a basis for $\varphi'(M, \bar{a}')$ over N_0 and I_1 is a basis for $\varphi'(M, \bar{b}')$ over N_1, then $\mu = k + |I_0| = k + |I_1|$. Since k is finite $|I_0| = |I_1|$. From this point on the proof is just like the uncountable case. This proves that every model of T' is homogeneous.

For M an uncountable model of T', M is $\aleph_0$−saturated (see Exercise 3.1.3) and homogeneous, hence M is saturated (by Corollary 2.2.6).

Some of the results in this section will be generalized later to arbitrary $\aleph_0$−stable (in fact, the more general superstable) theories. For example, a countable superstable theory which is not $\aleph_0$−categorical must have infinitely many countable models. This will be proved by establishing Proposition 3.4.1 for superstable theories. We will also show (in Section 5.4) that an $\aleph_0$−stable theory has a saturated model in every cardinality, not just the regular ones.

Historical Notes. All of the main theorems here are due to Baldwin and Lachlan [BL71], although our exposition owes a lot to Lascar [Las86].

Exercise 3.4.1. Suppose that T is an uncountably categorical theory, $N \subset M$ are models of T, and φ, ψ are strongly minimal formulas over N. Show that $\dim(\varphi(M)/N) = \dim(\psi(M)/N)$.

Exercise 3.4.2. Give a quick proof of the Baldwin-Lachlan Theorem assuming that there is a strongly minimal formula over $\emptyset$.

3.5 Introduction to ω−stable Groups

In this section we study ω−stable theories in which the universe is a group under some definable operation. The goal is to elucidate the degree to which the group-theoretic and stability-theoretic properties of these structures influence each other. The importance of this study lies both in the breadth of the class of groups with an ω−stable theory and the manner in which groups arise in a "geometrical" analysis of ω−stable theories. Results pertaining to this second point will be discussed in Sections 4.4 and 4.5.

Definition 3.5.1. *Let T be an ω−stable theory with universal domain $\mathfrak{C}$ such that $(\mathfrak{C}, \cdot)$ is a group for some definable binary operation $\cdot$. Then $(\mathfrak{C}, \cdot)$ is called an ω−*stable group. *Adopting more standard notation, ω−stable groups will usually be represented by G, H, G', etc.*

Remark 3.5.1. Let T be an ω–stable theory with universal domain $\mathfrak{C}$.

(i) For $(\mathfrak{C}, \cdot)$ to be an ω–stable group it is not necessary for $\cdot$ to be in the language, being a definable operation is enough. Also, there may be definable relations on $\mathfrak{C}$ which are not definable in the group language; i.e., the restriction of $\mathfrak{C}$ to a language containing only a function symbol for $\cdot$.

(ii) Let X be a definable subset of $\mathfrak{C}$ and $\cdot$ a definable operation on X such that $(X, \cdot)$ is a group. The restriction of $\mathfrak{C}$ to X contains no new definable relations, hence $(X, \cdot)$ is an ω–stable group in this restricted universe. Since the definable relations on X in $\mathfrak{C}$ are the same as the definable relations on X in the restriction there is no loss in calling X an ω–stable group without first restricting to X. From hereon, when referring to an ω–stable group G we always leave open the possibility that G is a definable subset of some larger theory.

(iii) The restriction to countable theories (and the consistent use of ω–stable over $\aleph_0$–stable) is purely a convention adopted by the authors in the area. Virtually anything proved here is true in an uncountable t.t. theory with the same justification.

(iv) Here, the term "ω–stable group" only applies to the universal domain of the relevant theory. This is nonstandard. Most authors call the model G an ω–stable group if there is a definable group operation on G and $Th(G)$ is ω–stable. However, we have found our terminology to be more appropriate for the presentation of the material in this book.

Here are some basic examples.

Example 3.5.1. (i) The universal domain of the theory of vector spaces over a fixed field is an ω–stable group.

(ii) The universal domain K of the theory of algebraically closed fields of a fixed characteristic is an ω–stable group under $+$, and $K \setminus \{0\}$ is an ω–stable group under $\cdot$.

(iii) Let M be the direct sum of $\aleph_0$ many copies of the group $\mathbb{Z}_4 = \mathbb{Z}/4\mathbb{Z}$, $T = Th(M)$ and G the universal domain of T. The reader can show that T is quantifier-eliminable, from which we conclude:

- $2G$ is a strongly minimal set. In fact, $2G$ is a vector space over the field with two elements and there are no definable relations on $2G$ except those defined in the vector space language.
- $MR(G) = 2$.
- T is totally categorical.

(iv) Consider the special case of an abelian group G in the language containing only the group operation $+$ and 0. Macintyre proved (in [Mac71a])

> **Theorem.** *$Th(G)$ is ω–stable if and only if G is of the form $D \oplus H$, where D is divisible and H is of bounded order.*

Given an arbitrary ω–stable abelian group G let G_0 be the restriction of G to a language containing only the group operation and the identity. By

Exercise 3.1.9 G_0 is also an ω−stable group. Thus, the theorem tells us the underlying group structure of G_0 *and* G. Namely, G is a direct sum of a divisible group and a group of bounded order.

Perhaps the most important class of ω−stable groups is the collection of algebraic groups over an algebraically closed field. Here is a summary of the critical concepts.

Let K be an algebraically closed field which we take to be the universal domain of its theory, for simplicity. Let $n < \omega$ and $K[\bar{x}]$ the ring of polynomials in $\bar{x} = (x_1, \ldots, x_n)$ over K. A set $V \subset K^n$ is called an *affine algebraic subset of* K^n if for some $p_1(\bar{x}), \ldots, p_k(\bar{x}) \in K[\bar{x}]$,

$$V = \{\, \bar{a} \in K^n : p_1(\bar{a}) = 0 \wedge \ldots \wedge p_k(\bar{a}) = 0 \,\}.$$

The affine algebraic subsets of K^n form the closed sets in a topology on K^n called the *Zariski topology* on K^n. This topology is *Noetherian*, meaning that it has the descending chain condition on closed sets. (If $V_i \subset K^n$, $i < \omega$, are affine algebraic sets then the ideal generated by the polynomials defining the V_i's is generated by finitely many such polynomials since $K[\bar{x}]$ is Noetherian. Thus, $\bigcap_{i<\omega} V_i$ is the intersection of finitely many of these sets.) Certainly, any affine algebraic set is definable. A Zariski closed set V is called *irreducible* if it cannot be written as $V_1 \cup V_2$, where $V_i \subsetneq V$, $i = 1, 2$, are Zariski closed sets. An irreducible Zariski closed subset of K^n is called an *affine algebraic variety* or simply an *affine variety.* A set $X \subset K^n$ is a *constructible set* if it is a boolean combination of affine algebraic subsets of K^n. The elimination of quantifiers for algebraically closed fields can be stated as: a subset of K^n is definable if and only if it is constructible.

Turning to groups, the set $M_n(K)$ of $n \times n$ matrices over K is definable ($M_n(K)$ is a subset of K^{n^2}). The operations of addition and matrix multiplication on $M_n(K)$ and the determinant function (from $M_n(K)$ into K) are definable over K. Thus, the *general linear group* $\mathrm{GL}_n(K) = \{a \in M_n(K) : \det(a) \neq 0\}$ of invertible $n \times n$ matrices over K under multiplication is definable. Also definable is $\mathrm{SL}_n(K)$, the set of elements of $\mathrm{GL}_n(K)$ with determinant 1. In fact, $M_n(K)$, $\mathrm{GL}_n(K)$ and $\mathrm{SL}_n(K)$ are all affine algebraic sets. Most commonly encountered subgroups of $\mathrm{GL}_n(K)$, such as the upper triangular matrices over K, are definable. What is not so obvious is the definability of the projective linear group $\mathrm{PGL}_n(K) = \mathrm{GL}_n(K)/Z$, where Z is the center of the general linear group. This will be verified after our discussion of T^{eq} in the next chapter.

A group H is an *affine algebraic group* if it is a subgroup of $\mathrm{GL}_n(K)$ (for some n) which is also an affine algebraic set. We will not take the time to define an "algebraic group over K", stating only that every affine algebraic group is an algebraic group and every algebraic group is definable. In fact, using much deeper model theory it can be shown that every connected (see below) definable group over K is definably isomorphic to an algebraic group (due to Hrushovski [Hru90b] and, in part, to van den Dries).

Note: $MR(\mathrm{GL}_n(K)) = n^2$, so any affine algebraic group has finite Morley rank.

The interest in ω–stable groups began with the study of the natural examples, for example, vector spaces, algebraically closed fields and affine algebraic groups over algebraically closed fields. In fact, much of what we know about ω–stable groups has been obtained by generalizing results about affine algebraic groups. Such a generalization is facilitated by replacing Zariski topology dimension theory by Morley rank independence. One such result, Zil'ber's Indecomposability Theorem, is proved below. In Zil'ber's Ladder Theorems he shows that any sufficiently complicated uncountably categorical theory contains definable groups and these groups have a significant influence on the structure of the definable subsets (see Section 4.4). These results have been generalized (by Hrushovski and others) to the extent that almost any problem in stability theory involves groups on some level. For instance, groups played a critical role in the proof of Vaught's conjecture for superstable theories of finite rank [Bue93].

Remember: When X is a definable set of Morley rank α, $\deg(X) = 1$ if and only if whenever $Y \subset X$ is definable in M and has Morley rank α, $MR(X \setminus Y) < \alpha$ (see Exercise 3.3.6). Also, when T is t.t. and $\bar{a}$ and $\bar{b}$ are interalgebraic over A, $MR(\bar{a}/A) = MR(\bar{a}\bar{b}/A) = MR(\bar{b}/A)$.

Definition 3.5.2. *Let G be a group, X a set and $\star$ a map from $G \times X$ into X. The triple $(G, X, \star)$ is a* group action *if*

(1) $e \star x = x$, *for e the identity of G and all $x \in X$, and*
(2) *for all g, $h \in G$ and $x \in X$, $(gh) \star x = g \star (h \star x)$.*

Often the $\star$ is dropped from the notation, the group action is denoted (G, X) and $g \star x$ is written gx.

Let (G, X) be a group action.

- *(G, X) is* faithful *if, whenever $g \in G \setminus \{1\}$, $gx \neq x$ for some $x \in X$.*
- *Given $0 \neq k < \omega$, (G, X) is* k–transitive *if for all $x_1, \ldots, x_k, y_1, \ldots, y_k \in X$ such that $x_i \neq x_j$ and $y_i \neq y_j$ for $1 \leq i < j \leq k$, there is a $g \in G$ such that $gx_i = y_i$ for $1 \leq i \leq k$. (G, X) is* sharply k–transitive *if the g in the previous sentence is unique. Equivalently, (G, X) is* k–transitive *and only 1 fixes k elements of X.*
- *The term* transitive *is used in place of 1–transitive and* regular *replaces sharply 1–transitive.*
- *For $x \in X$ let $stab(x)$, called the* stabilizer of x, *denote $\{ g \in G : gx = x \}$.*
- *The* orbit of $x \in X$ *is $Gx = \{ gx : g \in G \}$.*
- *If $Y \subset X$ and $g \in G$, the* translation of Y by g *is $gY = \{ gy : y \in Y \}$.*

Clearly, the stabilizer of a point is a subgroup of G and if $H = stab(x)$ and g, $h \in G$, then $gH = hH$ if and only if $gx = hx$. This defines a one-to-one correspondence between the left cosets of H and the orbit of x. Thus, letting G/H denote the set of left cosets of H in G, $|G/H| = |Gx|$; i.e., $[G : stab(x)] = |Gx|$.

We will see that an ω–stable group G acts on the types over G and the definability of types yields definable stabilizer subgroups. First, some terminology about ω–stable groups acting on types.

Let G be an ω–stable group and p a 1–type over the set A. Let p^{-1} denote $\{\varphi(x^{-1}) : \varphi \in p\}$. For a an element let $ap = \{\varphi(a^{-1} \cdot x) : \varphi(x) \in p\}$ and $pa = \{\varphi(x \cdot a^{-1}) : \varphi(x) \in p\}$, both types over $A \cup \{a\}$. Given $a \in A$, ap is also a type over A and, if $p = tp(b/A)$, $ap = tp(a \cdot b/A)$. Since any b and $a^{-1} \cdot b$ are interalgebraic over a the formulas $\varphi(x)$ and $\varphi(a^{-1} \cdot x)$ have the same Morley rank. In fact, since multiplication defines a function, $\varphi(x)$ and $\varphi(a^{-1} \cdot x)$ also have the same Morley degree (see the exercises). Thus, the types p and ap have the same Morley rank and degree. The types ap and pa are called the *left* and *right translates of p by a*, respectively. The term *translation* may be used in place of *left translation.*

Given an ω–stable group G, $a \in G$ and $p \in S_1(G)$, $ap \in S_1(G)$. In fact, left (or right) translation defines a group action of G on $S_1(G)$. By the above comments the action preserves Morley rank.

Definition 3.5.3. *Let G be an ω–stable group and $p \in S_1(G)$. The* left stabilizer of p, *denoted $stab(p)$, is the stabilizer of p under left translation; i.e., $stab(p) = \{a \in G : ap = p\}$. The* right stabilizer of p *is the stabilizer of p under right translation. By default, the term "stabilizer of p" refers to the left stabilizer of p.*

Lemma 3.5.1. *Let G be an ω–stable group.*

(i) Let $P \subset S_1(G)$ be the collection of all types of some fixed Morley rank. The action of G on P is definable in the sense that for all p, $q \in P$, there is a formula ψ over G such that

$$\text{for all } a \in G,\ ap = q \text{ if and only if } \models \psi(a).$$

There is a formula equivalent to ψ over any set A such that p and q are both definable over A.

(ii) Given $p \in S_1(G)$, $stab(p)$ is a definable subgroup of G and

$$MR(stab(p)) \leq MR(p).$$

Moreover, if p is definable over A, $stab(p)$ is definable over A.

(iii) If p, $q \in S_1(G)$ and q is a right translate of p then $stab(p) = stab(q)$.

Proof. (i) First notice that G acts on P since translation preserves Morley rank. Let p be in P and $\varphi \in p$ be a formula of Morley rank $\alpha = MR(p)$ and degree 1. If $a \in G$, then $\varphi(a^{-1} \cdot x)$ is also a formula of Morley rank α and degree $1 = \deg(p)$. Let q be another element of P. Since every element of P has Morley rank α, q is ap if and only if $\varphi(a^{-1} \cdot x) \in q$. Let $\varphi'(x, y) = \varphi(y^{-1} \cdot x)$ and let $\psi(y)$ be the formula defining $q \restriction \varphi'$. Then, for all $a \in G$, $q = ap$ if and only if $\models \psi(a)$. From an inspection of the proof we can choose ψ to be over any set A such that p and q are both definable over A.

(ii) As a special case of (i) notice that $\{a \in G : ap = p\} = stab(p)$ is definable. Furthermore, by the proof of (i), if p is definable over the finite set $A \subset G$ then $stab(p)$ is definable over A.

(iii) Left to the reader as a short exercise.

Notation. Given an ω–stable group G we may write ab for $a \cdot b$ when there is no danger of confusing ab with the pair of elements (a, b).

It is frequently handy to work with restrictions of elements of $S_1(G)$ when showing that one type is a translate of another. The necessary tool is

Corollary 3.5.1. *Let G be an ω–stable group, $p, q \in S_1(G)$ and $a \in G$. Suppose that p and q are definable over A. Then,*

(1) *$q = ap$ if and only if*

(2) *there is a b realizing $p \upharpoonright A$ such that b is independent from a over A, ab realizes $q \upharpoonright A$ and ab is independent from a over A.*

Proof. Interpolating a few more equivalences will make the proof easy.

Claim. (a) $q = ap$
$\iff$ (b) for all sets $B \supset A \cup \{a\}$, $q \upharpoonright B = ap \upharpoonright B$;
$\iff$ (c) there is a set $B \supset A \cup \{a\}$, $q \upharpoonright B = ap \upharpoonright B$.

That (a) implies (b) is clear from the definition of translation and (b) $\Longrightarrow$ (c) holds trivially. Suppose (c) holds and $B \supset A \cup \{a\}$ is such that $q \upharpoonright B = ap \upharpoonright B$. Let $MR(p) = \alpha$ and let $\varphi \in p \upharpoonright A$ be a formula of Morley rank α and degree 1. Since $ap \upharpoonright B = q \upharpoonright B$, $\alpha = MR(p \upharpoonright B) = MR(q \upharpoonright B)$ and $\varphi(a^{-1}x) \in q$. Since q is definable over A, $MR(q)$ must also be α. As in the proof of Lemma 3.5.1(i), ap is the unique element of $S_1(G)$ which contains $\varphi(a^{-1}x)$ and has Morley rank α. Thus $ap = q$, proving the claim.

Turning to the main assertion of the corollary, (1) implies (2) is just a matter of unraveling the notation. Now assume that a and b meet the conditions in (2). Let $B = A \cup \{a\}$. Since p and q are both definable over A and both b and ab are independent from a over A, b realizes $p \upharpoonright B$ and ab realizes $q \upharpoonright B$. Since $a \in B$, $tp(a \cdot b/B) = ap \upharpoonright B$; i.e., $ap \upharpoonright B = q \upharpoonright B$. By the claim $ap = q$, proving the corollary.

The most fundamental property of an ω–stable group is

Proposition 3.5.1. *An ω–stable group G has the descending chain condition on definable subgroups. That is, if $G = H_0 \supset H_1 \supset \ldots \supset H_i \supset \ldots$ is a chain of definable subgroups of G, there is an $n < \omega$ such that $H_m = H_n$ for all $m \geq n$.*

Proof. Let I be the collection of pairs (α, m) where α is an ordinal and $1 \leq m < \omega$. Well-order I lexicographically; i.e., $(\alpha, m) < (\beta, k)$ if $\alpha < \beta$ or $\alpha = \beta$ and $m < k$. Associate with any formula ψ the pair $(MR(\psi), \deg(\psi)) \in I$. Let $H \subset K$ be subgroups of G defined by ψ and φ, respectively. If $H \neq K$ there is an $a \in K$ such that $a\psi$ defines a coset $aH \neq H$. Since ψ and $a\psi$ have the same Morley rank and both imply φ, $(MR(\psi), \deg(\psi)) < (MR(\varphi), \deg(\varphi))$; i.e., $(MR(H), \deg(H)) < (MR(K), \deg(K))$. If $G = H_0 \supset H_1 \supset \ldots \supset H_i \supset \ldots$ is an infinite descending chain of subgroups, where $H_{i+1} \neq H_i$, for all $i < \omega$, then $\{ (MR(H_i), \deg(H_i)) : i < \omega \}$ is an infinite descending chain in $(I, <)$, which is impossible since this is a well-ordering.

Corollary 3.5.2. *An ω−stable group G has a minimal definable subgroup G^o of finite index in G. G^o is a normal subgroup of G and is definable over $\emptyset$.*

Proof. Let $\mathcal{H}$ be the collection of all definable subgroups of G having finite index in G. If H_1, $H_2 \in \mathcal{H}$ then $H_1 \cap H_2 \in \mathcal{H}$. Combining this with the previous proposition shows that $\bigcap \mathcal{H} = G^o$ is the intersection of finitely many elements of $\mathcal{H}$, hence G^o is definable. Given $g \in G$, $g^{-1}G^o g = g^{-1} \bigcap \mathcal{H} g = \bigcap \{ g^{-1}Hg : H \in \mathcal{H} \} = \bigcap \mathcal{H} = G^o$, hence G^o is normal. That G^o is definable without parameters is left to the exercises.

Definition 3.5.4. *Let G be an ω−stable group. The minimal definable subgroup of finite index in G is called the* connected component *of G, and is denoted G^o. We call G* connected *if $G = G^o$.*

Remark 3.5.2. Since G is the union of finitely many cosets of G^o, $MR(G^o) = MR(G)$. Also, G^o is itself connected as an ω−stable group.

Definition 3.5.5. *Let G be an ω−stable group. A 1−type p over G is a* generic type *if $MR(p)$ is maximal; i.e., $MR(p) = MR(G)$.*

Warning: p is not assumed to be a complete type in the definition.

The set P of generic types in $S_1(G)$ is nonempty and finite. Since translation preserves Morley rank, $GP = P$ (under the action of G on $S_1(G)$ by translation).

Lemma 3.5.2. *Let G be ω−stable and P the set of generics in $S_1(G)$. Then the action of G on P is transitive.*

Proof. Let p and q be any two elements of $S_1(G)$ and $\alpha = MR(p) = MR(q) = MR(G)$. The goal is to find a c such that $cp = q$. Let G_0 be a model of $Th(G)$ and note that p and q are definable over G_0 (since they are both free extensions of their restrictions to $\emptyset$). Let a and b be realizations of $p_0 = p \restriction G_0$ and $q_0 = q \restriction G_0$, respectively, which are independent over G_0. Let φ be a formula in q_0 of Morley rank α and degree 1, and let $c = ba^{-1}$. Using that c and b are interalgebraic over $G_0 \cup \{a\}$, $MR(c/G_0 \cup \{a\}) = MR(b/G_0 \cup \{a\}) = MR(q) = \alpha$. Since the Morley rank of $tp(c/G_0)$ cannot be $> \alpha$, c and a are independent over G_0. Similarly, c is independent from b over G_0. Since $ca = b$ Corollary 3.5.1 applies to show that $q = cp$, as desired.

This leads to the following linking of the group-theoretic and model-theoretic structure of an ω–stable group.

Corollary 3.5.3. *Let G be an ω–stable group. Then*
(i) $[G : G^o] = \deg(G)$.
(ii) $p \in S_1(G)$ is generic if and only if $stab(p) = G^o$.
(iii) G is connected if and only if $\deg(G) = 1$.

Proof. Let P be the generic types in $S_1(G)$, G_0 any model of $Th(G)$ and P_0 the generic elements of $S_1(G_0)$. Each element of P is the unique free extension of some element of P_0, hence $|P_0| = |P| = \deg(G)$. The reader should verify that when $tp(a/G_0)$ and $tp(b/G_0)$ are in P_0 and $aG^o \neq bG^o$, $tp(a/G_0) \neq tp(b/G_0)$. Thus, $[G : G^o] \leq \deg(G)$.

Now let p be an arbitrary element of P. Since the action of G on P is transitive one of the basic facts about group actions gives the equation:

$$[G : stab(p)] = |Gp| = |P| = \deg(G).$$

Thus, the definable subgroup $stab(p)$ has finite index in G. Since G^o is the minimal such group, $[G : stab(p)] \leq [G : G^o]$. We conclude both that $\deg(G) = [G : G^o]$ (i.e., (i) holds) and $stab(p) = G^o$, for any generic p. On the other hand, if $p \in S_1(G)$ and $stab(p) = G^o$ then $MR(p) \geq MR(G^o) = MR(G)$ (by Lemma 3.5.1(ii)), so p is a generic type, proving (ii).

Part (iii) follows immediately from (i). This proves the corollary.

Remark 3.5.3. An ω–stable group $G = G^o \cup a_1G^o \cup \ldots \cup a_nG^o$, where $\{1, a_1, \ldots, a_n\}$ form a complete set of representatives of the cosets of G^o in G. Each a_iG^o has degree 1 and the same Morley rank as G.

Remark 3.5.4. The above analysis of left translation proceeds in an identical manner for right translation. In particular, given an ω–stable group G, $p \in S_1(G)$ is generic if and only if the stabilizer of p with respect to right translation is G^o.

When F is a field, $F^* = F \setminus \{0\}$.

Corollary 3.5.4. *If F is an ω–stable field, then $(F, +)$ and $(F^*, \cdot)$ are both connected. Thus, F has degree* 1.

Proof. By Corollary 3.5.3, $(F, +)$ is connected if and only if $(F^*, \cdot)$ is connected if and only if $\deg(F) = 1$. Let H be a subgroup of $(F, +)$ of finite index. Let I be the ideal $\bigcap\{kH : k \in F^*\}$. By Proposition 3.5.1 there are $k_1, \ldots, k_n \in F^*$ such that $I = k_1H \cap \ldots \cap k_nH$, hence I is a nonzero ideal. Since F is a field, $I = F$. Thus, $H = F$ and $(H, +)$ is connected. This proves the corollary.

As an application of the previous Corollary 3.5.3 we prove that an ω-stable group has a "large" abelian subgroup.

Proposition 3.5.2. *An ω-stable group G contains an infinite definable abelian subgroup.*

The proof uses the following basic group-theoretic fact.

Lemma 3.5.3. *If all elements of a group G have finite order and all elements of $G \setminus \{1\}$ are conjugate, then $|G| \leq 2$.*

Proof. We may assume $G \neq \{1\}$. Let $g \in G \setminus \{1\}$ be of prime order p. By the conjugacy condition all nonidentity elements of G have order p. First suppose p is odd. Choose $h \in G$ such that $h^{-1}gh = g^{-1}$. Then for all n, $(h^{-1})^n gh^n = g^{(-1)^n}$. When $n = p$ this yields $g = g^{-1}$, a contradiction. Thus $p = 2$. By a standard exercise G is abelian, hence G has 2 elements.

Remember, for g an element of a group G the *centralizer of g* is $C(g) = \{ h : h^{-1}gh = g \}$. Let g^G denote the conjugacy class of g, $\{ h^{-1}gh : h \in G \}$.

Proof of Proposition 3.5.2. Suppose the proposition fails and G is a counterexample of minimal Morley rank α and Morley degree d. This minimality condition implies that every proper definable subgroup of G is finite. In particular G is connected, hence $d = 1$ by Corollary 3.5.3(i).

Let Z be the center of G, a proper definable subgroup. We will contradict the finiteness of Z by applying Lemma 3.5.3 to the group G/Z. For any $g \in G \setminus Z$ $C(g)$ is a proper definable subgroup of G, hence is finite. Since $g \in C(g)$, g must have finite order. There is a natural one-one correspondence between the conjugacy class g^G of g and the set of cosets $G/C(g)$. Since $C(g)$ is finite g^G must have rank α. If $g^G \neq h^G$, $g^G \cap h^G = \emptyset$, so the fact that G has degree 1 implies there is only one conjugacy class among the elements of $G \setminus Z$. By Lemma 3.5.3 G/Z contains at most two elements, contradicting that Z is finite. This proves the proposition.

Corollary 3.5.5. *A strongly minimal group is abelian.*

Definition 3.5.6. *An element a of an ω-stable group G is* generic over A *if $tp(a/A)$ is generic.*

It is frequently more appropriate to work with elements than types, calling for an equivalent definition and some additional results.

Lemma 3.5.4. *Let G be an ω-stable group. Then $a \in G$ is generic over A if and only if*

$$\forall b \in G(b \underset{A}{\perp} a \Longrightarrow b \underset{A}{\perp} b \cdot a).$$

Proof. First suppose a is generic over A and $b \in G$ is independent from a over A. Then $MR(G) = MR(a/A \cup \{b\}) = MR(b \cdot a/A \cup \{b\}) \leq MR(b \cdot a/A) \leq MR(G)$. Thus, $MR(b \cdot a/A \cup \{b\}) = MR(b \cdot a/A)$, as required.

Conversely, suppose ba is independent from b over A whenever b is independent from a over A. Let g be generic over $A \cup \{a\}$. As in the previous paragraph, ga is generic over $A \cup \{a\}$. By assumption, ga is independent from g over A, hence ga is generic over $A \cup \{g\}$. Since a is interalgebraic with ga over $A \cup \{g\}$, a is also generic over $A \cup \{g\}$. This proves the lemma.

Remark 3.5.5. Given an ω-stable group G and generic elements a and b, $aG^o = bG^o$ if and only if

(*) for all sets A, if a and b are generic over A, then $tp(a/A) = tp(b/A)$.

(See Exercise 3.5.3.)

Translation provides information not only about $S_1(G)$, but about the formulas over G. The following corresponds to and slightly strengthens Lemma 3.5.2. A definable subset X of an ω-stable group is *generic* if it has maximal Morley rank; i.e., X is defined by a generic formula.

Lemma 3.5.5. *Let G be an ω-stable group and $X \subset G$ definable. Then, X is generic if and only if there are left translates (or right translates) $X_0, \ldots, X_k$ of X such that $G = \bigcup_{i \leq k} X_i$.*

Proof. The proof is written for left translates; it is the same for right translates. Let $\alpha = MR(G)$. First assume $G = \bigcup_{i \leq k} X_i$, where $X_0, \ldots, X_k$ are left translates of X. Then one of the X_i's has Morley rank α, hence X has Morley rank α.

Conversely, suppose $MR(X) = \alpha$ and X is definable over the finite set A. Notice it suffices to prove the lemma with X replaced by some translate of X. The proof will be easier after performing a few reductions. Since $X = (a_1 G^o \cap X) \cup \ldots \cup (a_k G^o \cap X)$ for some $a_1, \ldots, a_k$, there is some a such that $X \cap aG^o$ has Morley rank α. Since $(a^{-1}X) \cap G^o$ has Morley rank α, we may as well assume that $MR(X \cap G^o) = \alpha$. Since G is the union of finitely many translates of G^o it suffices to show that G^o is the union of finitely many translates of $X \cap G^o$. So, we can take G to be connected. By Corollary 3.5.3, G has a unique generic type in $S_1(G)$. Since $MR(X) = \alpha$ this generic type contains the formula defining X; i.e.,

$$\text{whenever } \ a \in G \text{ is generic over } \ A, \ a \in X.$$

Let $\{ a_i : i < \omega \}$ be an independent set of elements generic over A and let b be an arbitrary element of G. There is an $i < \omega$ such that a_i is independent from b over A, hence a_i is generic over $A \cup \{b\}$. Then $a_i \cdot b$ is generic over A, hence $a_i \cdot b \in X$, equivalently $b \in a_i^{-1}X$. Thus, $G = \bigcup_{i<\omega} a_i^{-1}X$. By compactness, G is the union of finitely many translates of X, completing the proof.

Corollary 3.5.6. *Let G be a connected ω-stable group and $X \subset G$ a generic definable set. Then $X \cdot X = G$.*

Proof. Let X be A-definable and b an arbitrary element of G. In the proof of the lemma we found an element a, generic over A, such that $b \in a^{-1}X$. Since $MR(a^{-1}/A)$ is also α, $a^{-1} \in X$ (by the connectedness of G). Thus $b \in X \cdot X$.

As stated above, many of the results about ω-stable groups are obtained by generalizing proofs about algebraic groups. Zil'ber's Indecomposability Theorem is one such result. Chevalley proved the following about an algebraic group G over an algebraically closed field. Let X_i, $i \in I$, be a family of constructible (i.e., definable) subsets of G such that for each $i \in I$, the identity element e is in X_i and the Zariski closure $\bar{X}_i$ of X_i is irreducible. Then the subgroup H of G generated by the X_i's is Zariski closed, connected and $H = X_{i_1}^{\epsilon_1} \cdots X_{i_n}^{\epsilon_n}$ for some $i_1, \ldots, i_n \in I$ and $\epsilon_j = \pm 1$. (When $X \subset G$, $X^{+1} = X$ and $X^{-1} = \{x^{-1} : x \in X\}$.) In the general ω-stable context there is no topology, hence nothing exactly like Zariski closure or irreducibility. Zil'ber's substitute for irreducibility is the following.

Definition 3.5.7. *Given an ω-stable group G and $X \subset G$ definable, X is* indecomposable *if for any definable subgroup H of G, either $|X/H| = 1$ or X/H is infinite (where $X/H = \{xH : x \in X\}$).*

When X is a group it is indecomposable exactly when it is connected.

Theorem 3.5.1 (Zil'ber's Indecomposability Theorem). *Suppose that G is an ω-stable group of finite Morley rank and, for $i \in I$, X_i is an indecomposable definable subset of G containing the identity element e. Let H be the subgroup of G generated by $\bigcup_{i \in I} X_i$. Then H is definable, connected and for some $i_1, \ldots, i_n \in I$, $H = X_{i_1} \cdot \ldots \cdot X_{i_n}$.*

Proof. Let $\chi = \{Y : Y = \Pi_{j \in J} X_j,\ J \subset I \text{ is finite}\}$. Since each element of Y is definable and $MR(G)$ is finite there is an $X = X_{i_1} \cdot \ldots \cdot X_{i_k} \in \chi$ which has maximal Morley rank among the elements of χ. Let $MR(X) = m$. Let φ_i be the formula defining X_i, φ the formula defining X and $p \in S_1(G)$ a type containing φ with Morley rank m. Let H be the connected component of $stab(p)$. By Lemma 3.5.1 and Corollary 3.5.2 H is definable, hence to complete the proof it remains to show that $X_i \subset H$ for all $i \in I$, and $H = X_{i_1} \cdot \ldots \cdot X_{i_n}$, for some $i_1, \ldots, i_n \in I$. The first step is handled in

Claim. For all $i \in I$, $X_i \subset H$.

Suppose $X_i \not\subset H$. Since $e \in X_i \cap H$, $|X_i/H| > 1$, hence the indecomposability of X_i forces X_i/H to be infinite. Since H has finite index in $H^* = stab(p)$, X_i/H^* is also infinite. Let $\{a_j : j < \omega\} \subset X_i$ be such that $a_jH^* \neq a_lH^*$ for $j \neq l < \omega$. Elements b, $c \in G$ have the same coset with respect to $stab(p)$ if and only if $bp = cp$. Thus, $\{a_jp : j < \omega\}$ is an infinite collection of types of Morley rank m. However, each of these types contains the formula defining $X_i \cdot X$, which also has Morley rank m (by the maximality of m). This contradiction proves the claim.

Being a group H therefore contains not only X but the group generated by the X_i's. The Morley rank of H is $\leq m = MR(p)$ by Lemma 3.5.1(ii). Since $X \subset H$, $MR(H)$ must equal m, hence X is a generic subset of H. Since H is connected $X \cdot X = H$ (by Corollary 3.5.6). A fortiori, H is the group generated by the X_i's, proving the theorem.

This theorem yields the definability of some typical subgroups, for example, commutator subgroups. The *commutator* of elements a, b in a group G is the element $[a,b] = a^{-1}b^{-1}ab$. For A and B subsets of G, $[A,B]$ denotes the subgroup of G generated by $\{ [a,b] : a \in A,\ b \in B \}$. Notice that $G' = [G,G]$ is the minimal normal subgroup H of G such that G/H is abelian. A priori, there is no reason to think that G' is definable, however it follows from the next lemma that G' is definable when G is a connected group of finite Morley rank.

Lemma 3.5.6. *Let G be a group of finite Morley rank, H a connected definable subgroup of G and A any subset of G. Then the group $[A,H]$ is definable and connected. Moreover, there are finitely many elements $a_1, \ldots, a_n \in A$ such that $x \in [A,H]$ if and only if there are $h_1, \ldots, h_n \in H$ such that $x = [a_1,h_1] \cdots [a_n,h_n]$.*

Proof. For $a \in A$ let $X_a = \{ h^{-1}ah : h \in H \}$. Then $[A,H]$ is the group generated by $\bigcup\{ a^{-1}X_a : a \in A \}$. The indecomposability of $a^{-1}X_a$ is needed to satisfy the hypotheses of Zil'ber's Indecomposability Theorem. Since indecomposability is invaraint under translation it suffices to show that each set X_a is indecomposable. The first step towards this end is

Claim. X_a is indecomposable if $|X_a/K|$ is 1 or infinite for any definable subgroup K of G which is normalized by H; i.e, $h^{-1}Kh = K$ for all $h \in H$.

Let K be any definable subgroup of G and suppose $1 < |X_a/K| < \aleph_0$. Let K_0 be the group $\bigcap_{h \in H} h^{-1}Kh$ and notice that K_0 is normalized by H. We proceed to show that $1 < |X_a/K_0| < \aleph_0$. The group K_0 is the intersection of finitely many of the groups $h^{-1}Kh$, $h \in H$ (by Proposition 3.5.1) hence is definable. Given $h \in H$, $h^{-1}X_ah = X_a$, so $X_a/h^{-1}Kh$ is also finite. If K_1 and K_2 are any two subgroups such that X_a/K_i is finite (for $i = 1,2$) then $X_a/(K_1 \cap K_2)$ is also finite. Thus X_a/K_0 is finite. Since $|X_a/K| \leq |X_a/K_0|$, $1 < |X_a/K_0| < \aleph_0$, proving the claim.

Fix $a \in A$ and let K be any definable subgroup of G which is normalized by H. Let $H_0 = \{ x \in H : (x^{-1}ax)K = aK \}$. Since K is normalized by H

- H_0 is a subgroup of H and
- for x, $y \in H$, $x^{-1}axK = y^{-1}ayK$ if and only if $xH_0 = yH_0$.

Thus, when $X_a/K = [H : H_0]$ is finite, $|X_a/K| = 1$ (by the connectedness of H). This proves that X_a is indecomposable.

By Zil'ber's Indecomposability Theorem $[A, H]$ is definable, connected and has the form specified in the last sentence of the lemma.

Remark 3.5.6. Implicit in the previous lemma is a proof that $[A, H]$ is infinite or $\{1\}$ under the stated hypotheses.

In all of the above results G is the universal domain of its theory. Many of the results yield information about the subgroups of an arbitrary model G_0 of $Th(G)$ with little additional work. In many instances the condition used to define a subgroup H of G, when relativized to G_0, defines $H \cap G_0$. Here are two examples.

Corollary 3.5.7. *Let G be an ω–stable group, G_0 a model of $Th(G)$ and $p \in S_1(G_0)$. Then $H_0 = \{a \in G_0 : ap = p\}$ is a subgroup definable in G_0. In fact, for p' the unique free extension of p in $S(G)$ and $H = stab(p')$, $H_0 = H \cap G_0$.*

Proof. This is assigned as Exercise 3.5.4.

Corollary 3.5.8. *Let G be an ω–stable group and G_0 a model of $Th(G)$. Then $G^o \cap G_0$ is*

$$\bigcap \{H \subset G_0 : H \text{ is a subgroup of finite index definable in } G_0\}.$$

Proof. This is Exercise 3.5.5.

Remark 3.5.7. Let G be an ω–stable group, G_0 a model of $Th(G)$ and H a subgroup of G_0 definable in G_0. Suppose $H = \varphi(G_0)$. Then H is called *connected* if $\varphi(G)$ is connected. By the previous corollary, H is connected if and only if there is no proper subgroup K of H definable in G_0 and having finite index in H.

Corollary 3.5.9. *Let G be a group of finite Morley rank, G_0 a model of $Th(G)$, H a connected subgroup definable in G_0 and A any subset of G_0. Then the group $[A, H]$ is definable in G_0 and connected.*

Proof. Left to the reader in Exercise 3.5.6.

The following relative version of Zil'ber's Indecomposability Theorem is somewhat less elementary. First we need a definition of indecomposability that applies to sets definable in a model.

Definition 3.5.8. *Let G be an ω–stable group, G_0 a model of $Th(G)$ and X a set definable in G_0. Then X is* indecomposable *if for any subgroup H of G, definable in G, either $|X/H| = 1$ or X/H is infinite.*

Corollary 3.5.10 (Zil'ber's Indecomposability Theorem (relative)). *Let G be an ω–stable group of finite Morley rank and G_0 a model of $Th(G)$. For each $i \in I$ suppose X_i is an indecomposable set definable in G_0 which contains the identity e. Let H be the subgroup of G_0 generated by $\bigcup_{i \in I} X_i$. Then H is definable in G_0, connected and for some $i_1, \ldots, i_n \in I$, $H = X_{i_1} \cdot \ldots \cdot X_{i_n}$.*

Proof. For $i \in I$, let ψ_i be the formula defining X_i and $Y_i = \psi_i(G)$.

Claim. For $i \in I$, Y_i is indecomposable.

Suppose to the contrary that $K = \varphi(G, \bar{a})$ is a definable group and $|Y_i/K| = k$, where $1 < k < \omega$. The formula $\varphi(x, \bar{y})$ can be chosen so that, given $\bar{b}$ in G_0 satisfying $\exists x \varphi(x, \bar{y})$, $K_0 = \varphi(G_0, \bar{b})$ is a subgroup of G_0 and $|X_i/K_0| = k$. This contradicts the indecomposability of X_i to prove the claim.

By Zil'ber's Indecomposability Theorem the group H_0 generated by $\bigcup_{i \in I} Y_i$ is definable, connected and for some $i_1, \ldots, i_n \in I$, $H_0 = Y_{i_1} \cdot \ldots \cdot Y_{i_n}$. Then (using that H_0 is G_0–definable) $H_0 \cap G_0 = X_{i_1} \cdot \ldots \cdot X_{i_n} = H$ is a connected subgroup definable in G_0 and containing X_i for all $i \in I$. This proves the corollary.

By Lemma 3.5.6, whenever G is a connected group of finite Morley rank each element of the series of derived groups $G = G^{(0)} \supset G' \supset G'' \supset \ldots \supset G^{(n)} \supset \ldots$ is definable and connected (where $G^{(n+1)} = [G^{(n)}, G^{(n)}] = (G^{(n)})'$). Similarly, the series $G = G_0 \supset G_1 \supset G_2 \supset \ldots \supset G_n \supset \ldots$, where $G_{n+1} = [G, G_n]$, consists of connected definable groups. By Proposition 3.5.1 each of these descending chains of groups terminates.

For an arbitrary group G the statement "G is simple" involves quantification over subsets of G. While it is not possible to give a first-order set of axioms for the class of simple groups, when G is an ω–stable group of finite Morley rank being simple is an elementary property in the following sense.

Proposition 3.5.3. *Let G be a group of finite Morley rank.*

(i) If a model G_0 of $Th(G)$ is nonabelian and not simple then G_0 has a definable nontrivial normal subgroup.

(ii) If H is a model of $Th(G)$, then H is simple if and only if G is simple.

Proof. (i) Assume to the contrary that G_0 is nonabelian, not simple and has no definable nontrivial normal subgroup. Since $(G_0)^o$ is a definable normal subgroup G_0 must be connected. The center of G_0, call it Z, is a normal subgroup of G_0, definable in G_0. Since G_0 is nonabelian Z is a proper subgroup of G_0, hence $Z = \{1\}$. Let $x \neq 1$ be an element of a nontrivial normal subgroup N of G_0. Since $x \notin Z$, $H = [x, G_0]$ is not $\{1\}$, while it is definable in G_0 and connected by Lemma 3.5.9. Since H is contained in any normal subgroup containing x, $H \subset N$. The minimal normal subgroup H_0 containing H is the group generated by $\bigcup\{g^{-1}Hg : g \in G\}$. By the relative version of Zil'ber's Indecomposability Theorem (Corollary 3.5.10) H_0 is definable. Since $\{e\} \neq H_0 \subset N$, H_0 is a definable nontrivial normal subgroup of G, a contradiction which proves (i).

(ii) By (i) it suffices to show that when G contains a definable nontrivial normal subgroup, so does H. This is left as an exercise to the reader.

3.5.1 A Group Acting on a Strongly Minimal Set

This section is a study of an ω−stable group action (G, X) in which X is strongly minimal. This study will yield a result to be used later (Theorem 3.5.2) and show the strength of the hypotheses on (G, X). Many of the arguments are very group-theoretic and specific to this problem. However, the theorem is important enough to warrant a reasonably complete proof.

Notation. When G and H are groups and there is a an embedding $\theta : H \to \mathrm{Aut}(G)$, $G \rtimes H$ is the corresponding semidirect product of G and H. (The particular automorphism θ is generally understood from context.)

When K is a field $K^\times$ denotes the (multiplicative) group of units of K and K^+ denotes the additive group of K.

For G a group we write $G \rhd N$ when N is a normal subgroup of G.

Remark 3.5.8. In this section, when (G, X) is an ω−stable group action it is understood that multiplication on G and the action of G on X are $\emptyset$−definable operations.

Here are three examples of an ω−stable group acting on a strongly minimal set of increasing complexity (at least of increasing rank).

Example 3.5.2. Let G be a strongly minimal group. By Corollary 3.5.5 G is abelian, so additive notation is used for G. The most natural action on the strongly minimal set G is translation by G itself: $x \mapsto x + a$, $a \in G$. This is a regular action.

Now suppose G has an element of order > 2. Here is another action on G. The map $x \mapsto -x$ produces an embedding of $\mathbb{Z}_2$ into $\mathrm{Aut}(G)$. Furthermore $(a, 0)$ and $(a, 1)$, as elements of $G \rtimes \mathbb{Z}_2$, define bijections on G, $x \mapsto x + a$ and $x \mapsto -x + a$, respectively. This gives a faithful transitive group action of $G \rtimes \mathbb{Z}_2$ on G. Note that $G \rtimes \mathbb{Z}_2$ is nonabelian, has Morley rank 1 and the action is not regular.

Example 3.5.3. Let K be an algebraically closed field. Let G_{aff} be the group of affine transformations on K, $x \mapsto a + bx$, where $a \in K$ and $b \in K^\times$. Then G_{aff} acts sharply 2−transitively on K and $G \cong K \rtimes K^\times$. When K is the universal domain of its theory (G_{aff}, K) comprises a group of Morley rank 2 acting on a strongly minimal set.

Example 3.5.4. Let K be an algebraically closed field and $\mathbb{P}^1$ the projective line over K, thinking of $\mathbb{P}^1$ as the set of 2×1 column vectors over K, factored by the equivalence relation: $x \sim y \iff K^\times x = K^\times y$. The group $\mathrm{GL}_2(K)$ of invertible 2×2 matrices over K defines an action $*$ on $\mathbb{P}^1$. Note: for any $g, h \in \mathrm{GL}_2(K)$, $g * x = h * x$ for all $x \in \mathbb{P}^1$ if and only if $g = \lambda h$ for some $\lambda \in K^\times$. Let $\mathrm{PGL}_2(K)$ be the quotient of $\mathrm{GL}_2(K)$ by $\{\lambda \cdot 1 : \lambda \in K^\times\}$, where 1 is the identity of $\mathrm{GL}_2(K)$. The action of $\mathrm{PGL}_2(K)$ on $\mathbb{P}^1$ is faithful.

Note that this action is sharply 3–transitive and (when K is the universal domain) $\mathrm{PGL}_2(K)$ is a group of Morley rank 3 acting faithfully on a strongly minimal set.

What is quite surprising is that these are essentially the only examples of an ω–stable group action on a strongly minimal set.

Theorem 3.5.2. *Let (G, X) be an ω–stable transitive faithful group action with X strongly minimal. Then $MR(G) \leq 3$ and*

(1) *If $MR(G) = 1$, G^o acts regularly on X.*
(2) *If $MR(G) = 2$, there is a field K definable on X and the action of G on X is definably isomorphic to the affine action of $K \rtimes K^\times$ on K.*
(3) *If $MR(G) = 3$, there is a field K definable on $X \setminus \{a\}$ (for some $a \in X$) and the action of G on X is isomorphic to the action of $\mathrm{PGL}_2(K)$ on $\mathbb{P}^1$, the projective line over K.*

The bulk of this subsection is devoted to the proof of this theorem. The proof will require three additional propositions which will not be proved here. The first two represent a nontrivial amount of work.

Proposition 3.5.4. *A connected group of Morley rank 2 is solvable.*

Proof. This is Theorem 6 of [Che79].

Definition 3.5.9. *Given a group H acting on a set A, $B \subset A$ is H–*invariant *if $HB = B$; i.e., for all $h \in H$, $b \in B$, $hb \in B$.*

Proposition 3.5.5. *Let $\mathfrak{C}$ be the universal domain of a theory of finite Morley rank; G and A infinite definable abelian groups. Suppose there is a definable faithful action of G on A which induces an embedding of G into the automorphism group of A. Further suppose that A contains no infinite definable proper G–invariant subgroup. Then there is a definable field F such that the additive group of F is definably isomorphic to A and there is a definable embedding of G into the multiplicative group of F so that the action of G on A corresponds to multiplication in F.*

Remark 3.5.9. Here is a more precise statement of the conclusion of the proposition. The definable action $*$ of G on A is assumed to induce an embedding θ of G into $\mathrm{Aut}(A)$ by: $\theta(g)$ is the automorphism α such that $\alpha(a) = g*a$, for all $g \in G$, $a \in A$. (Here $\mathrm{Aut}(A)$ is simply the automorphism group of A as an abelian group.) Use θ to define $A \rtimes G$. Then the proposition yields a definable field F and a definable embedding from $A \rtimes G$ into $F^+ \rtimes F^\times$ which restricts to an isomorphism of A onto F^+.

Proposition 3.5.6. *Let T be an ω–stable theory, F an infinite definable field of finite Morley rank and S a definable group of field automorphisms of F. Then $S = \{1\}$.*

Finally we also make use of the following fact whose proof is omitted in favor of bigger fish.

Lemma 3.5.7. *Let G be an ω-stable group of Morley rank $k+1<\omega$. Suppose H is a definable subgroup with $MR(G/H)=1$. Then $MR(H)=k$.*

The restriction to transitive actions in the theorem does not eliminate many interesting cases:

Lemma 3.5.8. *Let (G,X) be an ω-stable faithful group action with X strongly minimal and G infinite.*
(i) If Y is a finite orbit, $|Y|\leq \deg(G)$.
(ii) There is an $\emptyset$-definable finite set $Y\subset acl(\emptyset)\cap X$ such that G acts transitively on $X\setminus Y$.

Proof. (i) Given $y\in Y$, $|Y|=[G:stab(y)]$. Thus, $stab(y)$ has finite index in G and $|Y|=[G:stab(y)]\leq[G:G^o]=\deg(G)$.

(ii) If each orbit is finite, $G^o\subset stab(x)$ for all $x\in X$, contradicting that the action is faithful and G is infinite. Thus, X contains an infinite orbit. Since an orbit is a definable set and X is strongly minimal, X contains only one infinite orbit and this orbit is $X\setminus Y$ for some finite Y. Since Y is the union of all finite orbits, Y is $\emptyset$-definable.

Lemma 3.5.9. *Let (G,X) be a transitive faithful group action with G abelian. Then the action is regular.*
If, in addition, (G,X) is ω-stable, $MR(G)=MR(X)$.

Proof. Let $a\in X$ and suppose $g\in G$ is such that $ga=a$. Let b be any element of X. By the transitivity of the action there is an $h\in G$ such that $ha=b$. Then $gb=gha=hga=ha=b$. Since the action is faithful, $g=1$, proving that the action is regular.

Suppose (G,X) is ω-stable, $a\in X$ and $g\in G$ is generic over a. By the regularity of the action $g\in dcl(a,ga)$. Thus, $MR(G)=MR(g)=MR(g/a)\leq MR(ga/a)\leq MR(X)$. Now suppose $b,c\in X$ are independent with $MR(b)=MR(c)=MR(X)$. By the transitivity of the action there is a $h\in G$ such that $hb=c$. Then $MR(X)=MR(c/b)\leq MR(h/b)\leq MR(G)$, completing the proof.

Notation. From here until the end of the subsection we assume (G,X) to be an ω-stable transitive faithful group action with X strongly minimal and G infinite, unless stated otherwise. These hypotheses may be repeated in key results to make later reference easier.

Lemma 3.5.10. *If $N\lhd G$ is infinite and definable then N acts faithfully and transitively on X.*

Proof. It is clear that the action of N on X is faithful. If N^o acts transitively on X then certainly N acts transitively on X, so we may as well assume N is connected. Suppose the orbit of x under N is finite. Then $Nx = \{x\}$ by Lemma 3.5.8(i). An arbitrary $y \in X$ is gx for some $g \in G$ (by the transitivity of the action of G) hence, since N is normal, $Ny = Ngx = gNx = \{gx\} = \{y\}$. In other words the elements of N fix every element of X, contradicting that the action is faithful and $N \neq \{1\}$. Thus, every orbit under N is infinite. Since X is strongly minimal and N is definable X can only contain one infinite orbit under N. That is, N acts transitively on X, as required.

Proof of Theorem 3.5.2(i). Suppose $MR(G) = 1$. Then G^o is a strongly minimal group, hence abelian by Corollary 3.5.5. By Lemma 3.5.10 the action of G^o on X is transitive, so the action is regular by Lemma 3.5.9, proving (i) of the theorem.

Proof of Theorem 3.5.2(ii). We assume now that $MR(G) = 2$. By Proposition 3.5.4 G^o is solvable. Since $G^o \lhd G$, G^o acts transitively on G (by Lemma 3.5.10). If G^o were abelian Lemma 3.5.9 would contradict that $MR(G^o) = 2$ and $MR(X) = 1$. Hence G^o is nonabelian. Let $A = (G^o)' = [G^o, G^o]$, a definable connected subgroup of G^o by Lemma 3.5.6. Since G^o is nonabelian $A \neq \{1\}$, hence A is infinite (see Remark 3.5.6). The solvability of G^o forces A to be a proper subgroup, so A must be strongly minimal. By Theorem 3.5.2(i) A acts regularly on X.

Fix $x \in X$ and let $G_x = \{g \in G : gx = x\}$, the stabilizer of x in G. For $g, h \in G$, $gG_x = hG_x$ if and only if $gx = hx$. Thus, the map $g \mapsto gx$ defines a bijection between G/G_x and X. Since $MR(G) = 2$ the only possibility for $MR(G_x)$ is 1.

Claim. Conjugation defines an embedding θ of G_x into $\mathrm{Aut}(A)$. Moreover, the action of G_x^o on A by conjugation is faithful and regular on $A \setminus \{1\}$.

Since A acts regularly on X, for any $g \in G$ there is a unique $a \in A$ such that $gx = ax$, hence $g \in aG_x$. Since A is a normal subgroup of G, conjugation defines a homomorphism θ of G_x into $\mathrm{Aut}(A)$. Suppose towards a contradiction that θ is not an embedding; i.e., there is a $g \neq 1$ in G_x which commutes with each element of A. Then given $y \in X$ there is an $a \in A$ with $y = ax$, hence $gy = gax = agx = ax = y$. In other words, g fixes each element of X, contradicting that G acts faithfully on X. Thus θ is an embedding of G_x into $\mathrm{Aut}(A)$.

Conjugation also defines a group action of G_x on A. The above argument shows this action to be faithful on $A \setminus \{1\}$. By part (i) of the theorem conjugation defines a faithful regular action of G_x^o on $A \setminus \{1\}$, proving the claim.

The embedding θ in the claim can be used to define $A \rtimes G_x$. There is a map ψ from G into $A \rtimes G_x$ defined by $\psi(g) = (a, h)$ if and only if $g = ah$.

(The first sentence in the proof of the claim shows that $G = A \cdot G_x$.) A routine verification shows ψ to be an isomorphism.

Since A has no infinite definable proper subgroups it is G_x^o–invariant. Since $MR(G_x) = 1$, G_x^o is abelian. Thus, Proposition 3.5.5 yields a definable field F and a definable embedding σ of $A \rtimes G_x^o$ into $F^+ \rtimes F^\times$ which restricts to an isomorphism of A onto F^+. This isomorphism guarantees that F has Morley rank 1. By Corollary 3.5.4, F is strongly minimal, hence the embedding of G_x^o into $F^\times$ is surjective.

The field structure can be transferred onto A as follows. Let $1'$ be any element of $A \setminus \{1\}$. For each $a \neq 1$ in A let a'' be the unique element of G_x^o taking $1'$ to a (which exists by the claim). Define a binary relation $\otimes$ on A by: $1 \otimes a = a \otimes 1 = 1$ for all $a \in A$ and $a \otimes b = (a'' \cdot b'')1'$, for $a, b \in A \setminus \{1\}$. The reader can show that $(A, \cdot, \otimes, 1, 1')$ is a field isomorphic to F (via a definable bijection). Since the action of A on X is regular there are also definable operations on X under which X is a field (definably) isomorphic to F. The action of A on X corresponds to the translations $x \mapsto x + a$, $a \in A$, while the action of G_x^o on X corresponds to the dilations $x \mapsto bx$, $b \in G_x^o$. (See Example 3.5.3.) Thus $A \rtimes G_x^o$ acts on X like the affine group of the field F. To finish the Morley rank 2 case we need only show

Claim. $G_x = G_x^o$.

Since G_x^o acts faithfully on A it acts regularly by Lemma 3.5.9. Fix $a \in A$. For any $g \in G_x$ there is an $h \in G_x^o$ such that $gag^{-1} = hah^{-1}$. Since $h^{-1}g \in S = \{ f \in G_x : faf^{-1} = a \}$ the claim will be proved once we show $S = \{1\}$.

Define $\oplus$ on G_x^o by

$$g_0 \oplus g_1 = g_2 \text{ if and only if } (g_0 a g_0^{-1})(g_1 a g_1^{-1}) = g_2 a g_2^{-1}).$$

Just as $(A, \cdot, \otimes)$ was a field isomorphic to F, $(G_x^o, \oplus, \cdot)$ is a field isomorphic to F.

The group S acts on G_x^o by conjugation. We claim this action defines an embedding of S into the group Γ of field automorphisms of $(G_x^o, \oplus, \cdot)$. Conjugation by $s \in S$ is clearly a bijection σ of G_x^o which is an automorphism of $\cdot$. Also, σ is an automorphism of $\oplus$ since

$$\begin{aligned}(sg_0s^{-1}asg_0^{-1}s^{-1})(sg_1s^{-1}asg_1^{-1}s^{-1}) &= (sg_0ag_0^{-1}s^{-1})(sg_1ag_1^{-1}s^{-1}) \\ &= sg_0ag_0^{-1}g_1ag_1^{-1}s^{-1}.\end{aligned}$$

Thus S is a definable group of automorphisms of the field $(G_x^o, \oplus, \cdot)$. By Proposition 3.5.6 $S = \{1\}$. This completes the proof of the claim and Theorem 3.5.2(ii).

For the cases when $MR(G) > 2$ we need the following two lemmas.

Lemma 3.5.11. *Suppose $MR(G) \geq 2$ and $x \in X$. Then G_x acts faithfully and transitively on a cofinite subset of X. Moreover G_x^o acts transitively on a cofinite subset of X.*

Proof. This follows immediately from Lemma 3.5.8 once we show that G_x is infinite. Suppose to the contrary that G_x is finite. Then for any $y \in X$ there are finitely many $g \in G$ such that $gx = y$. Thus, picking g to be a generic of G independent from x and $y = gx$, $MR(g) \leq MR(y/x) \leq 1$. This contradicts that $MR(G) \geq 2$ to prove the lemma.

Lemma 3.5.12. *Let $MR(G) = k + 1 \geq 3$. Suppose that any ω–stable transitive faithful action (H, Z) with Z strongly minimal and $MR(H) = k$ acts sharply k–transitively. Given $x \in X$ let Y be the unique infinite orbit under G_x^o. Then $X \setminus Y = \{x\}$ and G acts sharply $k + 1$–transitively on X.*

Proof. First observe

Claim. $MR(G_x) = k$.

Let $X \setminus Y = Z$ and $n = |Z|$. Define an equivalence relation $\sim$ on X by $a \sim b$ if and only if $G_a^o = G_b^o$. Any finite orbit of G_x^o contains a single element (by Lemma 3.5.8) so $G_z \supset G_x^o$ for any $z \in Z$. By the transitivity of the action of G, $MR(G_y) = MR(G_{y'})$ for all $y, y' \in X$. Thus, $G_z^o = G_x^o$ for all $z \in Z$; i.e., the $\sim$ –class of x is Z. For any $g \in G$ and $a \in X$, $G_{ga} = gG_a g^{-1}$, so $G_{ga}^o = gG_a^o g^{-1}$. Thus, $\sim$ is preserved by the action of G; i.e., $a \sim b$ if and only if $ga \sim gb$. Since G acts transitively on X every $\sim$ class contains n elements.

Claim. If $a \neq b \in Y$ then $a \not\sim b$.

Let H denote G_x^o and fix $a \neq b \in Y$. Since there is a definable one-to-one correspondence between X and the cosets of G_x in G, $MR(G/G_x) = 1$. By Lemma 3.5.7, $MR(G_x) = k$, hence H acts sharply k–transitively on Y. Thus, H_a acts sharply $(k-1)$–transitively on $Y \setminus \{a\}$. Let $\bar{c}$ be a sequence of $k-1$ distinct elements of $Y \setminus \{a\}$. By the sharpness of the action there is a $\bar{c}$–definable bijection between H_a and $(Y \setminus \{a\})^{k-1}$. Since $\deg(Y \setminus \{a\})^{k-1} = 1$ (by Exercise 3.5.10) H_a also has degree 1. Hence H_a is connected. Since $H_a \subset G_a$ and H_a is connected Exercise 3.5.7 forces H_a to be a subgroup of G_a^o. If $a \sim b$, $H_a \subset G_b$, hence any element of H_a fixes b. The action of H_a on $Y \setminus \{a\}$ is $(k-1)$–transitive, so this is impossible. This proves the claim.

Thus, each $\sim$ –class contains a single element and $X \setminus Y = \{x\}$. In particular G_x acts transitively and faithfully on $X \setminus \{x\}$. Since $MR(G_x) = k$, G_x acts sharply k–transitively on $X \setminus \{x\}$. Since this is true for each $x \in X$, G acts sharply $(k+1)$–transitively on X. This proves the lemma.

Corollary 3.5.11. *If G has finite Morley rank $k \geq 2$ then the action of G on X is sharply k–transitive.*

Proof. It follows from Theorem 3.5.2(ii) that a group of Morley rank 2 acting faithfully and transitively on a strongly minimal set acts sharply 2–transitively. From here the corollary follows by induction on k.

Proof of Theorem 3.5.2(iii). In this case $MR(G) = 3$. By the previous corollary G acts sharply 3–transitively on X. Fix a point in X and call it ∞. Then G_∞ acts sharply 2–transitively on $Y = X \setminus \{\infty\}$. By (ii) of the theorem there is a field K defined on Y. Moreover, if $0 \in Y$ is the zero of K, $G_{\infty,0}$ (= the set of elements of G fixing both ∞ and 0) is isomorphic (via a definable map) to the multiplicative group of K. Among other things, $G_{\infty,0}$ is strongly minimal.

Since the action of G on X is sharply 3–transitive there is a unique $\alpha \in G$ such that α maps the triple $(0, 1, \infty)$ to $(\infty, 1, 0)$. Since α^2 fixes $\{0, 1, \infty\}$, α^2 must be 1.

Claim. Conjugation by α defines an automorphism σ of $G_{\infty,0}$, $\sigma \neq 1$, $\sigma^2 = 1$ and $\sigma g = g^{-1}$ for all $g \in G_{\infty,0}$.

Given $g \in G_{\infty,0}$, $\alpha g \alpha^{-1}$ is also in $G_{\infty,0}$, hence conjugation by α defines an automorphism σ of $G_{\infty,0}$. Let $a \in X \setminus \{0, 1, \infty\}$ and $g \in G_{\infty,0}$ such that $g1 = a$. If $\sigma = 1$, then $\alpha g \alpha^{-1} = g$, hence $\alpha a = \alpha g 1 = g1 = a$, a contradiction since G acts faithfully and $\alpha \neq 1$. Since $\alpha^2 = 1$, $\sigma^2 = 1$. It remains to show that σ is inversion. Let $B = \{a \in G_{\infty,0} : \sigma a = a\}$ and $C = \{a \in G_{\infty,0} : \sigma a = a^{-1}\}$. Since $G_{\infty,0}$ is strongly minimal and B is a proper definable subgroup, B is finite. Consider the map $\tau : G_{\infty,0} \longrightarrow G_{\infty,0}$ defined by $\tau(x) = \sigma(x)x^{-1}$. Then, for any $x \in G_{\infty,0}$,

$$\sigma(\tau(x)) = \sigma^2(x)\sigma(x^{-1}) = x\sigma(x)^{-1} = \tau(x)^{-1},$$

so τ maps $G_{\infty,0}$ to C. If $\tau(x) = \tau(y)$, then $\sigma(xy^{-1}) = xy^{-1}$ and $x \in yB$. Since B is finite, x is algebraic over y. This shows that the kernel of τ is finite, hence C contains a generic. Thus, C is all of $G_{\infty,0}$, completing the proof of the claim.

In other words, σ is inversion on $G_{\infty,0}$. Given $a \in K^\times$, let $h \in G_{\infty,0}$ be such that $h1 = a$. Then $\alpha a = \alpha h 1 = h^{-1}\alpha 1 = h^{-1}1 = a^{-1}$. Thus, α acts like inversion on $K^\times$. It follows that G contains the group of automorphisms of $\mathbb{P}^1$ generated by all affine maps $x \mapsto cx+d$ and $x \mapsto x^{-1}$. Thus $\mathrm{PGL}_2(K)$ embeds into G. Since $\mathrm{PGL}_2(K)$ itself acts 3–transitively on X and the action of G on X is sharply 3–transitive, this embedding of $\mathrm{PGL}_2(K)$ into G is surjective. That is, the action of G on X is isomorphic to the action of $\mathrm{PGL}_2(K)$ on $\mathbb{P}^1$.

This proves Theorem 3.5.2(iii).

To complete the proof of Theorem 3.5.2 it remains to show that $MR(G) \leq 3$.

Claim. $MR(G) \neq 4$.

Suppose to the contrary that $MR(G) = 4$. By Lemma 3.5.12 G acts sharply 4–transitively on X. Fix two points ∞_1 and ∞_2 in X. Then G_{∞_1,∞_2} acts sharply 2–transitively on $X \setminus \{\infty_1, \infty_2\}$ so there is a field structure K

definable on $X \setminus \{\infty_1, \infty_2\}$. Moreover, the action of the multiplicative group of K (on itself) is isomorphic to the action of $G_{\infty_1,\infty_2,0} = H$.

There are σ_1, $\sigma_2 \in G$ such that σ_1 maps $(0, 1, \infty_1, \infty_2)$ to $(\infty_1, 1, 0, \infty_2)$ and σ_2 maps $(0, 1, \infty_1, \infty_2)$ to $(\infty_2, 1, \infty_1, 0)$. As in the proof of Theorem 3.5.2(iii), $\sigma_i H \sigma_i^{-1} = H$ and $\sigma_i \notin C(H)$ (= the centralizer of H) for $i = 1, 2$. Repeating the proof of the previous claim shows that $\sigma_i h \sigma_i^{-1} = h^{-1}$, for all $h \in H$. Let $\omega = \sigma_1\sigma_2$. For any $h \in H$, $\omega h = \sigma_1 h^{-1} \sigma_2 = h\sigma_1\sigma_2 = h\omega$. If $x \in K^\times \setminus \{1\}$ and $h \in H$ is such that $h1 = x$ then

$$\omega x = \omega h 1 = h\omega 1 = h1 = x.$$

Since $\omega 1 = 1$, ω is the identity on $K^\times$. It follows that $\omega = 1$, contradicting the fact that $\omega 0 = \infty_2$. This proves the claim.

Let k be the minimal natural number > 3 such that there is an ω–stable group G of Morley rank k and a faithful, transitive, ω–stable action of G on X. Given $x \in X$, G_x acts faithfully and transitively on a strongly minimal subset of X and $MR(G_x) = k - 1$. Thus, k must be 4, contradicting the claim. This proves that $MR(G) \leq 3$ when $MR(G)$ is finite.

To finish the proof we must show that any ω–stable group G acting faithfully and transitively on a strongly minimal set must have finite Morley rank. We know

$$\{1\} = \{g \in G : \forall x \in X,\ gx = x\} = \bigcap_{x \in X} \{g \in G : gx = x\}.$$

By the descending chain condition on subgroups there are $x_1, \ldots, x_n \in X$ such that $\{1\} = \{g \in G : gx_1 = x_1, \ldots, gx_n = x_n\}$. It follows that the map $g \mapsto (gx_1, \ldots, gx_x)$ is a one-to-one map of G into X^n. Thus $MR(G) \leq MR(X^n) = n$.

This completes the proof of Theorem 3.5.2.

3.5.2 $\bigwedge$–definable Groups and Actions

Occasionally (most notably in Section 4.5) a theorem giving the existence of a group will not immediately yield a definable group, but a group on the set of realizations of a type in the universe. We show in Theorem 3.5.3 that any such group in a t.t. theory is actually definable. The relevant definitions are as follows.

Definition 3.5.10. *Let $\mathfrak{C}$ be the universal domain of a theory. A subset X of $\mathfrak{C}$ is called* infinity-definable over A *(abbreviated $\bigwedge$–definable over A) if for some type p over A, $X = p(\mathfrak{C})$. X is $\bigwedge$–definable if it is $\bigwedge$–definable over some set A.*

Every definable subset of $\mathfrak{C}$ is $\bigwedge$–definable.

Given $\mathfrak{C}$ the universe of a t.t. theory and D an $\bigwedge$–definable set, specifically, $D = p(\mathfrak{C})$;

– $MR(D)$ and $\deg(D)$, the Morley rank and degree of D, are defined to be $MR(p)$ and $\deg(p)$, respectively.

Definition 3.5.11. *Let T be a complete theory.*

(i) We call $(G, \cdot)$ an $\bigwedge$−definable group over A in $\mathfrak{C}$ *if*

– *$(G, \cdot)$ is a group,*
– *G is a subset of $\mathfrak{C}$, $\bigwedge$−definable over A and*
– *there is a function f, definable over A in $\mathfrak{C}$, such that $f \restriction G \times G$ defines the binary operation $\cdot$ on G under which G is a group.*

(ii) Similarly, a group action $(G, \cdot, X, \star)$ is an $\bigwedge$−definable group action over A in $\mathfrak{C}$ *if $(G, \cdot)$ is an $\bigwedge$−definable group over A in $\mathfrak{C}$, X is a subset of $\mathfrak{C}$, $\bigwedge$−definable over A, and $\star$ is the restriction to $G \times X$ of an A−definable function.*

(iii) An $\bigwedge$−definable ω−stable group (action) is an $\bigwedge$−definable group (action) in an ω−stable theory.

Theorem 3.5.3. *An $\bigwedge$−definable ω−stable group is definable.*

Remark 3.5.10. Let G be an ω−stable $\bigwedge$−definable group over $\emptyset$. Since G may not be definable we cannot relativize the universe to G and retain a full description of the definable relations on G. That is, Proposition 3.3.3 may fail when relativizing to a set which isn't definable. So, we have to keep sight of the ambient theory.

We say $X \subset G^n$ is a *locally definable relation on* G if $X = Y \cap G^n$ for Y some definable n−ary relation.

Remark 3.5.11. The reader is asked to verify the following basic facts. Let G be an $\bigwedge$−definable ω−stable group over A. An element $a \in G$ is *generic over* $B \supset A$ if $MR(a/B) = MR(G)$. A type $p \in S(\mathfrak{C})$ is *generic* if p extends the type defining G and $MR(p) = MR(G)$.

(i) If $a, b \in G$ and a is generic over $B \cup \{b\}$, then $a \cdot b$ and $b \cdot a$ are generic over $B \cup \{b\}$.

(ii) For any formula $\psi(x, y)$ there is a formula $\theta(x$ such that for all a, $\models \theta(a)$ if and only if $\psi(a, y)$ is in every generic type.

The theorem will follow quickly from previously proved results and

Lemma 3.5.13. *Let G be an $\bigwedge$−definable ω−stable group over A. Then there is a definable group H such that G is a subgroup of H.*

Proof. Let $\cdot$ be a definable function whose restriction to $G \times G$ is the group operation on G. Let Φ be a set of definable sets such that $G = \bigcap \Phi$. Pick (by compactness) $X_0 \in \Phi$ such that for all $x, y, z \in X_0$, $x \cdot (y \cdot z) = (x \cdot y) \cdot z$ and $x \cdot 1 = x$. For any $Y \in \Phi$ let

$$X_Y = \{x \in X_0 : \text{ for all } \ y \in G \text{ generic over } \ x, \ x \cdot y \in Y\}.$$

Then

$$G = \bigcap \{ X_Y : Y \in \Phi \}.$$

(Simply because G is a group, $G \subset X_Y$, for all $Y \in \Phi$. If $x \in X_Y$ for each $Y \in \Phi$, then for any $y \in G$ generic over x, $x \cdot y \in G$ (because $G = \bigcap \Phi$). Since $x, y \in X_0$, $x = x \cdot 1 = x \cdot (y \cdot y^{-1}) = (x \cdot y) \cdot y^{-1} \in G$.) We can furthermore assume that for any finite $\Phi_0 \subset \Phi$ there is a $Y \in \Phi$ such that $Y \subset \bigcap \Phi_0$.

Claim. There is a $Z \in \Phi$ such that $X_0 \supset Z$ and for all $x, y \in X_Z$, $x \cdot y \in X_0$.

If $x \in X_{X_0}$ and $y \in G$, then $x \cdot y \in X_{X_0}$. (Let $z \in G$ be generic over $\{x, y\}$. Then $y \cdot z$ is generic over $\{x, y\}$ and $(x \cdot y) \cdot z = x \cdot (y \cdot z) \in X_0$.) Let $Z_1 \in \Phi$ be such that for all $x \in X_{X_0}$ and $y \in Z_1$, $x \cdot y \in X_0$. A set $Z \in \Phi$ such that $X_Z \subset X_{X_0} \cap Z_1$ satisfies the requirements of the claim.

Let $X_1 = X_Z$.

Claim. If $x \in X_1$ and $y \in G$ then $x \cdot y \in X_1$.

Simply because $x, y \in X_Z$, $x \cdot y \in X_0$. Choose a $z \in G$ generic over $\{x, y\}$. Then $y \cdot z \in G$ is generic over x. Since $x \in X_Y$, $x \cdot (y \cdot z) \in Z$. Thus, $(x \cdot y) \cdot z \in Z$, proving that $x \cdot y \in X_Z$ as required.

Let $X_2 = \{ y \in X_1 : \forall x (x \in X_1 \Rightarrow x \cdot y \in X_1) \}$. Then X_2 is a definable set closed under $\cdot$, and $\cdot$ is associative on X_2. We proved in the claim that $G \subset X_2$. Thus, the invertible elements of X_2 form a definable group containing G.

Proof of Theorem 3.5.3. Let G be an ω–stable $\bigwedge$–definable group over A. Let Φ be a collection of definable sets such that $G = \bigcap \Phi$. By Lemma 3.5.13 there is a definable group $H \supset G$. In fact, from the proof of the lemma we see that for any $X \in \Phi$ there is a definable group H_X, $X \supset H_X \supset G$. Hence, $G = \bigcap_{X \in \Phi} H_X$. The descending chain condition on definable groups in an ω–stable theory (Proposition 3.5.1) yields a finite $\Psi \subset \Phi$ such that $G = \bigcap_{X \in \Psi} H_X$. This proves the theorem.

Historical Notes. Proposition 3.5.1 is due to Macintyre [Mac71b]. Groups of finite Morley rank were studied by Zil'ber [Zil77b] (translated in [Zil91]) and [Zil77a] and independently by Cherlin [Che79]. The notion of a generic type came out of these papers, [CS80] and Poizat's [Poi83a]. Zil'ber's Indecomposability Theorem is found in [Zil77b]. Theorem 3.5.3 is due to Hrushovski [Hru90b, Theorem 2], although Poizat had earlier proved that an $\bigwedge$–definable group which is contained in a definable group is the intersection of its definable supergroups [Poi81].

Exercise 3.5.1. Let $\varphi(x)$ be a formula in an ω–stable group G and let a be an element. Show: $\varphi(x)$ and $\varphi(a^{-1} \cdot x)$ have the same Morley rank and degree.

Exercise 3.5.2. Verify that the connected component of an ω−stable group G is definable without parameters.

Exercise 3.5.3. Prove Remark 3.5.5.

Exercise 3.5.4. Prove Corollary 3.5.7. HINT: Use Corollary 3.5.1.

Exercise 3.5.5. Prove Corollary 3.5.8.

Exercise 3.5.6. Prove Corollary 3.5.9.

Exercise 3.5.7. Let G be an ω−stable group and H a definable subgroup of G. Assuming that H is connected show that $H \subset G^o$.

Exercise 3.5.8. Let T be an ω−stable theory and G, H infinite groups definable in the universal domain of T. Let $K = G \times H$. Then c is a generic of K if and only if $c = (a, b)$, where $a \in G$ and $b \in H$ are generics and a is independent from b.

Exercise 3.5.9. Give a proof of Proposition 3.5.3(ii).

Exercise 3.5.10. Let T be ω−stable and (H, Y) a definable group action, where Y is infinite. Suppose H acts sharply k−transitively on Y and H is connected. Then $Y, Y^2, \ldots, Y^k$ all have degree 1.

4. Fine Structure of Uncountably Categorical Theories

In the preceding chapter, for T a countable uncountably categorical theory, we solved problems concerning the number of models of T in a fixed cardinality. However, this study leaves many unanswered questions about uncountably categorical theories, and raises others. Here are a few such questions.

- In [Vau61] Vaught asked if an uncountably categorical theory can be finitely axiomatizable. (It was through Zil'ber's work on this problem that geometrical stability theory, the area in which the subject matter of this chapter belongs, was born.)
- Can we isolate a broad class of uncountably categorical theories which have a strongly minimal formula (or at least a formula of Morley rank 1) over $\emptyset$? (While working on the Baldwin-Lachlan Theorem we recognized that an easier proof would be possible in such theories.)
- Are there strongly minimal sets which are radically different from the examples given in Example 3.1.1?

What is surprising is that work on each of these questions has given insight into the others. The issues underlying this connection are the following imprecisely worded problems concerning the definable relations in models of uncountably categorical theories. Recall that algebraic closure restricted to the subsets of strongly minimal set defines a pregeometry.

(1) Find a natural and meaningful dividing line between "simple" pregeometries and "complex" pregeometries among those which occur as the pregeometry on a strongly minimal set.
(2) Prove that whenever the pregeometry on a strongly minimal set is simple, the Morley rank dependence relation on tuples is also simple in a meaningful way.

In order to formulate the properties which will meet these requirements we need the notion of M^{eq} (M a model), which is developed in the next section. In the expansion M^{eq} we have not only the elements of M but elements which act as names for the definable relations in M. This expansion is used in most of model theory today.

4.1 T^{eq}

Most of the theorems we have proved so far make few distinctions between tuples from a model and elements of the model, the standard hypothesis being: "Let $\bar{a}$ be a tuple from M and ...". It is a slight deficiency in our notion of a model that we cannot use more uniform terminology for elements of M and tuples from M. Another annoyance is the nonuniqueness of the parameters involved in defining sets. As we look more deeply at the relationships between the definable subsets of a model a natural question (in a t.t. theory) might be: Given a definable set $D = \varphi(\mathfrak{C}, \bar{a})$, what is the Morley rank of the type of the parameters used to define D? This is an ambiguous question since there may be $\bar{a}$ and $\bar{b}$ with $D = \varphi(M, \bar{a}) = \varphi(M, \bar{b})$ and $MR(\bar{a}) \neq MR(\bar{b})$. Both of these deficiencies are removed by expanding the model M to M^{eq}. Shelah calls the additional elements "imaginary elements", in analogy to numbers which we add to the reals to form the complex numbers. As with the complex numbers, the most efficient proof of a theorem about the real elements of a theory may involve imaginaries. This expansion is formulated using many-sorted logic.

It is common in mathematics for the universe of a structure to consist of several disjoint classes of elements. A simple example is a projective plane $\mathcal{P}$ which consists of a set P of "points", a set L of "lines" and a binary incidence relation ε between points and lines. In expressing properties of these planes variables are restricted to ranging over either points or lines. Adopt the convention that p, p', ... denote arbitrary points and l, l', ... denote lines. Then, one of the axioms for a projective plane can be stated as: $\forall p \forall p' \exists l (p \varepsilon l \wedge p' \varepsilon l)$. To formulate this plane as a model of a first-order language we would add unary predicates P_0 and L_0 and let the universe be the disjoint union of the interpretations of these two. In any useful formula involving the variable v we would have an occurrence of $P_0(v)$ or $L_0(v)$. The following approach offers a more natural formalization.

Let I be a nonempty set whose elements are called *sorts*. The logical symbols of *I–sorted logic* are the same as first-order logic, except that for each sort i there are variables v_1^i, v_2^i, ... of sort i (and each variable is tagged with a sort). An *I–sorted language* L consists of predicate, constant and function symbols. For each n–ary predicate symbol P there is an n–tuple of sorts $(i_1, \ldots, i_n)$ and P is said to be a predicate of sort $(i_1, \ldots, i_n)$. Similarly, a constant symbol is of a particular sort and the arguments of a function symbol have specified sorts. We leave it to the reader to define the terms and formulas of L. (For example, if P is a predicate symbol of sort $(i_1, \ldots, i_n)$ and $x_1, \ldots, x_n$ are variables of sorts $i_1, \ldots, i_n$, respectively, then $Px_1 \ldots x_n$ is an atomic formula.)

An I–sorted structure $\mathcal{M}$ consists of the following.

1. For each $i \in I$ there is a nonempty set M_i called the universe of sort I.
2. For each predicate symbol P of sort $(i_1, \ldots, i_n)$ there is a relation $P^{\mathcal{M}} \subset M_{i_1} \times \ldots \times M_{i_n}$.

3. For a constant symbol c of sort i there is an element $c^{\mathcal{M}}$ of M_i.
4. For each function symbol f of sort ... (the obvious clause).

The definitions of truth and satisfaction are the predictable ones, given that $\forall v^i$ means "for all elements of M_i." The submodel and elementary submodel relations are defined much like the 1–sorted versions, as are elementary maps and isomorphisms. (Ordinary first-order logic as described in Chapter 1 is called 1–sorted logic.)

Let T be a complete I–sorted theory. Given $\sigma = (i_1, \ldots, i_n)$ a sequence of sorts, $S_\sigma(\emptyset)$ denotes the set of complete types in a sequence of variables of sorts $i_1, \ldots, i_n$. In situations where we used $S_n(\emptyset)$ in a 1-sorted theory we will use $S_\sigma(\emptyset)$ in an I–sorted theory. $S(\emptyset)$ denotes $\bigcup_\sigma S_\sigma(\emptyset)$.

So far, we have only stated definitions and theorems for 1-sorted logic. However, everything we have done extends trivially to many-sorted logics. For example, the term categorical in λ is defined by exactly the same statement. We chose to work in 1-sorted logic only to simplify the notation. We will, however, freely apply past results to many-sorted theories and models.

It is possible to transform a many-sorted structure into an ordinary one-sorted structure much as we did above for projective planes. The reader is referred to [End72] for the details.

For L a language and T a theory in L, L^{eq} and T^{eq} are defined as follows. As before, we assume for notational simplicity alone that L and T are 1-sorted. Let $\mathcal{E}$ be the set of all formulas $E(\bar{x}, \bar{y})$ such that for some n and every model M of T, E defines an equivalence relation on M^n. Let $I = \{\, i_E : E \in \mathcal{E} \,\}$ be a collection of (distinct) sorts. For each $E \in \mathcal{E}$ let f_E be a function symbol taking n–tuples from the sort $i_=$ into the sort i_E. Finally, let L^{eq} be the I–sorted language which contains $\{\, f_E : E \in \mathcal{E} \,\}$ and for each element of L a corresponding element whose arguments are required to range over the sort $i_=$. (For example, if P is an n–ary relation symbol of L then L^{eq} contains a relation symbol P of sort $(i_=, \ldots, i_=)$, where there are n copies of $i_=$.) The axioms for T^{eq} are the axioms for T restricted to the sort $i_=$, together with all statements expressing: f_E is a surjective map of n–tuples from $i_=$ onto i_E such that $\forall \bar{x}\bar{y}(E(\bar{x}, \bar{y}) \longleftrightarrow f_E(\bar{x}) = f_E(\bar{y}))$. From hereon we will identify T with its copy on $i_=$ in T^{eq}.

Statements made in T^{eq} can always be reduced to statements in T. This is made precise in the following lemma, which is proved by induction on formulas (left to the reader).

Lemma 4.1.1. *For any formula $\varphi(v_0, \ldots, v_n)$ of L^{eq}, with v_j a variable of sort i_{E_j}, there is a formula $\varphi^*(\bar{w}_0, \ldots, \bar{w}_n)$ of L such that*

$$T^{eq} \models \forall \bar{w}_0 \ldots \bar{w}_n (\, \varphi(f_{E_0}(\bar{w}_0), \ldots, f_{E_n}(\bar{w}_n)) \longleftrightarrow \varphi^*(\bar{w}_0, \ldots, \bar{w}_n)\,).$$

Let T be a complete theory in L with universal domain $\mathfrak{C}$. Let $\mathfrak{C}^{eq}$ be an expansion of $\mathfrak{C}$ to a model of T^{eq}. (For E a formula defining an equivalence relation on n-tuples let $(\mathfrak{C}^{eq})_{i_E} = \mathfrak{C}^n/E(\mathfrak{C}) =$ the $E(\mathfrak{C})$-equivalence classes on $\mathfrak{C}^n$, and let f_E be the quotient map.) Notice that $\mathfrak{C}^{eq}$ is obtained from $\mathfrak{C}$ simply by closing under the functions of the language L^{eq}. This observation makes it clear that $\mathfrak{C}^{eq}$ is the unique model N of T^{eq} with $\mathfrak{C} = N_{i_=}$. Furthermore, an automorphism f of $\mathfrak{C}$ can be extended uniquely to an automorphism of $\mathfrak{C}^{eq}$. Given $A \subset \mathfrak{C}$ let A^{eq} denote the closure of A under the maps f_E, $E \in \mathcal{E}$.

Corollary 4.1.1. *Let T be a complete theory in L with universal domain $\mathfrak{C}$.*
(i) T^{eq} is complete.
(ii) Any relation on $\mathfrak{C}$ definable in $\mathfrak{C}^{eq}$ is definable in $\mathfrak{C}$.
(iii) $\mathfrak{C}^{eq}$ is a saturated model of T^{eq}.
(iv) T^{eq} is λ-stable if and only if T is λ-stable (for all $\lambda \geq |T|$).
(v) T^{eq} is t.t. if and only if T is t.t. Also, for φ a formula of T, $MR(\varphi)$, computed in T, is the same as $MR(\varphi)$, computed in T^{eq}.

Proof. (i) and (ii) follow immediately from Lemma 4.1.1.
(iii) To see that $\mathfrak{C}^{eq}$ is a saturated model let $A \subset \mathfrak{C}^{eq}$ have cardinality $< |\mathfrak{C}^{eq}|$ and $p \in S_1(A)$. Let B be a subset of $\mathfrak{C}$ of cardinality $\leq |A|+|T| < |\mathfrak{C}^{eq}|$ such that $A \subset B^{eq}$. Supposing that the variable in p ranges over the sort i_E there is a type q over B (in L) such that if $\bar{b}$ realizes q then $f_E(\bar{b})$ realizes p. Since $\mathfrak{C}$ is saturated, q (hence p) is realized in $\mathfrak{C}^{eq}$.
(iv) follows from much that same argument used to prove (iii). (v) is left to the reader.

Thus, we can use $\mathfrak{C}^{eq}$ as the universal domain of T^{eq}.
Not only does this expansion to $\mathfrak{C}^{eq}$ not add any new structure to $\mathfrak{C}$, but there is a one-to-one correspondence between the elementary submodels of the two models. The reader can verify that if M is an elementary submodel of $\mathfrak{C}$ then M^{eq} is an elementary submodel of $\mathfrak{C}^{eq}$. Conversely, if $N \prec \mathfrak{C}^{eq}$ then $N = M^{eq}$ for some elementary submodel M of $\mathfrak{C}$.

Definition 4.1.1. *Let T be a complete theory, possibly many-sorted, with universal domain $\mathfrak{C}$.*
(i) If D is a definable set in $\mathfrak{C}^n$ (for some n), d is called a name for D *if $f(D) = D \iff f(d) = d$, for all $f \in \mathrm{Aut}(\mathfrak{C})$.*
(ii) If every definable set has a name in $\mathfrak{C}$, we say that T has built-in imaginary elements.

Proposition 4.1.1. *Given a complete theory T, T^{eq} has built-in imaginaries.*

Proof. Let $\mathfrak{C}$ be the universal domain of T and $D = \varphi(\mathfrak{C}, \bar{a})$, where $\varphi(\bar{x}, \bar{y})$ is a formula of L. Let $E(\bar{y}, \bar{y}')$ be the equivalence relation: $E(\bar{y}, \bar{y}') \iff \forall \bar{x}(\varphi(\bar{x}, \bar{y}) \leftrightarrow \varphi(\bar{x}, \bar{y}'))$. Then, for all $\bar{b}$ and $\bar{c}$, $\models E(\bar{a}, \bar{b}) \iff \varphi(\mathfrak{C}, \bar{b}) = \varphi(\mathfrak{C}, \bar{c})$. An automorphism of $\mathfrak{C}^{eq}$ permutes the set D if and only if it fixes $\bar{a}/E$. Thus, T^{eq} has a name for every definable set in $\mathfrak{C}$. We leave it to the reader to show that if D is a definable subset of $(\mathfrak{C}^{eq})^n$, for some n, then there is also a name for D in $\mathfrak{C}^{eq}$. This proves the proposition.

This is the fundamental property of T^{eq} arising in most applications. Instead of "T has built-in imaginaries" we may say T *has imaginaries* or T *has elimination of imaginaries*. By the proposition, T^{eq} has imaginaries, for any complete theory T. However, we use this term even when the theory is not T^{eq} for some other theory. For example, when k is an algebraically closed field and we restrict k^{eq} to the structure whose sorts are the sets k^n, $n < \omega$, we obtain a theory with elimination of imaginaries. (This was proved by Poizat in [Poi83b]; see also [Hod93, 4.4.6].) Informally, the passage from $\mathfrak{C}$ to $\mathfrak{C}^{eq}$ is described as "adding names for definable sets".

Definition 4.1.2. *Let T be a complete theory with universal domain $\mathfrak{C}$. For A a set the* definable closure of A, *denoted $dcl(A)$ is* $\{a :$ for all b, $tp(b/A) = tp(a/A) \implies a = b\}$. *Sets B and C are* interdefinable over A *if $dcl(B \cup A) = dcl(C \cup A)$.*

Of course, $A \subset dcl(A) \subset acl(A)$. Note: $a \in dcl(A)$ if and only if there is a formula $\varphi(v)$ over A such that $\models \exists! v\varphi(v)$ and $\models \varphi(a)$.

Recall that a formula φ is almost over A if it has finitely many conjugates over A, up to equivalence. Thus, if φ is almost over A in $\mathfrak{C}^{eq}$ there are finitely many elements which are the names for the conjugates over A of $\varphi(\mathfrak{C})$. Continuing with this observation yields

Lemma 4.1.2. *Suppose that T has built-in imaginary elements.*

(i) d is a name for the definable set D if and only if D is definable over d and $d \in dcl(A)$ for any set A such that D is definable over A.

(ii) A formula φ is almost over a set A if and only if $\varphi(\mathfrak{C})$ is definable over $acl(A)$.

Proof. (i) Let d be a name for D. By Lemma 3.3.8(i), D is definable over d. Suppose that D is definable over A, and f is an automorphism of $\mathfrak{C}$ fixing A. Then $f(D) = D$, so $f(d) = d$, from which it follows that $d \in dcl(A)$. To prove the converse, let e be a name for D. Since D is definable over e, $d \in dcl(e)$. By the first part of the proof, $e \in dcl(d)$; i.e., $dcl(d) = dcl(e)$. Thus, d is a name for D.

(ii) Suppose that $\varphi(\mathfrak{C})$ is definable over $\bar{a} \subset acl(A)$. Since there are only finitely many possible images of $\bar{a}$ under automorphisms that fix A, there are only finitely many conjugates of φ over A.

Conversely, suppose that φ is almost over A and a is a name for $\varphi(\mathfrak{C})$. If f is an automorphism of $\mathfrak{C}$, $f(a)$ is a name for $f(\varphi(\mathfrak{C}))$. Thus, $\{f(a) : f \in \mathrm{Aut}(\mathfrak{C})$ fixes $A\}$ is finite, implying that $a \in acl(A)$.

This lemma is one indication of how working in $\mathfrak{C}^{eq}$ smooths out certain arguments. Intuitively, the parameters defining a formula which is almost over A are closely tied to A. However, to make this precise in the original theory we needed to introduce an equivalence relation over A having finitely many classes, using this to show, e.g., that when φ is almost over A and M is a model $\supset A$ there is a formula ψ over M equivalent to φ. If we work in $\mathfrak{C}^{eq}$ we simply observe that every model containing A also contains $acl(A)$, from which it is clear that a formula almost over A is equivalent to a formula over any model containing A.

When working in $\mathfrak{C}^{eq}$ we can also replace finite tuples by elements in most settings without changing the validity of an argument. For $\bar{a}$ a finite sequence let b be a name for $\bar{a}$ as a definable set over $\bar{a}$. Then, $dcl(\bar{a}) = dcl(b)$. Proving a property about a definable relation satisfied by $\bar{a}$ quickly reduces to proving a similar property about a formula satisfied by b. Along the same lines, proving a property of the definable subsets of $\mathfrak{C}^{eq}$ implicitly proves the same property for the definable relations on $\mathfrak{C}^{eq}$. In settings where we would have said "Given a tuple $\bar{a}$ from $\mathfrak{C}^{eq}$ ..." we will say "Given a in $\mathfrak{C}^{eq}$ ..." Other advantages of working in $\mathfrak{C}^{eq}$ will be uncovered in later applications.

From hereon, unless stated otherwise, we restrict our attention to theories with built-in imaginaries.

The term "T is a theory" will mean "T is a theory with built-in imaginary elements". Since T^{eq} of any theory has built-in imaginaries any theory appears as a sort in a theory with built-in imaginaries. If we want to know what a certain theorem says about an ordinary 1–sorted theory, when it is proved for theories with built-in imaginaries, we need only read off what the result says about a particular sort of the theory. In jargon this assumption is known as "working in T^{eq}" or "working in $\mathfrak{C}^{eq}$". The only time we may abandon this convention is when we are analyzing a natural example, such as a module or one of the theories built on equivalence relations. Then we may become sloppy and say, e.g., "Let T be the the theory of $(\mathbb{Z}, +)$." Even in this setting, where "element" means an element of $\mathfrak{C}$, we assume the elements of $\mathfrak{C}^{eq}$ are available in proofs.

This passage from ordinary theories into theories with imaginaries has the following effect on our standard examples. Suppose that $T = T_0^{eq}$, where T_0 is a theory of equivalence relations. Then for E one of the equivalence relations and a an element (of the right sort), $f_E(a)$ is an element of the universe.

Now suppose $T = T_0^{eq}$, where T_0 is a theory of (infinite) vector spaces, and V denotes the universal domain of T_0. Let W be a linear (hence definable) subspace of V^n. There is a sort of $\mathfrak{C}$ consisting of V^n/W. Since dimension and Morley rank are the same in a vector space we can write the expected identity, $MR(W) + MR(V^n/W) = MR(V^n)$, for these definable sets. In general, for G any group, G^{eq} contains the quotient of G^n by any definable normal subgroup. For example, when $G = \mathrm{GL}_n(K)$ (where K is some field) $\mathrm{PGL}_n(K)$, which is the quotient of G by its center, is definable in G^{eq}.

Historical Notes. All of this is by Shelah [She90], although T^{eq} was first treated as a many-sorted theory (in writing) by Makkai [Mak84].

4.1.1 Totally Transcendental Theories Revisited

In this subsection totally transcendental theories are studied further under the built-in imaginaries hypothesis. Previous results are restated to set the current viewpoint and to emphasize items particularly relevant to his chapter. Also, the proof of Theorem 3.3.1(i) is completed and a new tool (the canonical parameter) is introduced.

The first lemma is little more than a combination of previous results stated under the built-in imaginaries requirement.

Lemma 4.1.3. *Let $\mathfrak{C}$ be the universal domain of a t.t. theory, a an element and A a set. Then,*

(i) $tp(a/acl(A))$ is stationary.

(ii) Moreover, there is an $e \in dcl(A \cup \{a\}) \cap acl(A)$ such that $\deg(a/A \cup \{e\}) = 1$.

Proof. (i) Let $p^* \in S(\mathfrak{C})$ be a free extension of $tp(a/acl(A))$. By Theorem 3.3.1(ii), there is a defining scheme for p^* consisting of formulas almost over A. Any formula almost over A is equivalent to a formula over $acl(A)$, by Lemma 4.1.2(ii). Thus, p^* is definable over $acl(A)$. We conclude from Theorem 3.3.1(i) that $tp(a/acl(A))$ is stationary.

(ii) By Exercise 4.1.5, $tp(a/acl(A))$ is implied by $tp(a/dcl(A \cup \{a\}) \cap acl(A))$. Since the theory is t.t. there is a finite $B \subset dcl(A \cup \{a\}) \cap acl(A)$ such that $1 = \deg(a/dcl(A \cup \{a\}) \cap acl(A)) = \deg(a/B)$. In other words there is an $e \in dcl(A \cup \{a\}) \cap acl(A)$ such that $\deg(a/A \cup \{e\}) = 1$.

Lemma 4.1.4. *Let T be t.t., $p \in S(\mathfrak{C})$ and A a set. If p does not split over A then p is a free extension of $p \restriction A$ and $p \restriction A$ is stationary.*

Proof. Let $B = acl(A)$. By Lemmas 3.3.2(iii), $p \restriction B$ is a free extension of $p \restriction A$. Hence, to show that p is a free extension of $p \restriction A$ it suffices to show that p is a free extension of $p \restriction B$. Suppose, to the contrary, that for some b, $p \restriction (B \cup \{b\})$ is not a free extension of $p \restriction B$. Let $r = tp(b/B)$, which is stationary by the previous lemma, and let I be an infinite Morley sequence in r over B. Let a realize $p \restriction (B \cup I)$. Let J be a finite subset of I such that a is independent from I over $B \cup J$ and let $c \in I \setminus J$. Then c is independent from a over $B \cup J$, in fact, c is independent from a over B (by the transitivity of independence). Since p does not split over B, $tp(a/B \cup \{c\}) = p \restriction (B \cup \{c\})$ is conjugate to $p \restriction (B \cup \{b\})$ over B. Thus, a depends on c over B, a contradiction which proves that p is a free extension of $p \restriction A$.

Turning to the stationarity of $p \restriction A$, observe that $p \restriction A$ has a unique extension over B (since p does not split over A). Hence, if $q \in S(\mathfrak{C})$ is a free extension of $p \restriction A$, $q \supset p \restriction B$. Since $p \restriction B$ is stationary, q must be p. In other words, $p \restriction A$ is stationary, proving the lemma.

This completes the proof of Theorem 3.3.1(i).

In this chapter it is more natural to work with sets of realizations of types than types; i.e., $\bigwedge$-definable sets (see Definition 3.5.10).

It is worth restating some previously defined notions in an equivalent form involving definable sets. For any sets A and B, $A \triangle B$ denotes the symmetric difference of A and B.

Let $\mathfrak{C}$ be the universal domain of a totally transcendental theory. Let D be an $\bigwedge$-definable set, specifically, $D = p(\mathfrak{C})$.

- $MR(D)$ and $\deg(D)$, the Morley rank and degree of D, are defined to be $MR(p)$ and $\deg(p)$, respectively.

Now suppose D to be the definable set $\varphi(\mathfrak{C})$.

- D is called a *strongly minimal set* if φ is strongly minimal.
- D is a strongly minimal set if and only if every definable subset of D is finite or cofinite.
- D has Morley rank $\geq \alpha$ if for all $\beta < \alpha$ there are definable subsets X_i of D, for $i < \omega$, such that (a) $MR(X_i) \geq \beta$ and (b) $MR(X_i \cap X_j) < \beta$, for $i < j < \omega$.
- If D has Morley rank α, then the degree of D is the maximal k such that there are definable subsets $X_1, \ldots, X_k$ of D satisfying (a) $MR(X_i) = \alpha$, for $i = 1, \ldots, k$, and (b) $MR(X_i \cap X_j) < \beta$, for $1 \leq i < j \leq k$.

Let D be $\bigwedge$-definable over A and have Morley rank α. There may be elements of D which belong to A-definable sets of Morley rank $< \alpha$. For example, some elements of the universal domain of algebraically closed fields of characteristic 0 are in $acl(\emptyset)$, namely the algebraic elements. Motivated by the terminology used in algebraic geometry we attach the label "generic" to the elements of D having maximal Morley rank over A.

Definition 4.1.3. *Let $\mathfrak{C}$ be the universal domain of a totally transcendental theory, D a subset which is $\bigwedge$-definable over A, $B \supset A$ and $a \in D$. We call a* generic over B *if $MR(a/B) = MR(D)$; otherwise a is* nongeneric over B.

Remark 4.1.1. If G is an ω-stable group, A a set and $a \in G$, then a is generic over A in the sense of Definition 3.5.6 if and only if a is generic over A in the sense of Definition 4.1.3.

For example, if D is an $\emptyset$-definable strongly minimal set, $a \in D$ is generic over B if and only if $a \notin acl(B)$. For any $\bigwedge$-definable set D and set B, D contains an element generic over B (because every type in a t.t. theory has a free extension). Intuitively, "most" of the elements of D are generic over any set B. In fact, if X and Y are definable over B and $X \triangle Y$ contains only elements nongeneric over B, then X and Y are "almost equal". The "almost equal" relation between sets is explicitly defined as follows.

Definition 4.1.4. *Let $\mathfrak{C}$ be the universal domain of a totally transcendental theory, X an $\bigwedge$−definable set over A, Y an $\bigwedge$−definable set over B and $\alpha = \max\{MR(X), MR(Y)\}$. We write $X \sim^* Y$ if for all $a \in X \triangle Y$, $MR(a/A \cup B) < \alpha$. The restriction of $\sim^*$ to sets of degree 1 is denoted $\sim$. That is, if X, Y, A and B are as above and, additionally,* $\deg(X) = \deg(Y) = 1$, *$X \sim Y$ if for all $a \in X \triangle Y$, $MR(a/A \cup B) < \alpha$.*

Remark 4.1.2. The detailed verifications of the following are left to the reader. Let X, Y, A and B be as in the definition of $\sim^*$.

(i) If $X \sim^* Y$, then $MR(X) = MR(Y)$. (Suppose, to the contrary, that $MR(X) < MR(Y) = \alpha$. Then any element of Y generic over $A \cup B$ is in $X \triangle Y$; i.e., $MR(Y \setminus X) = \alpha$; contradiction.)

(ii) Suppose that $MR(X) = MR(Y) = \alpha$. The domains A and B play no active role in the definition. That is to say, for sets $A' \supset A$ and $B' \supset B$, $X \sim^* Y$ (over A and B) if and only if $X \sim^* Y$ (over A' and B'). (There is an $a \in X \triangle Y$ such that $MR(a/A \cup B) = \alpha$ if and only if there is an $a \in X \triangle Y$ such that $MR(a/A' \cup B') = \alpha$.)

(iii) If $X = p(\mathfrak{C})$ and $Y = q(\mathfrak{C})$ both have degree 1, then $X \sim Y$ if and only if p and q have the same free extension in $S(\mathfrak{C})$. (This follows quickly from (ii).)

(iv) $\sim$ is an equivalence relation on the $\bigwedge$−definable sets of degree 1.

Since $\mathfrak{C}$ has built-in imaginaries, the quotient set of any definable equivalence relation is a definable subset of $\mathfrak{C}$. This property was used to show that every definable set X in $\mathfrak{C}$ has a name in $\mathfrak{C}$; i.e., an element x such that for all $f \in \mathrm{Aut}(\mathfrak{C})$, $f(X) = X$ if and only if $f(x) = x$. While $\sim$ is not a definable equivalence relation we will show that for each $\sim$−class, $\mathfrak{C}$ contains an element that acts like a "name" for the class (formalized as follows).

Definition 4.1.5. *Let $\mathfrak{C}$ be the universal domain of a t.t. theory and let X be an $\bigwedge$−definable set of degree 1. An element $c \in \mathfrak{C}$ is a* canonical parameter *of X if*

$$\forall f \in \mathrm{Aut}(\mathfrak{C}),\ f(X) \sim X \text{ if and only if } f(c) = c.$$

If $X = p(\mathfrak{C})$ a canonical parameter of X is also called a canonical parameter of p.

Remark 4.1.3. (i) By Remark 4.1.2(iii), degree 1 sets $X = p(\mathfrak{C})$ and $Y = q(\mathfrak{C})$ are $\sim$−equivalent if and only if p and q have the same free extension in $S(\mathfrak{C})$. Thus, for $f \in \mathrm{Aut}(\mathfrak{C})$ and p^* the free extension of p in $S(\mathfrak{C})$, $f(X) \sim X$ if and only if $f(p^*) = p^*$. So, a canonical parameter of X is an element c such that

$$\forall f \in \mathrm{Aut}(\mathfrak{C}),\ f(p^*) = p^* \text{ if and only if } f(c) = c. \tag{4.1}$$

This equivalence is the key to the proof of the next theorem.

(ii) If X is a degree 1 set and c and d are both canonical parameters of X, then $dcl(c) = dcl(d)$. (If $f \in \mathrm{Aut}(\mathfrak{C})$ and $f(c) = c$, then $f(X) \sim X$ and $f(d) = d$. Thus, $d \in dcl(c)$. Similarly, $c \in dcl(d)$.)

(iii) While a degree 1 set will not have a unique canonical parameter, by virtue of (ii), any two such are interdefinable over $\emptyset$. Thus, it is common to say *the* canonical parameter instead of *a* canonical parameter.

Theorem 4.1.1. *Let $\mathfrak{C}$ be the universal domain of a t.t. theory and X an $\bigwedge$-definable set of degree 1. Then X has a canonical parameter.*

Proof. Suppose X is $p(\mathfrak{C})$ and let p^* be the free extension of p in $S(\mathfrak{C})$.

Claim. A canonical parameter for X exists if there is a formula ψ such that

$(*)$ $\forall f \in \mathrm{Aut}(\mathfrak{C})$, $f(p^*) = p^*$ if and only if $f(\psi) = \psi$.

By the previous remark an element c is a canonical parameter for X if it satisfies (4.1). If ψ satisfies $(*)$, a name c for $\psi(\mathfrak{C})$ satisfies (4.1), proving the claim.

The definability of types is the key to finding such a ψ. Let $\varphi(x, a)$ be a formula in p with $MR(\varphi(x, a)) = MR(p) = \alpha$ and $\deg(\varphi(x, a)) = 1$, where $\varphi(x, y)$ is over $\emptyset$. By the definability of types in t.t. theories (Theorem 3.3.1) there is a formula $\psi(y)$ such that for all $b \in \mathfrak{C}$, $\varphi(x, b) \in p$ if and only if $\models \psi(b)$.

Claim. For all $f \in \mathrm{Aut}(\mathfrak{C})$, $f(p^*) = p^*$ if and only if $f(\psi) = \psi$.

Let $f \in \mathrm{Aut}$ $(\mathfrak{C})$. First suppose that $f(p^*) = p^*$. Then

$$\begin{aligned} f(\psi(\mathfrak{C})) &= f(\{ b : \varphi(x, b) \in p^* \}) \\ &= \{ b : \varphi(x, b) \in f(p^*) \} \\ &= \{ b : \varphi(x, b) \in p^* \} = \psi(\mathfrak{C}). \end{aligned}$$

On the other hand, if $f(\psi(\mathfrak{C})) = \psi(\mathfrak{C})$, then $\varphi(x, f(a))$ is in p^* as well as in $f(p^*)$. Since $MR(p^*) = MR(\varphi(x, f(a))) = MR(f(p^*))$ and $\deg(\varphi(x, f(a))) = 1$, $f(p^*)$ must be p^*. This proves the claim and the theorem.

Corollary 4.1.2. *Let $\mathfrak{C}$ be the universal domain of a t.t. theory, $X = p(\mathfrak{C})$ a set of degree 1, p^* the free extension of p in $S(\mathfrak{C})$ and c the canonical parameter of X.*

(i) If p^* is definable over A, then $c \in dcl(A)$.

(ii) If p is over A, then $c \in dcl(A)$.

Proof. (i) Since p^* is definable over A, $f(p^*) = p^*$ for any $f \in \mathrm{Aut}(\mathfrak{C})$ which fixes A pointwise (by Theorem 3.3.1 and Lemma 3.1.8). Hence, $f(c) = c$ for any $f \in \mathrm{Aut}(\mathfrak{C})$ which fixes A pointwise. We conclude that $c \in dcl(A)$.

(ii) Since p has degree 1, $p^* \restriction A$ has degree 1. Thus, p^* is definable over A and we conclude from (i) that $c \in dcl(A)$.

Corollary 4.1.3. *Let $\mathfrak{C}$ be the universal domain of a t.t. theory, $X = r(\mathfrak{C})$ an $\bigwedge$-definable set of degree 1, $p^* \in S(\mathfrak{C})$ the free extension of r and c a canonical parameter of X.*

(i) If Y is an $\bigwedge$-definable set of degree 1 and $Y \sim X$, then c is a canonical parameter of Y.

(ii) p^* is definable over c.

(iii) There is a degree 1 formula $\varphi(v, c)$ over c such that p^* is the unique free extension of $\varphi(x, c)$. Moreover, if q is any type over c of degree 1 having p^* as a free extension, then for all $f \in \mathrm{Aut}(\mathfrak{C})$, $f(q(\mathfrak{C})) \sim q(\mathfrak{C})$ if and only if $f(q(\mathfrak{C})) = q(\mathfrak{C})$.

Proof. (i) Let $Y = r'(\mathfrak{C})$ and $f \in \mathrm{Aut}(\mathfrak{C})$. Since $Y \sim X$, p^* is the free extension of r' in $S(\mathfrak{C})$. Thus, $Y \sim f(Y)$ if and only if $p^* = f(p^*)$. Since $p^* = f(p^*)$ if and only if $c = f(c)$, c is a canonical parameter of Y.

(ii) Let $\psi(x, y)$ be a formula over $\emptyset$ and let $\theta(y)$ be a formula such that $\psi(x, a) \in p^* \iff\ \models \theta(a)$. If $f \in \mathrm{Aut}(\mathfrak{C})$ fixes c, $f(p^*) = p^*$, hence $f(\theta(\mathfrak{C})) = \theta(\mathfrak{C})$. In other words, θ is invariant under the automorphisms of $\mathfrak{C}$ which fix c. By Lemma 3.3.8(i), θ is equivalent to a formula over c.

(iii) Since p^* is definable over c, p^* is the unique free extension of $p^* \restriction c$. Hence, there is a formula $\varphi(v, c) \in p^* \restriction c$ with $MR(\varphi(v, c)) = MR(p^*)$ and $\deg(\varphi(v, c)) = 1$. Now let q be any type over c of degree 1 such that p^* is a free extension of q. For any $f \in \mathrm{Aut}(\mathfrak{C})$,

$$\begin{aligned} f(q(\mathfrak{C})) \sim q(\mathfrak{C}) &\iff f(p^*) = p^* \\ &\iff f(c) = c \\ &\iff f(q(\mathfrak{C})) = q(\mathfrak{C}), \end{aligned}$$

completing the proof.

Remark 4.1.4. Let X be an $\bigwedge$-definable set of degree 1 and c the canonical parameter of X. There is (over c) a definable Y of degree 1 such that $Y \sim X$ and for all $f \in \mathrm{Aut}(\mathfrak{C})$, $f(Y) \sim Y$ if and only if $f(Y) = Y$. In this way the set Y acts as a canonical representative for its $\sim$-class.

The next result only ties together numerous previous results to give easily referenced tools for later use.

Lemma 4.1.5. *Let $\mathfrak{C}$ be the universal domain of a t.t. theory, a an element and A a set.*

(i) Let B be an algebraically closed set containing A. Then, a is independent from B over A if and only if the canonical parameter c of $tp(a/B)$ is in $acl(A)$.

(ii) Let $p = tp(a/acl(A))$ and c the canonical parameter of p. Then, there is a Morley sequence I in p such that $c \in dcl(I)$.

Proof. (i) First notice that $tp(a/B)$ is stationary, by Lemma 4.1.3, hence it does have a canonical parameter. Let p^* be the unique free extension of $tp(a/B)$ in $S(\mathfrak{C})$. If a is independent from B over A, p^* is a free extension of $p^* \restriction A$, hence $c \in acl(A)$ by Corollary 4.1.2(i).

Conversely, if $c \in acl(A)$ then p^* is definable over $acl(A)$ (since p^* is definable over c). Thus, a is independent from B over A by Theorem 3.3.1.

(ii) Let p^* be the free extension of $tp(a/B)$ in $S(\mathfrak{C})$. By Corollary 3.3.3, there is a Morley sequence I in p such that p^* is definable over I. Thus $c \in dcl(I)$ by Corollary 4.1.2.

Corollary 4.1.4. *Let $\mathfrak{C}$ be the universal domain of a t.t. theory and a, b elements of $\mathfrak{C}$. There is a c such that*

(1) *$c \in acl(a)$,*
(2) *b is independent from a over c,*
(3) *$tp(b/c)$ is stationary, and*
(4) *there is a finite c–independent set $\{b_0, \ldots, b_n\}$ of realizations of $tp(b/c)$ such that $c \in dcl(b_0, \ldots, b_n)$.*

A final word about notation:

Notation. In this chapter we may state a result about $\emptyset$–definable sets in a t.t. theory, instead of A–definable sets for an arbitrary A. However, if $\mathfrak{C}$ is the universal domain of a t.t. theory and A is a finite set then $\mathfrak{C}_A$, the model with constants added to the language for the elements of A, is also t.t. Thus, a statement proved for the $\emptyset$–definable sets in an arbitrary t.t. theory is true of all definable sets in an arbitrary t.t. theory. (Except, of course, statements explicitly mentioning the parameters over which the set is defined.)

4.1.2 D^{eq} for a Strongly Minimal D

In subsequent sections much attention will be given to definable relations on a fixed definable set D and the canonical parameters of degree 1 relations on D, especially when D is strongly minimal. The elements of $\mathfrak{C}^{eq}$ most relevant to D are isolated in

Definition 4.1.6. *Let $\mathfrak{C}$ be the universal domain of a t.t. theory and let D be a set which is $\bigwedge$–definable over A. Then $D^{eq} = \{\, x \in \mathfrak{C}^{eq} : x \in dcl(D \cup A) \,\}$.*

Lemma 4.1.6. *Let D be an A–definable set in the universal domain of a t.t. theory and X a degree 1 definable relation on D. Then the canonical parameter of X is in D^{eq}.*

Proof. By Proposition 3.3.3 there is a $B \subset D$ such that X is definable over $A \cup B$. By Corollary 4.1.2(ii), $c \in dcl(A \cup B)$.

Let D be a strongly minimal set. Recall from Remark 3.1.4 that dimension on D satisfies

(Additivity) For $\bar{a}$ and $\bar{b}$ finite sequences from D,
$$\dim(\bar{a}\bar{b}) = \dim(\bar{a}/\bar{b}) + \dim(\bar{b}).$$

Since $\dim(\bar{a}) = MR(\bar{a})$ when $\bar{a}$ is a finite sequence from D (by Lemma 3.3.4), Morley rank on D satisfies the corresponding additivity condition. In fact, the elements of D^{eq} are tied closely enough to D to prove

Proposition 4.1.2. *Let $\mathfrak{C}$ be the universal domain of a t.t. theory and let D be a strongly minimal set, definable over A. Then for all $a, b \in D^{eq}$*

$$MR(ab/A) = MR(a/\{b\} \cup A) + MR(b/A).$$

Proof. Without loss of generality, $A = \emptyset$. Let $\bar{c}$ and $\bar{d}$ be finite sequences from D such that $a \in dcl(\bar{c})$ and $b \in dcl(\bar{d})$. Let $\bar{c}_0$ be a maximal subsequence of $\bar{c}$ which is independent from a and $\bar{c}_1 = \bar{c} \setminus \bar{c}_0$. By the maximality of $\bar{c}_0$, any $e \in \bar{c}_1$ is in $acl(a, \bar{c}_0)$. Hence, a and $\bar{c}_1$ are interalgebraic over $\bar{c}_0$. By Lemma 3.3.2(ii), $MR(a/\bar{c}_0) = MR(\bar{c}_1/\bar{c}_0)$. The sequence $\bar{c}$ can be chosen so that $\bar{c}_0$ is independent from $\{b, \bar{d}\}$. (Given $\bar{c}_0\bar{c}_1 = \bar{c}$ let $\bar{e}_0$ be a realization of $r = tp(\bar{c}_0/acl(\emptyset))$ independent from $\{a, b, \bar{d}\}$. Since r is stationary, $tp(\bar{e}_0/a) = tp(\bar{c}_0/a)$. Thus there is an $\bar{e}_1$ from D such that $a \in dcl(\bar{e}_0\bar{e}_1)$ and $\bar{e}_1 \subset acl(\bar{e}_0, a)$.) Similarly, for $\bar{d}_0$ a maximal subsequence of $\bar{d}$ which is independent from b and $\bar{d}_1 = \bar{d} \setminus \bar{d}_0$, b is interalgebraic with $\bar{d}_1$ over $\bar{d}_0$ and $MR(b/\bar{d}_0) = MR(\bar{d}_1/\bar{d}_0)$. Without loss of generality, $\bar{d}_0$ is independent from $\{a, \bar{c}, b\}$.

The following sequence of equations shows that $MR(ab) = MR(a/b) + MR(b)$. (The details are left to the reader.)

1. $MR(ab) = MR(ab/\bar{c}_0\bar{d}_0) = MR(\bar{c}_1\bar{d}_1/\bar{c}_0\bar{d}_0)$;
2. $MR(a/b) = MR(a/b\bar{d}_0) = MR(a/b\bar{d}) = MR(a/b\bar{d}\bar{c}_0)$ (since $\bar{c}_0$ is independent from $\{a, b, \bar{d}\}$) and $MR(a/b\bar{d}\bar{c}_0) = MR(\bar{c}_1/b\bar{d}\bar{c}_0) = MR(\bar{c}_1/\bar{d}\bar{c}_0)$;
3. $MR(b) = MR(b/\bar{d}_0\bar{c}_0) = MR(\bar{d}_1/\bar{d}_0\bar{c}_0)$;
4. $MR(\bar{c}_1\bar{d}_1/\bar{c}_0\bar{d}_0) = MR(\bar{c}_1/\bar{d}\bar{c}_0) + MR(\bar{d}_1/\bar{c}_0\bar{d}_0)$.

This proves the proposition.

When working with sequences from a strongly minimal set we prefer $\dim(-)$ over $MR(-)$ to emphasize the additivity property. Because of the previous proposition we can use $\dim(-)$ to denote Morley rank on D^{eq} and keep the property that $\dim(-)$ is additive where it is defined:

Definition 4.1.7. *Let $\mathfrak{C}$ be the universal domain of a t.t. theory and let D be a strongly minimal set. For $\bar{a}$ a finite sequence from D^{eq} we define* $\dim(\bar{a})$ *to be $MR(\bar{a})$.*

Historical Notes. All of this is by Shelah [She90], although T^{eq} was first treated as a many-sorted theory (in writing) by Makkai [Mak84].

Exercise 4.1.1. Prove Lemma 4.1.1.

Exercise 4.1.2. Show that $p \in S(A)$ (in $\mathfrak{C}$) has a unique extension over A^{eq} (in $\mathfrak{C}^{eq}$). Use this observation to show that when T is t.t. and p is a type, $MR(p)$ is the same, whether computed in T or T^{eq}.

Exercise 4.1.3. Prove that T^{eq} is quantifier eliminable whenever T is quantifier eliminable.

Exercise 4.1.4. Suppose that T has built-in imaginaries and A is a finite set. Show that there is an a and a formula $\varphi(x, a)$ such that $b \in A$ if and only if $\models \varphi(b, a)$.

Exercise 4.1.5. Let $\mathfrak{C}$ be the universal domain of a complete theory, A a set, a an element and $A' = acl(A) \cap dcl(A \cup \{a\})$. Show that $tp(a/A')$ implies $tp(a/acl(A))$. (We are working in mt^{eq} here.)

Exercise 4.1.6. Let T be the 1−sorted theory in a language with a single binary relation E saying that E is an equivalence relation with infinitely many infinite classes and no finite classes. Let $\mathfrak{C}/E$ denote the sort in $\mathfrak{C}^{eq}$ consisting of the E−classes of the elements of $\mathfrak{C}$. Prove that $\mathfrak{C}/E$ is strongly minimal. (HINT: Use automorphisms of $\mathfrak{C}^{eq}$.)

4.2 The Pregeometries on Strongly Minimal Sets

In this section we introduce the property, namely local modularity, which divides the "geometrically simple" and "geometrically complex" strongly minimal sets. This property will be defined in the context of arbitrary pregeometries.

Definition 4.2.1. *Let $(S, c\ell)$ be a pregeometry. The* localization *of S at $A \subset S$ is defined to be the pregeometry $(S, c\ell')$, where $c\ell'(X) = c\ell(X \cup A)$ for all $X \subset S$. (The reader can verify that S is indeed a pregeometry under $c\ell'$.) An* isomorphism *between $(S, c\ell)$ and another pregeometry $(S_0, c\ell_0)$ is simply a bijection f from S onto S_0 which respects the closure operators; i.e., $X \subset S$ is closed if and only if $f(X)$ is a closed subset of S_0. As usual, an isomorphism of a pregeometry onto itself is called an automorphism. S is said to be* homogeneous *if for any closed $A \subset S$ and $a, b \in S \setminus A$, there is an automorphism of S which is the identity on A and maps a to b.*

In the exercises the reader is asked to verify that the pregeometry on a strongly minimal set is homogeneous. Now to the more substantive definitions.

Definition 4.2.2. *Let $(S, c\ell)$ be a pregeometry.*

(i) S is trivial *if for all nonempty $X \subset S$, $c\ell(X) = \bigcup\{\, c\ell(a) : a \in X \,\}$.*

(ii) S is modular *if for all closed $X, Y \subset S$,*

$$\dim(X) + \dim(Y) = \dim(X \cup Y) + \dim(X \cap Y) \quad \textit{(Modularity Law)}. \tag{4.2}$$

(iii) S is projective *if S is nontrivial and for all $a, b \in S$ and $X \subset S$ such that $a \in c\ell(X \cup \{b\})$, there is a $c \in c\ell(X)$ such that $a \in c\ell(\{b, c\})$.*

(iv) S is locally modular (locally projective) *if for some $a \in S$ the localization of S at $\{a\}$ is modular (projective).*

(v) For any of the properties defined in (i)-(iv), a strongly minimal set $D = \varphi(\mathfrak{C})$ is said to have the property if the pregeometry associated to D has the property. Similarly with strongly minimal formulas and types containing strongly minimal formulas.

Remark 4.2.1. Let $(S, c\ell)$ be a pregeometry.

(i) It is easy to show that S possesses one of the properties defined above if and only if the geometry associated to S also has that property. (See Exercise 4.2.2.) Also, each of the properties is invariant under isomorphism (in the class of pregeometries).

(ii) If S is trivial then S is modular.

(iii) The Modularity Law is equivalent to

Any two closed subsets X and Y of S are independent over $X \cap Y$.

Proof. Without loss of generality, X and Y have finite dimension. By the additivity of dimension (see Exercise 3.1.8), $\dim(X \cup Y) = \dim(X/Y) + \dim(Y)$. X and Y are independent over $X \cap Y$ if and only if $\dim(X/Y) = \dim(X/X \cap Y)$. Thus, X and Y are independent over $X \cap Y$ if and only if

$$\begin{aligned} \dim(X \cup Y) &= \dim(X/X \cap Y) + \dim Y \\ &= \dim X - \dim(X \cap Y) + \dim Y. \end{aligned}$$

Example 4.2.1. (i) Let D be the universal domain of the theory in the empty language with only infinite models. Then D is a trivial strongly minimal set.

(ii) Let F be a division ring, V a vector space over F and $\langle - \rangle$ the linear span operator on V. Then, $S = (V, \langle - \rangle)$ is a modular pregeometry. If V has dimension ≥ 2, S is nontrivial and projective. The geometry associated to V, P, is called a *projective geometry over F*.

A remark about the dimension of P is in order. As a geometry, the dimension of P equals the dimension of V. Strictly in the context of projective geometries over a division ring, however, it is customary to define the dimension of P to be $\dim(V) - 1$ (or ∞, if $\dim(V) = \infty$). For example, a projective plane over $\mathbb{R}$ has dimension 2 as a real projective space, but dimension 3 as a geometry. In this book, $\dim(P)$ will always denote the dimension of P as a geometry (hence $\dim(P) = 3$ when P is a projective plane).

Turning to model-theoretic considerations, formulate V as a structure in the natural language of vector spaces and suppose that it is infinite. Let

$c\ell(-)$ be algebraic closure on V. It was proved earlier that $Th(V)$ is quantifier-eliminable, hence $c\ell(-) = \langle - \rangle$ and $Th(V)$ is strongly minimal. Thus, V (when it's the universal domain of its theory) is a modular and projective strongly minimal set.

(iii) Affine spaces provide examples of locally modular strongly minimal sets which are not modular, however it will take some time to formulate these structures as strongly minimal sets. Remember (from Definition 3.5.2) that a group action (G, X) is called *regular* if for each pair x, $y \in X$ there is a unique $g \in G$ such that $gx = y$. Notice that if X is a coset of the group G in a supergroup of G, then the group operation defines a regular group action of G on X.

Let V be a vector space of dimension ≥ 1 over a division ring F. Following [BM67], an *affine space derived from* V is defined to be a regular group action of V on a set P. For a fixed group G, if G acts regularly on both X and Y, then (G, X) and (G, Y) are isomorphic as group actions. Thus, all affine spaces derived from V are isomorphic. An *affine space over* F is an affine space derived from some vector space over F.

Let W be a vector space over F properly containing V, $a \in W$ and $A = a + V$. As stated above, (V, A) is a regular group action under $+$, hence an affine space derived from V. There are various ways to formulate an affine space as a structure in a first-order language. The most natural formulation is in a two-sorted language L^* with the symbols needed to define a vector space on the first sort and a binary operator $\star$ such that given v_1 in the first sort and v_2 in the second sort, $v_1 \star v_2$ is an element of the second sort. Then, interpreting the first sort by V, the second sort by A and $\star$ by the group action turns (V, A) into a structure for L^*. It is easy to show that the theory of $M = (V, A)$ is quantifier-eliminable in this language. From hereon suppose (V, A) is the universal domain of its theory in L^*.

The relations on V definable in M are simply those definable in the vector space V. For any $x \in A$ there is a bijection between V and A definable over x (see the definition of a regular group action). Thus, A is a strongly minimal set and the localization of A at any element x is isomorphic (as a pregeometry) to the pregeometry on V. Since V is modular we conclude that A is locally modular.

Claim. When $a \notin V$, $A = a + V$ is not modular.

Let $c\ell$ denote algebraic closure restricted to A and (in the proof of the claim) let $\dim(-)$ be dimension in the pregeometry $(A, c\ell)$. Let b be an element of A, x a nonzero element of V and c an element of A which is independent from $\{b, x + b\}$. Let $X = c\ell(b, x + b)$, $Y = c\ell(c, x + c)$ and notice that $\dim(X) = \dim(Y) = 2$ and $\dim(X \cup Y) = 3$. If $\dim(X \cap Y)$ were $2 = \dim(X)$ we would have $X = Y$, contradicting that $\dim(X \cup Y) = 3$. Thus $\dim(X \cap Y) \leq 1$. Since $x \in acl(X) \cap acl(Y)$ any element of $X \cap Y \setminus c\ell(\emptyset)$ is interalgebraic with x. By the elimination of quantifiers in the model (V, A), no element of

V is algebraic in an element of A. Thus, $\dim(X \cap Y) = 0$, proving that the modularity law (4.2) fails for X and Y. This proves the claim.

The (1−sorted) structure A' whose universe is A and whose definable relations are those definable in M is also known as an affine space over F. A natural 1−sorted language in which $Th(A')$ is quantifier-eliminable is specified as follows. Let V, W and A be the objects defined above. We need to find relations on A from which the vector space V and the action of V on A can be recovered. Replace the action of V on A by the ternary operation f:

$$f(x, y, z) = x + y - z, \quad \text{for all } x,\ y,\ z \in A.$$

The action of F on V induces a family of binary operators g_α, $\alpha \in F$, on A given by the rule:

$$g_\alpha(x, y) = \alpha x + (1 - \alpha)y, \quad \text{for all } x,\ y \in A.$$

It is left to the reader to see that A in the language $\{f, g_\alpha\}_{\alpha \in F}$ is quantifier-eliminable and has the same definable relations as A'. Note: The vector space V and its action on A are definable in $(A')^{eq}$. (See Exercise 4.2.3.)

(iv) (A strongly minimal set which is not locally projective.) Let K be the universal domain for the theory of algebraically closed fields of some fixed characteristic. It was noted previously that K is a strongly minimal set on which field-theoretic closure is the same as algebraic closure. To see that K is not even close to being locally projective, let $\{a, c_0, \ldots, c_n\}$ be an algebraically independent set of elements of K and let $b = c_0 a^n + c_{n-1} a^{n-1} + \ldots + c_1 a + c_0$. Not only is there no $d \in acl(c_0, \ldots, c_n) \cap K$ such that $b \in acl(a, d)$, but there is no set $\{d_0, \ldots, d_{n-1}\} \subset acl(c_0, \ldots, c_n) \cap K$ such that $b \in acl(a, d_0, \ldots, d_{n-1})$. In particular, this shows that no localization of K at a finite set is projective.

In the example the only nontrivial modular strongly minimal set (a vector space) is also projective. The next lemma shows that this is no accident.

Lemma 4.2.1. *A pregeometry (S, cl) is modular if and only if*

(P) *for all a, $b \in S$ and $X \subset S$ such that $a \in cl(X \cup \{b\})$, there is a $c \in cl(X)$ such that $a \in cl(\{b, c\})$.*

Thus, a pregeometry is projective if and only if it is nontrivial and modular.

Proof. The proof of this lemma is elementary but will serve to familiarize the reader with the definitions. First, suppose that S is modular, $X \subset S$ is closed and $a \in cl(X \cup \{b\})$. We need to find a $c \in X$ such that $a \in cl(\{b, c\})$. Let $Y = cl(\{a, b\})$ and assume, without loss of generality, that both a and b are not in X, hence $\dim(Y/X) = 1$. If $a \in cl(\{b\})$ we are done, so we can also assume that $\dim(Y) = 2$. By the Modularity Law on S, $\dim(X \cap Y) = 1$. Let c be an element of $(X \cap Y) \setminus cl(\emptyset)$. Since $b \notin X$, $c \notin cl(\{b\})$, hence $a \in cl(\{b, c\})$ by the exchange property.

Now suppose that S satisfies **(P)**.

Claim. For all closed $X, Y \subset S$, if $c \in c\ell(X \cup Y)$ then there are $a \in X$ and $b \in Y$ such that $c \in c\ell(\{a, b\})$.

This is proved by induction on $\dim(X \cup Y)$, which we can assume to be finite. Let $c \in c\ell(X \cup Y)$. Without loss of generality, there are $a \in Y$ and a closed $Z \subset Y$ such that $a \notin c\ell(X \cup Z)$, $Y = c\ell(Z \cup \{a\})$ and $c \notin c\ell(X \cup Z)$. Since S satisfies **(P)** and $c \in c\ell(X \cup Z \cup \{a\})$ there is a $b \in c\ell(X \cup Z)$ such that $c \in c\ell(\{a, b\})$. The conditions on Z force $\dim(X \cup Z)$ to be less than $\dim(X \cup Y)$, hence the inductive hypothesis yields $d \in X$ and $e \in Z$ such that $b \in c\ell(\{d, e\})$. Thus, $c \in c\ell(\{d, a, e\})$. Since a and e are both in the closed set Y the projectivity of S produces an element $f \in Y$ such that $c \in c\ell(\{d, f\})$. This proves the claim.

Assume, towards a contradiction, that S is not modular and let X and Y be closed subsets of S which are dependent over $X \cap Y = Z$. From this dependence we get a closed Y', $Z \subset Y' \subset Y$ and an $a \in Y$ such that $a \in c\ell(X \cup Y') \setminus Y'$. By the claim there are $b \in Y'$ and $c \in X$ such that $a \in c\ell(\{b, c\})$. Since $a \notin Y'$ the exchange property implies that $c \in c\ell(\{a, b\}) \subset Y$. Thus, $c \in Z \subset Y'$, contradicting that $a \notin Y'$. This proves the lemma.

A natural problem is: Characterize the infinite projective geometries which are (a) isomorphic to a strongly minimal set, or at least (b) isomorphic to the geometry associated to a strongly minimal set. In this introductory section only a fraction of what is known will be stated. The restrictions on the geometries are less stringent in part (b) of the problem, so it is discussed first. In the main example above we showed that any infinite projective geometry over a division ring F is the geometry associated to some model of a strongly minimal theory, namely a vector space over F. The following classical result (see, e.g., [Hal59]) shows the converse to be true (when the dimension of the strongly minimal set is sufficiently large).

Lemma 4.2.2. *Let P be a projective geometry of dimension ≥ 4 in which each closed set of dimension 2 contains at least 3 elements. Then P is isomorphic to projective geometry over some division ring F.*

Let D be a strongly minimal set such that the geometry associated to D is isomorphic to projective geometry P over a division ring F. The geometry P is derived from a vector space V as outlined in Example 4.2.1. It is natural to ask if V is $\emptyset$–definable in D^{eq}, or at least definable in D^{eq} over some set of parameters. This, and similar questions on representing strongly minimal sets using groups, will be investigated throughout this chapter.

A pregeometry $(S, c\ell)$ is *locally finite* if for all closed $X \subset S$ of finite dimension there is a finite $A \subset X$ such that $X = \bigcup\{c\ell(a) : a \in A\}$. (Thus, $(S, c\ell)$ is locally finite if in the associated geometry the closure of a finite set is finite.)

The strongest classical result about locally projective, locally finite geometries is

Lemma 4.2.3 (Doyen-Hubaut). *Let P be a locally projective, locally finite, geometry of dimension ≥ 4 in which all closed sets of dimension 2 have the same cardinality. Then P is affine or projective geometry over a finite field.*

Let D be a locally projective, locally finite, strongly minimal set and let P be the geometry associated to D. Then all closed sets of dimension 2 in P have the same cardinality because D (hence P) is homogeneous. Thus, P is affine or projective geometry over some finite field. Again, the problem of defining the relevant affine space or vector space in D^{eq} is difficult and will be discussed later.

4.2.1 Plane Curves

From a model-theoretic standpoint a deficiency of the definition of modularity is that it is stated in terms of closed sets, which are potentially undefinable objects. Our next goal is to find an equivalent of modularity which is a property of definable relations and rank instead of closed sets and dimension. This will make it easier to study modularity and local modularity with model-theoretic techniques.

Definition 4.2.3. *Let D be a strongly minimal set, definable over $\emptyset$ in the universal domain $\mathfrak{C}$ of a t.t. theory. A strongly minimal subset of D^2 is called a* plane curve *in D. If C and C' are plane curves in D we write $C \approx C'$, and say that C and C' are* equivalent curves, *if the symmetric difference of C and C' is finite. Slightly abusing the terminology, we identify a $\approx$ −class of plane curves and say that C and C' are the* same *plane curve if $C \approx C'$. A strongly minimal formula φ such that $\varphi(D)$ is a plane curve in D is also called a plane curve in D.*

Remark 4.2.2. Let D be a strongly minimal set as in the definition and C, C' plane curves in D. Then, $C \approx C'$ if and only if, in the notation of Definition 4.1.4, $C \sim C'$. The new notation is introduced only to emphasize that the relation will only be applied to plane curves. By Corollary 4.1.3, C and C' are considered to be the same plane curve if and only if they have the same canonical parameter. Furthermore, there is a plane curve $C_0 \approx C$ which acts as a canonical representative for the $\approx$ −class of C in the sense that, for all $f \in \mathrm{Aut}(\mathfrak{C})$, $f(C_0) \approx C_0$ if and only if $f(C_0) = C_0$. Often we will express the equivalence of C and C' by saying "C equals C' (up to a finite set)."

Example 4.2.2. (i) Let D be a trivial strongly minimal set defined over the empty set in the universal domain of a t.t. theory. Let C be a plane curve in D, defined over $A \subset D$, and let $(a, b) \in C \setminus acl(A)$. Since $tp(ab/A)$ is strongly minimal, $\{a, b, A\}$ cannot be independent, and this set cannot be pairwise independent since D is trivial. First suppose that $a \in acl(A)$. Then, the reader can verify that C equals $\{(c,d) \in D^2 : c = a\}$ (up to a finite set). Similarly, when $b \in acl(A)$. Now suppose that a depends on b; i.e., a and b are interalgebraic and (a, b) is independent from A over $\emptyset$. Thus, there is a strongly minimal set C' which is equal to C up to a finite set and has finitely many conjugates over $\emptyset$, equivalently, the canonical parameter of C is in $acl(\emptyset)$. The reader can verify that these are the only plane curves in D.

(ii) Let V be the universal domain of infinite vector spaces over some division ring F. Let C be a plane curve in V, defined over $A \subset V$. Then, up to a finite set, C is defined by a linear equation $f(x, y) = 0$ of the form $\alpha x + \beta y + \gamma_0 a_0 + \ldots + \gamma_n a_n = 0$, where $a_0, \ldots, a_n \in A$, $\alpha, \beta, \gamma_0, \ldots, \gamma_n \in F$ and the nullity of $f(x, y) = 0$ is 1. The element $b = \gamma_0 a_0 + \ldots + \gamma_n a_n$ is a canonical parameter of C.

Thus, any plane curve C in V is defined (up to a finite set) by an equation of the form $\alpha x + \beta y = b$, where α, $\beta \in F$ and $b \in V$.

(iii) Let K be the universal domain of algebraically closed fields of a fixed characteristic, and let C be a plane curve in K. Then, C is defined (up to a finite set) by an equation of the form $f(x, y) = 0$, where f is an irreducible polynomial over K. (This follows from the elimination of quantifiers and some basic facts about varieties found in, for example, [Har80, I.1.13].) Let $\{c_0, \ldots, c_n\}$ be algebraically independent. The equation $y = c_n x^n + c_{n-1} x^{n-1} + \ldots + c_1 x + c_0$ defines a plane curve C' whose canonical parameter is interdefinable with the set $\{c_0, \ldots, c_n\}$. In particular, for each $n < \omega$ there is a plane curve whose canonical parameter has dimension $= n$.

In the example each plane curve in a modular strongly minimal set is relatively simple in that its canonical parameter has dimension ≤ 1. Strongly minimal sets with this property deserve a special name.

Definition 4.2.4. *Let D be an A–definable strongly minimal set in a t.t. theory with universal domain $\mathfrak{C}$. D is called* linear *if for every plane curve C in D the canonical parameter of C has dimension (over A) ≤ 1. If $D = \varphi(\mathfrak{C})$ is linear, φ is also called linear.*

An algebraically closed field K (which is not locally modular) fails to be linear, in fact, for any $k < \omega$ there is a plane curve in K whose canonical parameter has dimension $\geq k$. (See (iii) in the previous example.)

The next lemma not only connects local modularity and linearity but shows that for D a strongly minimal set, if *any* localization of D is modular then D is locally modular.

Lemma 4.2.4. *Let D be a strongly minimal set over A^* in a t.t. theory with universal domain $\mathfrak{C}$. The following are equivalent.*

(1) *D is linear.*
(2) *D is locally modular.*
(3) *For some set $A \supset A^*$, the localization of D at A is modular.*

Proof. Without loss of generality, $A^* = \emptyset$.

(1) $\Longrightarrow$ (2). Assume D to be linear and let e be any element of $D \setminus acl(\emptyset)$. We need to show that the localization of D at e (denoted D_e) is modular. It suffices to show that D_e satisfies **(P)** of Lemma 4.2.1. To this end, let B be a subset of D and $a, b \in D$ such that $a \in acl(B \cup \{b, e\})$. To satisfy **(P)** we need a

($\diamond$) $d \in acl(B \cup \{e\}) \cap D$ such that $a \in acl(b, d, e)$.

If $b \in acl(B \cup \{e\})$ or $a \in acl(b, e)$ we are done. Thus we can assume that $b \notin acl(B \cup \{e\})$ and $\{a, b, e\}$ is independent. Letting $B' = acl(B \cup \{e\})$, $p = tp(ab/B')$ is strongly minimal. Let c be a canonical parameter of p. By the linearity of D, $\dim(c) \leq 1$. The element d satisfying ($\diamond$) is found via

Claim. (i) $a \in acl(b, c)$.

(ii) There is a $d \in D$ such that $acl(c, e) = acl(d, e)$.

From the data: c is the canonical parameter of p, $MR(b/B \cup \{e\}) = 1$ and $MR(ab/B \cup \{e\}) = 1$, we derive $MR(b/c) = MR(ab/c) = 1$, establishing (i). Since $a \in acl(b, c) \setminus acl(b)$, a depends on c over b. Combining this dependence with $\dim(c) \leq 1$ yields: $c \in acl(a, b)$, c is independent from b and c is independent from e. Since D is strongly minimal there is an automorphism f of the universe such that $f(c) = c$ and $f(b) = e$. Setting $d = f(a)$ yields an element meeting the requirements in (ii) and completes the proof of the claim.

Simply because c is the canonical parameter of a free extension of a type over $B \cup \{e\}$, $c \in acl(B \cup \{e\})$. Thus, $d \in acl(B \cup \{e\})$ and $a \in acl(b, d, e)$ (by the claim); i.e., ($\diamond$) holds for this d. This completes the proof that D_e satisfies **(P)**, hence D is locally modular.

(2) $\Longrightarrow$ (3). This case is trivially true.

(3) $\Longrightarrow$ (1). This case is proved in the two steps delineated in

Claim. (i) If the localization of D at some set B is linear, then D is linear.

(ii) A modular strongly minimal set is linear.

Suppose that D is not linear. Then, there are $(a, b) \in D^2$ and c such that $p = tp(ab/c)$ is strongly minimal, c is the canonical parameter of p and $\dim(c) > 1$. By applying an automorphism to (a, b, c) if necessary, we can require B to be independent from (a, b, c). The type $q = tp(ab/B \cup \{c\})$ is simply a free extension of p, hence q is strongly minimal with canonical parameter c. Since $\dim(c/B) > 1$, the localization of D at B is not linear, proving (i).

Turning to (ii) let D_0 be a modular strongly minimal set (in the universal domain of some t.t. theory). Let $a, b \in D_0$ and $C \subset D_0$ such that $p = tp(ab/C)$ is strongly minimal. First suppose $a \in acl(C)$. Then $1 = \dim(ab/C) = \dim(ab/C \cup \{a\}) = \dim(ab/a)$; i.e., the free extension of p over $C \cup \{a\}$ is definable over $acl(a)$. The canonical parameter of p is algebraic in a (by Lemma 4.1.5), hence has dimension ≤ 1. Similarly, if $b \in acl(C)$.

We are left with the case when a and b are not in $acl(C)$. Then $a \in acl(C \cup \{b\})$, so the modularity of D_0 yields a $c \in acl(C) \cap D_0$ such that $a \in acl(b, c)$. From $\dim(ab/C) = \dim(ab/C \cup \{c\}) = \dim(ab/c)$ we conclude as above that a canonical parameter of p has dimension ≤ 1 since it is algebraic in c. This completes the proof of (i), the claim and this final case of the lemma.

As a first application of the lemma we show that it is impossible for an uncountably categorical theory to contain both a locally modular strongly minimal set and a strongly minimal set which is not locally modular.

Lemma 4.2.5. *Let D_1 and D_2 be strongly minimal sets in the universal domain $\mathfrak{C}$ of an uncountably categorical theory. Then D_1 is locally modular if and only if D_2 is locally modular.*

Proof. Let M be an $\aleph_0$–saturated model over which both D_1 and D_2 are definable. Assume that D_1 is locally modular. By Lemma 4.2.4, for D any strongly minimal set over M, D is locally modular if and only if the localization of D over M is locally modular. Let D_i' be the localization of D_i over M (for $i = 1, 2$). Then, D_1' is locally modular and it suffices to show that D_2' is locally modular. Let a_1 be any element of $D_1' \setminus M$. By Exercise 3.3.18, there is an element $a_2 \in D_2$ which is interalgebraic with a_1 over M. It follows that the geometry associated to D_1' is isomorphic to the geometry associated to D_2'. Hence D_2' is locally modular (see Remark 4.2.1(i)). This proves the lemma.

A plane curve can be thought of as an element of the universe by identifying it with its canonical parameter. This identification supports the following concept.

Definition 4.2.5. *For D a strongly minimal set in a t.t. theory, a* definable ($\bigwedge$–definable) family of plane curves in *D is a definable ($\bigwedge$–definable) set X such that each element of X is the canonical parameter of a plane curve.*

Such families are common in the study of both vector spaces and algebraically closed fields. For instance, in Example 4.2.2(ii), the collection of equations $\{\, \alpha x + \beta y = b : \ b \in V \,\}$ (for fixed α, $\beta \in F$) is a definable family of plane curves (since b is the canonical parameter of $\alpha x + \beta y = b$). In Example 4.2.2(iii), where K denotes the universal domain of algebraically closed fields of a given characteristic, let C be the plane curve defined by $y = a_n x^n + a_{n-1} x^{n-1} + \ldots + a_1 x + a_0$, where $a_0, \ldots, a_n$ are arbitrary elements. Since $(a_0, \ldots, a_n)$ is the canonical parameter of C a definable family of plane curves is obtained by letting the coefficients vary over all $n+1$–tuples in K.

An elementary but fundamental fact about plane curves is

Lemma 4.2.6. *Let D be an $\emptyset$–definable strongly minimal set in a t.t. theory, C a plane curve in D, c the canonical parameter of C and a a generic of C.*

(i) The following are equivalent:

(1) $\dim(a) = 2$.
(2) $\dim(c) > 0$.
(3) *a depends on c.*
(4) *C is not contained in an $\emptyset$–definable set of Morley rank* 1.

(ii) If $\dim(a) = 2$, $\dim(c/a) = \dim(c) - 1$.

(iii) If $\dim(a) = 2$, *C may be chosen (from among the collection of equivalent plane curves) so that* $b \in C \implies \dim(b) = 2$.

(iv) Suppose that $k = \dim(c) > 0$ *and let* $\{a_1, \ldots, a_k\}$ *be a set of generics of C independent over c. Then,* $\{a_1, \ldots, a_k\}$ *is an independent set of generics of* D^2.

Proof. (i) Since a is a generic of a strongly minimal subset of D^2, $\dim(a/c) = 1$ and $\dim(a) \leq 2$. Thus, $\dim(a) = 2$ if and only if a depends on c. From here, part (i) follows from simple facts about canonical parameters.

(ii) Assuming that $\dim(a) = 2$, $\dim(ac) = \dim(a/c)+\dim(c) = 1+\dim(c)$. Also, $\dim(ac) = \dim(c/a)+\dim(a) = \dim(c/a)+2$, so $\dim(c/a) = \dim(c)-1$.

(iii) By (i) a depends on c. Thus, there is a formula $\varphi(x, c) \in p = tp(a/c)$ such that any $b \in \varphi(\mathfrak{C}, c)$ depends on c. We may chose $\varphi(x, c)$ to be strongly minimal, hence defining a plane curve C' equivalent to C. Then, $b \in C' \implies$ b depends on $c \implies \dim(b) = 2$ (by (i)).

(iv) Let $p \in S(c)$ be the (strongly minimal) type realized by generic elements of C. Since $\{a_1, \ldots, a_k\}$ is a Morley sequence in p and all Morley sequences in p are conjugate it suffices to find one Morley sequence in p of length k which is an independent set of generics of D^2. Let $I \subset C$ be an infinite Morley sequence in p. By Corollary 3.3.3, p is definable over I, hence $c \in acl(I)$. Let $B = \{b_1, \ldots, b_n\}$ be a minimal subset of I in which c is algebraic. By (i) each b_i is a generic of D^2. To complete the proof of (iv) we need only show

Claim. B is independent and $n = k$.

Assume, to the contrary, that B is dependent. Then b_n depends on $B' = \{b_1, \ldots, b_{n-1}\}$ (since B is indiscernible). Since $\dim(b_n) = 2$, the dependence of b_n on B' forces $\dim(b_n/B')$ to be 1. Since B is a Morley sequence in p, $\dim(b_n/B' \cup \{c\}) = 1$, hence b_n and c are independent over B'. Since $c \in acl(B)$, this independence forces c to be algebraic in B', contradicting the minimality assumption on B. Thus B is independent.

The following straightforward dimension calculation shows that $n = k$. We know that $\dim(c) = k$ and $\dim(B \cup \{c\}) = \dim(B) = 2n$ (since B is an independent set of generics of D^2 and $c \in acl(B)$). Furthermore, $\dim(B \cup \{c\}) = \dim(B/c)+\dim(c) = n+k$ (since B is a Morley sequence in a strongly minimal type over c). Thus $n = k$, completing the proof of (iv).

A linear strongly minimal set is relatively simple in that a localization at some element is modular. How does linearity effect the behavior of the collection of all plane curves simply as a family of subsets of D^2?

Lemma 4.2.7. *Let D be a linear strongly minimal set and C a plane curve in D with canonical parameter c.*

(i) If $a \in C$ and $\dim(a/c) = 1$, then $c \in acl(a)$.

(ii) If $C' \neq C$ is another plane curve in D, $\dim(C \cap C') \leq 2$.

Proof. (i) Certainly, this is true when $c \in acl(\emptyset)$. In the remaining case $\dim(c) = 1$ and $\dim(a) = 2$, hence $c \in acl(a)$ by Lemma 4.2.6(ii).

(ii) Let c' be a canonical parameter for C'. Since C and C' are distinct plane curves, $C \cap C' \subset acl(c, c')$. If $\dim(c) = \dim(c') = 0$ then $C \cap C' \subset acl(\emptyset)$, so (ii) holds in this case. Now suppose c or c' has dimension > 0, say $\dim(c) = 1$. Certainly, $\dim(C \cap C') \leq 1$ if $C \cap C' \subset acl(c)$, hence we can assume there is an $a \in C \cap C'$ with $\dim(a/c) = 1$. By Lemma 4.2.6(i), $\dim(a) = 2$ and c depends on a, hence $c \in acl(a)$. Applying the same lemma to the curve C' (which also contains a), $c' \in acl(a)$. Thus, $C \cap C' \subset acl(c, c') \subset acl(a)$, proving that $\dim(C \cap C') \leq 2$ in this final case.

Remark 4.2.3. When D is a linear strongly minimal set and C, C' are arbitrary plane curves in D, it is quite possible for $C \cap C'$ to be empty.

A collection of plane curves in a strongly minimal set is said to be *independent* if the corresponding collection of canonical parameters is independent.

The next lemma shows that the plane curves in a nonlinear strongly minimal set can have rather complicated intersections. First an example to illustrate this situation.

Example 4.2.3. Let D be the universal domain of algebraically closed fields of characteristic 0. Let C^* be the plane curve defined by the equation $y = ax + b$, where a and b are algebraically independent. Let X be the family of conjugates of C^* over $\emptyset$. If C is a plane curve in X defined by $y = a'x + b'$, then the pair (a', b') is a canonical parameter for C. Thus, each element of X has a canonical parameter of dimension 2. As a collection of subsets of D^2 the family X has the following two properties.

- If C and C' are independent elements of X then $\dim(C \cap C') \geq 2$. (This may fail in a linear strongly minimal set.)
- If a and b are independent generic elements of D^2, there is an element of X containing both a and b. (In the linear case no plane curve can contain an independent pair of generics of D^2.)
- If a is a generic of D^2 there are infinitely many elements of X containing a.

Lemma 4.2.8. *Suppose that D is an $\emptyset$–definable strongly minimal set in a t.t. theory containing the canonical parameter c^* of a plane curve C^* in D*

such that $\dim(c^*) = k > 1$. *Let* X *be the collection of conjugates of* C^* *over* $acl(\emptyset)$.

(i) *If* C *and* C' *are independent elements of* X, *then* $\dim(C \cap C') \geq 2$.

(ii) *If* a, b *is an independent pair of generics of* D^2 *there is a* $C \in X$ *containing both* a *and* b.

(iii) *For* $a \in D^2$ *generic,* $Y = \{ C \in X : a \text{ is a generic of } C \}$ *is infinite.*

Proof. (i) Since any two independent pairs of elements from X are conjugate over $acl(\emptyset)$, it suffices to find one independent pair of elements of X whose intersection has dimension ≥ 2. Let C be a generic of X with canonical parameter c, and let $a \in C$ be generic over c. Since $\dim(c) > 0$, $\dim(a) = 2$. By Lemma 4.2.6, $\dim(c/a) = k-1$. Let c' be a realization of $tp(c/\{a\} \cup acl(\emptyset))$ which is independent from c over a. Let C' be the element of X with canonical parameter c'. Since $a \in C' \cap C$ and $\dim(a) = 2$, to complete the proof of (i) it suffices to show that c' is independent from c over $\emptyset$. Since $\dim(c'/c) \geq k-1 > 0$, C' is distinct from C. Hence, $C \cap C'$ is finite, in particular $a \in acl(c, c')$. By the additivity of dimension, $\dim(c'ca) = \dim(c'/ca) + \dim(c/a) + \dim(a) = (k-1) + (k-1) + 2 = 2k$. Also, $\dim(c'ca) = \dim(a/c'c) + \dim(c'c) = \dim(c'c)$. Hence, $\dim(c'c) = 2k$, proving that c' and c are independent, as required.

(ii) Since all independent pairs of generics of D^2 are conjugate over $acl(\emptyset)$ and X is $\bigwedge$-definable over $acl(\emptyset)$ it suffices to find one $C \in X$ which contains an independent pair $\{a, b\}$ of generics of D^2. Since the canonical parameter of any $C \in X$ has dimension > 1 this follows directly from Lemma 4.2.6.

(iii) The proof that Y is infinite is left as an exercise to the reader. This proves the lemma.

Lemma 4.2.4 is such a basic result in the geometry of strongly minimal sets that from hereon it will be quoted tacitly. The term "linear" will be dropped in favor of the exclusive use of "locally modular".

Returning to the introductory discussion at the beginning of the chapter, it is local modularity that we will use as the dividing line between geometrically simple and geometrically complex strongly minimal sets. This choice for the dividing line is supported by the previous two lemmas and will be further justified in later sections. In these later sections we see that an uncountably categorical universal domain containing a strongly minimal set which is not locally modular is recognizably more complicated than one which does not.

How common are locally modular strongly minimal sets? Many of the examples of strongly minimal sets we've given so far are trivial, vector spaces or affine spaces. The next theorem suggests that this is not simply due to a lack of imagination; locally modular strongly minimal sets are the rule under some model-theoretic hypotheses.

Theorem 4.2.1 (Cherlin-Mills-Zil'ber). *A strongly minimal set in an* $\aleph_0$*-categorical theory is locally modular.*

For $\mathfrak{C}$ a universal domain a definable algebraically closed field K is called *pure* if every relation on K definable in $\mathfrak{C}$ is definable in the field language on K. Motivated by the known examples Zil'ber asked in [Zil84c]:

> Is there a strongly minimal set D which is not locally modular and does not have a definable pure algebraically closed field in D^{eq}?

This was answered affirmatively by Hrushovski in [Hru90a]:

Theorem 4.2.2. *There is a strongly minimal set D which is not locally modular such that D^{eq} does not contain an infinite definable group.*

Later (in Section 4.3.2) we will see that any nontrivial locally modular strongly minimal set D has a definable group in D^{eq} which is close to being a vector space.

In this chapter we only scratch the surface of what is known about strongly minimal sets. The reader is referred to Pillay's book [Pil] for further results.

Historical Notes. Local modularity, as a property of a strongly minimal set, was isolated by Zil'ber in [Zil80]. Lemma 4.2.4 is an alternate version of a theorem in Zil'ber's [Zil84a]. Theorem 4.2.1 was proved independently by Cherlin, Mills and Zil'ber; a good history can be found in [CHL85]. This result is an essential ingredient in the proof that a totally categorical theory is not finitely axiomatizable [CHL85].

Exercise 4.2.1. Prove Lemma 4.2.8(iii).

Exercise 4.2.2. Let S be a pregeometry. Show that S possesses one of the properties in Definition 4.2.2 if and only if the geometry associated to S possesses the property.

Exercise 4.2.3. Following the notation of the end of Example 4.2.1(iii), show that he vector space V and its action on A are definable in $(A')^{eq}$.

4.3 Global Geometrical Considerations

In this section we turn our attention from strongly minimal sets to the entire universe of an uncountably categorical theory. This study, which will occupy the remainder of the chapter, will be organized around the following admittedly vague questions. We begin with the premise that strongly minimal theories are the simplest uncountably categorical theories.

1. To what degree is every uncountably categorical universe built from strongly minimal sets?

2. If D_1 and D_2 are strongly minimal sets in an uncountably categorical universe, can we characterize the possible relations between elements from D_1 and elements from D_2? In other words, how much freedom do we have in specifying an uncountably categorical universe containing both D_1 and D_2?
3. If some strongly minimal set in the universe is locally modular, do we obtain sharper answers to the first two questions?

We first address Question 1 motivated by the behavior illustrated in the following examples.

Example 4.3.1. (i) This is a rather trivial example, but it defines what we consider to be the ideal situation. Let D be the universe of a strongly minimal theory and $X = D^n$ for some n. In this theory the set X (which has Morley rank n) can easily be decomposed in terms of strongly minimal sets. Explicitly, there are definable functions (the coordinate maps) $\pi_i : X \longrightarrow D$ $(1 \leq i \leq n)$ such that for any $a \in D$, a is in the definable closure of $(\pi_1(a), \ldots, \pi_n(a))$.

(ii) In this second example, the coordinatizing strongly minimal sets are a little harder to find. To begin with, let F be a field with more than two elements, V an infinite vector space over F and $X = V^2$. Let α and β be distinct nonzero elements of F. Define a binary relation $R(v, w)$ on X by the formula: for $a = (x_1, y_1)$, $b = (x_2, y_2) \in X$, $R(a, b) \Longleftrightarrow y_1 - \alpha x_1 = y_2 - \alpha x_2$. Similarly, $S(v, w)$ is defined on X by: for $a = (x_1, y_1)$, $b = (x_2, y_2) \in X$, $S(a, b) \Longleftrightarrow y_1 - \beta x_1 = y_2 - \beta x_2$. Let M be the model with universe X in a language consisting of two binary predicate symbols interpreted by R and S, respectively. The reader can show that $Th(M)$ is quantifier eliminable, uncountably categorical, the universal domain $\mathfrak{C}$ has Morley rank 2, and for any element a, $R(\mathfrak{C}, a)$ and $S(\mathfrak{C}, a)$ are strongly minimal. The sets of the form $R(\mathfrak{C}, a)$ and $S(\mathfrak{C}, a)$, as a ranges over elements of $\mathfrak{C}$, will be called curves of type R and type S, respectively. The canonical parameters of curves of type R form a strongly minimal set D_1 over $\emptyset$, and similarly the canonical parameters of curves of type S make up a strongly minimal set D_2. The sets D_1 and D_2 provide us with a coordinatization of the universe as follows. Let a be an element of $\mathfrak{C}$. There is a unique curve C_1 of type R containing a and a unique curve C_2 of type S containing a. Let $c_i \in D_i$ be the canonical parameter of C_i, for $i = 1, 2$. The axioms for T imply that a is the unique element of $C_1 \cap C_2$, hence $a \in dcl(c_1, c_2)$ and $(c_1, c_2) \in dcl(a)$. In this way the universe is decomposed into strongly minimal sets.

(iii) In this example (as in the previous two) the universe is the algebraic closure of a strongly minimal set (and some finite set of parameters). Here, however the coordinatizing strongly minimal sets are necessarily not $\emptyset$−definable. Let P be the projective plane over an infinite division ring F formulated in a language with a ternary relation symbol I and the interpretation: $I(a, b, c)$ if and only if a, b and c are collinear, for all a, b, $c \in P$. To keep the notation simple, assume P to be the universal domain of $Th(P)$.

Then, P has Morley rank 2 and for each $a \neq b \in P$, $I(P, a, b)$ (which is called a *line of* P) is strongly minimal. Let X be any line in P, ℓ the canonical parameter of X, and $a_1 \neq a_2$ two elements of $P \setminus X$. The following claim shows that P is almost strongly minimal.

Claim. For all $a \in P$ there are $x_1, x_2 \in X$ such that a and (x_1, x_2) are interdefinable over $A = \{\ell, a_1, a_2\}$.

For a given $a \in P$ let ℓ_i be the line containing a and a_i, for $i = 1, 2$. Let x_i be the element in the intersection of ℓ and ℓ_i. Since ℓ_i is the unique line containing a_i and x_i, and a is the unique element in the intersection of ℓ_1 and ℓ_2, a is in the definable closure of $A \cup \{x_1, x_2\}$. By a similar argument x_i is in the definable closure of $A \cup \{a\}$, proving the claim.

The next claim shows that there is no coordinatizing strongly minimal set over $acl(\emptyset)$.

Claim. There is no strongly minimal subset D of P^{eq} such that D is definable over $acl(\emptyset)$ (in P^{eq}) and for some $a \in P$, $acl(a) \cap D \neq acl(\emptyset)$.

A basic fact about any projective plane over a division ring is that its automorphism group is 2−transitive. In other words, for any $a_1 \neq a_2$ and $b_1 \neq b_2$ in P, there is an automorphism α of P such that $\alpha(a_i) = b_i$, for $i = 1, 2$. Suppose, to the contrary, that D is a strongly minimal subset of P^{eq} which is definable over $acl(\emptyset)$, and $a \in P$ is such that $acl(a) \cap D \neq acl(\emptyset)$. The 2−transitivity of $\text{Aut}(P)$ implies that $MR(a) = 2$. Hence, a cannot be algebraic in any $x \in (acl(a) \cap D) \setminus acl(\emptyset)$. This yields a $b \in P$, $b \neq a$, such that $x \in acl(b)$. If c in P is independent from a over $\emptyset$, then $acl(a) \cap acl(c) = acl(\emptyset)$. This contradicts the existence of an $\alpha \in \text{Aut}(P)$ such that $\alpha(a) = a$ and $\alpha(b) = c$, to prove the claim.

(iv) In this example, the universe can still be viewed as being constructed from strongly minimal sets, however, no finite collection of strongly minimal sets suffices. Let $M = \bigoplus_{i<\omega} (\mathbb{Z}_4)_i$, the direct sum of $\aleph_0$ copies of the additive group $\mathbb{Z}_4 = \mathbb{Z}/4\mathbb{Z}$. Let M^* be the universal domain of $Th(M)$. The theory of M^* is quantifier-eliminable and categorical in every infinite cardinality. By this quantifier-eliminability, $2M^*$ is a vector space over $\mathbb{Z}_2$ with no additional definable relations. In particular, $2M^*$ is a strongly minimal set. Furthermore, for each $a \in M^*$, $\{\, b \in M^* : 2b = 2a \,\} = a + 2M^*$, is strongly minimal. In this way, M^* is constructed from strongly minimal sets: Given $a \in M^*$, $2a$ is in the strongly minimal set $2M^*$ and a is in a strongly minimal set definable over $2a$. More globally, this could be written as $M^* = \bigcup_{x \in 2M^*} \{\, b \in M^* : 2b = x \,\}$, the union of a strongly minimal family of strongly minimal sets.

It is left as an exercise to the reader to see that there is *not* a collection of strongly minimal sets $D_1, \ldots, D_n$ over a finite $A \subset M^*$ such that every $a \in M^*$ is interalgebraic with a subset of $D_1 \cup \ldots \cup D_n$ over A. In this sense infinitely many strongly minimal sets are needed to construct M^*.

These examples help us formulate in a more specific way the question raised under 1 at the beginning of the section, and place limits on the possible answers. Let $\mathfrak{C}$ be the universal domain of an uncountably categorical theory. In the ideal situation there are (in $\mathfrak{C}^{eq}$) strongly minimal sets $D_1, \ldots, D_n$, with each D_i definable over $acl(\emptyset)$, such that each element a is interdefinable with a subset $C(a)$ of $D_1 \cup \ldots \cup D_n$. We think of $C(a)$ as a "set of coordinates for a with respect to $D_1, \ldots, D_n$". This ideal is attained in the first two examples, but fails in (iii) and (iv). In the projective plane we must settle for strongly minimal sets which are not definable over $acl(\emptyset)$. In the model M^* of (iv) no finitely many strongly minimal sets suffice to "coordinatize" the entire model. However, each $a \in M^*$ is interdefinable with the set $\{a, 2a\}$, and both $tp(2a)$ and $tp(a/2a)$ are strongly minimal. In other words, M^* is the union of a strongly minimal family of strongly minimal sets. These facts leave us with the hope that some useful reduction to strongly minimal sets will be possible. The final results (Proposition 4.3.2 and Corollary 4.3.4) will require a few preliminary definitions and lemmas.

In (i)–(iii) of the previous example, while the universe is not strongly minimal, it is the algebraic closure of a strongly minimal set over some finite set.

Definition 4.3.1. *The complete theory T is called* almost strongly minimal *if there is a strongly minimal set D, definable over a set A, such that $\mathfrak{C} = acl(D \cup A)$.*

Lemma 4.3.1. *The countable complete theory T is almost strongly minimal if and only if*

(∗) *there is a formula $\varphi(x, \bar{y})$ over $\emptyset$ and an isolated type $q \in S(\emptyset)$ such that for any model M there is an $\bar{a} \in q(M)$ such that $\varphi(x, \bar{a})$ is strongly minimal and $M = acl(\varphi(M, \bar{a}) \cup \{\bar{a}\})$.*

Proof. That (∗) implies T is almost strongly minimal is clear. Conversely, suppose T is almost strongly minimal, A is a set and $\psi(x, \bar{a})$ is a strongly minimal formula over A such that $\mathfrak{C} = acl(\psi(\mathfrak{C}, \bar{a}) \cup A)$. Let $D = \psi(\mathfrak{C}, \bar{a})$. By compactness we may take A to be finite. The proof is carried out in the following steps.

(a) For any strongly minimal formula $\theta(x)$ there is a set $B \supset A$ such that θ is over B and for any $b \in \theta(\mathfrak{C}) \setminus acl(B)$ there is a $c \in D$ interalgebraic with b over B.
(b) T is ω–stable.
(c) T is uncountably categorical.
(d) There is a sequence $\bar{b}$ realizing an isolated type, a strongly minimal formula $\varphi(x, \bar{b})$ and a set $B \supset \bar{b}$ such that $\mathfrak{C} = acl(\varphi(\mathfrak{C}, \bar{b}) \cup B)$.
(e) There is a sequence $\bar{c} \supset \bar{b}$ realizing an isolated type such that $\mathfrak{C} = acl(\varphi(\mathfrak{C}, \bar{b}) \cup \bar{c})$.

(a) First let B_0 be any set containing A and the parameters in θ. Let b be any element of $\theta(\mathfrak{C}) \setminus acl(B)$ and $\bar{d} \subset D$ such that $b \in acl(\bar{d} \cup A)$. Without loss of generality, $\bar{d}$ is of the form $d_0\bar{d}'$, where $d_0 \in D \setminus acl(B_0 \cup \bar{d}')$ and $b \notin acl(B_0 \cup \bar{d}')$. Let $B = B_0 \cup \bar{d}'$. Since d_0 is in a strongly minimal set over B and b depends on d_0 over B, d_0 and b are interalgebraic over B. All elements of $\theta(\mathfrak{C}) \setminus acl(B)$ realize the same type over B (since θ is strongly minimal), hence every element of $\theta(\mathfrak{C}) \setminus acl(B)$ is interalgebraic over B with an element of D, proving (a).

(b) Let M be a countable model of T containing A. For any a there is a $\bar{d} \subset D$ such that $a \in acl(M \cup \bar{d})$. Since D is strongly minimal $\{\, tp(\bar{d}/M) : \bar{d} \subset D \text{ is finite}\,\}$ is countable. Thus, $S_1(M)$ is countable. This proves that T is ω−stable.

(c) By Theorem 3.1.2 and (b) it suffices to show that T has no Vaughtian pair. Assume to the contrary that T has a Vaughtian pair. By Lemma 3.1.7 there is a Vaughtian pair (M, N) where $M \supset N$ are $\aleph_0$−saturated and $N \supset A$. For an arbitrary $a \in M \setminus N$ there is a $\bar{d} \subset \psi(M, \bar{a})$ such that $a \in acl(\bar{d} \cup A)$. Since $a \notin N$, $\bar{d} \not\subset N$, hence $\psi(M) \not\subset N$. By Corollary 3.1.2 there is a strongly minimal formula θ over N such that (M, N, θ) is a Vaughtian triple. Let c be an element of $\psi(M) \setminus N$. By (a) and the $\aleph_0$−saturation of N there is a set $B \subset N$ and a b satisfying θ such that b and c are interalgebraic over B. Then $b \in \theta(M) \subset N$, contradicting the fact that $c \notin N$ and proving (c).

(d) Since T is uncountably categorical there is a strongly minimal formula $\varphi(x, \bar{b})$, where $tp(\bar{b})$ is isolated. By (a) there is a set $B \supset \bar{b} \cup A$ such that $D \subset acl(\varphi(\mathfrak{C}, \bar{b}) \cup B)$. Thus, $\mathfrak{C} = acl(\varphi(\mathfrak{C}, \bar{b}) \cup B)$.

(e) By compactness there are formulas $\psi_0(x, \bar{y}_0, \bar{c}_0), \ldots, \psi_n(x, \bar{y}_n, \bar{c}_n)$ such that

(1) for all $\bar{d}_i \subset \varphi(\mathfrak{C}, \bar{b})$, $\psi_i(x, \bar{d}_i, \bar{c}_i)$ is algebraic and
(2) for any a there is an $i \leq n$ and a $\bar{d}_i \subset \varphi(\mathfrak{C}, \bar{b})$ such that $\models \psi_i(a, \bar{d}_i, \bar{c}_i)$.

Taking the disjunction of the ψ_i's gives one formula $\sigma(x, \bar{y}, \bar{c})$ such that

(∗) $\sigma(x, \bar{d}, \bar{c})$ is algebraic (for all $\bar{d} \subset \varphi(\mathfrak{C}, \bar{b})$) and for any a, $\models \sigma(a, \bar{d}, \bar{c})$ (for some $\bar{d} \subset \varphi(\mathfrak{C}, \bar{b})$).

There is a $\bar{b}$−definable set Z containing $\bar{c}$ such that (∗) holds for any $\bar{c}' \in Z$. Thus, (∗) holds for some $\bar{c}$ realizing an isolated type over $\bar{b}$, hence an isolated type over $\emptyset$.

This proves the lemma.

Corollary 4.3.1. *A countable almost strongly minimal theory is uncountably categorical.*

Note: There is an uncountably categorical theory which is not almost strongly minimal.See Example 4.3.1(iv).

Almost strongly minimal theories arise naturally in the study of ω−stable groups:

Proposition 4.3.1. *If G is a simple group of finite Morley rank, then $Th(G)$ is almost strongly minimal.*

Proof. Let D be a strongly minimal set in G. Let $\mathcal{C} = \{D \cap aQ : a \in G,\ Q$ is a definable subgroup of G and $D \cap aQ$ is infinite$\}$. Since D is strongly minimal, each element of $\mathcal{C}$ is cofinite in D. Thus, $\mathcal{C}$ is closed under finite intersections. By Proposition 3.5.1, $\bigcap \mathcal{C}$ is equal to some $D \cap aQ \in \mathcal{C}$. Clearly, $D \cap aQ$ is indecomposable. Let $g_0 \in D \cap aQ$ and $B = g_0^{-1}(D \cap aQ)$. Then B is indecomposable, strongly minimal and contains the identity 1.

Now let $\mathcal{B} = \{g^{-1}Bg : g \in G\}$ and $N =$ the group generated by $\bigcup \mathcal{B}$. Since each element of $\mathcal{B}$ is indecomposable and contains 1, Zil'ber's Indecomposability Theorem says that $N = B_1 \cdot \ldots \cdot B_k$ for some $B_1, \ldots, B_k \in \mathcal{B}$. Since N is normal (and not $\{1\}$) and G is simple, $G = B_1 \cdot \ldots \cdot B_k$. Thus, for some $g_1, \ldots, g_k \in G$, each element of G is of the form $g_1^{-1}b_1g_1 \cdot \ldots \cdot g_k^{-1}b_kg_k$, for some $b_1, \ldots, b_k \in B$. A fortiori, for $A = \{g_1, \ldots, g_k, g, a\}$, $G = acl(A \cup D)$. Thus, G is almost strongly minimal, proving the proposition.

Definition 4.3.2. *Let $\mathfrak{C}$ be the universal domain of a t.t. theory. A subset X of $\mathfrak{C}$, definable over A, is said to be* almost strongly minimal over A *if the restriction of $\mathfrak{C}$ to X is an almost strongly minimal theory. Equivalently, X is almost strongly minimal over A if there is a $B \subset X$ and a $D \subset X$ which is strongly minimal over $B \cup A$, such that $X \subset acl(D \cup B \cup A)$. A formula over A is almost strongly minimal if the set it defines is almost strongly minimal over A. A type over A is almost strongly minimal if it contains an almost strongly minimal formula over A.*

The next few results are used to show that elements in various relationships to almost strongly minimal sets are themselves elements of almost strongly minimal sets. The first lemma shows that we needn't be careful to choose a strongly minimal subset of X in verifying that X is almost strongly minimal.

Lemma 4.3.2. *Let $\mathfrak{C}$ be the universal domain of an uncountably categorical theory. Let X be a subset of $\mathfrak{C}$ definable over A such that for some $B \supset A$ there is an almost strongly minimal set D over B with $X \subset acl(D \cup B)$. Then, X is almost strongly minimal.*

Proof. To keep the notation simple suppose that $A = \emptyset$. We will prove the lemma in the case when D is strongly minimal, leaving the proof of the full result to the reader. By Corollary 3.1.2, there is a set B' and a strongly minimal set $D' \subset X$ definable over B'. In fact, by Proposition 3.3.3, we can require B' to also be a subset of X. Exercise 3.3.18 yields a set $C \supset B \cup B'$ such that $acl(D \cup C) = acl(D' \cup C)$. Thus, $X \subset acl(D' \cup C)$. Since $D' \subset X$, Proposition 3.3.3 can again be applied to find a $C' \subset X$ such that $X \subset acl(D' \cup C')$. This proves the lemma.

Note that any definable subset of an almost strongly minimal set is finite or almost strongly minimal (see Exercise 4.3.1).

In the next two results we see that almost strong minimality is preserved under finite unions and algebraic closure.

Lemma 4.3.3. *Let $\mathfrak{C}$ be the universal domain of an uncountably categorical theory and let $X_1, \ldots, X_n$ be almost strongly minimal subsets of $\mathfrak{C}$. Then, $X_1 \cup \ldots \cup X_n$ is almost strongly minimal.*

Proof. Simply from the definition, there are strongly minimal sets $D_i \subset X_i$, for $1 \leq i \leq n$, and a set A over which each D_i is definable such that $X_i \subset acl(D_i \cup A)$. Again quoting Exercise 3.3.18, we can take A to be large enough so that $D_i \subset acl(D_1 \cup A)$, for each i. By Lemma 4.3.2, $X_1 \cup \ldots \cup X_n$ is almost strongly minimal.

Lemma 4.3.4. *Let $\mathfrak{C}$ be the universal domain of an uncountably categorical theory. Let $A \subset B$ be sets, X an almost strongly minimal set over B, and a an element which is independent from B over A and algebraic in $X \cup B$. Then, a is an element of an almost strongly minimal set over A.*

Proof. Observe that we can, without loss of generality, take both A and B to be finite. Let $A' \supset A$ be a finite subset of $acl(A)$ such that $tp(a/A') = p$ is stationary.

Claim. There is a finite set $B' \supset A'$ and an almost strongly minimal set X' over B' such that $p(\mathfrak{C}) \subset acl(B' \cup X')$.

By Proposition 3.4.1 there is a k such that whenever $\{b_0, \ldots, b_k\}$ is independent over A', a is independent from some b_i over A'. Let $\{B_0, \ldots, B_k\}$ be a set of realizations of $tp(B/A')$ which is independent over A'. For $i \leq k$ there is an $f_i \in \mathrm{Aut}(\mathfrak{C})$ which maps B to B_i and is the identity on A'. Let $X_i = f_i(X)$, an almost strongly minimal set over B_i. Let a' be any realization of p. In the next paragraph we prove

$$a' \in acl(B_i \cup X_i). \tag{4.3}$$

For some i, a' is independent from B_i over A'. Since p is stationary, the unique free extension of p over $\mathfrak{C}$ does not split over A'. Thus, $a'B_i$ is conjugate over A' to aB. (Let f_1 be an automorphism of $\mathfrak{C}$ which fixes A' pointwise and maps a' to a. Then, $f_1(B_i)$ and B realize the same type over $A' \cup \{a\}$, hence there is an automorphism f_2 of $\mathfrak{C}$ which is the identity on $A' \cup \{a\}$ and maps $f_1(B_i)$ to B. The automorphism $f_2 f_1$ is the identity on A' and maps $a'B_i$ to aB.) By this conjugacy, $a' \in acl(B_i \cup X_i)$, as required.

To complete the proof of the claim let $B' = B_0 \cup \ldots \cup B_k$ and $X' = X_0 \cup \ldots \cup X_k$. By Lemma 4.3.3, X' is almost strongly minimal. For c an arbitrary realization of p, (4.3) applied with $a' = c$, shows that $c \in acl(B' \cup X')$. This proves the claim.

By a compactness argument there is a formula $\psi_0(v) \in p$ such that $\psi_0(\mathfrak{C}) \subset acl(B' \cup X')$. By Lemma 4.3.2, $\psi_0(\mathfrak{C})$ is almost strongly minimal. Let $\psi_0, \ldots, \psi_n$ be a list of the (finitely many) conjugates of ψ_0 over A. Then, $Y = \psi_0(\mathfrak{C}) \cup \ldots \cup \psi_n(\mathfrak{C})$ is definable over A, almost strongly minimal (by Lemma 4.3.3) and contains a, completing the proof of the lemma.

The reader should think of the next proposition in two parts. First, we find in $dcl(A \cup \{a\})$ an element of an almost strongly minimal set (which we think of as a "coordinate" for a over A). Secondly, there is a coordinate for a which significantly effects the relations between a and other elements of the universe. This result is central to our understanding of uncountably categorical theories.

Proposition 4.3.2. *Let $\mathfrak{C}$ be the universal domain of an uncountably categorical theory. Then for all sets A and $a \notin acl(A)$ there is a $d \in dcl(A \cup \{a\})$ such that d is an element of an almost strongly minimal set over A, and ad is dominated by d over A.*

Proof. To simplify the notation, take A to be the empty set. First we will find an element in $acl(a)$ (rather than $dcl(a)$) which meets the other requirements. Let M be an $\aleph_0$-saturated model which is independent from a. By Corollary 3.4.1, there is a strongly minimal set D over M and a sequence $\bar{c} = (c_1, \ldots, c_k)$ from D such that a is dominated by $\bar{c}$ over M and $\bar{c} \subset acl(M \cup \{a\})$. Let $B \subset M$ be a finite set such that $a\bar{c}$ is independent from M over B. (Hence, a is dominated by $\bar{c}$ over B and $\bar{c} \subset acl(B \cup \{a\})$.) By Corollary 4.1.4 there is an element b such that

(1) $b \in acl(a)$,
(2) a is independent from $B \cup \{\bar{c}\}$ over b, and
(3) $b \in acl(\bar{c}_0B_0, \ldots, \bar{c}_kB_k)$, for some set $\{\bar{c}_0B_0, \ldots, \bar{c}_kB_k\}$ of realizations of $tp(\bar{c}B/a)$ which is independent over a.

Claim. ab is dominated by b over $\emptyset$.

Let C be a set which is independent from b. It suffices to show that C' is independent from a for some conjugate C' of C over $\{a, b\}$, hence we can assume that C is independent from B over $\{a, b\}$. Since B is independent from $\{a, b\}$, B is independent from $C \cup \{b\}$. That is,

$$\{C, B, b\} \text{ is independent.} \tag{4.4}$$

By (2) and the fact that $\bar{c} \subset acl(B \cup \{a\})$, $\bar{c} \subset acl(B \cup \{b\})$. Thus, C is also independent from $\bar{c}$ over B. Since a is dominated by $\bar{c}$ over B, C is independent from a over B. (by (1)). By (4.4) and the transitivity of independence, C is independent from $B \cup \{a\}$. Combining this with (1) shows that C is independent from ab.

Claim. b is an element of an almost strongly minimal set over $\emptyset$.

Lemma 4.3.4 will be used to prove the claim. Let D_i be the strongly minimal set over B_i which is conjugate to D (for $i \leq k$), $B' = B_0 \cup \ldots \cup B_k$ and $X = D_0 \cup \ldots \cup D_k$. Then X is almost strongly minimal by Lemma 4.3.3. Since B_i realizes $tp(B/a)$, B_i is independent from a for each $i \leq k$. Since $\{B_0, \ldots, B_k\}$ is independent over a, $B' = B_0 \cup \ldots \cup B_k$ is independent from a (by the transitivity of independence). Thus, b is independent from B' (because $b \in acl(a)$). By (3), $b \in acl(B' \cup X)$, hence b belongs to an almost strongly minimal set over A, by Lemma 4.3.4.

Claim. There is a $d \in dcl(a)$ such that ad is dominated by d over $\emptyset$ and d is an element of an almost strongly minimal set over $\emptyset$.

Let X^* be the set of realizations of $tp(b/a)$ in $\mathfrak{C}$. Since X^* is finite there is a name d for X^* in $\mathfrak{C}$. Also, X^* is definable over a, hence $d \in dcl(a)$. Using: $b \in acl(A \cup \{d\})$ and $d \in dcl(A \cup \{a\})$, the reader can verify that ad is dominated by d over A. Finally, d is an element of an almost strongly minimal set over A since it is interalgebraic with a finite subset of an almost strongly minimal set.

This proves the proposition.

Remark 4.3.1. The most important part of the proposition is the existence of a "coordinate" for a from an almost strongly minimal set. However, that a is dominated by a coordinate d indicates the strength of the relationship between the two elements. The corollaries below make use of and reveal the ramifications of this domination relation.

Corollary 4.3.2. *Let a and b be elements of the universe of an uncountably categorical theory such that a depends on b. Then, there are $a' \in dcl(a)$ and $b' \in dcl(b)$ such that a' and b' belong to almost strongly minimal sets and a' depends on b'.*

Proof. By Proposition 4.3.2 there are $a' \in dcl(a)$ and $b' \in dcl(b)$ such that a' and b' belong to almost strongly minimal sets over $\emptyset$, a is dominated by a' over $\emptyset$ and b is dominated by b' over $\emptyset$. Since $a \not\perp b$ these domination relations force a' to be dependent on b', proving the corollary.

Corollary 4.3.3. *Let $\mathfrak{C}$ be the universal domain of an uncountably categorical theory. Then $\mathfrak{C}$ is almost strongly minimal or there are a and d such that $d \in dcl(a)$, $a \notin acl(d)$ and a is dominated by d over $\emptyset$.*

Proof. Suppose there are no elements a and d as in the statement. Then, by Proposition 4.3.2, for any $a \notin acl(\emptyset)$ there is a d interalgebraic with a such that d belongs to an almost strongly minimal set. Thus, any $a \notin acl(\emptyset)$ belongs to an almost strongly minimal set. By compactness there are $\emptyset$–definable sets $X_0, \ldots, X_k$ such that the sort of equality in $\mathfrak{C}$ is $X_0 \cup \ldots \cup X_k$ and each X_i is finite or almost strongly minimal. Thus, $\mathfrak{C}$ is almost strongly minimal.

Proposition 4.3.2 implies that the universe of an uncountably categorical theory is built from almost strongly minimal sets. This is formalized through the following concept. For α an ordinal, $C = \{ c_i : i < \alpha \}$ an indexed family, and $i \leq \alpha$, C_i denotes $\{ c_j : j < i \}$.

Definition 4.3.3. *Let $C = \{ c_i : i < \alpha \}$ be a sequence of elements in the universe of some complete theory. We call C an* almost strongly minimal construction *(*asm-construction, *for short) if for each $i < \alpha$, $tp(c_i/C_i)$ is almost strongly minimal or algebraic. A set A is* asm-constructible *if there is an enumeration of A which is an asm-construction.*

Remark 4.3.2. If $C = \{ c_i : i < \alpha \}$ and $C' = \{ c'_i : i < \alpha' \}$ are both asm-constructions, then the enumeration of $C \cup C'$ which lists C' after all the elements of C is also an asm-construction. Thus, the union of two asm-constructible sets is asm-constructible. In fact, the union of any number of asm-constructible sets is asm-constructible.

Corollary 4.3.4. *Let A be a set in the universe of an uncountably categorical theory. Then, $dcl(A)$ is asm-constructible.*

Proof. This proof is relegated to Exercise 4.3.2. It follows quickly from Proposition 4.3.2.

4.3.1 1–based Theories

We will return to asm-constructibility in arbitrary uncountably categorical theories in later sections, where the definable relations between different almost strongly minimal subsets of the universe are studied. In the remainder of this section the above results are extended assuming the theory contains a locally modular strongly minimal set.

A strongly minimal set D is modular if and only if, for all closed $X, Y \subset D$, X and Y are independent over $X \cap Y$. The following definition and Theorem 4.3.1 extend this property to uncountably categorical theories which contain a modular strongly minimal set.

Definition 4.3.4. *An uncountably categorical theory is called* 1–based *if for all subsets A and B of the universal domain $\mathfrak{C}$, A is independent from B over $acl(A) \cap acl(B)$.*

(As usual, if T is 1–based we also call $\mathfrak{C}$ 1–based.)

Lemma 4.3.5. *The following are equivalent for $\mathfrak{C}$ the universe of an uncountably categorical theory.*

(1) *$\mathfrak{C}$ is 1–based.*
(2) *For all $a \in \mathfrak{C}$ and sets A, a canonical parameter c of $tp(a/acl(A))$ is in $acl(a)$.*

Proof. First suppose $\mathfrak{C}$ to be 1–based, let $a \in \mathfrak{C}$ and $A \subset \mathfrak{C}$. Let $p \in S(\mathfrak{C})$ be the free extension of $tp(a/acl(A))$ and c a canonical parameter of p. Since $\mathfrak{C}$ is 1–based, p is a free extension of its restriction to $B = acl(a) \cap acl(A)$, hence $c \in acl(B) \subset acl(a)$, as desired. Now suppose that (2) holds. To prove that $\mathfrak{C}$ is 1–based it suffices to show that for all elements a and b, a is independent from b over $acl(a) \cap acl(b)$. For a arbitrary elements a and b let c be a canonical parameter of $tp(a/acl(b))$. Then, a is independent from b over c, $c \in acl(b)$ (because the relevant type is definable over $acl(b)$) and $c \in acl(a)$ (by (2)). Thus, a is independent from b over $acl(a) \cap acl(b)$, as required.

Remark 4.3.3. This equivalent definition explains the term "1–based". In Shelah's terminology, a type over $\mathfrak{C}$ is "based" on a set A if it is definable over A. An uncountably categorical theory is 1–based when, given a degree 1 type p and q the free extension of p in $S(\mathfrak{C})$, q is based on $acl(a)$ for any single a realizing p.

Theorem 4.3.1. *Given $\mathfrak{C}$ the universal domain of an uncountably categorical theory, $\mathfrak{C}$ is 1–based if and only if $\mathfrak{C}$ contains a locally modular strongly minimal set.*

The proof of this theorem will take several lemmas and propositions. Starting from the fact that the theorem is true on the restriction to a modular strongly minimal subset of the universe, we will prove the result for increasingly general sets.

First we take care of the easier direction of the biconditional:

Lemma 4.3.6. *Let $\mathfrak{C}$ be the universal domain of an uncountably categorical theory containing a strongly minimal set D which is not locally modular. Then, $\mathfrak{C}$ is not 1–based.*

Proof. For simplicity, suppose D is definable over $\emptyset$. Since D is not locally modular there is, by Lemma 4.2.4, a plane curve C in D such that a canonical parameter c of C has dimension $k > 1$. Let a be an element of C such that $\dim(a/c) = 1$. Suppose, towards a contradiction, that a and c are independent over $b \in acl(a) \cap acl(c)$. Since c is a canonical parameter of $tp(a/acl(c)) \supset tp(a/bc)$ and this type is a free extension of its restriction to b, Lemma 4.1.5(i) implies that $c \in acl(b)$. Hence, $c \in acl(a)$. Using the additivity of dimensions, $2 \geq \dim(a) = \dim(ac) = \dim(a/c) + \dim(c) = 1 + \dim(c)$. Since $\dim(c) > 1$ we have reached the contradiction which proves the lemma.

The next lemma shows that dependence between the elements of a modular strongly minimal set and other elements of the universe can only occur in a very simple way. The lemma implies that sets B and C are independent over $acl(B) \cap acl(C)$, when one of B or C is contained in a modular strongly minimal set (over $\emptyset$).

Lemma 4.3.7. *Let $\mathfrak{C}$ be the universal domain of an uncountably categorical theory, D an A–definable modular strongly minimal set. Then for all sets B,*

B and D are independent over $(D \cap acl(A \cup B)) \cup A$. Furthermore, for all sets $C \subset D$, B is independent from C over $acl(A \cup C) \cap acl(A \cup B)$.

Proof. Without loss of generality, both A and B are finite, and $D \cap acl(A \cup B) = D \cap acl(A)$. By taking A to be $\emptyset$ we can assume that $D \cap acl(B) = acl(\emptyset)$. We need to show that B and D are independent over $\emptyset$. Assuming, to the contrary, that B and D are dependent, there is a sequence $\bar{a}$ from D such that $\dim(\bar{a}/B) = \dim(\bar{a}) - 1$. As a consequence of Corollary 4.1.4 there is a set $B' \subset D$ such that

$$B' \underset{B}{\perp} \bar{a} \text{ and } B \underset{B'}{\perp} \bar{a}.$$

Thus, $\dim(\bar{a}/B') = \dim(\bar{a}) - 1$. By the modularity of D and Remark 4.2.1(iv), $\bar{a}$ and B' are independent over $D \cap acl(\bar{a}) \cap acl(B')$. Thus, there is a $c \in D \cap acl(\bar{a}) \cap acl(B')$ with $\dim(c) = 1$. Since B' and $\bar{a}$ are independent over B, $c \in acl(B)$. This contradicts the fact that $acl(B) \cap D = acl(\emptyset)$, to prove the first part of the lemma.

Turning to the furthermore clause, let $C' = D \cap acl(A \cup B)$. By the first part of the lemma, C is independent from B over $C' \cup A$. By the modularity of D, C and C' are independent over $acl(A \cup C) \cap acl(A \cup C')$. The transitivity of independence now implies that C and B are independent over $acl(A \cup C) \cap acl(A \cup B)$, completing the proof.

As a first application of this lemma we sharpen the picture of the relationship between two locally modular strongly minimal sets supplied by Lemma 4.2.5. (This corollary is not directly involved in the proof of the Theorem 4.3.1, however its central role in the theory justifies the digression.)

Corollary 4.3.5. *Let $\mathfrak{C}$ be the universal domain of an uncountably categorical theory and D_1, D_2 strongly minimal sets over $\emptyset$.*

(i) If D_1 and D_2 are both modular, then for all generic $b_1 \in D$ there is a $b_2 \in D_2$ which is interalgebraic with b_1.

(ii) Suppose that D_1 is locally modular, $a_1, b_1 \in D_1$ are independent generics and $b_2 \in D_2$ is generic. Then, there is an $a_2 \in D_2$ such that a_1 and a_2 are interalgebraic over $\{b_1, b_2\}$.

Proof. (i) Let M be a model. By Exercise 3.3.18, there are $a_i \in D_i \setminus M$, for $i = 1, 2$, such that a_1 and a_2 are interalgebraic over M. By Proposition 3.3.3, a_1 and a_2 are interalgebraic over $(D_1 \cap M) \cup (D_2 \cap M)$. Then, the modularity of D_1 and Lemma 4.3.7 yield a $b_1 \in D_1 \setminus acl(\emptyset)$ such that $b_1 \in acl((D_2 \cap M) \cup \{a_2\})$. By the same reasoning there is a $b_2 \in D_2 \setminus acl(\emptyset)$ which is algebraic in b_1. This proves the existence of *some* pair (b_1, b_2) satisfying the necessary conditions. However, all elements of $D_1 \setminus acl(\emptyset)$ realize the same type over $\emptyset$, so (i) holds.

(ii) This part follows immediately from (i) once we observe that for D any locally modular strongly minimal set (over $\emptyset$) and $a \in D \setminus acl(\emptyset)$, the localization of D at a is modular.

Lemma 4.3.8. *Let $\mathfrak{C}$ be the universal domain of an uncountably categorical theory which contains a locally modular strongly minimal set and let A be a subset of an almost strongly minimal set over C. Then for all B, A is independent from B over $acl(A \cup C) \cap acl(B \cup C)$.*

Proof. Without loss of generality, A is finite, and for simplicity, take C to be $\emptyset$. Let b be a canonical parameter of $tp(A/acl(B))$. Since $b \in acl(B)$ and A is independent from B over b, it remains only to show that $b \in acl(A)$. Let M be an $\aleph_0$-saturated model such that

(a) M is independent from $B \cup A \cup \{b\}$, and
(b) there is a modular strongly minimal set D, definable over M, and $\bar{c} \subset D$ such that A and $\bar{c}$ are interalgebraic over M.

By Lemma 4.3.7 there is a $d \in acl(\bar{c} \cup M) \cap acl(B \cup M)$ such that $\bar{c}$ and B are independent over $\{d\} \cup M$. In fact, A and B are independent over $\{d\} \cup M$ (since A and $\bar{c}$ are interalgebraic over M). Since $MR(A/B) = MR(A/B \cup M) = MR(A/B \cup M \cup \{d\}) = MR(A/M \cup \{d\})$, and b is a canonical parameter of $tp(A/acl(B))$, $b \in acl(M \cup \{d\})$. We can conclude that $b \in acl(A)$ using the facts: $d \in acl(M \cup \bar{c}) = acl(M \cup A)$, $b \in acl(M \cup \{d\})$, and $A \cup \{b\}$ is independent from M (by (a)). This proves the lemma.

Proof of Theorem 4.3.1. One direction of the "if and only if" is Lemma 4.3.6. Assume the universal domain contains a a locally modular strongly minimal set. Let A and B be sets and $C = acl(A) \cap acl(B)$. To prove the theorem we must show that A and B are independent over C. Without loss of generality, A is finite. By Proposition 4.3.2, there is a $d \in dcl(A \cup C)$ such that $A \cup \{d\}$ is dominated by d over C and d belongs to an almost strongly minimal set over C. Since $acl(A) \supset acl(C \cup \{d\})$, $acl(C \cup \{d\}) \cap acl(C \cup B)$ is also C. Thus, by Lemma 4.3.8, d is independent from B over C. Since $A \cup \{d\}$ is dominated by d over C, A is independent from B over C. This proves the theorem.

The following is due in various parts to Cherlin, Harrington, Lachlan and Zil'ber. It follows from Theorems 4.2.1 and 4.3.1. See [CHL85].

Corollary 4.3.6. *A totally categorical theory is 1−based.*

From this result Zil'ber, and later Cherlin, Harrington and Lachlan [CHL85], proved

Theorem 4.3.2. *A totally categorical theory is not finitely axiomatizable.*

(A considerable amount of work is required to prove the theorem from the preceding corollary.)

Corollary 4.3.7. *Let $\mathfrak{C}$ be the universal domain of a uncountably categorical theory and X an infinite definable subset of $\mathfrak{C}$. Then $\mathfrak{C}$ is 1−based if and only if the restriction of $\mathfrak{C}$ to X is 1−based.*

Proof. See Exercise 4.3.3.

The following definition and results help to round out our picture of 1–based theories by improving Proposition 4.3.2 and Corollary 4.3.4.

Definition 4.3.5. *Let $C = \{c_i : i < \alpha\}$ be a sequence of elements in the universal domain of a complete theory. We call C a* rank 1 construction *(*rk1-construction*) if for each $i < \alpha$, $MR(c_i/C_i) \le 1$. A set A is* rk1-constructible *if there is an enumeration of A which is a rk1-construction.*

Lemma 4.3.9. *Let $\mathfrak{C}$ be the universal domain of an uncountably categorical 1–based theory, A a set and $a \notin acl(A)$. Then there is a $c \in acl(A \cup \{a\})$ such that $MR(c/A) = 1$.*

Proof. By Proposition 4.3.2 it suffices to prove

Claim. Let a belong to an almost strongly minimal set over B. Then, there is a set $\bar{b} = \{b_0, \ldots, b_n\}$ such that $MR(b_i/B) \le 1$, for $i \le n$, and a is interalgebraic with $\bar{b}$ over B.

Let $M \supset B$ be an $\aleph_0$–saturated model which is independent from a over B. Let D be a strongly minimal set over M and $\bar{c} = \{c_0, \ldots, c_n\}$ a subset of D such that a is interalgebraic with $\bar{c}$ over M. Since the theory is 1–based, for each $i \le n$, there is $b_i \in acl(B \cup \{a\}) \cap acl(M \cup \{c_i\})$ such that a and $M \cup \{c_i\}$ are independent over $B \cup \{b_i\}$. Let $\bar{b} = \{b_0, \ldots, b_n\}$. For each i, $b_i \in acl(B \cup \{a\})$ and a is independent from M over B, hence $MR(b_i/B) = MR(b_i/M)$. Since $b_i \in acl(M \cup \{c_i\})$ and $MR(c_i/M) \le 1$, $MR(b_i/M) \le 1$ and b_i is interalgebraic with c_i over M. Because a is interalgebraic with $\bar{c}$ over M, a is interalgebraic with $\bar{b}$ over M, in fact, a is interalgebraic with $\bar{b}$ over B. The claim now follows from the fact that $MR(b_i/B) = MR(b_i/M) \le 1$, for each i. This proves the lemma.

Remark 4.3.4. The stronger version of the lemma with *acl* replaced by *dcl* is false. That is, there is a 1–based uncountably categorical theory containing a set A and $a \notin acl(A)$ such that there is no $c \in dcl(A \cup \{a\})$ with $MR(c/A) = 1$.

In [CHL85] the preceding lemma (in the totally categorical context) is called the Coordinatization Lemma.

Proposition 4.3.3. *Let A be a set in the universal domain of a 1–based uncountably categorical theory. Then $acl(A)$ is rk1-constructible.*

Proof. This is immediate by the previous lemma.

Remark 4.3.5. This finally gives us a reasonable picture of the manner in which the universal domain $\mathfrak{C}$ of a 1–based theory can be built from sets of Morley rank 1. For a any element of $\mathfrak{C}$ there is a set $\{c_0, \ldots, c_n\}$ interalgebraic with a such that $MR(c_i/C_i) \le 1$, for $i \le n$.

4.3.2 1−based Groups

This subsection is devoted to the study of definable groups in 1−based uncountably categorical theories. This examination will both illustrate the strength of the 1−based condition, and provide us with tools for later use in 1−based theories. A definition is needed to state the key result.

Definition 4.3.6. *Let G be an A−definable group in the universe of a complete theory. Let*

$$\mathcal{H} = \{ H : H \text{ is a subgroup of } G^n, \text{ for some } n, \text{ which is definable over } acl(A) \}.$$

G is called an abelian structure *if for every $n < \omega$, every definable subset of G^n is equal to a boolean combination of cosets of elements of $\mathcal{H}$.*

It is left to the exercises to show that a vector space is an abelian structure. It will be shown in Section 5.3.2 that a module, formulated in the natural language for modules over a particular ring, is an abelian structure. (In fact, later we will see that any locally modular strongly minimal group is not only an abelian structure, but essentially a vector space over some division ring.) An abelian structure has an abelian subgroup of finite index, supporting the use of the term “abelian”. (This is proved below in Corollary 4.3.12 in the context of uncountably categorical theories.)

It is not difficult to show directly that an algebraically closed field is not an abelian structure, although it also follows from the next theorem and Theorem 4.3.1.

Theorem 4.3.3. *Let G be an infinite definable group in the universal domain $\mathfrak{C}$ of an uncountably categorical theory. Then, G is an abelian structure if and only if $\mathfrak{C}$ is 1−based.*

The following type-oriented equivalent of being an abelian structure is easier to work with in proofs. We will only prove the lemma in the context of uncountably categorical theories, although it is true in a much broader setting. Remember, given an $\bigwedge$−definable group G and a set B, $S_n^G(B)$ denotes the set of complete n−types over B which extend the type defining G.

Lemma 4.3.10. *Let G be an A−definable group in the universal domain $\mathfrak{C}$ of an uncountably categorical theory. Then, G is an abelian structure if and only if*

(∗) *for any $n < \omega$ and $p \in S_n^G(G)$ there is a connected group $H \subset G^n$, definable over $acl(A)$, such that p is a left (or right) translate of the generic type of H.*

Proof. For simplicity, suppose $A = \emptyset$. In the proof we use the left translate version of (∗). The proof for right translates is the same.

First assume (∗) to be true. Fix an $n < \omega$ and let $\mathcal{H}_n = \{ H : H$ is a subgroup of G^n which is definable over $acl(\emptyset) \}$. We will prove by induction on Morley rank and degree that

(#) every definable subset X of G^n is equal to a boolean combination of cosets of elements of $\mathcal{H}_n$.

Let $\alpha = MR(G^n) + 1$ and $\omega^* = \omega \setminus \{0\}$. For any definable $X \subset G^n$, $(MR(X), \deg(X))$ is an element of the set of pairs $\alpha \times \omega^*$. Order $\alpha \times \omega$ lexicographically; i.e., for $\beta, \gamma < \alpha$ and $m, n \in \omega^*$, $(\beta, m) < (\gamma, n)$ if $\beta < \gamma$ or $\beta = \gamma$ and $m < n$. The induction will proceed using

$$\begin{aligned} &\text{if } X, Y \subset G^n \text{ are definable, } X \supset Y \neq \emptyset \text{ and } MR(X) = MR(Y), \\ &\qquad \text{then } (MR(X \setminus Y), \deg(X \setminus Y)) < (MR(X), \deg(X)). \end{aligned} \tag{4.5}$$

Let X be a B−definable subset of G^n. If $MR(X) = 0$, then X is a finite union of cosets of $\{0\}$, hence (#) is true in this case. Suppose that $MR(X) = \beta > 0$, $\deg(X) = k$, and (#) is true for any definable $Y \subset G^n$ with $(MR(Y), \deg(Y)) < (\beta, k)$. Let a be a generic element of X and $p \in S_n(G)$ a free extension of $tp(a/acl(B))$. Let $q \in S_n(G)$ and $g \in G^n$ be such that q is a generic type of an element H of $\mathcal{H}$ and $p = gq$. By Lemma 3.5.2 we can take q to be the generic of H^o, hence we may as well assume H is connected. Hence $\deg(H) = \deg(gH) = 1$ (by Corollary 3.5.3). Since a was chosen to be a generic of X, $\beta = MR(p) = MR(q) = MR(H)$. The formula defining gH is in p, hence $MR(X \cap gH) = \beta$ and (by (4.5)) $\deg(X \setminus gH) < \deg(X)$. Since $\deg(gH) = 1$, the same reasoning gives $MR(gH \setminus X) < \beta$. Thus, by induction, both $X \setminus gH$ and $gH \cap X = gH \setminus (gH \setminus X)$ are equal to a boolean combination of cosets of elements of $\mathcal{H}$. Since $X = (gH \cap X) \cup (X \setminus gH)$, we have proved that X is equal to a boolean combination of cosets of elements of $\mathcal{H}$.

Turning to the reverse implication, suppose G is an abelian structure and let $p \in S_n(G)$ have Morley rank β. Let $\mathcal{H} = \{ H : H$ is a subgroup of G^n which is definable over $acl(\emptyset) \}$.

Claim. There is a connected group $H \in \mathcal{H}$ and an $a \in G^n$ such that $MR(H) = \beta$ and the formula defining aH is in p.

Let $\varphi \in p$ be a formula of Morley rank β and degree 1; $X = \varphi(\mathfrak{C})$, which we can take to be a subset of G^n. A series of reductions will show that we can take X to be a coset of some element of $\mathcal{H}$. For $H, K \in \mathcal{H}$ and $a, b \in G^n$, if $Y = aH \cap bK \neq \emptyset$, then Y is a coset of $H \cap K$, also an element of $\mathcal{H}$. Thus X, which is equal to a boolean combination of elements of $\mathcal{H}$, can be written as a finite union of sets of the form $Y \setminus (Z_1 \cup \ldots \cup Z_n)$, where Y and $Z_1, \ldots, Z_n$ are cosets of elements of $\mathcal{H}$. If X is a finite union $Y_1 \cup \ldots \cup Y_k$ then some Y_i has Morley rank β. So, without loss of generality,

X is equal to $a_1H_1 \setminus (b_1K_1 \cup \ldots \cup b_nK_n)$, for some $H_1, K_1, \ldots, K_n \in \mathcal{H}$ and $a_1, b_1, \ldots, b_n \in G^n$. By the same reasoning we can require H_1 to be connected. Without loss of generality, $a_1H_1 \cap b_iK_i \neq \emptyset$, hence a coset of $H_1 \cap K_i$, for $1 \leq i \leq n$. Since H_1 is connected, $MR(H_1 \cap K_i) < MR(H_1)$, for $1 \leq i \leq n$, hence $MR(b_1K_1 \cup \ldots \cup b_nK_n) < MR(H_1)$. Since $MR(X) = \beta$, we conclude that $MR(H_1) = MR(a_1H_1) = \beta$, completing the proof of the claim.

With a and H as in the claim, let $q \in S_n(G)$ be the unique generic type of H. Then aq is the unique element of $S_n(G)$ having Morley rank β and containing the formula defining aH. Thus $p = aq$, completing the proof.

Corollary 4.3.8. *Let G be an $A-$definable abelian structure in the universal domain $\mathfrak{C}$ of an uncountably categorical theory. Let $p \in S_n(G)$ have Morley rank β and canonical parameter c. Then, there is a connected group $H \subset G^n$ of Morley rank β, definable over $acl(A)$, and an $a \in G^n$ such that the formula defining aH is in p and a name for aH is interdefinable with c over A.*

Proof. By the previous lemma there is a connected group $H \subset G^n$, definable over $acl(A)$, and an $a \in G^n$ such that the formula defining aH is in p. Let a^* be a name for aH. To show that a^* is interdefinable over A with c it suffices to prove

Claim. If f is an automorphism of $\mathfrak{C}$ which is the identity on A, then $f(p) = p$ if and only if $f(a^*) = a^*$.

Let $\mathrm{Aut}_A(\mathfrak{C})$ denote the set of automorphisms of $\mathfrak{C}$ which are the identity on A. The formula ψ over a^* defining aH has Morley rank β, degree 1, and is in p. Thus, if $f \in \mathrm{Aut}_A(\mathfrak{C})$ and $f(a^*) = a^*$, then $f(p) = p$ (since p is the unique extension of ψ or Morley rank β in $S_n(G)$). Now suppose $f \in \mathrm{Aut}_A(\mathfrak{C})$ and $f(p) = p$. Then, $f(\psi) \in p$, hence $aH \cap f(aH)$ has Morley rank β. The connected group H cannot have a proper definable subgroup of Morley rank β, so $H = f(H)$. Consequently, $aH = f(aH)$ and $f(a^*) = a^*$, completing the proof of the claim and the corollary.

Corollary 4.3.9. *Let G be an $A-$definable abelian structure in the universal domain $\mathfrak{C}$ of an uncountably categorical theory. Let $X \subset G^n$ be definable and*

$$\mathcal{H} = \{ H : H \text{ is a subgroup of } G^n \text{ definable over } acl(A) \}.$$

Then X is equal to a boolean combination of cosets of elements of $\mathcal{H}$.

Proof. See Exercise 4.3.5.

We are now in a position to prove the easy direction of Theorem 4.3.3.

Lemma 4.3.11. *Let $\mathfrak{C}$ be the universal domain of an uncountably categorical theory which contains an infinite definable abelian structure G. Then $\mathfrak{C}$ is 1−based.*

Proof. Suppose, to the contrary, that $\mathfrak{C}$ is not 1–based. Let G be definable over A. Since G is infinite there is a strongly minimal $D \subset G$ definable over some $B \supset A$. Since $\mathfrak{C}$ is not 1–based, D is not locally modular (by Theorem 4.3.1). By Lemma 4.2.4, D has a plane curve C such that the dimension (over B) of the canonical parameter c of C is > 1. Let a be a generic element of C over c, $p \in S(\mathfrak{C})$ the free extension of $tp(a/c)$, and observe that c is a canonical parameter of p. Since G is an abelian structure there is (by Corollary 4.3.8) a connected strongly minimal subgroup H of G^2, definable over $acl(A)$, and a $b \in G$ such that $C \cap bH$ is also strongly minimal and c is interalgebraic over A with a name b^* for bH. Since a is a generic of C, $a \in bH$, hence b^* and c are in $acl(A \cup \{a\})$. This contradicts Lemma 4.2.8(ii), completing the proof.

The proof of Theorem 4.3.3 will be complete once we have shown

Proposition 4.3.4. *Let $\mathfrak{C}$ be the universal domain of an uncountably categorical theory which contains an infinite definable group G, and assume $\mathfrak{C}$ is 1–based. Then, G satisfies*

(*) *for any $n < \omega$ and $p \in S_n^G(\mathfrak{C})$ there is a connected group $H \subset G^n$ definable over $acl(A)$ such that p is a translate of the generic type of H.*

A reasonable amount of work, found in subsequent lemmas, is required to prove the proposition. Most of the work revolves around stabilizers of types, introduced in Section 3.5. Indeed, the group H appearing in (*) is the stabilizer of p. For G an ω–stable group, $p \in S_n^G(G)$ can also be viewed as an element of $S_1^{G^n}(G)$. Thus, the facts proved in Section 3.5 about 1–types over an ω–stable group G extend transparently to $S_n^G(G)$, for any $n < \omega$. Also, facts proved about subgroups of G extend immediately to subgroups of G^n.

Remember, when H is an infinite group defined in the universal domain $\mathfrak{C}$ of an uncountably categorical theory then $\mathfrak{C}$ is 1–based if and only if the restriction to H is also 1–based. (See Corollary 4.3.7.)

Stabilizers enter our proof via

Lemma 4.3.12. *Let G be an ω–stable group, $p \in S_n(G)$ and $S = stab(p)$. Then*

(i) $MR(S) \leq MR(p)$, and

(ii) if $MR(S) = MR(p)$, p is a translate of a generic type of S and S is connected.

Proof. (i) This was proved in Lemma 3.5.1(ii).

(ii) Let A be a finite set over which p is definable and remember that S is a definable group over A. Let G' be a saturated model containing A, a a realization of $p \restriction G' = p'$ and g an element of S generic over $G' \cup \{a\}$. Since $tp(a/G' \cup \{g\}) = p \restriction (G' \cup \{g\})$, ga also realizes $p \restriction (G' \cup \{g\})$. By assumption, $MR(S) = MR(p)$, hence $MR(g/G' \cup \{a\}) = MR(p)$. Since g

and ga are interdefinable over $G' \cup \{a\}$, $MR(ga/G' \cup \{a\}) = MR(p)$; i.e., ga and a are G'–independent. Thus, $tp(ga/G' \cup \{a\}) = p \restriction (G' \cup \{a\})$.

Claim. S is connected.

Assuming that S is not connected there is an element g' of S, generic over $G' \cup \{a\}$, such that $tp(g'/G') \neq tp(g/G')$. Repeating the above argument, $g'a$ is also a realization of $p \restriction (G' \cup \{a\})$. An automorphism f of G which is the identity on $G' \cup \{a\}$ and maps ga to $g'a$ must map g to g'. This contradiction proves the claim.

Claim. Given $r \in S(G)$ the generic of S, p' is a right translate of $r \restriction G'$.

Since G' is saturated there is a $b \in G'$ realizing $p \restriction A$. Since S is connected, $r \restriction A$ has Morley degree 1 (by Corollary 3.5.3). Then, $tp(g/G' \cup \{a\}) = r \restriction (G' \cup \{a\})$, which is a free extension of $r \restriction A$, does not split over A (by Theorem 3.3.1(i)). Thus, $tp(a/A \cup \{g\}) = tp(b/A \cup \{g\})$ and $tp(gb/A \cup \{g\})$ is also $p \restriction (A \cup \{g\})$. Repeating the first paragraph of the proof for b instead of a, gb realizes p'. Drawing these facts together, for $r' = r \restriction G' = tp(g/G')$, $p' = r'b$, proving the claim.

By the second claim and Lemma 3.5.1, p is a right translate of the generic of S in $S(G)$, completing the proof.

Lemma 4.3.13. *Let $\mathfrak{C}$ be the universal domain of an uncountably categorical theory which contains an infinite definable group G, and assume $\mathfrak{C}$ is 1–based. Let $p \in S_n^G(\mathfrak{C})$ and $S = stab(p)$. Then, $MR(S) = MR(p)$ and S is definable over $acl(\emptyset)$.*

Proof. Without loss of generality, $n = 1$. Let c be the canonical parameter of p and let $\{a, a'\}$ be a Morley sequence over c in $p \restriction c$. Let $x = a'a^{-1}$ and notice that $MR(x) \geq MR(x/c) \geq MR(x/\{c, a\}) = MR(a'/\{c, a\}) = MR(p)$. Now let $g \in G$ be generic over $c \cup \{a, a'\}$ and let $q = pg$. Since right translates have the same stabilizer (Lemma 3.5.1(iii)), $stab(q) = S$. Proving that $MR(S) = MR(p) = MR(q)$ has been reduced to verifying that $x \in stab(q) = S$.

Since a and a' realize $p \restriction \{c, g\}$, ag and $a'g$ realize $q_0 = q \restriction \{c, g\}$. Since g is generic over $\{c, a, a'\}$, ag is generic over $\{c, a, a'\}$, hence x is independent from ag. The canonical parameter d of q is in $acl(ag)$ (since $\mathfrak{C}$ is 1–based) hence x is independent from $\{ag, d\}$. Furthermore, $a'g = xag$ by Corollary 3.5.1. Since q is the unique free extension of $tp(ag/c)$ we conclude that $x \in stab(q) = S$, completing the proof that $MR(S) = MR(p)$.

It remains to show that S is definable over $acl(\emptyset)$. Presently we only know S is definable over c, by the formula $\sigma(x, c)$, let's say. By Lemma 4.3.12(ii) S is connected, so we can take σ to have degree 1. Let $q_1 = q \restriction d$. Recapping what was proved above,

(a) d and c are independent.
(b) If $e \in S$ is generic over c and b realizes $q \restriction d$ and is generic over $\{e, c\}$ then $e \cdot b$ realizes q_1.

(c) For any d' realizing the type of d over $acl(\emptyset)$ independent from c, the conjugate of q_1 over d' is $pg' \upharpoonright d'$ for some g'.

We will show that $\sigma(v, c)$ is equivalent to any conjugate of itself over $acl(\emptyset)$.

Let c' be any realization of $tp(c/acl(\emptyset))$. Choose d' realizing $tp(d/acl(\emptyset))$ independent from $\{c, c'\}$. Let q_1' be the conjugate of q_1 over d' and b a realization of q_1' independent from $\{c, c'\}$ over d'. Let $q_1'' \in S_n^G(\mathfrak{C})$ be the unique free extension of q_1'. By (c) and Corollary 3.5.1, q_1' is a right translate of p, hence $S = \sigma(G)$ is $stab(q_1'')$. Since $stab(q_1'')$ is definable over d', $\sigma(v, c)$ is equivalent to a formula over d'. By the stationarity of types over $acl(\emptyset)$, c' and c have the same type over d', hence $\sigma(v, c')$ is equivalent to the same formula over d'. This proves the equivalence of $\sigma(v, c)$ and $\sigma(v, c')$, hence $\sigma(v, c)$ is equivalent to a formula over $acl(\emptyset)$.

This proves the lemma.

Combining Lemmas 4.3.10, 4.3.12 and 4.3.13 completes the proof of Theorem 4.3.3.

The following comes in handy when proving facts about the definable subsets of an abelian structure. (It is standard to use additive notation in an abelian structure.)

Corollary 4.3.10. *Let G be a $1-$based uncountably categorical group, $a \in G$ and p the unique free extension of $p' = tp(a/acl(\emptyset))$. There is a connected group S definable over $acl(\emptyset)$ (namely, the stabilizer of p) such that for any realization b of p',*

(1) *$b - a \in S$, and*
(2) *if b is independent from a, $b - a$ is a generic of S.*

Proof. Let S be the stabilizer of p, which by Lemma 4.3.13 is connected, definable over $acl(\emptyset)$ and has Morley rank $\alpha = MR(p)$.

Part (2) will be proved first (in a round about way). Let b be a realization of p' independent from a. Given $c \in S$ generic over a, $c + a$ realizes p'. Let c be a generic element of S which is independent from a. Since c and $c+a$ are interalgebraic over a, $MR(c+a/a) = MR(c/a) = \alpha$; i.e., $c+a$ is independent from a. Since p' is stationary, b and $c+a$ have the same type over $acl(\emptyset) \cup \{a\}$. Thus, $tp(b-a/acl(\emptyset) \cup \{a\}) = tp(c/acl(\emptyset))$, hence $b-a$ is a generic of S, proving (2).

Now assume only that b realizes p'. Let d be a realization of p independent from both a and b. By (2), both $a - d$ and $d - b$ are generic elements of S. Thus, $a - b = a - d + d - b$ is in S, completing the proof of the corollary.

As stated earlier, the most basic example of an abelian structure is an infinite vector space. In fact, it will be shown later that any module (formulated in the natural language for modules over a fixed ring) is an abelian structure. The next group of results investigates the degree to which every (uncountably categorical) abelian structure is a module, culminating in a proof that a strongly minimal abelian structure is (nearly) a vector space over a division ring of definable endomorphisms.

Corollary 4.3.11. *Let G be a $1-$based uncountably categorical group. Then, any connected definable subgroup of G^n is definable over $acl(\emptyset)$.*

Proof. Let H be a connected definable subgroup of G^n and let $p \in S_n(G)$ be the generic type of H. H is the stabilizer of p (by Corollary 3.5.3), hence H is definable over $acl(\emptyset)$ (by Lemma 4.3.13).

Definition 4.3.7. *A group is called* abelian-by-finite *if it has a definable abelian subgroup of finite index.*

Corollary 4.3.12. *A $1-$based uncountably categorical group G is abelian-by-finite.*

Proof. If G^o is abelian, G is abelian-by-finite, so we may take G to be connected. We will show that $Z(G)$ = the center of G, has finite index in G hence is all of G.

For $a \in G$ let $H_a = \{ (g, a^{-1}ga) : g \in G \}$, a definable subgroup of G^2.

Claim. H_a is connected.

Let $n = MR(G)$. If $(g,h) \in H_a$, h is interalgebraic with g over a. Hence, $MR(H_a)$ is also n. Suppose K is a definable subgroup of H_a of finite index, and let K_0, K_1 be the projections of K onto the first and second coordinates, respectively. Then, $n = MR(K) = MR(K_0) = MR(K_1)$, so, by the connectedness of G, $K_0 = G$. For any $g \in G$ there is a unique $x \in H_a$ whose first coordinate is x. Thus, K must be all of H_a, proving the claim.

By Lemma 4.3.11, H_a is definable over $acl(\emptyset)$, for any $a \in G$. A compactness argument shows that $\{ H_a : a \in G \}$ is some finite set of groups $\{H_{a_1}, \ldots, H_{a_k}\}$. For $a, b \in G$, $aZ(G) = bZ(G)$ if and only if $H_a = H_b$, hence $Z(G)$ has finite index in G. Since G is connected we conclude that G is abelian, as desired.

Definition 4.3.8. *Let G be a group which is $\bigwedge-$definable over A. Then, G^- denotes $acl(A) \cap G$. If B and C are subsets of G or elements of G, we write $B =^* C$ if $B + G^- = C + G^-$.*

With notation as in the definition, G^- is a subgroup of G, which is definable exactly when it is finite. The equivalence relation $=^*$ is simply the inverse image of equality under the quotient map from G into G/G^-.

Showing that a strongly minimal abelian structure is close to being a vector space requires the introduction of definable homomorphisms, accomplished as follows.

Definition 4.3.9. *Let G_0 and G_1 be $A-$definable groups (in the universal domain $\mathfrak{C}$ of a complete theory). A subgroup H of $G_0 \times G_1$ is called a $*-$homomorphism of G_0 into G_1 if*

– *H is definable,*
– *the projection of H onto the first coordinate is all of G_0, and*
– *$\{ a \in G_1 : (0,a) \in H \} = K$ is finite.*

*H is a $*-$*endomorphism of G_0 *if it is a $*-$homomorphism of G_0 into G_0. H is a $*-$*isomorphism of G_0 onto G_1 *if the projection of H onto the second coordinate is G_1 and $\{ a \in G_0 : (a,0) \in H \}$ is finite.*

With notation as in the definition, the $*-$homomorphism H is the graph of a definable homomorphism $\sigma_H : G_0 \longrightarrow G_1/K$, and, if H is $B-$definable, K is also $B-$definable. This homomorphism will also be called a $*-$homomorphism of G_0 into G_1. For $a \in G_0$, $\sigma_H(a)$ denotes the appropriate coset of K (hence a finite subset of G_1). Several elementary results and definitions are collected in

Definition 4.3.10. *Let G, H and K be $\emptyset-$definable abelian groups in the universal domain of a complete theory. Let $\mathcal{A} = \{ \sigma : \sigma$ is a $*-$homomorphism from G into $H \}$ and $\mathcal{B} = \{ \sigma : \sigma$ is a $*-$homomorphism from H into $K \}$. Addition on $\mathcal{A}$ is defined by the rule:*

$$\text{for } \sigma, \tau \in \mathcal{A} \text{ and } a \in G,\ (\sigma + \tau)(a) = \sigma(a) + \tau(a),$$

with the + on the right-hand side denoting addition on sets.

Multiplication between $\mathcal{A}$ and $\mathcal{B}$ is defined by:

$$\text{for } \sigma \in \mathcal{B},\ \tau \in \mathcal{A} \text{ and } a \in G,\ \sigma \cdot \tau(a) = \sigma(\tau(a)).$$

(We will largely be interested in multiplication when $G = H = K$.)

For $\sigma, \tau \in \mathcal{A}$ we write $\sigma =^* \tau$ if the graphs of σ and τ are $=^*$ as subsets of $G \times H$; i.e., $\sigma =^* \tau$ if for all $a \in G$, $\sigma(a) =^* \tau(a)$.

Let $\mathrm{Hom}^*(G,H) = \mathcal{A}/=^*$; *i.e.,* $\mathrm{Hom}^*(G,H)$ *is the set of equivalence classes of elements of $\mathcal{A}$ with respect to the equivalence relation $=^*$. Let* $\mathrm{End}^*(G)$ *denote* $\mathrm{Hom}^*(G,G)$. *The + operation extends to* $\mathrm{Hom}^*(G,H)$ *and $\cdot$ extends to* $\mathrm{End}^*(G)$ *in the obvious ways (for example, $(\sigma/=^*) + (\tau/=^*) = (\sigma+\tau)/=^*$). An element of* $\mathrm{Hom}^*(G,H)$ *is also called a $*-$homomorphism from G into H and an element of* $\mathrm{End}^*(G)$ *is called a $*-$endomorphism of G.*

Most statements made below involving a $*-$homomorphism σ remain valid after replacing σ by any $*-$homomorphism $\tau =^* \sigma$. This excuses the abuse of calling an element of $\mathrm{Hom}^*(G,H)$ a $*-$homomorphism. If G and H are $\emptyset-$definable groups, $a \in G$ and $\alpha \in \mathrm{Hom}^*(G,H)$ we write $\alpha(a) =^* b$ if there is a $*-$homomorphism σ such that α is $\sigma/=^*$ and $\sigma(a) =^* b$.

Remark 4.3.6. Let G and H be $\emptyset-$definable abelian groups in the universal domain of a complete theory. Suppose that σ is a $*-$isomorphism from G onto H and let $S \subset G \times H$ be the graph of σ. Let S^{-1} denote the inverse of S as a binary relation. Then, S^{-1} is the graph of a $*-$isomorphism τ from H into G and $\tau \cdot \sigma$ is a $*-$endomorphism of G which is $=^*$ the identity on G.

The straightforward proof of the following lemma is left to the reader.

Lemma 4.3.14. *Let G and H be definable abelian groups in the universal domain of a complete theory. Under the operations $+$ and $\cdot$ defined above,*
(i) $\mathrm{Hom}^*(G, H)$ *is an abelian group, and*
(ii) $\mathrm{End}^*(G)$ *is a ring.*

Definition 4.3.11. *For G and H definable abelian groups,* $\mathrm{Hom}^*(G, H)$ *is called the* group of $*$–homomorphisms from G into H *and* $\mathrm{End}^*(G)$ *is the* $*$–endomorphism ring of G.

If G and H are $\emptyset$–definable abelian groups, $a \in G$ and $\alpha \in \mathrm{Hom}^*(G, H)$, then $b =^* \alpha(a) \implies b \in acl(a)$. The next proposition shows (surprisingly) that all algebraic closure in a generic element of a connected 1–based group is witnessed by $*$–homomorphisms.

Proposition 4.3.5. *(i) Let G and H be $\emptyset$–definable groups in a 1–based uncountably categorical theory with G connected. Let A be a set, a an element of G generic over A and $b \in acl(A \cup \{a\}) \cap H$. Then, there is a $*$–homomorphism σ from G into H such that σ is definable over $acl(\emptyset)$ and $\sigma(a) =^* b'$ for some d, for some d independent from A with $MR(d) = MR(b)$. Furthermore, if $A = \emptyset$ we may take d to be b.*

(ii) Suppose, in addition, that $G = G_0 \times \ldots \times G_n$ and $a = (a_0, \ldots, a_n)$, where G_i is a connected group definable over $acl(\emptyset)$ and $a_i \in G_i$, for $i \le n$. Then, there are $\sigma_i \in$ $\mathrm{Hom}^*(G, H)$*, for $i \le n$, such that $\sigma(a) =^* \sum_{i \le n} \sigma_i(a_i)$.*

Proof. (i) Let $\bar{G}$ be the $\emptyset$–definable group $G \times H$. Let $p = tp((a, b)/acl(A))$, $X = p(\mathfrak{C})$, and S the stabilizer of the unique free extension of p in $S(\bar{G})$.

Claim. S is the graph of a $*$–homomorphism σ from G into H, definable over $acl(\emptyset)$.

Let $K = \{y : (0, y) \in S\}$. Since $b \in acl(A \cup \{a\})$, K is finite. For (a', b') an element of X A–independent from (a, b), $(a, b) - (a', b') \in S$ by Corollary 4.3.10. Hence, the projection of S onto the first coordinate contains a generic element, namely $a - a'$. Since G is connected, the projection of S onto the first coordinate must be all of G. Since the stabilizer of any type in a 1–based group is definable over $acl(\emptyset)$, σ is definable over $acl(\emptyset)$, proving the claim.

Since a is independent from A any $d \in \sigma(a)$ is independent from A. Moreover, $MR(d) = MR(b)$ since p is a translate of the generic type of S.

Now suppose $A = \emptyset$.

Claim. There is an element $-c =^* b$ such that $-c \in \sigma(a)$.

There is an element c such that $(-a, c) \in S$. Since $(-a, c) = (0, b+c) - (a, b)$, $(0, b+c)$ and (a, b) have the same coset with respect to S. Since S is definable over $acl(\emptyset)$ the difference of any two realizations of $q = tp((0, b + c)/acl(\emptyset))$

is in S. Since the group K is finite, the set of realizations of q is finite; i.e., $b+c \in H^-$. This completes the proof of the claim and (i) of the proposition.

(ii) Remember from Exercise 3.5.8 that a is a generic of G if and only if a_i is a generic of G_i (for $i \leq n$) and $\{a_0, \ldots, a_n\}$ is independent. Let S and σ be defined as in the proof of (i), bearing in mind that S is now a subgroup of $G_0 \times \ldots \times G_n \times H$. For $i \leq n$, let $S_i = \{ (x, y) : (0, \ldots, 0, x, 0, \ldots, 0, y) \in K \}$, where x is in the coordinate corresponding to G_i. As in the proof of (i), for each $i \leq n$, S_i is the graph of a $*-$homomorphism σ_i from G_i into H. It is easily verified that $\sum_{i \leq n} \sigma_i(a_i) =^* \sigma(a)$, proving the proposition.

Corollary 4.3.13. *Let G be a $1-$based uncountably categorical group. Any element of* End*(G) *is definable over $acl(\emptyset)$.*

(This corollary follows immediately from the preceding proposition.)

Theorem 4.3.4. *Let G be a $1-$based strongly minimal group.*

(i) $R =$ End$^*(G)$ is a division ring.

(ii) Let $b, a_0, \ldots, a_n \in G$ and suppose that b depends on $\{a_0, \ldots, a_n\}$. Then, there are $\alpha_0, \ldots, \alpha_n \in R$ such that $b =^* \sum_{i \leq n} \alpha_i a_i$.

Proof. (i) Let $\sigma \in R$ be nonzero. Both $K = ker(\sigma)$ and $H =$ the range of σ are $\emptyset-$definable subgroups of G. Since G is strongly minimal it has no infinite proper definable subgroup. Since σ is nonzero, this implies that K is a finite subgroup of G^- and $H = G$. Thus, the inverse of σ (as a relation on $G \times G$) is the graph of $*-$endomorphism of G. In other words, every nonzero element of G is invertible.

(ii) Since G is strongly minimal, $b \in acl(a_0, \ldots, a_n)$. There are $\alpha_i \in R$, $i \leq n$, such that $b =^* \sum_{i \leq n} \alpha_i a_i$ (by Proposition 4.3.5(ii)) completing the proof.

The following definition is a natural consequence of the theorem.

Definition 4.3.12. *A $1-$based strongly minimal group G is called a $*-$*vector space. *If $R \subset$* End*(G) *is a division ring of $*-$endomorphisms of G. Then G is called an $R - *-$*vector space.

An $R - *-$vector space G, for $R =$ End$^*(G)$, falls short of being a (quantifier-eliminable) vector space only in two ways.

- For $\sigma \in R =$ End$^*(G)$ and $a \in G$, $\sigma(a)$ may be a finite subset of G containing more than one element.
- G^- may contain a nonzero element.

Given $\sigma \in R$ and $a \in G$, $\sigma(a) \subset a + G^-$, hence σ induces an endomorphism of the group G/G^-. Moreover, G/G^- is an $R-$vector space. When $G^- = \{0\}$ this observation and Theorem 4.3.4 yield

Corollary 4.3.14. *Let G be a $1-$based strongly minimal such that $G^- = \{0\}$. There is a division ring R of endomorphisms of G, each definable over $acl(\emptyset)$, such that G is an $R-$vector space and every definable relation on G is equivalent to a boolean combination of $R-$linear relations.*

Proof. Left to the reader in Exercise 4.3.6.

Corollary 4.3.15. *The pregeometry on a $1-$based strongly minimal group G is projective.*

Proof. (We know simply from Theorem 4.3.1 that the pregeometry on G is locally projective.) Let $a, b, c_0, \ldots, c_n \in G$ be such that $a \in acl(b, c_0, \ldots, c_n)$. By Theorem 4.3.4 there are $\beta, \gamma_0, \ldots, \gamma_n \in \mathrm{End}^*(G)$ such that $a =^* \beta b + \gamma_0 c_0 + \ldots + \gamma_n c_n$. Any $d =^* \gamma_0 c_0 + \ldots + \gamma_n c_n$ is an element of $acl(c_0, \ldots, c_n) \cap G$ such that $a \in acl(b, d)$. This proves the projectivity of G.

Finally, we see that in the context of a $1-$based uncountably categorical theory strongly minimal groups are unique, up to $*-$isomorphism.

Corollary 4.3.16. *Let G and H be $\emptyset-$definable strongly minimal groups in the universal domain of a $1-$based uncountably categorical theory. Then, there is a $*-$isomorphism σ from G onto H which is definable over $acl(\emptyset)$.*

Proof. By Corollary 4.3.15, G and H are modular strongly minimal sets. Corollary 4.3.5 yield $a \in G \setminus G^-$ and $b \in H \setminus H^-$ which are interalgebraic over $\emptyset$. There is a $*-$homomorphism σ from G into H, definable over $acl(\emptyset)$, with $\sigma(a) =^* b$, by Proposition 4.3.5(i). Since G and H are strongly minimal, σ must be a $*-$isomorphism.

Historical Notes. Proposition 4.3.2 is more or less due to Shelah [She90, III.5]. In Zil'ber's early writings he worked with the condition "$\mathfrak{C}$ does not contain a definable pseudoplane". This property developed into a statement about canonical parameters in [CHL85]. Our main result, Theorem 4.3.1, is equivalent to one by Zil'ber in [Zil84a] and [Zil84b], and in the totally categorical context, implicit in [CHL85]. A generalization of the theorem, with up to date definitions, is found in [Bue86]. A weak version of Theorem 4.3.3 can be extracted from Zil'ber's writings. In its present form the theorem was first proved (independently) by Hrushovski and Pillay [HP87]. Proposition 4.3.5 and related results are due to Hrushovski in [Hru87].

Exercise 4.3.1. Show that any definable subset of an almost strongly minimal set is finite or almost strongly minimal.

Exercise 4.3.2. Prove Corollary 4.3.4.

Exercise 4.3.3. Prove Corollary 4.3.7.

Exercise 4.3.4. Prove that a vector space is an abelian structure.

Exercise 4.3.5. Prove Corollary 4.3.9

Exercise 4.3.6. Proof Corollary 4.3.14.

Exercise 4.3.7. Let G and H be $\emptyset$–definable strongly minimal groups in the universal domain of a 1–based uncountably categorical theory. Show that $\mathrm{End}^*(G) \cong \mathrm{End}^*(H)$ (as rings).

4.4 Automorphism Groups of Constructions

Let $\mathfrak{C}$ be the universal domain of an uncountably categorical theory. We proved that $\mathfrak{C}$ is asm-constructible, in fact, for any $a \in \mathfrak{C}$ there are $c_0, \ldots, c_n$, with $c_n = a$, such that $tp(c_i/\{c_0, \ldots, c_{i-1}\})$ is almost strongly minimal. If $\mathfrak{C}$ is also 1–based it is rk1-constructible. In this way $\mathfrak{C}$ is decomposed in terms of strongly minimal sets. In this section the structure gleaned from this decomposition is strengthened by describing, for X_1 and X_2 two almost strongly minimal subsets of $\mathfrak{C}$, the definable relations on $X_1 \times X_2$. We will see that (among other things) it always possible to choose the almost strongly minimal sets in a construction (like the one above) to be closely bound to one another, in a sense to be made precise momentarily. First a few motivating examples.

Example 4.4.1. (i) Let D be a definable set (over $\emptyset$) in the universal domain $\mathfrak{C}$ of a complete theory. A definable $X \subset D^{eq}$ is contained in $dcl(D)$. Any definable relation on $X \cup D$ reduces to a definable relation on D (in a way the reader is left to formalize). Notice that the condition $Y \subset dcl(D)$ is equivalent to "any $f \in \mathrm{Aut}(\mathfrak{C})$ which is the identity on D is also the identity on Y." Here the definable set X is "tightly bound" to D.

(ii) Let k_0 be an algebraically closed field of characteristic 0 and k_1 a proper elementary submodel. Let L be the language of fields together with a unary predicate P and $M_0 = (k_0, k_1)$ the model in L where k_1 interprets P and k_0 is the universe. Let (k^*, ℓ^*) be the universal domain of $Th(M_0)$. The relationship between k^* and the definable subset ℓ^* is described classically with the Galois group of k^* over l^*; i.e., the group of field automorphisms of k^* which fix ℓ^* pointwise. Below we use such automorphism groups to describe the relationships between two definable sets.

(iii) Let M be the abelian group $\bigoplus_{i<\omega}(\mathbb{Z}_4)_i$ and M^* the universal domain of $Th(M)$ (see Example 4.3.1(iv)). Let $V = 2M^*$, a a generic of M^* and $H = a + V$, which is also definable over $2a \in V$. For any $b \in H$ there is an automorphism of M^* which is the identity on V and maps a to b. Since $H \subset dcl(V \cup \{b\})$ for any $b \in H$, there is no nontrivial automorphism of M^* which fixes $V \cup \{b\}$ pointwise. In other words, the group G_0 of all automorphisms of M^* which fix V pointwise acts regularly on H. Let $G = \{\sigma : \sigma = \tau \restriction H$ for some $\tau \in G_0\}$.

Claim. $G \cong (V, +)$.

For $c \in V$ let τ_c be defined by: $\tau_c(x) = x + c$, for all $x \in H$. Observe that τ_c is in G. Fixing $a \in H$, any $\sigma \in G$ is determined by $\sigma(a)$; i.e., if $b = \sigma(a) = \sigma'(a)$, where $\sigma' \in G$, then $\sigma' = \sigma$. Since any $b \in H$ is $a + c$ for some $c \in V$, every $\sigma \in G$ is τ_c for some $c \in V$. Moreover, $\tau_c \cdot \tau_d = \tau_{d+c}$. This proves the claim.

(iv) Let P be the universal domain of the theory of a projective plane over an algebraically closed field, say the complex numbers. (P is formulated in a 2–sorted language with a single binary relation ϵ. The first sort in P is the set of "points" of P, the second the set of "lines" of P and $x\epsilon\ell$ is read "x lies on ℓ".) Let ℓ_1 and ℓ_2 be names for two distinct lines, D_i the set of points on ℓ_i, for $i = 1, 2$. Let $G_0 = \{\sigma \in \text{Aut}(P) : \sigma \restriction (D_1 \cup \{\ell_1, \ell_2\}) = \text{ the identity }\}$ and $G = \{\sigma \restriction D_2 : \sigma \in G_0\}$. The reader is asked to show the following in Exercise 4.4.1.

(a) For any $a_1 \neq a_2$ and $b_1 \neq b_2$ in $D_2 \setminus D_1$ there is a $\sigma \in G$ such that $\sigma(a_1) = b_1$ and $\sigma(a_2) = b_2$.
(b) Given $a_1 \neq a_2$ in $D_2 \setminus D_1$, $D_2 \subset dcl(D_1 \cup \{a_1, a_2\})$.

In group action terminology the action of G on $D_2 \setminus D_1$ is sharply 2–transitive. (It is 2–transitive because there is only one orbit in the set of distinct pairs from $D_2 \setminus D_1$. It is sharply 2–transitive because (by (b)) the σ in (a) is unique.)

The condition stated intuitively as "D_2 is closely bound to D_1" is formalized in

Definition 4.4.1. *Let $\mathfrak{C}$ be the universal domain of a complete theory. Let D_1 be an A–definable subset of $\mathfrak{C}$ and D_2 a subset of $\mathfrak{C}$ definable over $B \subset D_1 \cup A$. D_2 is said to be* finitely generated over $D_1 \cup A$ *if there are:*

(1) *a finite $\bar{b} \subset D_2$, and*
(2) *a function f, definable over $B \cup \bar{b}$, taking D_1^n onto D_2, for some n.*

When (1) and (2) hold $\bar{b}$ is called a fundamental generator of D_2 over $D_1 \cup A$ *and f is called the* generating function of D_2 over $D_1 \cup A$.

In Example 4.4.1(i) X is finitely generated over D with fundamental generator $\emptyset$. In Example 4.4.1(ii) k^* is not finitely generated over ℓ^*. The coset H of Example 4.4.1(iii) is finitely generated over V; any $b \in H$ is a fundamental generator with generating function $+$. Finally the projective line D_2 in Example 4.4.1(iv) is finitely generated over $D_1 \cup \{\ell_1, \ell_2\}$ with any pair of distinct points of $D_2 \setminus D_1$ as fundamental generator. It is left to the reader to describe the corresponding generating function.

Remark 4.4.1. Let $\mathfrak{C}$ be the universal domain of a complete theory, D_1 an $\emptyset$–definable subset of $\mathfrak{C}$ and D_2 a subset of $\mathfrak{C}$ definable over $B \subset D_1$. Let $\mathfrak{C}$, D_1 and D_2 be as in the definition, with D_1 $\emptyset$–definable (for simplicity).

(i) If D_2 is finite it is finitely generated over D_1.

(ii) If D_2 is finitely generated over D_1 there is a finite $\bar{b} \subset D_2$ such that $D_2 \subset dcl(D_1 \cup \bar{b})$. Thus, if D_1 is almost strongly minimal, D_2 is also almost strongly minimal (by Lemma 4.3.2).

(iii) If $D_2 \subset D_1^{eq}$ there is a definable function f taking D_1^n onto D_2, for some n. Hence, D_2 is finitely generated over D_1 with $\emptyset$ as a fundamental generator and generating function f.

(iv) Suppose that D_2 is finitely generated over D_1 and let f, B, n and $\bar{b}$ witness this as in the definition. Then, there is a B–definable $Y \subset D_1^{eq}$ and a $B \cup \bar{b}$–definable bijection g between D_2 and Y.

(This shows that there is little difference between a finitely generated set and an element of D_1^{eq}, although parameters outside of D_1^{eq} may be needed to define it.) To prove this fact let $E(x, y)$ be the equivalence relation on D_1^n defined over $B \cup \bar{b}$ by the rule: for all $\bar{x}$, $\bar{y} \in D_1^n$, $E(\bar{x}, \bar{y}) \iff f(\bar{x}) = f(\bar{y})$. Let Y be the set of equivalence classes of E and g the obvious $B \cup \bar{b}$–definable bijection from Y onto D_2 derived from f. Since E is a definable relation on D_1 there is a $B' \subset D_1$ such that E is B'–definable. Hence, $Y \subset D_1^{eq}$.

(v) If D_2 is finitely generated over $D_1 \cup A$ and D_3 is finitely generated over $D_2 \cup B$, then D_3 is finitely generated over $D_1 \cup A \cup B$. (The proof is left to the reader in Exercise 4.4.2.)

(vi) Let D_1 be A–definable and D_2 definable over $A \cup D_1$. Suppose there are: $\bar{b} \subset D_2$, a definable $X \subset D_1^n$ (for some n) and a $(A \cup \bar{b})$–definable function f taking X onto D_2. It is easy to find from f a function defined on all of D_1^n, hence D_2 is finitely generated over $D_1 \cup A$.

In Example 4.4.1(iii), where a is a generic of M^* and $H = a + V$, $\{2a, a\}$ defines an asm-construction of a. Here, H is not only a strongly minimal set over $2a$, but is finitely generated over V (= the strongly minimal set containing $2a$). We will show later that for $\mathfrak{C}$ the universal domain of an uncountably categorical theory and $b \in \mathfrak{C}$, there is an asm-construction $c_0, \ldots, c_n$ of b where c_i is an element of an almost strongly minimal set X_i, definable over $C_i = \{c_0, \ldots, c_{i-1}\}$, such that, for $1 \leq i \leq n$, X_i is finitely generated over $X_{i-1} \cup C_i$. Thus, we can gain more detailed information about an uncountably categorical theory through the relation "D_2 is finitely generated over D_1".

The definable relations holding between the elements of two definable sets are best studied with the following object.

Definition 4.4.2. *Let $\mathfrak{C}$ be the universal domain of a complete theory. Let D_1 be an A–definable subset of $\mathfrak{C}$ and D_2 a subset of $\mathfrak{C}$, $\bigwedge$–definable over $D_1 \cup A$. A map $\sigma : D_2 \longrightarrow D_2$ is an* automorphism of D_2 over $D_1 \cup A$ *if σ is the restriction to D_2 of an element of* Aut($\mathfrak{C}$) *which is the identity on $D_1 \cup A$. The collection of all automorphisms of D_2 over $D_1 \cup A$ is a group denoted* Aut($D_2/D_1 \cup A$). *When D_2 is definable and finitely generated over $D_1 \cup A$,* Aut($D_2/D_1 \cup A$) *is called the* binding group of D_2 over $D_1 \cup A$. *When $A = \emptyset$ it is omitted.*

In Example 4.4.1(i), $\mathrm{Aut}(X/D)$ is trivial and in (ii), $\mathrm{Aut}(k^*/\ell^*)$ is $\mathrm{Gal}(k^*/\ell^*)$, the Galois group of k^* over ℓ^*. In the third example, $\mathrm{Aut}(H/V) \cong (V, +)$. In (i) and (iii), with D_1 and D_2 the relevant definable sets, D_2 is finitely generated over D_1. This degree of control over the relations between the elements of D_2 and D_1 is reflected in the simplicity of $\mathrm{Aut}(D_2/D_1)$. In both (i) and (iii) $\mathrm{Aut}(D_2/D_1)$ is a definable group in the following sense.

Definition 4.4.3. *Let $\mathfrak{C}$ be the universal domain of a complete theory. Let D_1 be an $A-$definable subset of $\mathfrak{C}$ and D_2 a subset of $\mathfrak{C}$, $\bigwedge-$definable over $D_1 \cup A$. We say that $\alpha \in G = \mathrm{Aut}(D_2/D_1 \cup A)$ is definable if α agrees with a definable function g_α on D_2. In this case α is identified with a name for g_α. G is called* definable *if every element of G is definable and (G, D_2) is a definable group action.*

Remark 4.4.2. In the definition, when each $\alpha \in G$ is definable $G \subset \mathfrak{C}$ since we identify a definable function with its name. Remember: (G, D_2) is a definable group action if G and D_2 are definable sets and both the group operation and the action of G on D_2 are definable.

The goal of this section is the following set of "Ladder Theorems" by Zil'ber. The first two are improvements of Corollary 4.3.4 and Proposition 4.3.3, respectively.

Theorem 4.4.1 (Main Ladder Theorem). *Let $\mathfrak{C}$ be the universal domain of an uncountably categorical theory and a an element. Then there is a sequence $a_0, \ldots, a_{n-1}, a_n = a$ and definable sets $D_0, \ldots, D_n$ such that for $i \leq n$ and $A_i = \{a_0, \ldots, a_{i-1}\}$,*

(1) *$a_i \in dcl(a)$;*
(2) *$a_i \in D_i$;*
(3) *D_0 is $\emptyset-$definable and almost strongly minimal; D_i is finite or almost strongly minimal and D_i is definable over A_i;*
(4) *D_i is finitely generated over $D_0 \cup \ldots \cup D_{i-1}$ (when $i > 0$);*
(5) *$G_i = \mathrm{Aut}(D_i/D_0 \cup \ldots \cup D_{i-1})$ is definable.*

Theorem 4.4.2 (1−based Ladder Theorem I). *Let $\mathfrak{C}$ be the universal domain of a 1−based uncountably categorical theory and a an element. Then there is a sequence $a_0, \ldots, a_{n-1}, a_n = a$ and definable sets $D_0, \ldots, D_n$ such that for $i \leq n$ and $A_i = \{a_0, \ldots, a_{i-1}\}$,*

(1) *$a_i \in acl(a)$;*
(2) *$a_i \in D_i$;*
(3) *D_i is finite or strongly minimal and D_i is definable over A_i;*
(4) *D_i is finitely generated over $D_0 \cup \ldots \cup D_{i-1}$ (when $D_0 \cup \ldots \cup D_{i-1}$ is infinite);*
(5) *$G_i = \mathrm{Aut}(D_i/D_0 \cup \ldots \cup D_{i-1})$ is definable, has Morley rank ≤ 1 and is abelian-by-finite (when $D_0 \cup \ldots \cup D_{i-1}$ is infinite).*

An infinite definable abelian group G in a universal domain is called *minimal abelian* if there is no infinite definable subgroup of G.

Theorem 4.4.3 (Simple Ladder Theorem). *Let $\mathfrak{C}$ be the universal domain of an uncountably categorical theory and a an element. There is a sequence of definable sets $D_0, \ldots, D_n$ such that for all $i \leq n$*

(1) $a \in D_n$.
(2) *D_i is finite or almost strongly minimal and finitely generated over $D_0 \cup \ldots \cup D_{i-1}$ (when $D_0 \cup \ldots \cup D_{i-1}$ is infinite).*
(3) *If $D_0 \cup \ldots \cup D_{i-1}$ is infinite, $G_i = \text{Aut}(D_i/D_0 \cup \ldots \cup D_{i-1})$ is definable. When G_i is infinite it is simple or minimal abelian.*

Theorem 4.4.4 (1–based Ladder Theorem II). *Let $\mathfrak{C}$ be the universal domain of a 1–based uncountably categorical theory and a an element. Then there is a sequence $a_0, \ldots, a_{n-1}, a_n = a$ and definable sets $D_0, \ldots, D_n$ such that for $i \leq n$ and $A_i = \{a_0, \ldots, a_{i-1}\}$,*

(1) $a_i \in acl(a)$;
(2) $a_i \in D_i$;
(3) *D_i is finite or strongly minimal and D_i is definable over A_i;*
(4) *D_i is finitely generated over $D_0 \cup \ldots \cup D_{i-1}$ (when $D_0 \cup \ldots \cup D_{i-1}$ is infinite);*
(5) *When $D_0 \cup \ldots \cup D_{i-1}$ is infinite both $G_i = \text{Aut}(D_i/D_0 \cup \ldots \cup D_{i-1})$ and the action of G_i on D_i are definable over $D_0 \cup \ldots \cup D_{i-1}$. Moreover, when G_i is infinite it is strongly minimal (and abelian).*

The 1–based Ladder Theorem I will follow rather quickly from the Main Ladder Theorem using Proposition 4.3.3. The Simple Ladder Theorem says that in the sequence of almost strongly minimal sets we can choose $G_i = \text{Aut}(D_i/D_0 \cup \ldots \cup D_{i-1})$ to be finite, minimal abelian, or simple if we are willing to sacrifice other properties; namely, that $dcl(a) \cap D_i$ is nonempty and D_i is definable over a.

With notation as in the Main Ladder Theorem, $\{a_0, \ldots, a_n\}$ is an asm-construction of a. The existence of a sequence satisfying (1)–(3) was proved in Corollary 4.3.4. The object of this section is to obtain an asm-construction with the additional properties specified in (4)–(6). Certain results leading up to Corollary 4.3.4 (Proposition 4.3.2, for one) will be redone in this section to emphasize different points and increase the scope of the methods.

The first major result of the section indicates when we can expect $\text{Aut}(D_2/D_1)$ to be definable.

Theorem 4.4.5 (Binding Group Theorem). *Let $\mathfrak{C}$ be the universal domain of a t.t. theory, D_1 an $\emptyset$–definable set and D_2 a D_1–definable set which is finitely generated over D_1. Then $\text{Aut}(D_2/D_1)$ is a definable group.*

The proof of this theorem involves the notion of the type of an element over a definable set. When $\mathfrak{C}$ is a universal domain, a an element and X an

$\emptyset$–definable subset of $\mathfrak{C}$ the type of a over X is, by fiat, not a type since X is not a set. However, many properties of $tp(a/X)$ reduce to properties of types over sets by the following lemma.

Lemma 4.4.1. *Let $\mathfrak{C}$ be the universal domain of a t.t. theory T, a an element and X an $\emptyset$–definable subset of $\mathfrak{C}$.*

(i) There is a type r over a such that $tp(b/X) = tp(a/X)$ if and only if b realizes r.

(ii) r is equivalent to a type over a subset of X of cardinality $\leq |T|$.

(iii) There is $X_0 \subset X$ of cardinality $\leq |T|$ such that $tp(a/X_0)$ implies $tp(a/X)$. In fact, for any set A there is a $Y \subset X$ of cardinality $\leq |T| + |A|$ such that $(*)$ if A is conjugate to B over Y there is an elementary map from A to B which is the identity on X. (The notation $tp(A/Y) \models tp(A/X)$ will be used as shorthand for $(*)$.)

(iv) If $tp(b/X) = tp(a/X)$ there is an automorphism of $\mathfrak{C}$ which maps a to b and is the identity on X.

(v) There is a formula $\rho(x)$ over a implied by r such that any b realizing $tp(a) \cup \{\rho(x)\}$ realizes r.

Proof. (i) Let $\varphi(x, \bar{y})$ be a formula over $\emptyset$ and $E_\varphi(x, x')$ the $\emptyset$–definable equivalence relation expressing:

$$\text{for all } \bar{y} \text{ from } X\,(\,\varphi(x, \bar{y}) \leftrightarrow \varphi(x', \bar{y})\,).$$

Letting $\Xi(x, x') = \{\,E_\varphi(x, x') : \varphi \text{ is a formula over } \emptyset\}$ and $r = \Xi(x, a)$ produces a type meeting the requirements of (i).

Turning to (ii), since $\mathfrak{C}$ is assumed to be t.t., $tp(a/X)$ is definable over X (by Lemma 3.3.11). Thus, given a formula $\varphi(x, \bar{y})$ over $\emptyset$ there is a formula $\psi_\varphi(\bar{y})$ over $\bar{b}_\varphi \subset X$ such that

$$\text{for all } \bar{y} \text{ from } X(\models \varphi(a, \bar{y}) \text{ if and only if } \models \psi_\varphi(\bar{y})\,).$$

Then, for φ any formula over $\emptyset$, $E_\varphi(x, a)$ is equivalent to the $\bar{b}_\varphi$–definable relation:

$$\text{for all } \bar{y} \text{ from } X(\,\varphi(x, \bar{y}) \longleftrightarrow \psi_\varphi(\bar{y})\,).$$

There are $|T|$ many sets of the form $\bar{b}_\varphi$, so we have proved (ii).

(iii) This is immediate by (ii).

(iv) Since X has the same cardinality as $\mathfrak{C}$ (when it is infinite) we cannot simply use the homogeneity of $\mathfrak{C}$ to find such an automorphism. Instead an automorphism of $\mathfrak{C}$ is constructed using

Claim. There is a chain of elementary maps f_α, $\alpha < \kappa = |\mathfrak{C}|$, such that for all α,

(1) $f_\alpha \restriction X$ is the identity on X;
(2) $f_\alpha(a) = b$;
(3) for all $c \in \mathfrak{C}$ there are β, $\gamma < \kappa$ such that c is in the domain of f_β and c is in the range of f_γ.

To begin let f_0 be the elementary map which is the identity on X and takes a to b. The detailed construction of the chain will be left to the reader. The essential features are contained in the proof of

($\sharp$) If f is an elementary map defined on $X \cup A$ (for some set A) and $c \in \mathfrak{C}$, then there is an elementary map g extending f which is defined on $X \cup A \cup \{c\}$.

By (iii) there is a set $Y \subset X$ such that $tp(A \cup \{c\}/Y) \models tp(A \cup \{c\}/X)$. Since $tp(f(A)/Y) = tp(A/Y)$ there is a d such that $tp(f(A) \cup \{d\}/Y) = tp(A \cup \{c\}/Y)$. Since the type of $A \cup \{c\}$ over Y implies its type over X the map g which extends f and takes c to d is elementary. This proves ($\sharp$) and the claim.

To complete the proof we need only observe that $g = \bigcup_{\alpha<\kappa} f_\alpha$ is an automorphism of $\mathfrak{C}$ which is the identity on X and takes a to b.

(v) Let $\Xi' = \{ E_i(x, x') : i < |T| \}$ be a set of formulas obtained from $\Xi(x, x')$ by closing under finite conjunctions. Since any formula implied by $\Xi'(x, a)$ is implied by $E_i(x, a)$ for some i, there is an i such that $(MR(\Xi'(x, a)), \deg(\Xi'(x, a))) = (MR(E_i(x, a)), \deg(E_i(x, a)))$. Let $p = tp(a)$.

Claim. Any b realizing $p \cup \{E_i(x, a)\}$ also realizes r.

Assuming the claim to fail there is a $j \neq i$ such that $p \cup \{E_i(x, a)\}$ does not imply $E_j(x, a)$. Let b be a realization of $p \cup \{E_i(x, a)\}$ such that $\not\models E_j(b, a)$. Then $\Xi'(x, a)$ and $\Xi'(x, b)$ are extensions of $p \cup \{E_i(x, a)\}$ which are contradictory and have the same Morley rank and degree (since they are conjugate). This contradicts that $E_i(x, a)$ has the same Morley rank and degree as $\Xi'(x, a)$, proving the claim and completing the proof of the lemma.

Before getting to the proof of the Binding Group Theorem we show that when "finitely generated" is replaced by "finite" the proof needs no assumption other than the completeness of the theory. The proof of the lemma helps to motivate certain steps in the proof of the Binding Group Theorem.

Lemma 4.4.2. *Let $\mathfrak{C}$ be the universal domain of a complete theory, D_1 a definable set and D_2 a finite D_1-definable set. Then $G = \mathrm{Aut}(D_2/D_1)$ is a D_1-definable group and the action of G on D_2 is also D_1-definable.*

Proof. The proof is clear after a few moments thought but we may as well think aloud. First observe that there is a (finite) set $A \subset D_1$ such that $G = \mathrm{Aut}(D_2/A)$. Let $\bar{D}$ be the set of all enumerations of D_2. Identify $\alpha \in G$ with $d_\alpha = \{ (\bar{c}, \alpha(\bar{c})) : \bar{c} \in \bar{D} \}$ and let $\bar{G} = \{ d_\alpha : \alpha \in G \}$. Let $\mathrm{Aut}_A(\mathfrak{C})$ denote the set of automorphisms of $\mathfrak{C}$ which fix A pointwise. If $\beta \in \mathrm{Aut}_A(\mathfrak{C})$ then $\beta(d_\alpha) = d_{\beta\alpha\beta^{-1}}$, hence $\bar{G}$ is invariant under the elements of $\mathrm{Aut}_A(\mathfrak{C})$. By Lemma 3.3.8(i), $\bar{G}$ is $A-$definable. Define $\cdot$ on $\bar{G}$ by: $d_\alpha \cdot d_\beta = d_{\alpha\cdot\beta}$ (for $\alpha \in G$). Arguing as above, $\cdot$ is invariant under the elements of $\mathrm{Aut}_A(\mathfrak{C})$ hence $\cdot$ is also $A-$definable. This proves that the group $\bar{G}$ (which we identify with

G) is A–definable. The action of $\bar{G}$ on D_2 is defined by: $d_\alpha * x = \alpha(x)$. If $\beta \in \mathrm{Aut}_A(\mathfrak{C})$, then

$$(\forall x, y \in D_2)(\forall d_\alpha \in \bar{G})(\ d_\alpha * x = y \iff \beta(d_\alpha) * \beta(x) = \beta(y)\ ).$$

Thus, $*$ is A–definable. We conclude that through map $\alpha \mapsto d_\alpha$ from G onto $\bar{G}$ we can identify the action of G on D_2 with $(\bar{G}, \cdot, *)$.

Proof of Theorem 4.4.5 (Binding Group Theorem). Let D_2 be B–definable for $B \subset D_1$ finite. Let $\bar{b}$ be a fundamental generator of D_2 over D_1 with generating function $f(y_1, \ldots, y_n, \bar{z})$; i.e., $f(D_1^n, \bar{b}) \supset D_2$. Let $\psi_0(\bar{z})$ be a formula in $tp(\bar{b}/B)$ such that for any $\bar{c}$ satisfying ψ_0, $f(y_1, \ldots, y_n, \bar{c})$ is a function mapping D_1^n onto D_2. Let $\bar{c} \in \psi_0(\mathfrak{C})$ realize an isolated type in $S(B)$. By Lemma 4.4.1(v), $tp(\bar{c}/D_1)$ is isolated by some formula ψ. Let $X = \psi(\mathfrak{C})$. To prove the theorem it suffices to show:

Claim. There are $\tau : X \longrightarrow \mathrm{Aut}(D_2/D_1)$ and $\bar{c}$–definable operations,

$$\cdot : X \times X \longrightarrow X \text{ and } * : X \times D_2 \longrightarrow D_2$$

such that $*$ defines an action of the group $(X, \cdot)$ on D_2 and τ is an isomorphism of the group action $(X, \cdot, *)$ onto $\mathrm{Aut}(D_2/D_1)$.

Let $\theta(x, x', \bar{y}, \bar{y}')$ be a formula (over B) defining the relation:

$$\bar{y}, \bar{y}' \in X,\ x, x' \in D_2 \text{ and } \exists \bar{z} \in D_1^n(\ x = f(\bar{z}, \bar{y}) \wedge x' = f(\bar{z}, \bar{y}')\ ).$$

Let $\alpha \in \mathrm{Aut}(D_2/D_1)$ and suppose $\alpha(\bar{c}) = \bar{c}'$. Then for all $x, x' \in D_2$, $\theta(x, x', \bar{c}, \bar{c}') \iff x' = \alpha(x)$, so α is a definable map which we denote $\beta_{\bar{c}'}$. Notice that $\beta_{\bar{c}'}$ is the unique element of $\mathrm{Aut}(D_2/D_1)$ which takes $\bar{c}$ to $\bar{c}'$. Since ψ isolates a complete type over D_1 every $\bar{d} \in X$ realizes $tp(\bar{c}/D_1)$. Hence for any $\bar{d} \in X$ there is an $\alpha \in \mathrm{Aut}(D_2/D_1)$ such that $\alpha(\bar{c}) = \bar{d}$.

With these facts in hand we can define the necessary mappings $\cdot$ and $*$. Let τ be the bijection from X onto $\mathrm{Aut}(D_2/D_1)$ such that $\tau(\bar{d})$ is the unique $\gamma \in \mathrm{Aut}(D_2/D_1)$ such that $\beta_{\bar{d}} = \gamma$. Define the binary operation $\cdot$ on X by: $\beta_{\bar{d}\cdot\bar{e}} = \beta_{\bar{d}}\beta_{\bar{e}}$. Define $* : X \times D_2 \longrightarrow D_2$ by: $\bar{d} * a = \tau(\bar{d})(a)$. Using the formula $\theta(x, x', \bar{y}, \bar{y}')$ a routine argument shows that $\cdot$ and $*$ are both $\bar{c}$–definable. Furthermore, τ is a group action isomorphism of $(X, \cdot, *)$ onto $\mathrm{Aut}(D_2/D_1)$. This proves the claim, hence the theorem.

Remark 4.4.8. There may be many definable group actions isomorphic to $\mathrm{Aut}(D_2/D_1)$; i.e., many binding groups of D_2 over D_1. In the proof we picked $\bar{c}$ to be any element satisfying ψ_0 and realizing an isolated type over B. A different isolated completion of ψ_0 would lead to a different binding group. The set of fundamental generators X used as the universe of the binding group will be called the *special set of fundamental generators.*

The proof of the Binding Group Theorem finds, for any $\bar{c} \in X$ (a special set of fundamental generators), a copy of the binding group defined on X over $\bar{c}$. In the following corollary we show that while the action of the binding group generally needs a parameter from X there is a single B–definable group that works for all $\bar{c}$.

Corollary 4.4.1. *Let $\mathfrak{C}$ be the universal domain of a t.t. theory, D_1 an $\emptyset$–definable set and D_2 a B–definable set, where $B \subset D_1$, which is finitely generated over D_1. Let $\bar{c}$ be a fundamental generator of D_2 over D_1 such that $r = tp(\bar{c}/D_1)$ is isolated and let $X = r(\mathfrak{C})$. For each $\bar{c} \in X$ let $(G_{\bar{c}}, \cdot_{\bar{c}}, *_{\bar{c}})$ denote the copy of the binding group definable over $\bar{c}$, and let $\tau_{\bar{c}}$ denote the isomorphism of $G_{\bar{c}}$ onto* $\mathrm{Aut}(D_2/D_1)$ *(as group actions on D_2). Then there is an B–definable group $(G, \circ)$ and a formula $\epsilon(x, \bar{y})$ such that*

(1) *$G \subset D_1^{eq}$.*
(2) *For each $\bar{c} \in X$, $\epsilon(x, \bar{c})$ defines an isomorphism $\epsilon_{\bar{c}}$ of $(G, \circ)$ onto $(G_{\bar{c}}, \cdot_{\bar{c}})$.*
(3) *For each $\bar{c} \in X$ let $\pi_{\bar{c}} = \tau_{\bar{c}}\epsilon_{\bar{c}}$, an isomorphism of $(G, \circ)$ onto* $\mathrm{Aut}(D_2/D_1)$. *Let $\star_{\bar{c}}$ be the definable action of G on D_2 given by: $g \star_{\bar{c}} x = \epsilon_{\bar{c}}(g) *_{\bar{c}} x = \pi_{\bar{c}}(g)x$ (for $g \in G$ and $x \in D_2$). Hence $\pi_{\bar{c}}$ is an isomorphism of $(G, \circ, \star_{\bar{c}})$ onto* $\mathrm{Aut}(D_2/D_1)$ *as group actions.*
(4) *For each $\bar{c} \in X$, $\star_{\bar{c}}$ induces a regular group action of G on X.*
(5) *If $\gamma \in$* $\mathrm{Aut}(D_2/D_1)$, *$\bar{c} \in X$ and $\bar{d} = \gamma(\bar{c})$ (also an element of X) then for all $g \in G$, $\pi_{\bar{d}}(g) = \gamma\pi_{\bar{c}}(g)\gamma^{-1}$.*

Proof. Let $\bar{c} \in X$. Since $X \subset D_2^k$ (for some k) X is finitely generated over D_1. In fact, there is a $\bar{c}$–definable function $f_{\bar{c}}$ mapping D_1^m (for some m) onto X. By Remark 4.4.1(iii) there is a B–definable $G \subset D_1^{eq}$ and a $B \cup \bar{c}$–definable bijection $\epsilon_{\bar{c}}$ mapping G onto X. Since $G_{\bar{c}}$ is defined on X there is definable binary operation $\circ$ on G such that $\epsilon_{\bar{c}}$ is an isomorphism of $(G, \circ)$ onto $(G_{\bar{c}}, \cdot_{\bar{c}})$. Since all elements of X realize the same type over D_1 (hence the same type over D_1^{eq}) $\epsilon_{\bar{d}}$ is an isomorphism of $(G, \circ)$ onto $(G_{\bar{d}}, \cdot_{\bar{d}})$ for any $\bar{d} \in X$. This proves (1) and (2).

There is really nothing to prove in (3), its role being solely to set notation and viewpoint. Turning to (4) remember that X is a subset of D_2^k, hence $\star_{\bar{c}}$ defines an action of G on X. Since all elements of X have the same type over D_1 the action is transitive. For any $\bar{c} \in X$, $X \subset dcl(D_1 \cup \bar{c})$, hence only the identity of G can fix $\bar{c}$. In other words $\star_{\bar{c}}$ defines a regular action.

(5) Let $g \in G$ and $g \star_{\bar{c}} \bar{c} = \bar{e}$. Then by definition of $\star_{\bar{c}}$, $\pi_{\bar{c}}(g)$ is the unique element of $\mathrm{Aut}(D_2/D_1)$ taking $\bar{c}$ to $\bar{e}$. Since γ is in $\mathrm{Aut}(D_2/D_1)$, $g \star_{\bar{d}} \bar{d} = \gamma\bar{e}$. That is, $\pi_{\bar{d}}(g)$ is the unique element of $\mathrm{Aut}(D_2/D_1)$ taking $\bar{d}$ to $\gamma\bar{e}$. From here it is easy to see that $\pi_{\bar{d}}(g) = \gamma\pi_{\bar{c}}(g)\gamma^{-1}$.

Remark 4.4.4. This corollary gives us the picture of binding groups most useful in applications. Specifically, $G \subset D_1^{eq}$ is a definable group and there are

– a uniformly definable family of group actions $\{\star_{\bar{c}} : \bar{c} \in X\}$ and
– a family of maps $\{\pi_{\bar{c}} : \bar{c} \in X\}$ such that

for each $\bar{c} \in X$, $\pi_{\bar{c}}$ is an isomorphism of $(G, \circ, \star_{\bar{c}})$ onto $\mathrm{Aut}(D_2/D_1)$ (as group actions on D_2). From now on the term "binding group" refers to this copy of $\mathrm{Aut}(D_2/D_1)$ contained in D_1^{eq}.

The Binding Group Theorem allows us to apply all of our knowledge of ω–stable groups to binding groups. In particular, when $\mathfrak{C}$ is a 1–based uncountably categorical theory the binding group is abelian-by-finite. The strength of this fact will be discussed later in the context of the Ladder Theorems.

The applicability of the Binding Group Theorem depends on the existence of "many" sets which are finitely generated over a fixed set. The following result is the key in the context of uncountably categorical theories.

Theorem 4.4.6. *Let $\mathfrak{C}$ be the universal domain of an uncountably categorical theory, D an infinite A–definable set and a an element not in $acl(A)$. Then there is a $b \in dcl(A \cup \{a\}) \setminus acl(A)$ such that b is an element of an A–definable set which is finitely generated over $D \cup A$.*

The bulk of the proof of this theorem will be done in the context of a t.t. theory satisfying an additional condition (which is always true in an uncountably categorical theory).

Definition 4.4.4. *Let $\mathfrak{C}$ be the universal domain of a t.t. theory, D an A–definable set and Y $\bigwedge$–definable over A. Then Y is* foreign to D over *A if for any set $B \supset A$ and any $a \in Y$ which is generic over B, a is independent from $D \cup B$ over A. For q a type over A, q is foreign to D if $q(\mathfrak{C})$ is foreign to D over A.*

Notice the potential asymmetry in the foreign relation; Y may be foreign to D while D is not foreign to Y. This is possible because over a set B we test for independence using an arbitrary subset of X and a *generic element* of Y.

Theorem 4.4.6 will follow quickly from

Proposition 4.4.1. *Let $\mathfrak{C}$ be the universal domain of a t.t. theory, D an A–definable set, $p = tp(a/acl(\emptyset))$ and $Y = p(\mathfrak{C})$. If Y is not foreign to X there is a $b \in dcl(A \cup \{a\}) \setminus acl(A)$ such that b is an element of an A–definable set which is finitely generated over $D \cup A$.*

The proof of the proposition will be split between two results involving the following concept.

Definition 4.4.5. *Let $\mathfrak{C}$ be the universal domain of a t.t. theory, D an A–definable set and Y $\bigwedge$–definable over A. Then Y is said to be D–*internal *over A if for all $a \in Y$ there is a $B \supset A$ such that a is generic over B and $a \in dcl(B \cup D)$. If q is a type over A and $q(\mathfrak{C})$ is $D-$internal over A we also call q $D-$internal over A.*

Remark 4.4.5. Let D be an $\emptyset-$definable set in the universal domain of a t.t. theory, and Y $\bigwedge-$definable over $\emptyset$. The proofs of the following observations are left to the reader.

(i) When Y is finitely generated over D, Y is $D-$internal.

(ii) If Y is $D-$internal any conjugate of Y over $\emptyset$ is $D-$internal.

(iii) $p \in S(acl(\emptyset))$ is $D-$internal if for *some* a realizing p there is a B such that a is independent from B and $a \in dcl(B \cup D)$.

(iv) If $tp(a/acl(\emptyset))$ is $D-$internal and $b \in dcl(a)$, then $tp(b/acl(\emptyset))$ is $D-$internal.

(v) If $tp(a_i/acl(\emptyset))$ is $D-$internal for $i \leq n$, and b is the name for $\{a_0, \ldots, a_n\}$, then $tp(b/acl(\emptyset))$ is $D-$internal.

Notation. An $\bigwedge-$definable set X over A which is the set of realizations of a complete type over A is called a *locus over A*. Given an element a, the *locus of a over A* is the set of realizations of $tp(a/A)$. Note: the locus of a over A is the orbit of a under the automorphisms of $\mathfrak{C}$ which fix A.

Lemma 4.4.3. *Let $\mathfrak{C}$ be the universal domain of a t.t. theory, D an infinite $A-$definable set and Y a set $D-$internal over A such that Y is a locus over $acl(A)$. Then there is an $A-$definable set $X \supset Y$ such that X is finitely generated over $D \cup A$.*

Proof. Without loss of generality, $A = \emptyset$. The proof proceeds through the following steps.

(a) Let $a^* \in Y$ be generic over $\bar{b}^*$ such that $a^* = f(\bar{d}, \bar{b}^*)$ for some definable function $f(\bar{x}, \bar{b}^*)$ and $\bar{d} \subset D$. Let $q = tp(\bar{b}^*/acl(\emptyset))$. Then for all $\bar{b}'$ realizing q and $a' \in Y$ generic over $\bar{b}'$, $a' = f(\bar{d}', \bar{b}')$ for some $\bar{d}' \subset D$. Without loss of generality, $f(\bar{x}, \bar{b}^*)$ is defined on all of D^k for some k.

(b) Let $B = \{\bar{b}_i : i < \omega\}$ be a Morley sequence in q. Then for any $a \in Y$ there is a $\bar{b}_i \in B$ such that $a = f(\bar{d}, \bar{b}_i)$ for some $\bar{d} \subset D$.

(c) There is an $n < \omega$ such that

$$(\forall a \in Y)(\exists i \leq n)(\exists \bar{d} \subset D)(\ a = f(\bar{d}, \bar{b}_i)\ ).$$

(d) For $\bar{b} = \bar{b}_0 \cup \ldots \cup \bar{b}_n$ there is a single $\bar{b}-$definable function $g(\bar{z}, \bar{b})$ such that for any $a \in Y$, $a = g(\bar{d}, \bar{b})$ for some $\bar{d} \subset D$.

(e) There is a definable set $X \supset Y$ such that the condition in (d) is true with Y replaced by X.

Proofs:

(a) Every type in $S(\emptyset)$ is stationary, hence for any $\bar{b}'$ realizing q and $a' \in Y$ generic over $\bar{b}'$, $tp(a'\bar{b}') = tp(a\bar{b}^*)$, which is sufficient.

(b) Let a be any element of Y. By Corollary 3.3.1 there is a $\bar{b}_i \in B$ which is independent from a. Then, $a = f(\bar{d}, \bar{b}_i)$ for some $\bar{d} \subset D$.

(c) By (b) and a compactness argument there is such an n.

(d) This step is accomplished with a simple trick for producing one definable function from finitely many. Without loss of generality, $n > 1$. Define the function $g(\bar{z}, \bar{b})$ on D^{nk} so that for all $\bar{d}_0 \ldots, \bar{d}_n \in D^k$, $g(\bar{d}_0 \ldots \bar{d}_n, \bar{b}) = f(\bar{d}_i, \bar{b}_i)$, where i is the minimal index such that $f(\bar{d}_i, \bar{b}_i) \neq f(\bar{d}_j, \bar{b}_j)$ for all $j \neq i$, if one exits, and $i = n$, otherwise.

To verify that $g(\bar{z}, \bar{b})$ maps onto Y let $a \in Y$, $i \leq n$ and $\bar{d}_i \in D^k$ such that $a = f(\bar{d}_i, \bar{b}_i)$. To obtain $\bar{d}_0, \ldots, \bar{d}_n$ such that $a = g(\bar{d}_0, \ldots, \bar{d}_n, \bar{b})$ it suffices to find $\bar{d}_j$ (for $j \neq i$) such that $f(\bar{d}_l, \bar{b}_l) = f(\bar{d}_{l'}, \bar{b}_{l'})$ for all $l, l' \neq i$. Let $c \in Y$ be generic over $\{\bar{b}, \bar{d}_i, a\}$. Then, for each $j \neq i$ there is a $\bar{d}_j$ such that $c = f(\bar{d}_j, \bar{b}_j)$. This proves (d).

(e) Letting $X = g(D^{nk}, \bar{b})$ meets the requirement.

This proves the lemma.

The reader should compare the following lemma and its proof to Proposition 4.3.2.

Lemma 4.4.4. *Let D be an $A-$definable set in a t.t. theory and Y a locus over $acl(A)$. If Y is not foreign to D over A there is a $b \in dcl(A \cup \{a\}) \setminus acl(A)$ such that $p = tp(b/acl(A))$ is $D-$internal over A.*

Proof. Without loss of generality, $A = \emptyset$. Let $\bar{b}_0$ be a finite set independent from a and $\bar{d}_0 \subset D$ such that a depends on $\bar{d}_0$ over $\bar{b}_0$. By Corollary 4.1.4 there is an element c such that

(1) $c \in acl(a)$,
(2) a is independent from $\bar{b}_0\bar{d}_0$ over c, and
(3) $c \in dcl(\bar{d}_0\bar{b}_0, \ldots, \bar{d}_k\bar{b}_k)$, for some set $B = \{\bar{d}_0\bar{b}_0, \ldots, \bar{d}_k\bar{b}_k\}$ which is a Morley sequence over a in $tp(\bar{b}_0\bar{d}_0/acl(a))$.

Let $\bar{b} = \bar{b}_0 \ldots \bar{b}_k$, $\bar{d} = \bar{d}_0 \ldots \bar{d}_k$ and $q = tp(c/acl(\emptyset))$. Since a is independent from $\bar{b}_0$ and B is a Morley sequence over a, a is independent from $\bar{b}$. Thus, c is independent from $\bar{b}$. Since $\bar{d} \subset D$ and $c \in dcl(\bar{b}\bar{d})$, q is $D-$internal.

To obtain a realization of a $D-$internal type which is in $dcl(a)$ instead of only $acl(a)$ let b be a name for the (finite) set of conjugates of c over a. Since any conjugate of q is $D-$internal, b is a finite set of elements each realizing a $D-$internal type over $acl(\emptyset)$. By Remark 4.4.5, $p = tp(b/acl(\emptyset))$ is $D-$internal. This proves the lemma.

Proof of Proposition 4.4.1. The proposition follows immediately from the combination of Lemma 4.4.3 and Lemma 4.4.4.

Proof of Theorem 4.4.6. Without loss of generality, $A = \emptyset$. Let Y be the locus of a over $acl(\emptyset)$. It suffices (by Proposition 4.4.1) to show that Y is not foreign to D. Let M be a countable saturated model and b an element of Y generic over M. Then there is a $c \in D$ such that $tp(c/M)$ is strongly minimal (by Corollary 3.1.2) and $c \in acl(M \cup \{b\})$ (by Exercise 3.3.18). Thus, Y is not foreign to D.

Our first reward is a proof of the Main Ladder Theorem.

Proof of Theorem 4.4.1. Without loss of generality, $a \notin acl(\emptyset)$. By Proposition 4.3.2 there is an element $a_0 \in dcl(a)$ such that a_0 is in an $\emptyset$–definable almost strongly minimal set D_0. Now suppose $a_0, \ldots, a_i$ and $D_0, \ldots, D_i$ have been defined to satisfy (1)–(5) up to i. If $a \in acl(A_i)$ let $a_i = a$ and end the construction. Otherwise there is an $a_i \in dcl(A_i \cup \{a\}) \setminus acl(A_i)$ and an A_i–definable set D_i such that $a_i \in D_i$ and D_i is finitely generated over $D_0 \cup \ldots \cup D_{i-1}$ (by Theorem 4.4.6). Since $A_i \subset dcl(a)$, $a_i \in dcl(a)$. By the Binding Group Theorem (Theorem 4.4.5) $G_i = \mathrm{Aut}(D_i/D_0 \cup \ldots \cup D_{i-1})$ is definable, proving the theorem.

Proof of Theorem 4.4.2. The most important additional tool in this proof is Lemma 4.3.9, which says

($\sharp$) for any set A and $a \notin acl(A)$ there is a $c \in acl(A \cup \{a\})$ such that $MR(c/A) = 1$.

This fact is augmented with the following to obtain sets which are strongly minimal in addition to having Morley rank 1. (This is just a restatement of Lemma 4.1.3(ii).)

($\sharp\sharp$) For any a and finite set A there is an $e \in dcl(A \cup \{a\}) \cap acl(A)$ such that $\deg(a/A \cup \{e\}) = 1$.

Let a be any element of the universal domain. The choice of elements a_i and sets D_i proceeds as follows through several cases. The construction ends at the first step in which a_i is set to a. After defining these objects we will prove the necessary properties of the binding groups.

Case 1. $a \in acl(A_i)$. Let $a_i = a$ and D_i be the set of realizations of $tp(a/A_i)$.

Case 2. $a \notin acl(\emptyset)$ and $i = 0$. By ($\sharp$) there is a $c \in acl(a)$ such that $MR(c/\emptyset) = 1$. If $tp(c)$ is strongly minimal let $a_0 = c$ and D_0 be an $\emptyset$–definable strongly minimal set containing c. If, on the other hand, $\deg(c) > 0$ choose $e \in dcl(c) \cap acl(\emptyset)$ such that $\deg(c/e) = 1$ (by ($\sharp\sharp$)). In this case we let $a_0 = e$, $D_0 =$ the set of realizations of $tp(e)$, $a_1 = c$ and D_1 a strongly minimal set over a_0 which contains c.

Case 3. $a \notin acl(A_i)$ and $D_0 \cup \ldots \cup D_{i-1}$ is infinite. By ($\sharp$) there is a $c \in acl(A_i \cup \{a\})$ such that $MR(c/A_i) = 1$. Since $\bar{D} = D_0 \cup \ldots \cup D_{i-1}$ is infinite, Theorem 4.4.6 yields a $c' \in dcl(A_i \cup \{c\}) \setminus acl(A_i)$ such that c' belongs to an A_i–definable set which is finitely generated over $\bar{D}$. Thus, we may as well require c to belong to an A_i–definable set of Morley rank 1 which is finitely generated over $\bar{D}$.

If $tp(c/A_i)$ is strongly minimal we let $a_i = c$ and D_i an A_i–definable strongly minimal set which contains c and is also finitely generated over $\bar{D}$. If $\deg(c/A_i) > 1$ we interpose another element of $acl(a)$ as follows. By (♯♯) there is an $e \in dcl(A_i \cup \{c\}) \cap acl(A_i)$ such that $\deg(c/A_i \cup \{e\}) = 1$; i.e., $tp(c/A_i \cup \{e\})$ is strongly minimal. Let $a_i = e$ and D_i the (finite) set of realizations of $tp(e/A_i)$. Let $a_{i+1} = c$ and D_{i+1} an A_{i+1}–definable strongly minimal set. Notice that D_{i+1} is finitely generated over $D_0 \cup \ldots \cup D_i$.

The reader should observe that the described cases encompass all possibilities (until $a_n = a$ and the construction terminates). It remains to show that (when $D_0 \cup \ldots \cup D_{i-1}$ is infinite)

(♭) $G_i = \mathrm{Aut}(D_i/D_0 \cup \ldots \cup D_{i-1})$ is definable over $D_0 \cup \ldots \cup D_{i-1}$ and has Morley rank ≤ 1.

When D_i is finite this is true by Lemma 4.4.2. Suppose D_i is infinite. That G_i is definable over $D_0 \cup \ldots \cup D_{i-1}$ is simply by Corollary 4.4.1. Let X be a special set of fundamental generators for D_i over $D_0 \cup \ldots \cup D_{i-1}$ and recall that $MR(G_i) = MR(X)$, which we have assumed is > 0. Since $D_0 \cup \ldots \cup D_{i-1}$ is infinite one of $D_0, \ldots, D_{i-1}$ is strongly minimal. By Lemma 4.4.5(ii), for any $a \in D_i \setminus acl(A_i)$, $D_i \subset acl(D_0 \cup \ldots \cup D_{i-1} \cup \{a\})$. Since X is a subset of D_i^k for some k, and all elements of X realize the same type over $D_0 \cup \ldots \cup D_{i-1}$, $MR(X) = MR(a/D_0 \cup \ldots \cup D_{i-1}) \leq 1$. This proves (♭) and completes the proof of the theorem.

We turn now to the Simple Ladder Theorem, which will follow rather quickly from

Proposition 4.4.2. *Let $\mathfrak{C}$ be the universal domain of a t.t. theory, D_1 an infinite $\emptyset$–definable set and D_2 a set which is finitely generated over D_1 and definable over $B \subset D_1$. Let G be a binding group of D_2 over D_1.*

(i) Suppose that $B \subset C \subset D_1$ and f is a C–definable function from D_2 onto a set F. Let H be $\{h \in G : h \text{ is the identity on } F\}$. Then H is a C–definable normal subgroup of G. Furthermore, $H = \{1\}$ if and only if $D_2 \subset dcl(D_1 \cup F)$.

(ii) Conversely, let H be a definable normal subgroup of G. Then there is a definable set F, finitely generated over D_1 such that for any $\bar{c} \in X$, $\mathrm{Aut}(D_2/D_1 \cup F) = \pi_{\bar{c}}(H)$ and $\mathrm{Aut}(F/D_1) = \mathrm{Aut}(D_2/D_1)/\pi_{\bar{c}}(H)$. If H is B–definable then we can take F to be the set of realizations of an isolated type over D_1.

Proof. For the statement of (i) to make sense the reader must observe that the action of G on D_2 extends in a unique way to an action of G on $D_2 \cup F$. Let X be the special set of fundamental generators of D_2 over D_1. In the proof we freely draw on the notation used in Corollary 4.4.1. In particular,

G is a definable group in D_1^{eq} and for each $\bar{c} \in X$, $\star_{\bar{c}}$ defines an action of G on D_2 and a regular action of G on X.

(i) For any $\bar{c} \in X$ let $\varphi_{\bar{c}}(x)$ be the formula $x \in G \wedge (\forall y \in F)(x \star_{\bar{c}} y = y)$. Then, $H = \varphi_{\bar{c}}(\mathfrak{C})$, hence H is definable over $C \cup \{\bar{c}\}$. Since F is C–definable and the elements of X all have the same type over D_1^{eq}, $\varphi_{\bar{c}}$ is equivalent to $\varphi_{\bar{d}}$ for all $\bar{c}$, $\bar{d} \in X$. It follows that H is C–definable. The reader should verify that H is normal.

If $D_2 \subset dcl(D_1 \cup F)$ then any $h \in H$ must be the identity on D_2; i.e., $H = \{1\}$. On the other hand, if $a \in D_2$ and $a \notin dcl(D_1 \cup F)$ there is a $b \neq a$ realizing $tp(a/D_1 \cup F)$. By Lemma 4.4.1(iv) there is an $h \in G$ which maps a to b and fixes every element of $D_1 \cup F$; that is, $H \neq \{1\}$.

(ii) The set F will be the quotient of X by some D_1–definable equivalence relation.

Claim. Let Y be a D_1–definable set such that each $\star_{\bar{c}}$ defines an action of G on Y. For each $\bar{c} \in X$ define an equivalence relation $E_{\bar{c}}$ on Y by:

$$E_{\bar{c}}(x, y) \text{ if and only if } \exists \gamma \in H(\gamma \star_{\bar{c}} x = y).$$

Then for all $\bar{c}$, $\bar{d} \in X$, $E_{\bar{c}}$ is equivalent to $E_{\bar{d}}$.

Note: For $\bar{c} \in X$ and $x \in Y$, $E_{\bar{c}}(\mathfrak{C}, x) = \pi_{\bar{c}}(H)x$. Remember that the action of $\mathrm{Aut}(D_2/D_1)$ on X is regular. Pick $\bar{c}$, $\bar{d} \in X$ and let $\gamma \in \mathrm{Aut}(D_2/D_1)$ be such that $\bar{d} = \gamma\bar{c}$. By Corollary 4.4.1, for any $g \in G$, $\pi_{\bar{d}}(g) = \gamma \cdot \pi_{\bar{c}}(g) \cdot \gamma^{-1}$. Since H is normal $\pi_{\bar{d}}(H) = \gamma \cdot \pi_{\bar{c}}(H) \cdot \gamma^{-1} = \pi_{\bar{c}}(H)$. Thus, $E_{\bar{d}}(\mathfrak{C}, x) = \pi_{\bar{d}}(H)x = \gamma \cdot \pi_{\bar{c}}(H) \cdot \gamma^{-1}x = \pi_{\bar{c}}(H)x = E_{\bar{c}}(\mathfrak{C}, x)$, proving the claim.

Now apply the claim with $X = Y$. Let E be the equivalence relation such that for all x, $y \in X$, $E(x, y)$ holds if and only if there is a $\bar{c} \in X$ such that $E_{\bar{c}}(x, y)$. Then E is D_1–definable. Let F be the set of E–classes of elements of X. Since $F \subset dcl(X \cup D_1)$ any element of $\mathrm{Aut}(D_2/D_1)$ extends uniquely to an element of $\mathrm{Aut}(F/D_1)$.

Claim. For any $\bar{c} \in X$, $H = \{ g \in G : g \star_{\bar{c}} x = x \text{ for all } x \in F \}$.

Fix $\bar{c} \in X$ and $\star_{\bar{c}}$ as an action of H on F. It is immediate from the definition of E that any $g \in H$ is the identity on F. Conversely, suppose that $g \star_{\bar{c}} x = x$ for all $x \in F$. Let $\bar{d}$ be any element of X and $\bar{e} = g \star_{\bar{c}} \bar{d}$. Since g fixes every element of F (under $\star_{\bar{c}}$) $\bar{d}$ and $\bar{e}$ have the same type over F. Then $\bar{d}$ and $\bar{e}$ must be E–equivalent (since F is the set of E–classes), hence there is an $h \in H$ with $\bar{e} = h \star_{\bar{c}} \bar{d}$. Since the action of G on X is regular we conclude that $g = h$, proving the claim.

Since $\mathrm{Aut}(D_2/F \cup D_1) = \{\gamma \in \mathrm{Aut}(D_2/D_1) : \gamma \text{ is the identity on } F\}$, the claim proves that $\mathrm{Aut}(D_2/F \cup D_1) = \pi_{\bar{c}}(H)$, for any $\bar{c} \in X$.

Clearly, any element of $\mathrm{Aut}(F/D_1)$ extends to an element of $\mathrm{Aut}(D_2/D_1)$. Thus, the natural embedding of $\mathrm{Aut}(D_2/D_1)$ into $\mathrm{Aut}(F/D_1)$ is surjective. The kernel of this embedding is $\pi_{\bar{c}}(H)$ hence

$$\mathrm{Aut}(F/D_1) = \mathrm{Aut}(D_2/D_1)/\pi_{\bar{c}}(H).$$

Finally notice that when H is B–definable, so is the equivalence relation E. Remember that $F = X/E$. Since X is B–definable and all elements of X realize the same complete type over D_1, the same is true of F. This proves the proposition.

Proof of Theorem 4.4.3. To begin let $D_0, \ldots, D_l$ be a sequence of almost strongly minimal sets and $a_0, \ldots, a_l = a$ a sequence of elements meeting the requirements (1)–(5) of Theorem 4.4.1. Suppose, for example, that $G_1 = \mathrm{Aut}(D_1/D_0)$ is infinite, nonsimple and not minimal abelian. Let H be a definable normal subgroup of G. By Proposition 4.4.2 there is a D_0–definable set F, finitely generated over D_0, such that $\mathrm{Aut}(F/D_0) = G_1/H$ and $\mathrm{Aut}(D_1/D_0 \cup F) = H$. Replace the original sequence $D_0, D_1, \ldots, D_l$ by $D_0, F, D_1, \ldots, D_l$. Continuing this process produces a sequence of sets (in finitely many steps) satisfying (1)–(3) in the statement of the theorem.

A much more refined picture can be obtained when the theory is 1–based. Recall that a definable group in a 1–based theory is abelian-by-finite. Thus, in a 1–based theory the connected component of any binding group is abelian. The first part of the next lemma shows the strength of this condition.

Lemma 4.4.5. *Let $\mathfrak{C}$ be the universal domain of a 1–based uncountably categorical theory, D_1 an infinite $\emptyset$–definable set and D_2 a set, finitely generated over D_1 and definable over $B \subset D_1$. Let X be a special set of fundamental generators of D_2 over D_1, $\bar{c} \in X$, $(G, \cdot, \star_{\bar{c}})$ the binding group of D_2 over D_1 (presented as in Corollary 4.4.1) and $\pi_{\bar{c}}$ the isomorphism of $(G, \cdot, \star_{\bar{c}})$ onto* $\mathrm{Aut}(D_2/D_1)$. *Suppose G is abelian.*

(i) There is a B–definable action $\star$ of G on D_2 such that for all $\bar{c} \in X$, $\star = \star_{\bar{c}}$.

(ii) Let $a \in D_2$ and Y the set of realizations of $tp(a/D_1)$. Then $Y \subset dcl(D_1 \cup \{a\})$.

Proof. (i) For each $\bar{d} \in X$, $\pi_{\bar{d}}$ is a group action isomorphism, hence $g \star_{\bar{d}} x = \pi_{\bar{d}}(g)x$, for all $g \in G$ and $x \in D_2$. Let $\bar{d}$ and $\bar{e}$ be arbitrary elements of X. There is a $\gamma \in \mathrm{Aut}(D_2/D_1)$ such that $\bar{e} = \gamma(\bar{d})$ and, more to the point, $\pi_{\bar{e}}(g) = \gamma\pi_{\bar{d}}(g)\gamma^{-1}$. Since G is abelian we conclude that for all $g \in G$ and $x \in D_2$, $g \star_{\bar{d}} x = g \star_{\bar{e}} x$. Since the elements of X realize an isolated type over D_1 there is a D_1–definable action $\star$ of G on D_2 such that $\star = \star_{\bar{d}}$ for all $\bar{d} \in X$.

(ii) Simply because G is $\mathrm{Aut}(D_2/D_1)$, Y is the orbit of a under the action of G. Since $G \subset dcl(D_1)$ and the action of G on D_2 is definable over D_1, $Y \subset dcl(D_1 \cup \{a\})$, as needed to prove the lemma.

Proof of Theorem 4.4.4. Combining Theorem 4.4.2 with Proposition 4.4.2 will prove the theorem. For $\mathfrak{C}$ as hypothesized and a an arbitrary element let $a'_0, \ldots, a'_l = a$ and $D'_0, \ldots, D'_l$ satisfy all of the requirements of Theorem 4.4.2. We will find sets $D_0, \ldots, D_n$ and elements $a_0, \ldots, a_n = a$ satisfying the additional requirements of this theorem. These a_i and D_i will be chosen so that the a'_j and D'_j are among them. Suppose $a_0, \ldots, a_{i-1}$ and $D_0, \ldots, D_{i-1}$ have been found satisfying the conditions of the theorem "up to $i-1$" and let $\bar{D} = D_0 \cup \ldots \cup D_{i-1}$. Suppose j is minimal so that D'_j is not among $D_0, \ldots, D_{i-1}$. If D'_j is finite let $a_i = a'_j$ and note that (5) holds for $G_i = \mathrm{Aut}(D_i/\bar{D})$ by Lemma 4.4.2. Now suppose D'_j to be strongly minimal, in which case $H = \mathrm{Aut}(D'_j/\bar{D})$ has Morley rank 1. As a group H is definable over A_i by Corollary 4.4.1. If H is strongly minimal, let $D_i = D'_j$, $a_i = a'_j$ and $G_i = H$. Assuming that $\deg(H) > 1$ let G be the connected component of H, a strongly minimal normal subgroup of H which is A_i–definable. By Proposition 4.4.2 there is a definable set F such that

- F is finitely generated over $\bar{D}$,
- F is A_i–definable and the elements of F realize the same complete type over $\bar{D}$,
- $\mathrm{Aut}(D'_j/\bar{D} \cup F) \cong G$ and
- $\mathrm{Aut}(F/\bar{D}) \cong (H/G)$.

Since G has finite index in H, F is finite. Let $D_i = F$, $D_{i+1} = D'_j$, a_i any element of F and $a_{i+1} = a'_j$. Then $G_i = H/G$ is finite and $G_{i+1} = G$ is strongly minimal.

That a strongly minimal group is abelian is proved in Corollary 3.5.5. We proved in Lemma 4.4.5(i) that group action of G_i on D_i is definable over $D_0 \cup \ldots \cup D_{i-1}$ (since G_i is abelian). This proves the theorem.

Recipe. I'm sure you've worked up quite an appetite by now. After a long day of mathematics there is nothing like a big plate of lasagna. This recipe was given to me by Philipp Rothmaler in exchange for a preprint of [Bue87].

First we need a sauce bolognaise. Quickly brown 3/4 lb. of ground beef with a large chopped onion. Add salt by taste and remove most of the grease. Add 2 – 3 big chopped tomatoes, 2 – 3 tablesp. of tomato paste and the spices thyme, oregano, basil, black pepper, paprika and minced garlic (by taste). Cook under low heat until the tomatoes are saucy. (This could take quite a while; have a glass of wine and start the next section.)

When the sauce bolognaise is nearly finished it is time for the sauce bechamel. In a small sauce pan melt 3 tablesp. of butter and stir in 1 – 2 tablesp. of flour to make a smooth paste. Gradually add 1 cup of cold milk under low heat, stirring until it thickens. Add salt to taste and a few pinches of nutmeg.

The sauces, uncooked *(sic)* lasagna noodles and Mozzarella cheese are layered in a backing dish as follows. In the bottom of the dish put a thin layer of sauce bechamel, a layer of noodles and more bechamel on top. Then comes the bolognaise, Mozzarella, noodles, bechamel, bolognaise, etc. End the layering with a lot of Mozzarella on top. Cook at 350 for 30 minutes.

Historical Notes. With few exceptions the results in this section are due to Zil'ber. They originally appeared in various papers, but are compiled in [Zil93]. Binding groups are called liaison groups by some authors, most notably Poizat (see [Poi87]).

Exercise 4.4.1. Prove (a) and (b) in Example 4.4.1(iv).

Exercise 4.4.2. Prove: If D_2 is finitely generated over $D_1 \cup A$ and D_3 is finitely generated over $D_2 \cup B$, then D_3 is finitely generated over $D_1 \cup A \cup B$.

4.5 Defining a Group from a Pregeometry

The canonical example of a nontrivial modular strongly minimal set is a vector space. In fact, for any nontrivial modular strongly minimal set D there is a vector space V such that the geometry associated to D is isomorphic (as a geometry) to the geometry associated to V. In this section we show (roughly) how to find V as a definable group in D^{eq} from the pregeometry D. More precisely, from a configuration of points, that can always be found in a nontrivial modular strongly minimal set, a definable strongly minimal group is constructed. By Theorem 4.3.4 this definable strongly minimal group is a $*$−vector space. We will also analyze configurations of points leading to definable fields. This will lead to a characterization of the strongly minimal sets D containing a definable field in D^{eq}.

The configuration of points alluded to above is defined as follows. Note that the elements involved are not assumed to be from a strongly minimal set.

Definition 4.5.1. *Let $\mathfrak{C}$ be the universal domain of an uncountably categorical theory. A 6−tuple of elements $Q = (a_1, a_2, a_3, b_1, b_2, b_3)$ is called an* algebraic quadrangle *if the following hold for any $\{i, j, k\} = \{1, 2, 3\}$ and $\ell_{ijk} = \{b_i, a_j, a_k\}$.*

(1) *Q is pairwise independent and no element of Q is in $acl(\emptyset)$.*
(2) *$a_j \in acl(b_i, a_k)$.*
(3) *$b_i \in acl(b_j, b_k)$.*
(4) *b_i is interalgebraic with the canonical parameter of $tp(a_j a_k / acl(b_i))$.*
(5) *For $\{i', j', k'\} = \{1, 2, 3\}$, ℓ_{ijk} is independent from $\ell_{i'j'k'}$ over $\ell_{ijk} \cap \ell_{i'j'k'}$.*

For A a set and Q a 6−tuple the notion Q is an algebraic quadrangle over *A is defined with the obvious adjustments in (1)–(5).*

Remark 4.5.1. There are many variations on the above definition. All are known under the general heading of "Zil'ber's configuration", after Boris Zil'ber who first isolated the notion and proved a variant of the following theorem.

Remark 4.5.2. If Q is an algebraic quadrangle and A is independent from Q, then Q is an algebraic quadrangle over A. See Exercise 4.5.1.

The roles of the a_i's and b_i's is symmetric in the definition. Given an algebraic quadrangle $(a_1, a_2, a_3, b_1, b_2, b_3)$ and π a permutation of $\{1, 2, 3\}$,

$$(a_{\pi(1)}, a_{\pi(2)}, a_{\pi(3)}, b_{\pi(1)}, b_{\pi(2)}, b_{\pi(3)})$$

is also an algebraic quadrangle.

The main theorem of the section is

Theorem 4.5.1. *Let $\mathfrak{C}$ be the universal domain of an uncountably categorical theory and $Q = (a_1, a_2, a_3, b_1, b_2, b_3)$ an algebraic quadrangle. There is a finite set A independent from Q and $A-$definable sets X and G satisfying:*

(1) *There is a generic of X interalgebraic with a_1 over A and* $\deg(X) = 1$.
(2) $MR(G) = MR(b_2)$.
(3) *G is a connected definable group and there is a definable faithful transitive group action of G on X.*

Given an arbitrary uncountably categorical theory it is not at all clear that $\mathfrak{C}$ contains an algebraic quadrangle. However, we will see that a nontrivial strongly minimal set in 1−based theory contains an algebraic quadrangle, quickly leading to

Theorem 4.5.2. *Let $\mathfrak{C}$ be the universal domain of a $1-$based uncountably categorical theory and D a nontrivial strongly minimal set over $\emptyset$. There is a finite set A and a strongly minimal group G definable over A such that a generic of G is interalgebraic over A with an element of $D \setminus A$.*

As an application of Theorems 4.5.1 and 3.5.2 we offer Theorem 4.5.3, which does not *a priori* have anything to do with groups.

Definition 4.5.2. *Let D be a $\emptyset-$definable strongly minimal set. Then D is said to be* pseudomodular *if there is a $k < \omega$ such that whenever $X \cup \{a, b\} \subset D$ and $a \in acl(X \cup \{b\})$, there is a $Z \subset acl(X) \cap D$ of cardinality $\leq k$ such that $a \in acl(Z \cup \{b\})$.*

Remark 4.5.3. A strongly minimal set D is pseudomodular if and only if there is a k such that $MR(c/\emptyset) \leq k$, for c the canonical parameter of a plane curve of D. For this reason some authors say pseudolinear instead of pseudoprojective.

A modular strongly minimal set is pseudomodular with $k = 1$, while an algebraically closed field is not pseudomodular. In fact,

Theorem 4.5.3. *A pseudomodular strongly minimal set is locally modular.*

Throughout the section we assume the ambient theory to be uncountably categorical, although many of the results hold in much greater generality.

An algebraic quadrangle will not lead directly to a definable group action, but to an $\bigwedge$−definable collection of maps between $\bigwedge$−definable sets. Obtaining a definable group action from this collection of maps requires the following study.

4.5.1 Germs of Definable Functions

Here a definable function is identified with a name for the defining formula in $\mathfrak{C}^{eq}$. In this way a definable function is considered to be an element of the universe.

Throughout this section the theory is assumed to be uncountably categorical

(although we may restate this assumption to make results easier to reference).

Remark 4.5.4. (i) Let R be an A−definable binary relation on the universe and $X \subset dom(R)$ a locus over A such that the restriction of R to X defines a function. Then there is an A−definable function f which agrees with R on X.

(ii) Let g be an A−definable function and $a \in dom(g)$ such that $tp(a/A)$ is stationary. Then $tp(g(a)/A)$ and $tp(ag(a)/A)$ are stationary. (The proof is left as Exercise 4.5.2.)

Definition 4.5.3. *Let $\mathfrak{C}$ be the universal domain of an uncountably categorical theory, A a set and X, Y infinite loci over A such that* $\deg(X) = \deg(Y) = 1$. *An element g is a* generic map of X to Y *if g is a definable function and for all $a \in X$ generic over g, $g(a) \in Y$ and $\{a, g, g(a)\}$ is pairwise independent over A.*

When g is a generic map of X into Y we may also say g maps X to Y generically.

Remark 4.5.5. In the definition all elements of X and Y are generic over A since X and Y are loci over A. The assumption that X and Y are infinite is made only to eliminate trivial cases.

Remark 4.5.6. Let A be a set and X, Y loci over A such that $\deg(X) = \deg(Y) = 1$. Let g be a definable function.

(i) If a and b are elements of X generic over g, then $tp(g(a)/A) = tp(g(b)/A)$ and this type is stationary. Thus, g maps X into Y generically if and only if for *some* $a \in X$ generic over g, $g(a) \in Y$ and $\{a, g, g(a)\}$ is pairwise independent over A.

(ii) If g maps X to Y (generically) and $c \in Y$ is generic over g then there is some $b \in X$ generic over g such that $g(b) = c$. Thus, g maps onto the elements of Y generic over g.

(iii) Suppose $a \in X$, $b \in dcl(a,c) \cap Y$ and $\{a,b,c\}$ is pairwise independent over A. Then there is an $h \in dcl(c)$ which is a generic map from X into Y and takes a to b. (See Exercise 4.5.4.)

Let V be an algebraic variety. Morphisms g and h are called generically equal if there is an open set U on which g and h are both defined and agree. Being generically equal defines an equivalence relation on the "local morphisms" of V. The class of a morphism g under generic equality is the "germ of g".

Definition 4.5.4. *Let $\mathfrak{C}$ be the universal domain of an uncountably categorical theory, A a set and X, Y infinite loci over A such that* $\deg(X) = \deg(Y) = 1$. *Let g, h be generic maps of X into Y. We say g is* generically equal to h on X *if for all $a \in X$ generic over $\{g,h\}$, $g(a) = h(a)$. The set X is omitted from the term when it is clear from context.*

Lemma 4.5.1. *Let X and Y be infinite loci over $\emptyset$, g a generic map of X into Y and B a set. If $a \in X$ is generic over $B \cup \{g\}$ then $g(a)$ is generic over $B \cup \{g\}$.*

Proof. See Exercise 4.5.3.

Lemma 4.5.2. *Let X, Y and Z be degree 1 infinite loci over A. Let g map X generically to Y and h map Y generically to Z. Then $h \circ g$ maps X generically to Z.*

Proof. Without loss of generality, $A = \emptyset$. Let $a \in X$ be generic over $\{g,h\}$.

Claim. (i) $g(a)$ is generic over h.
(ii) $\{a, h \circ g, (h \circ g)(a)\}$ is pairwise independent.

(i) By Lemma 4.5.1.
(ii) By (i), $h(g(a))$ exists and is an element of Z such that $\{g(a), h, h(g(a))\}$ is pairwise independent. Again by Lemma 4.5.1, $h(g(a))$ is independent from $\{g,h\}$ and independent from a. Thus, $\{a, h \circ g, (h \circ g)(a)\}$ is pairwise independent, proving the claim.

Let $a \in X$ be generic over $\{g,h\}$ and $b \in X$ be generic over $h \circ g$. Then a is also generic over $h \circ g$, so a and b have the same type over $h \circ g$. Thus, $h \circ g(b)$ is defined, an element of Z and $\{b, h \circ g, (h \circ g)(b)\}$ is pairwise independent. In other words, $h \circ g$ maps X generically to Z.

Generic equality is not generally an equivalence relation, however

Lemma 4.5.3. *Let A be a set and X, Y infinite loci over A such that* $\deg(X) = \deg(Y) = 1$. *Let g be a generic map of X into Y and Z a B-definable set of generic maps of X into Y which contains g, where $B \supset A$. Then there is a B-definable equivalence relation $\sim$ such that for all $h, k \in Z$, h is generically equal to k if and only if $h \sim k$.*

Proof. Let $p \in S(\mathfrak{C})$ be the unique free extension of the type over A realized by the elements of X. Since $\deg(X) = 1$, p is definable over B. Since Z is B–definable there is a formula $f(x, z)$ over B such that for all $c \in Z$, $f(x, c)$ is a generic map of X into Y with name c. Let $\zeta(v)$ be the formula over B defining Z and $\varphi(x, y, z)$ the formula $\zeta(y) \wedge \zeta(z) \wedge (f(x, y) = f(x, z))$. Let $\theta(y, z)$ be the formula over B defining p_φ (where p is a type in x). Then, for all c, $d \in Z$, the following are equivalent:

- c is generically equal to d,
- $\forall a \in X$ generic over $\{c, d\}$, $f(a, c) = f(a, d)$,
- $\varphi(x, c, d) \in p$,
- $\models \theta(c, d)$.

This proves the lemma.

In algebraic geometry a germ of morphisms is an equivalence class X of generically equal morphisms. This germ is identified with a generic map g by defining $g(a)$ to be $h(a)$ for any h in X such that a is in the domain of h and a is generic over h. In other words, g is a canonical representative of X. We define a germ similarly except we must be sensitive to the fact that generic equality is definable only when restricted to a definable family of generic maps.

Definition 4.5.5. *Let A be a set and X, Y infinite loci over A such that* $\deg(X) = \deg(Y) = 1$. *A generic map g of X into Y is called a* germ *if*

(∗) *for all h realizing $tp(g/A)$, if h is generically equal to g on X then $h = g$.*

The next two lemmas give the existence of germs and useful tools for working with them.

Lemma 4.5.4. *Let A be a set and X, Y infinite loci over A such that* $\deg(X) = \deg(Y) = 1$. *Let h be a generic map from X to Y, $a \in X$ generic over h and $p \in S(\mathfrak{C})$ the unique free extension of $tp(ah(a)/A \cup \{h\})$.*

(i) Given a generic map g from X into Y, g is generically equal to h if and only if for any $b \in X$ generic over g, p is also the unique free extension of $tp(bg(b)/A \cup \{g\})$.

(ii) The canonical parameter c of p is interdefinable with a generic map from X into Y which is generically equal to h.

Proof. Without loss of generality, $A = \emptyset$. That $tp(ah(a)/h)$ is stationary and hence has a unique free extension in $S(\mathfrak{C})$ is Remark 4.5.4(ii).

(i) See Exercise 4.5.5.

(ii) Note, $c \in dcl(h)$ so a is generic over $\{c, h\}$. The restriction of p to c is stationary and $\{a, h(a)\}$ is independent from h over c.

Claim. $h(a) \in dcl(a, c)$.

Assuming the claim to fail there is a $b \neq h(a)$ realizing $tp(h(a)/c, a)$ which is independent from h over $\{c, a\}$, hence ab is independent from h over c. Thus, $tp(ab/c, h) = tp(ah(a)/c, h)$, so $b = h(a)$. This contradiction proves the claim.

By Remark 4.5.6(iii) there is a $g \in dcl(c)$ which is a generic map from X to Y with $g(a) = h(a)$. Since a is generic over $\{g, h\}$ we conclude that g is generically equal to h. By (i) p is definable over g, hence $c \in dcl(g)$. This proves the lemma.

Lemma 4.5.5. *Let A be a set and X, Y infinite loci over A such that* $\deg(X) = \deg(Y) = 1$. *Given a generic map g of X into Y the following are equivalent.*

(1) *g is a germ.*
(2) *For any generic map h of X into Y which is generically equal to g there is an $a \in X$ generic over h such that g is a canonical parameter of $tp(ah(a)/A \cup \{h\})$.*
(3) *For some set I of generics of X such that $I \cup \{g\}$ is A–independent,*

$$g \in dcl(\{(a, g(a)) : a \in I\}).$$

Proof. Without loss of generality $A = \emptyset$.

(2) $\Longrightarrow$ (1) Let $a \in X$ be generic over g. Let $p \in S(\mathfrak{C})$ be the unique free extension of $tp(ag(a)/g)$. Let g' be a realization of $tp(g)$ which is generically equal to g. By Lemma 4.5.4(i) p is also the unique free extension of $tp(bg'(b)/g')$, for any $b \in X$ generic over g'. Thus, any $\alpha \in \mathrm{Aut}(\mathfrak{C})$ that maps g to g' also maps p to itself. By (2) g is a canonical parameter of p, hence $\alpha(g) = g$. Thus, $g' = g$; i.e., g is a germ.

(1) $\Longrightarrow$ (3) Assume g is a germ and $I \subset X$ is an infinite g–independent set of elements generic over g. Let $J = \{ (a, g(a)) : a \in I \}$ and suppose g' realizes $tp(g/J)$. Since I is infinite there is an $a \in I$ generic over $\{g, g'\}$. Since g' and g have the same type over J, $g'(a) = g(a)$. Then $g' = g$ (because g is a germ), hence $g \in dcl(J)$.

(3) $\Longrightarrow$ (2) Let $I \subset X$ be a set of elements generic over g such that $g \in dcl(\{ (a, g(a)) : a \in I \}$. Without loss of generality, I is finite. By Lemma 4.5.4(ii) there is a map h generically equal to g which is a canonical parameter of $p_0 = tp(ag(a)/g)$, for some $a \in I$. It remains to show that $g \in dcl(h)$. Let $J = \{ (b, g(b)) : b \in I \}$. Since every element of J realizes p_0 and J is g–independent, J is a Morley sequence in p_0 over g. Since h is a canonical parameter of p_0, J is also a Morley sequence in p_0 over h. In particular, $tp(J/h)$ is stationary and J is independent from g over h. From $g \in dcl(J)$ we conclude that $g \in dcl(h)$.

Corollary 4.5.1. *Let X and Y be infinite loci over $\emptyset$ such that* $\deg(X) = \deg(Y) = 1$.

(i) Given a generic map g of X into Y there is a germ $h \in dcl(g)$ which is generically equal to g.

(ii) If R is an $\bigwedge-$definable set (over $\emptyset$) of generic maps there is an $\emptyset-$definable function γ such that for any $f \in R$, $\gamma(f)$ is a germ generically equal to f. Let R' be the $\bigwedge-$definable set $\gamma(R)$. We can choose γ so that for all h, $k \in R'$, if h is generically equal to k, then $h = k$.

Proof. (i) Combine Lemmas 4.5.4 and 4.5.5.
(ii) See Exercise 4.5.6.

Corollary 4.5.2. *Let X be an infinite loci over $\emptyset$ of degree 1. If g is a germ defined generically on X there is an $\emptyset-$definable set Z such that Z is a set of generic maps on X and for all h, $k \in Z$, if h and k are generically equal then $h = k$.*

Proof. Let $q = tp(g/A)$. There is a formula $\theta \in q$ such that any h satisfying θ is a map defined generically on X. By the Definability Lemma there is a formula $\sigma(y, z)$ over A such that whenever $\models \theta(h) \wedge \theta(k)$, $\models \sigma(h, k)$ if and only if

$$\forall a \in X \text{ generic over } \{h, k\},\ h(a) = k(a) \iff h = k.$$

Any pair of realizations of q satisfies $\sigma(y, z)$. The existence of Z now follows by compactness.

Remark 4.5.7. Given X and Y as usual, there may well be distinct germs g, h mapping X generically to Y which are generically equal. By the same token, when k maps X generically to Y there may be more than one germ in $dcl(k)$ which is generically equal to k.

Definition 4.5.6. *Let $\mathfrak{C}$ be the universe of an uncountably categorical theory, A a set and X, Y loci over A such that* $\deg(X) = \deg(Y) = 1$. *The set of germs from X into Y is denoted $\mathcal{O}(X,Y)$. Let $\mathcal{O}(X,X) = \mathcal{O}(X)$ and $\mathcal{O}^i(X,Y)$ the set of invertible elements of $\mathcal{O}(X,Y)$.*

Let Z be a degree 1 locus over A. With notation as in the definition, composition maps $\mathcal{O}(X,Y) \times \mathcal{O}(Y,Z)$ into $\mathcal{O}(X,Z)$ in that, given $g \in \mathcal{O}(X,Y)$ and $h \in \mathcal{O}(Y,Z)$, there is a germ in $\mathcal{O}(X,Z)$ generically equal to $h \circ g$. (So, in fact, the composition of g and h followed by the operation of taking a germ generically equal to the $h \circ g$ maps (g, h) to an equivalence class of generically equal germs.) In this sense $\mathcal{O}(X)$ is closed under composition. From hereon, when dealing with germs, composition will be denoted by $\cdot$ instead of $\circ$.

Our goal is to find a definable group $G \subset \mathcal{O}(X)$ acting on a definable set $X_0 \supset X$. As an intermediate step we find an $\bigwedge-$definable group contained in $\mathcal{O}(X)$. The naive way to find a group contained in $\mathcal{O}(X)$ which is at least $\mathrm{Aut}(\mathfrak{C})-$invariant is to close some locus $R = r(\mathfrak{C}) \subset \mathcal{O}^i(X)$ under inversion and composition. While this will yield a group H there is no reasons to think it is $\bigwedge-$definable unless there is a finite k such that each element of H is $h_1^{\epsilon_1} \cdot \ldots \cdot h_k^{\epsilon_k}$, where $h_i \in R$ and $\epsilon_i = \pm 1$, for $i = 1, \ldots, k$. We will show that there is such a bound k (and H is $\bigwedge-$definable) if R has generic composition (see Definition 4.5.7).

While not every germ is invertible we do have right cancelation:

Lemma 4.5.6. *Let X, Y and Z be degree 1 infinite loci over $\emptyset$, $h \in \mathcal{O}(X, Y)$ and g_1, $g_2 \in \mathcal{O}(Y, Z)$. If $g_1 \cdot h$ is generically equal to $g_2 \cdot h$ then g_1 is generically equal to g_2. Thus $g_1 \in acl(g_1 \cdot h, h)$.*

Proof. Let $a \in X$ be generic over $\{g_1, g_2, h\}$. Since $g_1(h(a)) = g_2(h(a))$ and $h(a)$ is generic over $\{g_1, g_2\}$ (by Lemma 4.5.1), g_1 is generically equal to g_2.

Lemma 4.5.7. *Let H be an $\bigwedge-$definable semigroup in an uncountably categorical theory which has right cancelation. Then H is a group.*

Proof. Recall from Exercise 3.3.15 that (*) given $\varphi(u, v)$ a formula and $A = \{ a_i : i < \omega \}$ a set such that $\models \varphi(a_i, a_j)$ if and only if $i \leq j$, A must be finite. Given an $a \in H$ we must find a $b \in H$ such that $ba = 1$. By compactness it suffices to show that for any definable $X \supset H$ (on which $\cdot$ is defined) there is a $b \in X$ such that $ba = 1$. Pick an arbitrary definable $X \supset H$. Without loss of generality, $\cdot$ is defined on $X \times X$ and satisfies the right cancelation law on X. Let $X_1 \subset X$ be a definable set such that $H \subset X_1$ and for all x, $y \in X_1$, $x \cdot y \in X$. Let $u|v$ denote the formula $(\exists w \in X_1)(w \cdot u = v)$. For $m \leq n < \omega$, $a^m | a^n$. By (*) there are $m < n$ such that $a^n | a^m$. Using right cancelation on X we get a $b \in X$ such that $b \cdot a = 1$, completing the proof.

Definition 4.5.7. *Let X be a degree 1 infinite locus over A and $R \subset \mathcal{O}(X)$ a degree 1 infinite locus over A. We say R has* generic composition *if for g, $h \in R$ independent, $\{g \cdot h, g, h\}$ is pairwise independent and $g \cdot h \in R$.*

One preliminary lemma before getting to the main result involving generic composition (which is essential to the proof of Theorem 4.5.1).

Lemma 4.5.8. *Let X be a degree 1 infinite locus over A and $R \subset \mathcal{O}(X)$ a degree 1 infinite locus over A with generic composition. Then for all $f, g, h \in R$ there are $j, k \in R$ such that $f \cdot g \cdot h = j \cdot k$.*

Proof. Let g_2 be an element of R independent from $\{f, g, h\}$. Since R has generic composition there is a $g_1 \in R$ such that $g = g_1 \cdot g_2$ and $\{g, g_1, g_2\}$ is pairwise independent. By Lemma 4.5.6 $g_1 \in acl(g, g_2)$. Thus, f is independent from g_1 over $\{g, g_2\}$. Since g_2 is independent from $\{f, g\}$, f is independent from g_1 over g. From the independence of g and g_1 we derive the independence of f and g_1. Let $j = f \cdot g_1$ and $k = g_2 \cdot h$. Since R has generic composition both j and k are in R. The equation $f \cdot g \cdot h = j \cdot k$ completes the proof.

Theorem 4.5.4. *Let X be an infinite locus of degree 1 over A in an uncountably categorical theory and $R \subset \mathcal{O}(X)$ an infinite locus of degree 1 over A with generic composition. There is an $A-$definable group $H \subset \mathcal{O}(X)$ which is connected and has R as its set of generic elements.*

Proof. Let $H_0 = R \cup \{1\}$ and $A = \emptyset$. Let

$$H_0' = \{ f : f \text{ is a generic map on } X \text{ and } f = g \cdot h \text{ for some } g, h \in H_0 \}.$$

Since H_0' is $\bigwedge$-definable Corollary 4.5.1 can be applied to find an $\emptyset$-definable function γ and an $\bigwedge$-definable set $H = \gamma(H_0')$ such that

- for all $f \in H_0'$, $\gamma(f)$ is a germ generically equal to f and
- $h = k$ whenever $h, k \in H$ are generically equal.

Without loss of generality, $\gamma(f) = f$ whenever $f \in H_0$; i.e., $R \subset H$. The key properties of H are highlighted in

Claim. (i) If $f, g \in H$ agree generically on X then $f = g$.
(ii) H is closed under multiplication.
(iii) H has right cancelation.

(i) is part of the definition of H. For (ii) let $f, g \in H$. There are $f_i, g_i \in H_0$, for $i = 1, 2$, such that f is a germ generically equal to $f_1 \cdot f_2$ and g is a germ generically equal to $g_1 \cdot g_2$. By Lemma 4.5.8 there are $h_1, h_2 \in H_0$ such that $f_1 f_2 g_1 g_2 = h_1 h_2$. There is a unique $h \in H$ generically equal to $h_1 h_2$, which we set equal to $f \cdot g$. H has right cancelation by Lemma 4.5.6, completing the proof of the claim.

From Lemma 4.5.7 we conclude that H is a group. By Theorem 3.5.3, H is not only $\bigwedge$-definable, but definable. Since H is a group each element of H is invertible. As a consequence

(*) whenever $A \subset H$, $a \in A$ and $b \in R$ is independent from A, $b \cdot a$ is interdefinable with b over A, hence $MR(b \cdot a/A) = MR(b/A) = MR(R)$.

It remains to show that H is connected and R is the set of generics of H.

Claim. If $a \in H$ and $b \in R$ is independent from a, then $b \cdot a$ is in R.

Let c and d be elements of R such that $a = c \cdot d$ and $\{c, d\}$ is independent from b. Since R has generic composition, $b \cdot c$ is an element of R. By (*), $b \cdot c$ is generic over $\{c, d\}$. Thus, $(b \cdot c) \cdot d$ is an element of R; i.e., $b \cdot a \in R$.

Let a be a generic of H and $b \in R$ generic over a. Then $b \cdot a$ is generic; it is also an element of R by the claim, hence the elements of R are generic. For $b, c \in R$ independent, $b \cdot c^{-1}$ is a generic in the connected component of H by basic facts about generics. Moreover, $b \cdot c^{-1} \in R$ by the claim. Thus, R is the set of generics in the connected component of H. Since H^o is closed under multiplication and every element of H is a product of elements of $R \cup \{1\}$, $H^o = H$. This proves the theorem.

We now make the jump from a definable group of generic maps on an $\bigwedge$-definable set to a definable group action.

Proposition 4.5.1. *Let X be a degree 1 locus over $\emptyset$ in an uncountably categorical theory and $G \subset \mathcal{O}^i(X)$ a connected $\emptyset$-definable group. Then there is a definable group action $(G, X_0, \star)$ for some definable X_0 such that*

- *the action of G on X_0 is faithful and transitive;*

– *X can be identified with a subset of X_0;*
– *for any $g \in G$ and $x \in X$ generic over g, $g \star x = g(x)$.*

Proof. We begin by addressing the problem of the elements of G only being defined generically on X.

Claim. There is an $\bigwedge$–definable set Y and a definable operation $\star$ such that

– $(G, Y, \star)$ is a faithful transitive group action,
– X can be identified with a subset of Y, $GX = Y$, and
– for $g \in G$ and $x \in X$ generic over g, $g \star x = g(x)$.

Consider the set Z of pairs (g, a), where $g \in G$ and $a \in X$. Define an equivalence relation $\sim$ on Z by: $(g, a) \sim (g', a')$ if and only if for every (some) $h \in G$ generic over $\{g, a, g', a'\}$, $(hg)a = (hg')a'$. ($hg \in \mathcal{O}(X)$ and a is generic over hg, so $(hg)a$ is defined.) As usual, by the Definability Lemma, $\sim$ is the restriction to Z of an $\emptyset$–definable equivalence relation. Let $[g, a]$ denote the $\sim$ –class of $(g, a) \in Z$ and Y the set of equivalence classes. We claim that given $g_0 \in G$ and $(g, a) \in Z$, if $(g, a) \sim (g', a')$, then $(g_0 g, a) \sim (g_0 g', a')$. (Given $h \in G$ generic over $\{g_0, g, g', a, a'\}$, hg_0 is generic over $\{g, g', a, a'\}$, hence $(hg_0)ga = (hg_0)g'a'$.) Thus, the operation $\star$ given by $g' \star [g, a] = [g'g, a]$ defines an action of G on Y. We may take $\star$ to be the restriction to $G \times Y$ of a definable operation.

That the map $a \mapsto [1, a]$ is an embedding of X into Y is clear since the elements of G are invertible germs. The definition of Y shows that any $y \in Y$ is $g \star x$ for some $x \in X$ and $g \in G$. It follows quickly that $(G, Y, \star)$ is a faithful transitive group action, completing the proof of the claim.

Let Y_0 be an $\emptyset$–definable set containing Y such that

– $\star$ is defined on $G \times Y_0$,
– for all $x \in Y_0$ and $g, h \in G$, $g \star (h \star x) = (gh) \star x$, and
– $g \star x = x \implies g = 1$.

Let $\theta(v)$ be the formula such that $\models \theta(a)$ if and only if for $x \in X$ generic over a, $(\exists g \in G)(g \star x = a)$. Since all elements of X are in the same orbit under the action of G, $\theta(\mathfrak{C}) \supset X$. Let $Y_1 = Y_0 \cap \theta(\mathfrak{C})$ and $X_0 = GY_1$. The reader can verify that $(G, X_0, \star)$ satisfies all of the conditions of the proposition.

Corollary 4.5.3. *Let X be a degree 1 locus over $\emptyset$ in an uncountably categorical theory and $R \subset \mathcal{O}(X)$ a degree 1 locus with generic composition. Then there is an $\emptyset$–definable group action $(H, X_0, \star)$ such that*

– *$H \subset \mathcal{O}(X)$ which is connected and has R as its set of generic elements;*
– *the action of H on X_0 is faithful and transitive;*
– *X can be identified with a subset of X_0;*
– *for any $g \in H$ and $x \in X$ generic over g, $g \star x = g(x)$.*

Proof. Simply combine Theorem 4.5.4 and Proposition 4.5.1.

Of course, this corollary is useless unless we can find a locus of germs with generic composition. Any instance in which we can prove such a locus exists is a special case of the next proposition.

Proposition 4.5.2. *Let X, Y and Z be loci over $acl(\emptyset)$ in an uncountably categorical theory and suppose there are $f \in \mathcal{O}(X,Y)$ and $g \in \mathcal{O}(Y,Z)$, both invertible, such that $\{f, g, g \cdot f\}$ is pairwise independent and $MR(f)$, $MR(g) < \omega$. Then there is a locus (over $acl(\emptyset)$) $R \subset \mathcal{O}(X)$ of invertible germs such that R has generic composition and $MR(R) = MR(f)$.*

Proof. Since $\{f, g, g \cdot f\}$ is pairwise independent and each element of the set is algebraic in the other two (by the invertibility of f and g) $MR(f) = MR(g) = MR(g \cdot f) = \alpha$. Let F be the locus of f over $acl(\emptyset)$, G the locus of g over $acl(\emptyset)$ and H the locus of $g \cdot f$ over $acl(\emptyset)$. Let f_0 be an element of F generic over f and R the locus of $f_0^{-1} \cdot f$ over $acl(\emptyset)$. One preliminary claim before showing that R has generic composition:

Claim. For independent k, $l \in R$ there is an independent $\{m_0, m_1, m_2, m_3\} \subset F$ such that $k = m_0^{-1} \cdot m_1$ and $l = m_2^{-1} \cdot m_3$.

If $\{f_0, f_1, f_2, f_3\} \subset F$ is independent then $k' = f_0^{-1} \cdot f_1$ and $l' = f_2^{-1} \cdot f_3$ are independent elements of R. The claim follows from the conjugacy over $acl(\emptyset)$ of all independent pairs in R.

Thus, to prove generic composition in R it suffices to show

Claim. Given $\{f_0, f_1, f_2, f_3\} \subset F$ independent there are $f_4, f_5 \in F$ such that $f_0^{-1} \cdot f_1 \cdot f_2^{-1} \cdot f_3 = f_4^{-1} \cdot f_5$ and $\{f_0^{-1} \cdot f_1, f_2^{-1} \cdot f_3, f_4^{-1} \cdot f_5\}$ is pairwise independent.

Let $g \in G$ be generic over $\{f_0, f_1, f_2, f_3\}$. As a first observation:

$$\{g \cdot f_0, g \cdot f_1, f_0^{-1} \cdot f_1\} \text{ is pairwise independent.} \tag{4.6}$$

(Since $\{g, f_0, f_1\}$ is independent $MR(g \cdot f_0 / f_0^{-1} \cdot f_1) \geq MR(g \cdot f_0 / \{g, f_0, f_1\}) = MR(g \cdot f_0 / \{g, f_0\}) = MR(g \cdot f_0)$. That is, $g \cdot f_0$ is independent from $f_0^{-1} \cdot f_1$. Similarly $g \cdot f_1$ is independent from $f_0^{-1} \cdot f_1$ and $g \cdot f_0$ is independent from $g \cdot f_1$.)

Write $(f_0^{-1} \cdot f_1)$ as $(g \cdot f_0)^{-1} \cdot (g \cdot f_1)$, where $(g \cdot f_0)^{-1}$ is independent from $(g \cdot f_1)$ by (4.6). From

- $(f_0^{-1} \cdot f_1)$, $(f_2^{-1} \cdot f_3) \in R$;
- $(g \cdot f_0)$, $(g \cdot f_1) \in H$;
- $\bar{a}_1 = \{f_0^{-1} \cdot f_1, g \cdot f_0\}$ is independent and
- $\bar{a}_2 = \{f_2^{-1} \cdot f_3, g \cdot f_1\}$ is independent;

we conclude that $tp(\bar{a}_1/acl(\emptyset)) = tp(\bar{a}_2/acl(\emptyset))$. Thus there is an $h \in H$ independent from $g \cdot f_1$ such that $f_2^{-1} \cdot f_3 = (g \cdot f_1)^{-1} \cdot h$. It is routine to verify the independence of $g \cdot f_0$ from $\{g \cdot f_1, f_2^{-1} \cdot f_3\}$, hence $(g \cdot f_0)^{-1}$ is independent from h. So, by the conjugacy over $acl(\emptyset)$ of $\{(g \cdot f_1)^{-1}, h\}$ and $\{(g \cdot f_0)^{-1}, h\}$,

$(g \cdot f_0)^{-1} \cdot h$ is equal to $f_4^{-1} \cdot f_5$, for some independent $f_4, f_5 \in F$. The pairwise independence of $\{f_0^{-1} \cdot f_1, f_2^{-1} \cdot f_3, f_4^{-1} \cdot f_5\}$ follows from a rank calculation like that done at the beginning of the proof. This proves the claim and completes the proof that R is a locus of invertible germs with generic composition. The reader should show that any $f_0^{-1} \cdot f_1 \in R$ can be written as $l^{-1} \cdot m$ for some $l, m \in H$ with $\{l, m, f_1\}$ independent. Thus, $MR(f_0^{-1} \cdot f_1) = \alpha$. This proves the proposition.

Corollary 4.5.4. *Let X and Y be infinite loci of degree 1 over $\emptyset$. Suppose there is an invertible germ in $\mathcal{O}(X,Y)$ and there is an $n < \omega$ such that $MR(f) \leq n$ for all invertible $f \in \mathcal{O}(X,Y)$. Then there is a locus (over $acl(\emptyset)$) $R \subset \mathcal{O}(X)$ of invertible germs which has generic composition.*

Proof. Let $g \in \mathcal{O}(X,Y)$ be an invertible germ whose type over $\emptyset$ has maximal Morley rank, C the locus of g over $acl(\emptyset)$ and $m = MR(C)$. Note: any invertible $f \in \mathcal{O}(X)$ (or $\mathcal{O}(Y)$) realizes a type of Morley rank $\leq m$ over $\emptyset$. (Without loss of generality, g is independent from f. Then $g \cdot f \in \mathcal{O}(X,Y)$ is interalgebraic with f over g, hence $n = MR(g) \geq MR(g \cdot f)$.) Let $h \in C$ be generic over g. Since $h^{-1} \cdot g$ is an invertible germ in $\mathcal{O}(X)$, $MR(h^{-1} \cdot g) \leq m$. A rank calculation shows that $\{g, h^{-1}, h^{-1} \cdot g\}$ is pairwise independent. Proposition 4.5.2 can be applied to find the locus R.

The main application of Proposition 4.5.2 is

Proposition 4.5.3. *Let X and Y be loci of degree 1 over $\emptyset$ in a 1−based uncountably categorical theory and suppose there is an invertible germ in $\mathcal{O}(X,Y)$. Then $\mathcal{O}(X)$ contains a connected group $\bigwedge$−definable over $acl(\emptyset)$ and having Morley rank $MR(X)$.*

Using existing results the proof will follow quickly from

Lemma 4.5.9. *Let X and Y be infinite loci of degree 1 over $\emptyset$ in a 1−based uncountably categorical theory and $g \in \mathcal{O}(X,Y)$. Then $MR(g) = MR(Y)$.*

Proof. Let $a \in X$ be generic over g, $b = g(a)$ and recall that $\{g, a, b\}$ is pairwise independent simply because g is a generic map. By Lemma 4.5.4(ii) and the 1−basedness of the theory $g \in acl(a, b)$, hence g is interalgebraic with b over a. Thus, $MR(g) = MR(g/a) = MR(b/a) = MR(b) = MR(Y)$, proving the lemma.

Proof of Proposition 4.5.3. Since there is an invertible germ in $\mathcal{O}(X,Y)$, $MR(X) = MR(Y)$. By Lemma 4.5.9 any invertible germ in $\mathcal{O}(X,Y)$, $\mathcal{O}(Y,X)$ or $\mathcal{O}(X)$ realizes a type of Morley rank $MR(Y)$. Then, given invertible f, $g \in \mathcal{O}(X,Y)$ independent, a standard rank calculation shows that $\{f^{-1}, g, f^{-1} \cdot g\}$ is pairwise independent. By Proposition 4.5.2, $\mathcal{O}(X)$ contains a locus R over $acl(\emptyset)$ with generic composition with $MR(R) = MR(f)$. There is an $acl(\emptyset)$−definable connected group $G \subset \mathcal{O}(X)$ which has R as its set of generic elements (by Theorem 4.5.4). Noting that $MR(G) = MR(X)$ completes the proof.

4.5.2 Getting a Group from an Algebraic Quadrangle

In this section Theorem 4.5.1 and its corollaries are proved. Theorem 4.5.4 reduces the problem to finding in $\mathcal{O}(X)$ for some X a locus of germs (with a special relationship to a_1) which has generic composition. The theme is to successively replace the original algebraic quadrangle by a "nicer" quadrangle until (many of) the algebraic closure relations in the quadrangle are instances of definable closure. A definition is needed to state the key result. Remember that every theory in this section is assumed to be uncountably categorical.

The following illustrates the relationship between algebraic quadrangles and group actions.

Remark 4.5.8. Let K be an algebraically closed field and G the group of affine transformations on K (see Example 3.5.3). Let $h,\ g \in G$ be independent generics and $a \in X$ generic over $\{h, g\}$. Then $(a, h(a), g^{-1}h(a), h, g^{-1}, g^{-1}h)$ is an algebraic quadrangle. (The verification is left to the reader.)

Definition 4.5.8. *Let A be a set and $Q = (a_1, a_2, a_3, b_1, b_2, b_3)$, $Q' = (a'_1, a'_2, a'_3, b'_1, b'_2, b'_3)$ algebraic quadrangles over A. Then Q is* interalgebraic with Q' over A *if for all $1 \le i \le 3$, a_i is interalgebraic with a'_i over A and b_i is interalgebraic with b'_i over A.*

Proposition 4.5.4. *Given an algebraic quadrangle $Q = (a_1, a_2, a_3, b_1, b_2, b_3)$ there is a finite set A independent from Q and $Q' = (a'_1, a'_2, a'_3, b'_1, b'_2, b'_3)$ an algebraic quadrangle over A such that*

(1) *Q and Q' are interalgebraic over A,*
(2) *a'_1 and a'_3 are interdefinable over $A \cup \{b'_2\}$, and*
(3) *a'_2 and a'_3 are interdefinable over $A \cup \{b'_1\}$.*

The proposition will be proved in several stages, finding progressively "nicer" algebraic quadrangles over increasingly large sets of parameters. To simplify the notation we will replace at each stage the original algebraic quadrangle Q by the nicer one and absorb the parameters into the language.

Lemma 4.5.10. *If $Q = (a_1, a_2, a_3, b_1, b_2, b_3)$ is an algebraic quadrangle and for each $1 \le i \le 3$, a'_i is interalgebraic with a_i over $\emptyset$ and b'_i is interalgebraic with b_i over $\emptyset$, then $Q' = (a'_1, a'_2, a'_3, b'_1, b'_2, b'_3)$ is an algebraic quadrangle.*

Proof. The proof quickly reduces to showing that, for instance, b'_3 is interalgebraic with the canonical parameter of $tp(a'_1a'_2/acl(b'_3))$. This is not difficult using that $\{a'_1, a'_2, b_3\}$ is pairwise independent, a'_1 is interalgebraic with a'_2 over b'_3, and the corresponding fact is true in Q. See Exercise 4.5.7.

Part of the definition of an algebraic quadrangle is that the ℓ_{ijk}'s are independent over their intersections. Using the independence of other sets we can show in addition

Lemma 4.5.11. *Let $Q = (a_1, a_2, a_3, b_1, b_2, b_3)$ be an algebraic quadrangle and $\{i, j, k\} = \{1, 2, 3\}$. Then $\{b_j, b_k\}$ is independent from $\{a_j, a_k\}$ over b_i.*

Proof. The proof is a two line exercise left to the reader.

The next lemma will see extensive use.

Lemma 4.5.12. *Let $Q = (a_1, a_2, a_3, b_1, b_2, b_3)$ be an algebraic quadrangle and $1 \leq i \leq 3$. A realization a of $tp(a_i/Q \setminus \{a_i\})$ is interalgebraic with a_i over $\emptyset$. Thus, letting e be a name for the (finite) set of realizations of $tp(a_i/Q \setminus \{a_i\})$, e is interalgebraic with a_i and $e \in dcl(Q \setminus \{a_i\})$.*

Proof. Without loss of generality, $i = 3$.

Claim. a and a_3 are interalgebraic over b_1 and interalgebraic over b_2.

Since a_2 and a_3 are interalgebraic over b_1, a_2 and a are interalgebraic over b_1. Thus, a and a_3 are interalgebraic over b_1. Similarly, a is interalgebraic with a_3 over b_2.

Let c be the canonical parameter of $tp(aa_3/acl(b_1b_2))$. Since a_3 is independent from $\{b_1, b_2\}$ and $a \in acl(a_3, b_1)$, aa_3 is independent from $\{b_1, b_2\}$ over b_1. Thus, $c \in acl(b_1)$. Similarly, $c \in acl(b_2)$. Since b_1 is independent from b_2, $c \in acl(\emptyset)$, hence aa_3 are interalgebraic over $\emptyset$.

It is clear from the first part of the lemma that e is interalgebraic with a_i. Since e is the name of a set definable over $Q \setminus \{a_i\}$, $e \in dcl(Q \setminus \{a_i\})$, completing the proof.

Lemma 4.5.13. *Let $Q = (a_1, a_2, a_3, b_1, b_2, b_3)$ be an algebraic quadrangle. There are b_1', a_2', a_3' and a finite set A such that*

(1) *A is independent from Q;*
(2) *$Q' = (a_1, a_2', a_3', b_1', b_2, b_3)$ is an algebraic quadrangle interalgebraic with Q over A; and*
(3) *$a_2' \in dcl(a_3', b_1', A)$.*

Proof. Let d_2 be a realization of $tp(b_2/acl(\emptyset))$ which is independent from Q. Since b_2 is independent from $\{b_1, a_2, a_3\}$,

$$tp(d_2/\{b_1, a_2, a_3\}) = tp(b_2/\{b_1, a_2, a_3\}),$$

hence there are c_1, d_3 so that $Q_0 = (c_1, a_2, a_3, b_1, d_2, d_3)$ realizes $tp(Q/acl(\emptyset))$. Let a_2' be a name for the finite set of realizations of $tp(a_2/Q_0 \setminus \{a_2\})$. Then $a_2' \in dcl(Q_0 \setminus \{a_2\})$ and a_2' is interalgebraic with a_2 over $\emptyset$ by Lemma 4.5.12.

Now fix $A = \{d_2\}$ as a set of parameters, $b_1' = \{b_1, d_3\}$ and $a_3' = \{a_3, c_1\}$. Let $Q' = (a_1, a_2', a_3', b_1', b_2, b_3)$. Then Q' is an algebraic quadrangle over A, interalgebraic with Q over A, and $a_2' \in dcl(A \cup \{a_3', b_1'\})$ as desired.

Lemma 4.5.14. *Let $Q = (a_1, a_2, a_3, b_1, b_2, b_3)$ be an algebraic quadrangle in which $a_2 \in dcl(a_3, b_1)$. Then there are b_2' interalgebraic with b_2 and a_1' interalgebraic with a_1 such that $a_1' \in dcl(a_3, b_2')$.*

Proof. Since $tp(a_1/a_3b_2)$ is algebraic there is a b_2' interalgebraic with b_2 so that $tp(a_1/a_3b_2')$ implies $tp(a_1/\{a_3\} \cup acl(b_2))$. Using Lemma 4.5.11 it follows that

$$tp(a_1/\{a_3, b_2'\}) \text{ implies } tp(a_1/\{a_3, b_1, b_2', b_3\}). \tag{4.7}$$

Claim. If a realizes $tp(a_1/\{a_3, b_2'\})$ then a_1 and a are interalgebraic.

Given a realizing $tp(a_1/\{a_3, b_2'\})$, a also realizes $tp(a_1/\{a_3, b_1, b_2', b_3\})$, by (4.7). Since $a_2 \in dcl(a_3, b_1)$, a realizes $tp(a_1/\{a_2, a_3, b_1, b_2', b_3\})$. The 6−tuple $(a_1, a_2, a_3, b_1, b_2', b_3)$ forms an algebraic quadrangle, so Lemma 4.5.12 forces a_1 and a to be interalgebraic as claimed.

Let a_1' be the (finite) set of realizations of $tp(a_1/\{a_3, b_2'\})$, which is hence in $dcl(a_3, b_2')$. By the claim a_1' is interalgebraic with a_1, proving the lemma.

Lemma 4.5.15. *Let $Q = (a_1, a_2, a_3, b_1, b_2, b_3)$ be an algebraic quadrangle in which $a_2 \in dcl(a_3, b_1)$ and $a_1 \in dcl(a_3, b_2)$. Then there is a d_3 independent from Q and there is $Q' = (a_1', a_2', a_3', b_1', b_2', b_3')$ an algebraic quadrangle over d_3, interalgebraic with Q over d_3 such that*

(1) a_1' and a_3' are interdefinable over $\{b_2', d_3\}$, and
(2) a_2' and a_3' are interdefinable over $\{b_1', d_3\}$.

Proof. First let $a_3' \in dcl(Q \setminus \{a_3\})$ be interalgebraic with a_3 (which exists by Lemma 4.5.12). Let d_3 be a realization of $tp(b_3/acl(\emptyset))$ which is independent from Q. Find d_1 and c_2 so that $tp(c_2d_1d_3/acl(b_2, a_1, a_3)) = tp(a_2b_1b_3/acl(b_2, a_1, a_3))$. Note that $a_3' \in dcl(a_1, c_2, d_1, b_2, d_3)$. Let $a_1' = (a_1, c_2)$ and $b_2' = (b_2, d_1)$. Summarizing, we have a_1', a_3', b_2' and d_3 so that

- $Q_0 = (a_1', a_2, a_3', b_1, b_2', b_3)$ is an algebraic quadrangle over d_3 interalgebraic with Q over d_3,
- $a_3' \in dcl(a_1', b_2', d_3)$, and
- $a_1' \in dcl(a_3', b_2', d_3)$.

Similarly we find elements d_2 and c_1 so that $tp(c_1d_2d_3/acl(b_1a_2a_3')) = tp(a_1b_2b_3/acl(b_1a_2a_3'))$. Let $a_2' = (a_2, c_1)$ and $b_1' = (b_1, d_2)$. Drawing together the accumulated properties:

- $Q' = (a_1', a_2', a_3', b_1', b_2', b_3)$ is an algebraic quadrangle over d_3 interalgebraic with Q over d_3,
- $a_3' \in dcl(a_1', b_2', d_3)$,
- $a_1' \in dcl(a_3', b_2', d_3)$,
- $a_3' \in dcl(a_2', b_1', d_3)$, and
- $a_2' \in dcl(a_3', b_1', d_3)$.

This proves the lemma.

Proof of Proposition 4.5.4. Combine Lemmas 4.5.13, 4.5.14 and 4.5.15.

A quadrangle with this amount of definable closure produces a group of germs acting generically on the locus of any of the a_i's over $acl(\emptyset)$:

Proposition 4.5.5. *Let $Q = (a_1, a_2, a_3, b_1, b_2, b_3)$ be an algebraic quadrangle in which a_2 is interdefinable with a_3 over b_1 and a_1 is interdefinable with a_3 over b_2. Let X be the locus of a_1 over $acl(\emptyset)$. Then there is a connected group $G \subset \mathcal{O}(X)$, definable over $acl(\emptyset)$, such that $MR(G) = MR(b_i)$.*

Proof. Let X, Y and Z be the loci over $acl(\emptyset)$ of a_1, a_2 and a_3, respectively. Since a_1 and a_3 are interdefinable over b_2 and $\{a_1, a_3, b_2\}$ is pairwise independent there is $f \in acl(b_2)$ which is the germ of an invertible generic map from X into Z with $f(a_1) = a_3$. By Lemma 4.5.5 f is the canonical parameter of $tp(a_1a_3/f)$, which is also the canonical parameter of $tp(a_1a_3/acl(b_2))$. From one clause in the definition of an algebraic quadrangle f is interalgebraic with b_2.

Similarly let g be an invertible germ in $\mathcal{O}(Z, Y)$ such that $g(a_3) = a_2$ and g is interalgebraic with b_1. Then $g \cdot f$ is an invertible germ from X to Y, a_1 is generic over $g \cdot f$ and $g \cdot f(a_1) = a_2$.

Claim. $g \cdot f$ is interalgebraic with b_3.

The germ $g{\cdot}f$ is definable over $\{b_1, b_2\}$ and a_1 is generic over $\{b_1, b_2\}$, hence $g{\cdot}f$ is interdefinable with the canonical parameter c of $p = tp(a_1a_2/acl(b_1, b_2))$ by Lemma 4.5.5. Since $b_3 \in acl(b_1, b_2)$ p is also $tp(a_1a_2/acl(b_1, b_2, b_3))$. Since Q is an algebraic quadrangle a_1 is interalgebraic with a_2 over b_3, thus p is the unique free extension of $tp(a_1a_2/acl(b_3))$. Hence both c and $g \cdot f$ are not only algebraic in b_3 but interalgebraic with b_3 as claimed.

By the claim and the pairwise independence of $\{b_1, b_2, b_3\}$ $\{f, g, g \cdot f\}$ is pairwise independent. By Proposition 4.5.2 and Theorem 4.5.4 there is a connected group $G \subset \mathcal{O}(X)$, definable over $acl(\emptyset)$, with $MR(G) = MR(f) = MR(b_i)$.

Proof of Theorem 4.5.1. Let $Q = (a_1, a_2, a_3, b_1, b_2, b_3)$ be the hypothesized algebraic quadrangle. By Proposition 4.5.4 there is a finite set A' independent from Q and an algebraic quadrangle $Q' = (a_1', a_2', a_3', b_1', b_2', b_3')$ over A' such that

(1) Q and Q' are interalgebraic over A',
(2) a_1' and a_3' are interdefinable over $A' \cup \{b_2'\}$, and
(3) a_2' and a_3' are interdefinable over $A' \cup \{b_1'\}$.

Proposition 4.5.5 yields a connected group G of germs acting generically on X' = the locus of a_1' over $acl(A')$ which is definable over $acl(A')$ and has Morley rank $= MR(b_2')$. By Proposition 4.5.1 there are:

– a finite $A \subset acl(A')$;

– an A–definable degree 1 set $X \supset X'$;
– an A–definable transitive group action of G on X.

This proves the theorem.

Theorem 4.5.2 will follow from a slightly more general result stated momentarily. An uncountably categorical theory with universal domain $\mathfrak{C}$ is *trivial* if for all A there is no set $\{A_0, A_1, A_2\}$ which is pairwise A–independent but not A–independent. Note: When $\mathfrak{C}$ is strongly minimal this definition agrees with the earlier definition of a trivial strongly minimal set. The set of elements $X = \{a_0, a_1, a_2\}$ is an *algebraic triangle over* A if X is pairwise A–independent and for each $i \leq 2$, $a_i \in acl(A \cup X \setminus \{a_i\}) \setminus acl(A)$.

Theorem 4.5.5. *An nontrivial* 1*–based uncountably categorical theory contains an infinite definable group.*

This theorem follows from the next two results.

Lemma 4.5.16. *Let $\mathfrak{C}$ be the universal domain of a* 1*–based uncountably categorical theory and A, A_0, A_1 and A_2 sets such that $\{A_0, A_1, A_2\}$ is pairwise A–independent but not A–independent. Then there are $a_i \in acl(A_i \cup A)$, for $i \leq 2$, such that $\{a_0, a_1, a_2\}$ is an algebraic triangle over A.*

Proof. Without loss of generality each A_i is finite and $A = \emptyset$. Find $a_0 \in acl(A_0) \cap acl(A_1 \cup A_2)$ so that A_0 is independent from $A_1 \cup A_2$ over a_0. Also choose $a_1 \in acl(A_1) \cap acl(A_0 \cup A_2)$ with A_1 independent from $A_0 \cup A_2$ over a_1 and $a_2 \in acl(A_2) \cap acl(A_0 \cup A_1)$ with A_2 independent from $A_0 \cup A_1$ over a_2. The pairwise independence of $\{A_0, A_1, A_2\}$ forces $\{a_0, a_1, a_2\}$ to be pairwise independent. Since $a_0 \in acl(A_1 \cup A_2)$ and A_1 is independent from $\{a_0\} \cup A_2$ over a_1, $a_0 \in acl(\{a_1\} \cup A_2)$. Continuing this reasoning $a_0 \in acl(a_1, a_2)$. By the symmetric roles of the a_i's in this proof, $a_1 \in acl(a_0, a_2)$ and $a_2 \in acl(a_0, a_1)$, proving the lemma.

Proposition 4.5.6. *Let $\mathfrak{C}$ be the universal domain of a* 1*–based uncountably categorical theory containing an algebraic triangle $P = \{c_0, c_1, c_2\}$. Then there is a finite set A, independent from P, an A–definable connected group G, an A–definable set X and an A–definable transitive action of G on X such that c_1 is interalgebraic over A with a generic of X and $MR(G) = MR(X)$.*

Proof. An algebraic quadrangle containing P is found as follows. First rename the elements of P as $b_2 = c_0$, $a_1 = c_1$ and $a_3 = c_2$. Let $b_1 a_2$ be a realization of $tp(b_2 a_1 / a_3)$ independent from $b_2 a_1$ over a_3. Let b_3 be the canonical parameter of $tp(a_1 a_2 / acl(b_1, b_2))$. We will show that $Q = (a_1, a_2, a_3, b_1, b_2, b_3)$ is an algebraic quadrangle.

Claim. b_2 is interalgebraic with the canonical parameter of $tp(a_1 a_3 / acl(b_2))$.

The canonical parameter c of $tp(a_1a_3/acl(b_2))$ is in $acl(b_2)$ and b_2 is independent from a_1a_3 over c. Since $b_2 \in acl(a_1, a_3)$, $b_2 \in acl(c)$ as claimed.

Since $tp(b_1a_2/a_3) = tp(b_2a_1/a_3)$, b_1 is interalgebraic with the canonical parameter of $tp(a_2a_3/acl(b_1))$. The element b_3 was chosen as the canonical parameter of $tp(a_1a_2/acl(b_3))$. The remaining steps in the verification that Q is an algebraic quadrangle are organized in

(a) $\{a_1, a_2, b_3\}$ is an algebraic triangle.
(b) $\{b_1, b_2, b_3\}$ is pairwise independent.
(c) $\{b_1, b_2, b_3\}$ is an algebraic triangle.
(d) For $\{i, j, k\} = \{i', j', k'\} = \{1, 2, 3\}$, ℓ_{ijk} is independent from $\ell_{i'j'k'}$ over $\ell_{ijk} \cap \ell_{i'j'k'}$ (where $\ell_{ijk} = \{b_i, a_j, a_k\}$).

(a) The a_3–independence of a_1b_2 and a_2b_1 forces a_1 and a_2 to be independent from b_1b_2. Since $b_3 \in acl(b_1, b_2)$, $\{a_1, a_2, b_3\}$ is pairwise independent. a_1 and a_2 are interalgebraic over b_3 because these elements are interalgebraic over b_1b_2. The theory is 1–based so $b_3 \in acl(a_1, a_2)$, proving (a).

(b) Again the selection of the elements $\{a_1, a_2, a_3, b_1, b_2\}$ yields the independence of b_1 from a_1a_2 and b_2 from a_1a_2. Thus $\{b_1, b_2, b_3\}$ is pairwise independent.

(c) a_2 is independent from $b_1b_2b_3$ and $b_1 \in acl(b_2, b_3, a_2)$, so $b_1 \in acl(b_2, b_3)$. Similarly $b_2 \in acl(b_1, b_3)$. It has already been noted that $b_3 \in acl(b_1, b_2)$, hence $\{b_1, b_2, b_3\}$ is an algebraic triangle.

(d) The cases not explicitly verified above are left to the reader.

Thus, Q is an algebraic quadrangle. By Theorem 4.5.1 there are:

- a finite set A independent from Q;
- an A–definable set X of degree 1 containing a generic interalgebraic with a_1 over A;
- an A–definable connected group G and A–definable transitive action of G on X with $MR(G) = MR(b_2)$.

Since $MR(b_2) = MR(c_0) = MR(c_1) = MR(a_1) = MR(X)$, $MR(G) = MR(X)$. This proves the proposition.

Proof of Theorem 4.5.5. This follows immediately from Lemma 4.5.16 and Proposition 4.5.6.

Completing our applications to 1–based theories we have:

Proof of Theorem 4.5.2. Being nontrivial D contains a finite B and $\{c_0, c_1, c_2\}$ which is an algebraic triangle over B. By Proposition 4.5.6 there is a finite $A \supset B$ and a connected A–definable group G of Morley rank 1. A connected group of Morley rank 1 is strongly minimal. Since the theory is uncountably categorical we can choose A large enough so that an element of $G \setminus acl(A)$ is interalgebraic over A with an element of $D \setminus acl(A)$.

Corollary 4.5.5. *Given a nontrivial locally modular strongly minimal set D there is a finite set A and a strongly minimal group G over A such that a generic of G is interalgebraic over A with an element of $D \setminus acl(A)$ and G is a $*-$vector space over some division ring F. Thus the geometry associated to D_A is projective geometry over F.*

Proof. The existence of G and its relationship to D is simply by Theorem 4.5.2. By Theorem 4.3.4 G is a $*-$vector space over the division ring $F = \text{End}^*(G)$. The geometry associated to G is a projective geometry over F. The relation of being interalgebraic over A defines a one-to-one correspondence between the elements of the geometry associated to G and the geometry associated to D_A. In other words the geometry associated to G is isomorphic to the geometry associated to D_A, completing the proof.

Remark 4.5.9. A more sophisticated series of arguments shows that when D is locally modular and nonmodular there is a strongly minimal group definable over $acl(\emptyset)$, an $\emptyset-$definable equivalence relation E with finite classes and an $acl(\emptyset)-$definable regular action of G on the strongly minimal set $D' = \{ a/E : a \in D \}$. Thus the geometry associated to D' (which is also the geometry associated to D) is affine geometry over the vector space G/G^-. See [Hru87].

Our final installment in this study of defining groups is Theorem 4.5.3. This is proved by assuming to the contrary the theory contains a pseudomodular strongly minimal set which is not locally modular, proving the theory contains an infinite definable field, and that this leads directly to a contradiction.

Lemma 4.5.17. *Let D be a strongly minimal set such that for some A there are $A-$definable operations $+$ and $\cdot$ under which D is a field. Then D is not pseudomodular.*

Proof. This follows quickly from Example 4.2.2(iii).

Lemma 4.5.18. *Let D be a strongly minimal set, A a finite set, $a \in D \setminus acl(A)$ and $a' \in acl(A \cup \{a\}) \cap D'$, for D' an $A-$definable strongly minimal set. Then D is pseudomodular if and only if D' is pseudomodular.*

Proof. See Exercise 4.5.8.

Proof of Theorem 4.5.3. Suppose to the contrary that D is pseudomodular, not locally modular, and $k > 1$ is the maximum Morley rank of a plane curve in D. Let $a_1, a_3 \in D$ and b_2 be such that $tp(a_1a_3/b_2)$ is strongly minimal, b_2 is the canonical parameter of this type and $MR(b_2) = k$. Let a_2b_1 be a realization of $tp(a_1b_2/a_3)$ independent from a_1b_2 over a_3. Let b_3 be the canonical parameter of $p = tp(a_1a_2/acl(b_1, b_2))$. Since p is strongly minimal b_3 is the name for a plane curve in D hence $MR(b_3) \leq k$.

Claim. $Q = (a_1, a_2, a_3, b_1, b_2, b_3)$ is an algebraic quadrangle.

As a first step $b_3 \in acl(b_1, b_2)$ because it is the canonical parameter of a type over $acl(b_1, b_2)$. From the a_3–independence of a_1b_2 and a_2b_1, a_1a_3 is independent from $b_1b_2b_3$ over b_2. Since $a_3 \in acl(b_1, b_3, a_1)$, a_1a_3 is independent from $b_1b_2b_3$ over b_1b_3. Thus $b_2 =$ the canonical parameter of $tp(a_1a_3/b_2)$ is in $acl(b_1, b_3)$. In other words, b_2 and b_3 are interalgebraic over b_1. From this relation and $MR(b_3) \leq k$ we conclude that $MR(b_3) = k$ and $\{b_1, b_2, b_3\}$ is pairwise independent. Similarly $b_1 \in acl(b_2, b_3)$. The remaining steps in showing that Q is an algebraic quadrangle are left to the reader.

By Theorem 4.5.1 there is a finite set A, an $a' \in acl(A \cup \{a_2\})$ which is a generic of an A–definable strongly minimal set D' and an A–definable group G acting transitively on D' such that $MR(G) = MR(b_2)$. By Lemma 4.5.18 D' is pseudomodular, while there is a definable field structure on D' (perhaps with extra parameters) by Theorem 3.5.2. This contradicts Lemma 4.5.17 to prove the theorem.

Historical Notes. Algebraic quadrangles were developed by Zil'ber in his proof that a totally categorical theory is not finitely axiomatizable. His most up to date treatment is found in [Zil93]. The proof given here is based on the more general results proved by Hrushovski. One source for this material is Bouscaren's article in [NP89]. It is also found in [BH]. A more complete set of results can be found in [Pil]. Theorem 4.5.3 was first proved (using methods different from those here) by Buechler and Hrushovski [Bue91].

Exercise 4.5.1. Prove Remark 4.5.2

Exercise 4.5.2. Prove Remark 4.5.4

Exercise 4.5.3. Prove Lemma 4.5.1

Exercise 4.5.4. Prove Remark 4.5.6(iii)

Exercise 4.5.5. Prove Lemma 4.5.4(i).

Exercise 4.5.6. Prove Corollary 4.5.1.

Exercise 4.5.7. Let $\mathfrak{C}$ be the universal domain of an uncountably categorical theory and $\{a_1, a_2, b\}$ a pairwise independent set such that $a_1 \in acl(a_2, b)$ is and b is interalgebraic with the canonical parameter of $tp(a_1a_2/acl(b))$. Suppose a_i' is interalgebraic with a_i, for $i = 1, 2$. Prove that b is interalgebraic with the canonical parameter of $tp(a_1'a_2'/acl(b))$.

Exercise 4.5.8. Prove (ii) of Lemma 4.5.18

5. Stability

In this chapter we state and prove the basic definitions and theorems relevant to all stable theories. The first section contains the most fundamental material. Here a freeness relation (see Definition 3.3.1) called forking independence is developed which agrees with Morley rank independence on a t.t. theory. Many of the theorems proved earlier for t.t. theories can be generalized to stable theories, the class of theories on which forking independence exists.

Sections 5.1 to 5.3 contain material which anyone working in stable theories must know. The first-time reader should feel free to skip the proofs in Section 5.5, although it is important to know the statements of the results found there. The forking independence relation is analyzed more deeply in Section 5.6. A class of types (namely those having weight 1) is isolated on which a well-behaved dimension theory exists.

5.1 Stability

Here we define a broad class of theories (called the stable theories) on which there is a freeness relation satisfying the conditions specified in Definition 3.3.1. As with t.t. theories, the freeness relation is defined via a rank (more accurately, a family of ranks). Intuitively, each of these ranks could be described as "Morley rank relative to a finite set of formulas". The overall goal of the section is to develop the relevant ranks and notion of freeness, prove the definability of types in stable theories and relate its existence to the number of types over sets.

Remember: Every complete theory discussed is assumed to have built-in imaginaries.

5.1.1 Ranks and Definability

Writing the formula φ in the form $\varphi(\bar{x}, \bar{y})$ indicates that the free variables in φ are in $\bar{x}\bar{y}$, $\bar{x}$ should be regarded as a sequence of free variables in the usual sense, but $\bar{y}$ is a placeholder for a sequence of parameters. For example, a general quadratic polynomial in x can be written as $\varphi(x, abc) = ax^2 + bx + c$, where a, b and c range over the possible coefficients. We call $\bar{x}$ the *object variables* and $\bar{y}$ the *parameter variables* in φ. When Δ is a set of formulas we write $\Delta = \Delta(\bar{x})$ when the object variables of any $\varphi \in \Delta$ are in $\bar{x}$. Following the conventions previously adopted for theories with built-in imaginaries, we will usually drop the bars from the variables and just write, e.g., $\Delta = \Delta(x)$.

When $\Delta = \Delta(x)$ is a set of formulas and A is a set we call the type p over A a *Δ–type* if each formula in p is of the form $\varphi(x, a)$ or $\neg\varphi(x, a)$ for some $a \in A$ and $\varphi(x, y) \in \Delta$. A Δ–type p is called *complete* if for all $a \in A$ and $\varphi(x, y) \in \Delta$, $\varphi(x, a)$ or $\neg\varphi(x, a)$ is in p. When $\Delta = \{\varphi\}$, for some formula φ, a Δ–type is called a *φ–type*. Let $S_\Delta(A)$ denote the set of complete Δ–types over A, $S_\varphi(A) = S_{\{\varphi\}}(A)$. When p is a type over A in the variable x, $p \restriction \Delta$ denotes the Δ–type

$$\{\varphi(x, a) \in p : \varphi(x, y) \in \Delta\} \cup \{\neg\varphi(x, a) \in p : \varphi(x, y) \in \Delta\}.$$

Our notion of freeness will be defined with the following class of ranks.

Definition 5.1.1. *Let T be a complete theory, $\Delta = \Delta(x)$ a set of formulas over $\emptyset$ and S the elements of $S(\mathfrak{C})$ in the variable x. For φ a formula in x and α an ordinal (or -1) the relation $R_\Delta(\varphi) = \alpha$, is defined as follows by recursion.*

(1) $R_\Delta(\varphi) = -1$ *if φ is inconsistent;*
(2) $R_\Delta(\varphi) = \alpha$ *if*

$$\{p \restriction \Delta : p \in S,\ \varphi \in p \text{ and } \neg\psi \in p \text{ for all formulas } \psi \text{ with } R_\Delta(\psi) < \alpha\}$$

is finite and nonempty.

For p any type in x, $R_\Delta(p)$ is defined to be

$$\inf\{R_\Delta(\varphi) : \varphi \text{ is implied by } p\}.$$

(Thus, for $p \in S$, $R_\Delta(p)$ is $\inf\{R_\Delta(\varphi) : \varphi \in p\}$.) The relation $R_\Delta(p) = \alpha$ is read the Δ–rank of p is α. *If there is no α with $R_\Delta(p) = \alpha$ we write $R_\Delta(p) = \infty$ and say that the Δ–rank of p does not exist.*

By convention, we only write $R_\Delta(p)$ when there is an x such that $\Delta = \Delta(x)$, Δ is a set of formulas over $\emptyset$ and p is a type in x.

As with Morley rank, Δ–rank is preserved under conjugacy. The rank $R_\Delta(-)$ is what Shelah calls $R(-, \Delta, \aleph_0)$ (see [She90, p.21]).

The following is little more than a restatement of Lemma 3.3.1.

Lemma 5.1.1. *Let T be a complete theory, $\Delta = \Delta(x)$ a set of formulas over $\emptyset$, p a type in x, S the set of elements of $S(\mathfrak{C})$ in x and α an ordinal.*

(i) If φ is a formula in x and Δ contains $x = y$, $R_\Delta(\varphi) = 0$ if and only if φ is algebraic.

(ii) $R_\Delta(p) = \alpha$ *if and only if there is a formula* φ *implied by* p *such that* $\{q \restriction \Delta : q \in S, \varphi \in q \text{ and } R_\Delta(q) = \alpha\}$ *is finite, nonempty and equal to* $\{q \restriction \Delta : q \in S, q \supset p \text{ and } R_\Delta(q) = \alpha\}$.

(iii) If $R_\Delta(p) = \alpha$ *there is a* $q \in S$ *such that* $q \supset p$ *and* $R_\Delta(q) = \alpha$.

(iv) If $p \in S$ *and* $R_\Delta(p) = \alpha$ *there is a* $\varphi \in p$ *such that* $p \restriction \Delta$ *is the only element of* $\{q \restriction \Delta : q \in S, \varphi \in q \text{ and } R_\Delta(q) \geq \alpha\}$.

(v) $R_\Delta(p) \geq \alpha$ *if and only if for all* $\beta < \alpha$ *and all* φ *implied by* p, $\{q \restriction \Delta : q \in S, \varphi \in q \text{ and } R_\Delta(q) \geq \beta\}$ *is infinite.*

(vi)

$$R_\Delta(\varphi) \text{ is the least ordinal } \alpha \text{ such that} \quad (5.1)$$
$$\{q \restriction \Delta : q \in S, \varphi \in q \text{ and } R_\Delta(p) \geq \alpha\} \text{ is finite.}$$

(vii) If $\Gamma(x) \supset \Delta(x)$, $R_\Gamma(p) \geq R_\Delta(p)$. *When* Γ *is the set of all formulas with object variable* x, $R_\Gamma(p) = MR(p)$. *Thus* $MR(p) \geq R_\Delta(p)$.

Proof. (i) If φ is algebraic, then the set of complete Δ–types over $\mathfrak{C}$ consistent with φ is finite, hence $R_\Delta(\varphi) = 0$. Suppose φ is nonalgebraic, satisfied by the distinct elements a_i, $i < \omega$. For each $i < \omega$, $\{x = a_i\}$ extends to a complete Δ–type r_i over $\mathfrak{C}$ consistent with φ. Then $\{r_i : i < \omega\}$ consists of infinitely many contradictory Δ–types consistent with φ, proving that $R_\Delta(\varphi) > 0$.

Each of (i)–(vi) is proved like the corresponding part in Lemma 3.3.1. The proof of part (vii) is assigned as Exercise 5.1.2.

We will tacitly assume that any finite set $\Delta(x)$ of formulas under consideration contains $x = y$. This ensures that a formula has Δ–rank 0 exactly when it is algebraic.

Definition 5.1.2. *Let* Δ *be a set of formulas in* x *and* p *a type in* x *with* $R_\Delta(p) = \alpha < \infty$. *Then the* Δ–multiplicity of p, *denoted* $\mathrm{Mult}_\Delta(p)$, *is the maximum* m *such that there are* $q_1, \ldots, q_m \in S(\mathfrak{C})$ *with* $q_i \supset p$, $R_\Delta(q_i) = \alpha$, *for* $1 \leq i \leq m$, *and* $i \neq j \implies q_i \restriction \Delta \neq q_j \restriction \Delta$. *(Equivalently, the* Δ*–multiplicity of* p *is the maximal* m *such that there are* m *complete* Δ*–types* $r_1, \ldots, r_m$ *over* $\mathfrak{C}$ *such that* $R_\Delta(p \cup r_i) = R_\Delta(p)$, *for* $1 \leq i \leq m$.*) Let* $(R, \mathrm{Mult})_\Delta(p)$ *denote the pair* $(R_\Delta(p), \mathrm{Mult}_\Delta(p))$.

Similar to the behavior of Morley rank and degree, for any set of formulas Δ and any type p there is a formula φ implied by p such that $(R, \mathrm{Mult})_\Delta(\varphi) = (R, \mathrm{Mult})_\Delta(p)$. Repeating the argument in Remark 3.3.1, when p is closed under finite conjunctions there is a $\varphi \in p$ such that $(R, \mathrm{Mult})_\Delta(\varphi) = (R, \mathrm{Mult})_\Delta(p)$.

Notation. If Δ is a finite set of formulas, p is a type and $X = p(\mathfrak{C})$, $(R, \mathrm{Mult})_\Delta(X) = (R, \mathrm{Mult})_\Delta(p)$.

Definition 5.1.3. *A complete theory* T *is called* stable *if for all* $p \in S(\mathfrak{C})$ *and all finite* Δ, $R_\Delta(p) < \infty$.

For any finite set of formulas Δ and any formula φ, $R_\Delta(\varphi) \le MR(\varphi)$, hence a totally transcendental theory is stable. Here are some more examples.

Example 5.1.1. Let $L = \{E_i : i < \omega\}$ and T be T_0^{eq}, where T_0 is the theory in L saying that each E_i is an equivalence relation with only infinite classes, $E_0(x,y)$ is $x = x$ and for every $i \ge 0$, E_{i+1} refines each E_i–class into two E_{i+1}–classes. Below, x denotes a variable in the sort of the equivalence relations. Let $\Delta = \{x = y, E_{i_1}, \ldots, E_{i_j}\}$ where $i_1 < \ldots < i_j$. Given a formula φ the number of nonalgebraic complete Δ–types over $\mathfrak{C}$ consistent with φ is equal to the number E_{i_j}–classes consistent with φ. Thus $R_\Delta(p) \le 1$ for any $p \in S(\mathfrak{C})$ in the variable x. If $\Gamma(x)$ is any finite set of formulas there is some set $\Delta = \{x = y, E_{i_1}, \ldots, E_{i_j}\}$ such that every Γ–type is a Δ–type (by elimination of quantifiers). Thus, for any $p \in S(\mathfrak{C})$, $R_\Gamma(p) \le 1$. Again using that T has elimination of quantifiers, for z a variable of any sort and $\Gamma(z)$ a finite set of formulas, $R_\Gamma(z = z) < \omega$. This proves the stability of T. The reader should notice that T is not totally transcendental.

Example 5.1.2. Alter the last example by requiring each E_i–class to be refined into *infinitely* many E_{i+1}–classes. Suppose $\Delta = \{x = y, E_0, \ldots, E_i\}$.

Claim. (i) If $j \ge i$ and a is an element, $(R, \mathrm{Mult})_\Delta(E_j(x,a)) = (1,1)$.
(ii) If $j < i$ and a is an element, $R_\Delta(E_j(x,a)) = i - j$.

All nonalgebraic completions of $E_j(x,a)$ over $\mathfrak{C}$ have the same Δ–type, so (i) holds.

(ii) First let $j = i - 1$. Then $Q = \{q \restriction \Delta : q \in S(\mathfrak{C}),\ E_{i-1}(x,a) \in q$ and $\neg\varphi \in q$ for all φ with $R_\Delta(\varphi) < 2\}$ is contained in $\{q \restriction \Delta : q \in S(\mathfrak{C}),\ E_{i-1}(x,a) \in q$ and $\neg E_i(x,b) \in q$ for all $b \in \mathfrak{C}\} = P$. Since $|P| = 1$, Q has cardinality $< |\mathfrak{C}|$, so $R_\Delta(E_{i-1}(x,a)) \le 2$. Since $E_{i-1}(x,a)$ is contained in infinitely many elements of $S_\Delta(\mathfrak{C})$ of Δ–rank 1, $R_\Delta(E_{i-1}(x,a)) = 2$.

The previous paragraph can be generalized to a downward induction which proves (ii) for all $j < i$, proving the claim.

Since T is quantifier-eliminable the claim can be used to show that for any formula $\varphi(x)$ and finite set Γ there is a set $\Delta = \{x = y, E_0, \ldots, E_i\}$ such that $R_\Gamma(\varphi) \le R_\Delta(\varphi)$ and $R_\Delta(\varphi) \le i$. Thus, T is stable. Moreover, for every $i < \omega$, there is a finite $\Delta(x)$ such that $R_\Delta(x = x) \ge i$.

This example also shows that the Δ–rank of a formula depends quite heavily on Δ. As Δ becomes larger the Δ–rank of $x = x$ increases without a finite bound.

Example 5.1.3. Let M be an infinite module over a ring R formulated in the natural language for R–modules. We will show in Corollary 5.3.4 that $T = Th(M)$ is stable.

Definition 5.1.4. *Let T be a stable theory.*

(i) We say that $p \in S(A)$ does not fork over $B \subset A$ if for all finite Δ, $R_\Delta(p) = R_\Delta(p \restriction B)$. When p does not fork over B, p is called a nonforking *extension of $p \restriction B$. The negation of nonforking is* forking.

(ii) A type q over A (perhaps incomplete) is said to fork over $B \subset A$ *if every* $p \in S(A)$ *containing* q *forks over* B.

(iii) For A, B and C sets we say that A is forking independent from B over C *and write* $A \underset{C}{\perp} B$ *if for all finite tuples a from A,* $tp(a/B \cup C)$ *does not fork over C. The negation of forking independent is* forking dependent *and is denoted* $A \underset{C}{\not\perp} B$. *We usually shorten these terms to "independent" or "dependent" since it is clear from context that we mean "forking independent" "forking dependent".*

Remark 5.1.1. In a t.t. theory we have already adopted "independent" to mean Morley rank independent. We will show in Corollary 5.1.4 however that Morley rank independence and forking independence are equivalent relations in a t.t. theory, eliminating the apparent conflict. Below we also redefine other terms (like "stationary") later showing the equivalence (in a t.t. theory) of this property with the one defined earlier.

Remark 5.1.2. If T is stable and $p \in S(A)$ forks over $B \subset A$ there is a formula $\varphi \in p$ such that $p \restriction B \cup \{\varphi\}$ forks over B. (Find a finite Δ such that $R_\Delta(p) < R_\Delta(p \restriction B)$ and a $\varphi \in p$ such that $R_\Delta(p) = R_\Delta(\varphi)$.)

Definition 5.1.5. *A collection of sets* $\mathcal{A}$ *is called* independent over B *or* B–independent *if each* $A \in \mathcal{A}$ *is independent from* $\bigcup(\mathcal{A} \setminus \{A\})$ *over* B.

Remark 5.1.3. Conditions (1), (2), (3), (5) and (7) in the definition of a freeness relation (Definition 3.3.1) hold for forking independence in a stable theory.

Verifications. Finite character and monotonicity (1) are clear, as is transitivity of independence (3). Since Δ–rank is invariant under automorphisms of $\mathfrak{C}$ so is independence (i.e., (5) holds). For any complete p and finite Δ there is a $\varphi \in p$ with $R_\Delta(p) = R_\Delta(\varphi)$. Thus, there is a set $B \subset dom(p)$ of cardinality $\leq |T|$ such that p does not fork over B, proving (2). If Δ contains the formula $x = y$ and $p \in S(\mathfrak{C})$, then $R_\Delta(p) = 0$ if and only if p is algebraic. Thus, $b \notin acl(A)$ implies that b depends on b over A; i.e., reflexivity (7) holds.

One part of (6) also holds: If $p \in S(A)$ and Δ is finite, $\{q \restriction \Delta : q \in S(\mathfrak{C})$ is a nonforking extension of $p\}$ is finite. Thus, the number of nonforking extensions of p in $S(\mathfrak{C})$ is $\leq 2^{|T|}$. What is not clear is that every complete type has at least one nonforking extension in $S(\mathfrak{C})$. This existence result as well as symmetry (4) will require real effort to verify.

Lemma 5.1.2. *Given a stable T and set A, any* $p \in S(acl(A))$ *does not fork over A.*

Proof. See Exercise 5.1.3.

Definition 5.1.6. *In a stable theory a complete type is called* stationary *if it has a unique nonforking extension. When p is complete, does not fork over* $A \subset dom(p)$ *and* $p \restriction A$ *is stationary we say that p is* stationary over A.

The following slightly extends Definition 3.3.7.

Definition 5.1.7. *Let T be a complete theory, $\Delta = \Delta(x)$ a set of formulas and $p \in S_\Delta(B)$. Then p is* definable over A *if for all formulas $\varphi(x,y) \in \Delta$, there is a formula $\psi(y)$ over A such that for all $b \in B$,*

$$\varphi(x,b) \in p \iff \models \psi(b).$$

Definability and nonforking are linked with

Theorem 5.1.1 (Definability Theorem). *Let T be stable and $p \in S(\mathfrak{C})$. Then p does not fork over A if and only if p is definable over $acl(A)$.*

The definability of nonforking extensions will be proved first. The analogue of this result for t.t. theories was proved using Morley sequences. There is a similar proof in this setting, however the alternative argument given here gives more insight into the properties of Δ–rank. Order lexicographically the collection $I = \{ (\beta,k) : \beta \text{ is an ordinal and } 1 \leq k < \omega \}$; i.e., $(\beta,k) < (\gamma,l)$ if $\beta < \gamma$ or $\beta = \gamma$ and $k < l$.

Lemma 5.1.3. *Let T be a complete theory, Δ a finite set of formulas, α an ordinal and $m < \omega$.*

(i) For any formula $\varphi(x,y)$ there is a set of formulas $\Gamma(y)$ such that for all a, $(R,\mathrm{Mult})_\Delta(\varphi(x,a)) \geq (\alpha,m)$ if and only if a realizes Γ.

(ii) For $\varphi(x)$ a formula over A with $(R,\mathrm{Mult})_\Delta(\varphi) = (\alpha,m)$ and $\delta(x,y) \in \Delta$ there is a formula $\psi(y)$ over A such that for all b, $\models \psi(b)$ if and only if $(R,\mathrm{Mult})_\Delta(\varphi(x) \wedge \delta(x,b)) = (\alpha,m)$.

Proof. The proof of the following preliminary fact is left to the reader.

Claim. Given $m > 1$ and α an ordinal,

$$(R,\mathrm{Mult})_\Delta(\varphi(x)) \geq (\alpha,m)$$

if and only if

there are $m_1, m_2 \geq 1$ with $m_1 + m_2 = m$, $\delta \in \Delta$ and b such that
- $(R,\mathrm{Mult})_\Delta(\varphi(x) \wedge \delta(x,b)) \geq (\alpha,m_1)$ and
- $(R,\mathrm{Mult})_\Delta(\varphi(x) \wedge \neg\delta(x,b)) \geq (\alpha,m_2)$.

Part (i) is proved by induction on the pairs (β,k), where β is an ordinal and $1 \leq k < \omega$. The minimal element of the order I is $(0,1)$ and $(R,\mathrm{Mult})_\Delta(\varphi(x,a)) \geq (0,1)$ exactly when $\models \exists x\varphi(x,a)$. Assume (i) holds for all elements of I less than (α,m). For $(\beta,n) < (\alpha,m)$ let $\Gamma_{(\beta,n)}(y)$ be a set of formulas such that for all a, $(R,\mathrm{Mult})_\Delta(\varphi(x,a)) \geq (\beta,n)$ if and only if a realizes $\Gamma_{(\beta,n)}$. First suppose that $m = 1$. Note:

$$(R, \mathrm{Mult})_\Delta(\varphi(x, a)) \geq (\alpha, 1) \text{ if and only if}$$
$$\forall \beta < \alpha, \forall n(\ (R, \mathrm{Mult})_\Delta(\varphi(x, a)) \geq (\beta, n)\ ).$$

Thus, $(R, \mathrm{Mult})_\Delta(\varphi(x, a)) \geq (\alpha, 1)$ if and only if a realizes $\bigcup_{\beta<\alpha, n<\omega} \Gamma_{(\beta,n)} = \Gamma_{(\alpha,1)}$. Supposing that $m > 1$ the claim yields m_1, $m_2 \geq 1$ with $m_1 + m_2 = m$, $\delta \in \Delta$ and b such that $(R, \mathrm{Mult})_\Delta(\varphi(x, a) \wedge \delta(x, b)) \geq (\alpha, m_1)$ and $(R, \mathrm{Mult})_\Delta(\varphi(x, a) \wedge \neg\delta(x, b)) \geq (\alpha, m_2)$. Let Γ_1 and Γ_2 be sets of formulas such that for all a and b,

- $(R, \mathrm{Mult})_\Delta(\varphi(x, a) \wedge \delta(x, b)) \geq (\alpha, m_1)$ if and only if ab realizes $\Gamma_1(y, z)$ and
- $(R, \mathrm{Mult})_\Delta(\varphi(x, a) \wedge \neg\delta(x, b)) \geq (\alpha, m_2)$ if and only if ab realizes $\Gamma_2(y, z)$.

Let

$$\Theta_{(m_1,m_2)}(y) = \{\ \exists z(\bigwedge \Gamma_1'(y, z) \wedge \bigwedge \Gamma_2'(y, z)) : \Gamma_1' \subset \Gamma_1,\ \Gamma_2' \subset \Gamma_2 \text{ finite}\ \},$$

a set of formulas which holds if and only if $\Gamma_1(y, z) \cup \Gamma_2(y, z)$ is consistent. Let $\Gamma(y)$ be a set of formulas such that a realizes $\Gamma(y)$ if and only if for some m_1, $m_2 > 1$ with $m_1 + m_2 = m$, a realizes $\Theta_{(m_1,m_2)}(y)$. This type Γ satisfies the requirements.

(ii) For all b and $\delta \in \Delta$,

$$\begin{aligned} (R, \mathrm{Mult})_\Delta(\varphi(x) \wedge \delta(x, b)) &= (\alpha, m) \\ \iff (R, \mathrm{Mult})_\Delta(\varphi(x) \wedge \neg\delta(x, b)) &< (\alpha, 1). \end{aligned}$$

(If $(R, \mathrm{Mult})_\Delta(\varphi(x) \wedge \neg\delta(x, b)) < (\alpha, 1)$, then any r in $P = \{p \upharpoonright \Delta : p \in S(\mathfrak{C}),\ \varphi \in p \text{ and } R_\Delta(p) \geq \alpha\}$ contains $\delta(x, b)$. Since $|P| = m$, $(R, \mathrm{Mult})_\Delta(\varphi(x) \wedge \delta(x, b)) = (\alpha, m)$. The converse follows immediately from the claim.)

By compactness and (i) there is a formula $\psi(y)$ over A such that $\models \psi(b)$ if and only if $R_\Delta(\varphi(x) \wedge \neg\delta(x, b)) < \alpha$. Thus, $\models \psi(b)$ if and only if $(R, \mathrm{Mult})_\Delta(\varphi(x) \wedge \delta(x, b)) = (\alpha, m)$.

Lemma 5.1.4. *Let T be stable, Δ is a finite set of formulas and $p \in S_\Delta(\mathfrak{C})$ such that $R_\Delta(p \cup \{\gamma\}) = R_\Delta(\gamma)$ for some formula γ over A. Then p is definable over $acl(A)$.*

Proof. Let $q = p \cup \{\gamma\}$, $\alpha = R_\Delta(q)$ and notice that $\mathrm{Mult}_\Delta(q) = 1$. Let $\xi(x)$ be a formula implied by q such that $(R, \mathrm{Mult})_\Delta(q) = (R, \mathrm{Mult})_\Delta(\xi) = (\alpha, 1)$. In other words, p is the only $p_0 \in S_\Delta(\mathfrak{C})$ such that $(R, \mathrm{Mult})_\Delta(p_0 \cup \{\xi\}) = (\alpha, 1)$. By Lemma 5.1.3(ii), for any $\delta(x, y) \in \Delta$ there is a formula $\psi(y)$ such that for all b, $\models \psi(b)$ if and only if $(R, \mathrm{Mult})_\Delta(\xi(x) \wedge \delta(x, b)) = (\alpha, 1)$. Thus, $\delta(x, b) \in p$ if and only if $\models \psi(b)$.

It remains to show that ψ is equivalent to a formula over $acl(A)$, By Lemma 4.1.2, it suffices to show that ψ is almost over A. Since $R_\Delta(p \cup \{\gamma\}) = R_\Delta(\gamma)$ and γ is over A, $P = \{\ r \upharpoonright \Delta : r \text{ is conjugate over } A \text{ to } q\}$ is finite. Furthermore, any formula conjugate over A to ψ defines some element of P. Since formulas defining the same Δ-type are equivalent, ψ is almost over A.

One direction of the Definability Theorem follows immediately:

Lemma 5.1.5. *If T is stable and $p \in S(\mathfrak{C})$ does not fork over A, then p is definable over $acl(A)$.*

Proving the other direction of the Definability Theorem, the Symmetry Lemma and the existence of nonforking extensions all involve the order property.

Definition 5.1.8. *Let T be complete and $\varphi(x, y)$ a formula. Then φ has the* order property *if there are sets of elements $\{ a_i : i < \omega \}$ and $\{ b_i : i < \omega \}$ such that $\models \varphi(a_i, b_j)$ if and only if $i \leq j < \omega$. We say that T has the* order property *if there is a formula of T with the order property.*

It can be shown that T is stable if and only if T does not have the order property. In fact, some authors take "T does not have the order property" to be the definition of stable. One direction of the equivalence is

Lemma 5.1.6. *A stable theory T does not have the order property.*

Proof. Suppose that T is stable. For Δ finite, A a set and p in $S_\Delta(A)$ there is a formula φ implied by p such that $R_\Delta(p) = R_\Delta(\varphi)$. Furthermore, for any such φ there are finitely many $r \in S_\Delta(A)$ with $R_\Delta(r \cup \{\varphi\}) = R_\Delta(\varphi)$. A formula implied by a Δ−type over A is implied by a finite Δ−type over A. There are $\leq |A| + \aleph_0$ finite Δ−types over A, so

$$|S_\Delta(A)| \leq |A| + \aleph_0. \tag{5.2}$$

Now assume, towards a contradiction, that there are $\varphi(x, y)$, $\{ a_i : i < \omega \}$ and $\{ b_i : i < \omega \}$ such that $\models \varphi(a_i, b_j)$ if and only if $i \leq j < \omega$. We will contradict (5.2) for $\Delta = \{\varphi\}$. Let $(Y, \leq)$ be a dense linear order without endpoints which has a dense subset X of cardinality $\kappa < |Y|$. (Note: κ must be infinite.) Let $C = \{ c_i : i \in Y \}$ and $D = \{ d_i : i \in X \}$ be sets of constant symbols and Φ the set of sentences $\{ \varphi(c_i, d_j) : i \in Y, j \in X$ and $i \leq j \} \cup \{ \neg\varphi(c_i, d_j) : i \in Y, j \in X$ and $i > j \}$. Compactness proves the consistency of Φ since for any finite $\Psi \subset \Phi$ the constants appearing in Ψ can be interpreted by some of the a_i's and b_j's to obtain a model of it. Thus, without loss of generality, C and D are subsets of the universe. However, the density of X forces each c_i to have a different φ−type over D, contradicting (5.2) since $|Y| > |X| \geq \aleph_0$. This proves the lemma.

Lemma 5.1.7. *Let T be stable, $\varphi(x, y)$ a formula over $\emptyset$, and $\varphi'(y, x)$ the formula φ with y as the object variable and x as the parameter variable. Suppose that $p(x) \in S_\varphi(\mathfrak{C})$ and $q(y) \in S_{\varphi'}(\mathfrak{C})$ are definable over A and consistent with p_0, $q_0 \in S(A)$, respectively. Then for all a realizing p_0 and b realizing q_0, $\varphi(x, b) \in p$ if and only if $\varphi'(y, a) \in q$; i.e.,*

$$\varphi(x, b) \in p \text{ if and only if } \varphi(a, y) \in q.$$

Proof. Assume to the contrary there are a' realizing p_0 and b' realizing q_0 such that $\neg\varphi(x, b') \in p$ and $\varphi(a', y) \in q$. By the definability over A of p and q, for all a realizing p_0, $\varphi(a, y) \in q$ and similarly for realizations of q_0. That is,

$$\forall a \text{ realizing } p_0,\ \forall b \text{ realizing } q_0(\ \neg\varphi(x, b) \in p \text{ and } \varphi(a, y) \in q\ ).$$

Define sets of elements $\{\, a_i : i < \omega \,\}$ and $\{\, b_i : i < \omega \,\}$ as follows. Assuming that a_i and b_i have been defined for $i < k$ let a_k realize the restriction of p to $A \cup \{\, a_i : i < k \,\} \cup \{\, b_i : i < k \,\}$ and b_k realize the restriction of q to $A \cup \{\, a_i : i \leq k \,\} \cup \{\, b_i : i < k \,\}$. Then, $\models \varphi(a_i, b_j)$ if and only if $i \leq j$. This contradicts Lemma 5.1.6, to prove the lemma.

Lemma 5.1.8. *Given $A = acl(A)$, Δ a finite set of formulas and $p \in S(A)$, there is at most one $q \in S_\Delta(\mathfrak{C})$ which is definable over A and consistent with p.*

Proof. Assume to the contrary that p is consistent with $q, q' \in S_\Delta(\mathfrak{C})$, both definable over A, and for some formula $\varphi \in \Delta$, $\varphi(x, b) \in q$ and $\neg\varphi(x, b) \in q'$. Let $r_0(y) = tp(b/A)$. Let φ' be the formula φ with object variable y and parameter variable x. Let $r(y) \in S_{\varphi'}(\mathfrak{C})$ be such that $R_{\varphi'}(r(y) \cup r_0(y)) = R_{\varphi'}(r_0(y))$. By Lemma 5.1.4, r is definable over A. Applying Lemma 5.1.7 to both q and q',

- for all a realizing p, $\varphi(a, y) \in r$ and,
- for all a realizing p, $\neg\varphi(a, y) \in r$.

This contradiction proves the lemma

From here we can quickly prove the existence of nonforking extensions, the other direction of the Definability Theorem and the Symmetry Lemma.

Corollary 5.1.1. *If T is stable and $p \in S(A)$ there is a $q \in S(\mathfrak{C})$, $q \supset p$, such that q does not fork over A.*

Proof. Let $p \in S(A)$. Every extension of p in $S(acl(A))$ is a nonforking extension (by Lemma 5.1.2) so may as well assume that $A = acl(A)$. For any finite set of formulas Δ there is a $q_\Delta \in S_\Delta(\mathfrak{C})$ with $R_\Delta(q_\Delta \cup p) = R_\Delta(p)$. By Lemma 5.1.4, q_Δ is definable over A, hence (by Lemma 5.1.8) q_Δ is the unique complete Δ–type over $\mathfrak{C}$ such that $R_\Delta(q_\Delta \cup p) = R_\Delta(p)$. If $\Gamma \supset \Delta$ are finite sets of formulas, q_Γ contains a Δ–type which is definable over A, hence $q_\Gamma \supset q_\Delta$. Thus, $q = \bigcup_\Delta q_\Delta$ is a nonforking extension of p in $S(\mathfrak{C})$.

Proof of the Definability Theorem (Theorem 5.1.1). One direction of the theorem was already proved in Lemma 5.1.5. Conversely, given $p \in S(A)$ suppose $q \in S(\mathfrak{C})$ extends p and is definable over $acl(A)$. Let p' be the restriction of q to $acl(A)$. In the proof of Corollary 5.1.1 we show that p' has a unique extension in $S(\mathfrak{C})$ which is definable over $acl(A)$ and this extension does not fork over $acl(A)$. Since p' is a nonforking extension of p (by Lemma 5.1.2) we have proved that q is a nonforking extension of p.

The symmetry of forking dependence follows easily from Lemma 5.1.7:

Corollary 5.1.2 (Symmetry Lemma). *If T is stable then for all sets A, B and C,*

$$A \underset{C}{\perp} B \Longrightarrow B \underset{C}{\perp} A.$$

Proof. Assuming the lemma to be false there are elements a and b and a set C such that a is independent from b over C, but b depends on a over C. These same relations hold when we replace C by $acl(C)$, so we may as take C to be algebraically closed. Let $q_0(y) = tp(b/C)$ and $\varphi(a, y) \in tp(b/C \cup \{a\})$ a formula such that $q_0 \cup \{\varphi(a, y)\}$ forks over C. Let $p \in S(\mathfrak{C})$ be a nonforking extension of $tp(a/C \cup \{b\})$ which by transitivity of independence also does not fork over C. Let $q \in S(\mathfrak{C})$ be a nonforking extension of q_0. By Lemma 5.1.5 and the fact that C is algebraically closed, both p and q are definable over C. Since $\varphi(x, b) \in p$, Lemma 5.1.7 implies that $\varphi(a, y) \in q$, contradicting that q does not fork over C.

This completes the proof of

Corollary 5.1.3. *In a stable theory forking independence is a freeness relation.*

The following corollaries all follow easily from the Definability Theorem and a couple other key results above. Stating them rounds out our picture of forking independence.

Corollary 5.1.4. *Suppose that T is a t.t. theory. Then $p \in S(\mathfrak{C})$ does not fork over A if and only if $MR(p) = MR(p \restriction A)$. A type is stationary if and only if it has Morley degree* 1.

Proof. Left to the reader in the exercises.

Corollary 5.1.5. *Let T be stable.*

(i) Every $p \in S(acl(A))$ is stationary.

(ii) If $p \in S(\mathfrak{C})$ is definable over A, then p does not fork over A.

(iii) If $p \in S(\mathfrak{C})$ is stationary over A, then p is definable over A.

Proof. (i) This follows immediately from the Definability Theorem and Lemma 5.1.8.

(ii) Assuming that $p \in S(\mathfrak{C})$ is definable over A, p does not fork over $acl(A)$ by the Definability Theorem. Since every $q \in S(acl(A))$ does not fork over A, p does not fork over A by the transitivity of independence.

(iii) Simply unraveling the definitions shows that if $q \in S(\mathfrak{C})$, $f \in \mathrm{Aut}(\mathfrak{C})$, φ is a formula over $\emptyset$ and ψ defines $q \restriction \varphi$, then $f(\psi)$ defines $f(q) \restriction \varphi$. If

$f \in \mathrm{Aut}(\mathfrak{C})$ fixes A then p and $f(p)$ are both nonforking extensions of $p \restriction A$, hence $p = f(p)$. Thus, given a formula φ and ψ which defines $p \restriction \varphi$, ψ is equivalent to $f(\psi)$ for any $f \in \mathrm{Aut}(\mathfrak{C})$ fixing A. By Lemma 3.3.8, ψ is equivalent to a formula over A.

Remark 5.1.4. In a stable theory it is possible to find sets $B \supset A$ and $p \in S(B)$ which is definable over A but forks over A. (Compare this with (ii) of the previous corollary.) However, when $p \in S(M)$, M a model and p is definable over $A \subset M$, p does not fork over A (see Exercise 5.1.7).

Remark 5.1.5. Let T be stable and $p \in S(A)$ stationary. Then for any finite set of formulas Δ, $\mathrm{Mult}_\Delta(p) = 1$. (The proof is assigned as Exercise 5.1.8.)

Corollary 5.1.6. *If T is stable, then any $p \in S(A)$ is definable over A.*

Proof. Let $q \in S(\mathfrak{C})$ be a nonforking extension of p. By the Corollary 5.1.1 there is a defining scheme for q, hence for p, consisting of formulas over $acl(A)$. By Lemma 3.3.11, p is definable over A.

As with t.t. theories, this definability of types yields

Corollary 5.1.7. *Let T be stable and D a subset of $\mathfrak{C}$ definable over A. Then, for any k and $H \subset D^k$ definable over $\mathfrak{C}$ there is a $B \subset D$ such that H is definable over $A \cup B$.*

(The proof of the corollary is the same as that giving Proposition 3.3.3.)

A related property (stated for t.t. theories as Corollary 3.3.7) is:

Lemma 5.1.9. *Let T be stable, M a model, φ a formula over $A \subset M$ and a a tuple form $\varphi(\mathfrak{C})$. Then $tp(a/\varphi(M) \cup A)$ implies $tp(a/M)$.*

Proof. Let b be a tuple from M and $p = tp(b/\varphi(M) \cup A)$. For $\psi(x, y)$ a formula over $\emptyset$, p_ψ is defined by some formula $\psi'(y)$ over $\varphi(M) \cup A$ (by Corollary 5.1.6). We claim that ψ' also defines $tp(b/\varphi(\mathfrak{C}) \cup A)$. (Otherwise, there is tuple c from $\varphi(\mathfrak{C})$ such that $\models \neg(\psi'(c) \leftrightarrow \psi(b, c))$. Since M is a model there is such a c in M, a contradiction.)

Now let a' be a realization of $tp(a/\varphi(M) \cup A)$, b, $\psi(x, y)$ and $\psi'(y)$ as above. Then $\models \psi'(a')$ (since $\psi' \in tp(a/\varphi(M) \cup A)$), so $\models \psi(b, a')$, as required to prove the lemma.

The following set of results refines our knowledge of the defining scheme of a type.

Definition 5.1.9. *Let T be stable.*

(i) The stationary complete types p and q are called parallel, *denoted $p \parallel q$, if they have the same nonforking extension in $S(\mathfrak{C})$. A stationary type p is* based *on a set A if there is a $q \in S(A)$ parallel to p, in which case we let $p|A$ denote q.*

(ii) A type p is called a strong type over A *if $p \in S(acl(A))$. For any a, $tp(a/acl(A))$ is called the* strong type of a over A *and is denoted $stp(a/A)$, simply writing $stp(a)$ when $A = \emptyset$.*

Remark 5.1.6. (i) Unraveling the definitions, when p is a stationary complete type and $r \in S(\mathfrak{C})$, the unique nonforking extension of p, is definable over A, $p|A = r \restriction A$. Do not confuse $|$ with $\restriction$.

(ii) At least in regard to terminology we identify types which have the same set of realizations. If $p \in S(A)$ is stationary it has a unique extension q in $S(acl(A))$, hence is equivalent to q. Thus any stationary type may be called a strong type. Many of the properties we prove of strong types hold for all types in a fixed parallelism class. Indeed, Hrushovski (in [Hru86]) defines a strong type to be an equivalence class of types under parallelism.

In a t.t. theory when X is a degree 1 $\emptyset$–definable set, $a \in X$ and $MR(a/A) = MR(X)$ we say "a is generic over A" (see Definition 4.1.3). When T is stable, $p \in S(A)$ is stationary and $B \supset A$, we say "a is generic over A" if a realizes p and a is independent from B over A (in other words a realizes $p|B$).

The following elegant notation for the defining scheme of a strong type (due to Harrington and promoted by Hrushovski) greatly improves the readability of some proofs.

Notation. Let T be stable, $p \in S(A)$ a strong type in the variable x, $q = p|\mathfrak{C}$, $\varphi(x, y)$ a formula over $\emptyset$ and $\psi(y)$ the formula over A defining $p \restriction \varphi$. If b is any element then $\models \psi(b)$ if and only if $\varphi(x, b) \in q$ if and only if $\models \varphi(a, b)$ whenever a is a realization of p generic over b. We will denote $\psi(y)$ by $(d_p x)\varphi(x, y)$, which is read "for generic x realizing p, $\varphi(x, y)$ holds."

Suppose T is stable, p is stationary and based on A and a realizes $p|A$. Equivalent ways of describing the relationships between φ, p and $(d_p x)\varphi(x, y)$ are:

– If $a \underset{A}{\perp} b$, $\models \varphi(a, b) \iff \models (d_p x)\varphi(x, b)$.

– If $\models \varphi(a, b)$, then $a \underset{A}{\perp} b \iff \models (d_p x)\varphi(x, b)$.

– If $\models (d_p x)\varphi(x, b)$, then $a \underset{A}{\perp} b \iff \models \varphi(a, b)$.

Since all formulas defining $q \restriction \varphi$ are equivalent, $(d_p x)\varphi(x, y)$ is uniquely determined and $p' \parallel p \implies (d_{p'} x)\varphi(x, \mathfrak{C}) = (d_p x)\varphi(x, \mathfrak{C})$. The variable of the type is not always the first one appearing in the formula. We will write $(d_r y)\varphi(x, y)$ for the formula in x defining $r \restriction \varphi'$, where $\varphi'(y, x) = \varphi(x, y)$ and $r = r(y)$.

Let $p \in S(A)$ be a type in a stable theory and $X = p(\mathfrak{C})$. Define an equivalence relation $\sim$ on X by: $a \sim b$ if for all $B \supset A$ such that a and

b are both generic over B, $tp(a/B) = tp(b/B)$. Notice that $a \sim b$ if and only if $tp(a/acl(A)) = tp(b/acl(A))$; i.e., the $\sim$ −class of a is the locus of a over $acl(A)$. This equivalence relation is expressible entirely in terms of A−definable equivalence relations as follows. Let $FE(A)$ denote the set of equivalence relations which are definable over A and have finitely many classes. The elements of $FE(A)$ are called *finite equivalence relations over A*. Then,

Lemma 5.1.10. *Let T be a complete theory. For all sets A and elements a, b,*

$$tp(a/acl(A)) = tp(b/acl(A))$$

if and only if $\models E(a,b)$, for all $E \in FE(A)$ (of the appropriate sort).

Proof. If $E \in FE(A)$ then the name e for the E−class of b is in $acl(A)$ and there is a formula $\eta(x,e)$ defining this class. Thus, if $tp(a/acl(A)) = tp(b/acl(A))$, $\models E(a,b)$, for all $E \in FE(A)$. Conversely, let $\varphi(x,y)$ be a formula over A such that $\exists x\varphi(x,y)$ is algebraic. The equivalence relation $E(x,x')$ defined by: $\forall y(\varphi(x,y) \leftrightarrow \varphi(x',y))$, is in $FE(A)$. Thus, assuming that $\models E(a,b)$ for all E in $FE(A)$, $tp(a/acl(A)) = tp(b/acl(A))$. This proves the lemma.

Corollary 5.1.8. *Let T be stable.*

(i) Suppose $p \in S(\mathfrak{C})$ does not fork over A and a realizes $p \restriction acl(A)$. Then $p \restriction A \cup \{a\}$ is stationary, hence p is definable over $A \cup \{a\}$.

(ii) If p, $p' \in S(\mathfrak{C})$ are both nonforking extensions of some $q \in S(A)$, then p and p' are conjugate over A.

(iii) When $a \not\perp_A b$, there is a formula $\psi(x,b) \in tp(a/A \cup \{b\})$ such that any c satisfying $\psi(x,b)$ depends on b over A.

Proof. (i) Let $q = p \restriction A \cup \{a\}$ and b be a realization of q. Then $\models E(b,a)$, for all $E \in FE(A)$, hence (by Lemma 5.1.10) $tp(b/acl(A)) = tp(a/acl(A))$. Thus, q is stationary and p is definable over $A \cup \{a\}$ (by Corollary 5.1.5(iii)).

(ii) Let a and a' be realizations of $p \restriction acl(A)$ and $p' \restriction acl(A)$, respectively. Then p is the unique nonforking extension of q containing $\{ E(x,a) : E \in FE(A) \}$ and p' is the unique nonforking extension of q containing $\{ E(x,a') : E \in FE(A) \}$. Since a and a' both realize q there is an automorphism f of $\mathfrak{C}$ which pointwise fixes A and takes a to a'. Then $f(\{ E(x,a) : E \in FE(A) \}) = \{ E(x,a') : E \in FE(A) \}$, so $f(p) = p'$.

(iii) Let $p = tp(b/A)$ and $q = stp(b/A)$. By the Symmetry Lemma, b depends on a over A, so there is a formula $\varphi(a,y) \in tp(b/A \cup \{a\})$ such that $p(y) \cup \{\varphi(a,y)\}$ forks over A. Since $\varphi(a,y)$ is not in the unique nonforking extension of q in $S(\mathfrak{C})$, a cannot satisfy the formula $(d_q y)\varphi(x,y)$. By (i) q is based on $A \cup \{b\}$, hence $(d_q y)\varphi(x,y)$ is equivalent to a formula over $A \cup \{b\}$. The formula $\psi(x) = \varphi(x,b) \wedge \neg(d_q y)\varphi(x,y)$ is a formula over $A \cup \{b\}$ such that $\models \psi(c) \implies c \not\perp_A b$, proving (iii).

Recalling Definition 5.1.4(ii), a (possibly incomplete) type p over B forks over $A \subset B$ if whenever c realizes p, c depends on B over A. Part (iii) of the corollary can be reworded as: If $p \in S(B)$ forks over $A \subset B$, there is a $\varphi \in p$ which forks over A.

The following direct consequence of definability is called the Open Mapping Theorem because it asserts that a certain map between topologies is open. The reader is referred to [Las86] or [Bal88] for an explanation.

Lemma 5.1.11 (Open Mapping Theorem). *Suppose that T is stable, $B \supset A$ and $\varphi(x)$ is a formula over B. Then there is a formula $\psi(x)$ over A such that $p \in S(A)$ has a nonforking extension containing $\varphi(x)$ if and only if $\psi \in p$.*

Proof. Without loss of generality, $B \setminus A$ contains a single element b. Let $q = stp(b/A)$ as a type in y and $\psi_0(x) = (d_q y)\varphi(x,y)$. Then, given a independent from b over A, $\models \psi_0(a) \iff \models \varphi(a,b)$. Let $\psi_0 = \psi_0(x,e)$, where $\psi_0(x,z)$ is over A, $e \in acl(A)$ and $\exists x\psi_0(x,z)$ isolates $tp(e/A)$. Let $\psi(x) = \exists z\psi_0(x,z)$. If a is independent from b and $\models \varphi(a,b)$ then $\models \psi_0(a,e)$, so $\models \psi(a)$. In other words, $\psi(x) \in tp(a/A)$. Now suppose $p \in S(A)$ and $\psi \in p$. Then $p \cup \{\psi_0(x,e')\}$ is consistent for some e'. Since $\exists x\psi_0(x,z)$ isolates a complete type over A, $p \cup \{\psi_0(x,e)\}$ is consistent. Since $e \in acl(A)$ there is a nonforking extension p' of p over $acl(A) \cup \{b\}$ containing $\psi_0(x,e)$. That is, for some a realizing p which is independent from b over A, $\psi_0(x,e)$. Thus, $\models \varphi(a,b)$ as needed to complete the proof.

Using this lemma we can generalize Lemma 3.3.10.

Corollary 5.1.9. *Let T be stable, $p \in S(A)$ nonisolated and $q \supset p$ is an isolated complete type. Then q forks over A.*

Proof. Suppose, to the contrary, that $q \in S(B)$ does not fork over A and $\varphi(x)$ isolates q. Let $\psi \in p$ be such that $r \in S(A)$ has a nonforking extension containing $\varphi(x)$ if and only if $\psi \in r$. Any $p' \in S(A)$ containing ψ has an extension $q' \in S(B)$ containing φ. Since q is the only element of $S(B)$ containing φ, $q \restriction A = p$ is the only element of $S(A)$ containing ψ; i.e., ψ isolates p. This contradiction proves the corollary.

Canonical parameters were introduced for t.t. theories in Section 4.1.1. The canonical parameter of a degree 1 type has much in common with the name for a formula. Recall that $\ulcorner\varphi\urcorner$ denotes the name for the formula φ. (Thus, for every automorphism f of $\mathfrak{C}$, $f(\varphi(\mathfrak{C})) = \varphi(\mathfrak{C})$ if and only if $f(\ulcorner\varphi\urcorner) = \ulcorner\varphi\urcorner$.)

Definition 5.1.10. *Let T be stable and p be a stationary type in x. Let*

$$D(p) = \{\, \ulcorner(d_p x)\varphi(x,y)\urcorner : \varphi(x,y) \text{ is a formula over } \emptyset \,\}.$$

The canonical base *of p, denoted $Cb(p)$, is $dcl(D(p))$.*

Clearly, if p is a stationary type definable over A then $Cb(p) \subset dcl(A)$.

Lemma 5.1.12 (Canonical Bases). *Let T be stable and p a stationary type.*

(i) p is based on $Cb(p)$.

(ii) If the stationary type q is parallel to p then $Cb(p) = Cb(q)$.

(iii) If $p \in S(\mathfrak{C})$, $Cb(p)$ is the largest set C such that for all $f \in \mathrm{Aut}(\mathfrak{C})$, $f(p) = p$ if and only if f fixes C pointwise.

Proof. (i) Clearly p is definable over $D(p)$, hence p is based on $Cb(p)$.

(ii) For $r = p|\mathfrak{C} = q|\mathfrak{C}$, $D(p) = D(r) = D(q)$, so $Cb(p) = Cb(q)$.

(iii) Let C be a maximal set such that for all $f \in \mathrm{Aut}(\mathfrak{C})$, $f(p) = p$ if and only if f fixes C pointwise. If $f \in \mathrm{Aut}(\mathfrak{C})$ and $f(p) = p$ then $f((d_p x)\varphi(x, y)) = (d_p x)\varphi(x, y)$, for all φ. Hence f fixes $Cb(p) = dcl(D(p))$ pointwise and (since $C = dcl(C)$) $Cb(p) \subset C$. To prove that $C \subset Cb(p)$, suppose $f \in \mathrm{Aut}(\mathfrak{C})$ fixes $Cb(p)$ pointwise. Since p is definable over $Cb(p)$, $f(p) = p$, hence f fixes C. Thus $C \subset dcl(Cb(p)) = Cb(p)$, completing the proof.

Remark 5.1.7. (i) Some authors take the property proved in (iii) of the previous lemma as the definition of a canonical base. Indeed, most of the properties we prove about canonical bases follow directly from this condition. Our definition makes it clear that each stationary type has a canonical base.

(ii) In Exercise 5.1.18 the reader is asked to show that for $p \in S(\mathfrak{C})$, $Cb(p) = dcl(C)$ if and only if for all $f \in Aut(\mathfrak{C})$, $f(p) = p$ if and only if f fixes C pointwise.

Corollary 5.1.10. *Let T be t.t., p a stationary type and c a canonical parameter of p. Then $Cb(p) = dcl(c)$.*

Remark 5.1.8. Requiring a canonical base to be definably closed guarantees the maximality condition in Lemma 5.1.12(iii). When defining the canonical parameter of a stationary type p in a t.t. theory we sacrificed this uniqueness in favor of p having a canonical parameter which is an element of the universe. Given an arbitrary stable theory and stationary p there may not be an element $c \in Cb(p)$ such that $dcl(c) = Cb(p)$. (Examples are given below.) See also Exercise 5.1.11.

Example 5.1.4. To illustrate this notion we examine canonical bases in an arbitrary theory of equivalence relations. Let $L = \{ E_i : i \in I \}$ be a collection of binary relations, T_0 a complete 1-sorted theory in L saying that each E_i is an equivalence relation and let $T = T_0^{eq}$. Then T has elimination of quantifiers regardless of the relationships between the E_i's axiomatized in T_0. Let $p \in S(\mathfrak{C})$ be a type in the same sort as the equivalence relations. By the elimination of quantifiers p is implied by

$$\{ E_i(x, a) : i \in I,\ a \in \mathfrak{C} \text{ and } E_i(x, a) \in p \} \cup$$
$$\{ \neg E_i(x, a) : i \in I,\ a \in \mathfrak{C} \text{ and } \neg E_i(x, a) \in p \}.$$

The data needed to determine p is:

(1) $\mathcal{E}(p) = \{ E_i : E_i(x, a) \in p \text{ for some } a \}$, and
(2) for $E_i \in \mathcal{E}(p)$, the set $e_i(p)$ which is the E_i–class of the elements b such that $E_i(x, b) \in p$.

(Given $q \in S(\mathfrak{C})$, if $\mathcal{E}(q) = \mathcal{E}(p)$ and $e_i(q) = e_i(p)$, for all $E_i \in \mathcal{E}(q)$, then $q = p$.) How can this data be used to find the canonical base of p or sufficiently, a set C such that $f(p) = p$ if and only if $f(C) = C$, for all $f \in \mathrm{Aut}(\mathfrak{C})$?

Claim. For all $f \in \mathrm{Aut}(\mathfrak{C})$, $f(p) = p$ if and only if f pointwise fixes $C(p) = \{ c_i : c_i \text{ is a name for } e_i(p), \text{ for } E_i \in \mathcal{E}(p) \}$.

Clearly, if $f \in \mathrm{Aut}(\mathfrak{C})$ and $f(p) = p$ then f pointwise fixes $C(p)$. Suppose, conversely, that $f \in \mathrm{Aut}(\mathfrak{C})$ pointwise fixes $C(p)$. By (1) and (2), $f(p) = p$ if and only if $\mathcal{E}(p) = \mathcal{E}(f(p))$ and $e_i(p) = e_i(f(p))$. For any $g \in \mathrm{Aut}(\mathfrak{C})$, $\mathcal{E}(g(p)) = \mathcal{E}(p)$. Let c_i be the element of $\mathfrak{C}$ which is a name for the equivalence class $e_i(p)$. We see, then, that

$$f(p) = p \iff e_i(f(p)) = e_i(p) \iff f(c_i) = c_i, \quad \text{for all } E_i \in \mathcal{E}(p).$$

This proves the claim.

Thus, $Cb(p) = dcl(C(p))$.

Pick T_0 to be the theory expressing that each E_i, $i < \omega$, has infinitely many class, E_{i+1} refines E_i and each E_i class contains infinitely many E_{i+1}–classes. The reader should find a $p \in S(\mathfrak{C})$ such that there is no $c \in Cb(p)$ with $Cb(p) = dcl(c)$.

Besides filling out our picture of the nonforking extensions of a complete type the following illustrates how canonical bases are used in proofs.

Lemma 5.1.13. *If T is stable and $p \in S(\mathfrak{C})$, then p does not fork over A if and only if p has $< |\mathfrak{C}|$ conjugates over A.*

Proof. Let $\mathrm{Aut}_A(\mathfrak{C})$ denote the set of automorphisms of $\mathfrak{C}$ which fix A pointwise. Given $f \in \mathrm{Aut}(\mathfrak{C})$, $f(p)$ is the unique nonforking extension of $f(p|Cb(p))$. Moreover, if $f, g \in \mathrm{Aut}(\mathfrak{C})$ and $f(Cb(p)) = g(Cb(p))$ then $f(p|Cb(p)) = g(p|Cb(p))$. Thus, there is a one-to-one correspondence between $\{ f(p) : f \in \mathrm{Aut}_A(\mathfrak{C}) \}$ and $\{ f(Cb(p)) : f \in \mathrm{Aut}_A(\mathfrak{C}) \}$.

Suppose p does not fork over A. Then $Cb(p) \subset acl(A)$, $\{ f(Cb(p)) : f \in \mathrm{Aut}_A(\mathfrak{C}) \}$ has cardinality $\leq 2^{|T|} < |\mathfrak{C}|$, so $\{ f(p) : f \in \mathrm{Aut}_A(\mathfrak{C}) \}$ has cardinality $< |\mathfrak{C}|$. If, on the other hand, $\{ f(p) : f \in \mathrm{Aut}_A(\mathfrak{C}) \}$ has unbounded cardinality, then so does $\{ f(Cb(p)) : f \in \mathrm{Aut}_A(\mathfrak{C}) \}$. Thus $Cb(p) \not\subset acl(A)$ and p must fork over A.

5.1.2 Stability and the Number of Types

We saw in the case of totally transcendental theories that there is a tight connection between the number of complete types over sets and the existence of ranks (a countable complete theory is t.t. if and only if it is $\aleph_0$–stable).

In this subsection we establish a similar connection for stable theories. The main result is

Proposition 5.1.1. *For T a complete theory and Δ a finite set, the following are equivalent.*

(i) *There is a $p \in S(\mathfrak{C})$ with $R_\Delta(p) = \infty$.*
(ii) *There is a $p \in S(\mathfrak{C})$ with $R_\Delta(p) \geq \omega$.*
(iii) *For every infinite cardinal λ there is a set A of cardinality λ such that $|S_\Delta(A)| = 2^\lambda$.*
(iv) *For some infinite set A, $|S_\Delta(A)| > |A|$.*

Proof. Fix T and Δ throughout the proof. Recall that R_φ denotes $R_{\{\varphi\}}$ when φ is a single formula. Most of the work is contained in

Claim. Suppose there is a $p \in S(\mathfrak{C})$ with $R_\Delta(p) \geq \omega$. Then for every infinite cardinal λ there is a set A of cardinality λ such that $|S_\Delta(A)| = 2^\lambda$.

First note that $R_\Delta(x = x) \geq \omega$. The proof of the claim involves the following sets of formulas. (Remember: ${}^\alpha 2$ is the set of functions from α into $2 = \{0, 1\}$ and ${}^{\alpha>}2 = \bigcup_{\beta<\alpha} {}^\beta 2$. Also, for ψ a formula, $\psi^0 = \psi$ and $\psi^1 = \neg\psi$.)

> Let α be an ordinal and, for each $s \in {}^{\alpha>}2$, let c_s be a new constant symbol. Given a formula φ, let $\Gamma(\varphi, \alpha)$ is $\{\varphi(x_\tau, c_{\tau\restriction i})^{\tau(i)} : \tau \in {}^\alpha 2,\ i < \alpha\}$.

An easy induction on α shows that $\Gamma(\varphi, \alpha)$ is consistent whenever $R_\varphi(x = x) \geq \alpha$.

Subclaim. There is a formula $\varphi \in \Delta$ such that $R_\varphi(x = x) \geq \omega$.

For $m \leq \omega$ let $\mathcal{W}_m$ be the collection of all sets of formulas W of the form

$$W = \{\varphi_{\tau\restriction i}(x_\tau, a_{\tau\restriction i})^{\tau(i)} : \tau \in {}^m 2,\ i < m\}$$
$$\text{for new constants } a_{\tau\restriction i} \text{ and } \varphi_{\tau\restriction i} \in \Delta.$$

We also require that for all $m \leq \omega$ and for all $\sigma, \tau \in {}^m 2$, if $\sigma \restriction i = \tau \restriction i$ and $\varphi_{\sigma\restriction i} = \varphi_{\tau\restriction i}$ then $a_{\sigma\restriction i} = a_{\tau\restriction i}$. As with the $\Gamma(\varphi, \alpha)$'s, $R_\Delta(x = x) \geq m$ implies the consistency of some element of $\mathcal{W}_m$. When $m < \omega$ every element of $\mathcal{W}_{m+1}$ contains an element of $\mathcal{W}_m$ (after renaming the variables). There are consistent elements of $\mathcal{W}_m$ for arbitrarily large finite m (since $R_\Delta(x = x) \geq \omega$). Since each $\mathcal{W}_m$ (for $m < \omega$) is finite there is a consistent element of $\mathcal{W}_\omega$. This yields $2^{\aleph_0}$ many Δ–types over some countable set A. A simple counting argument (using the finiteness of Δ) produces a $\varphi \in \Delta$ for which there are $2^{\aleph_0}$ many φ–types over A. Since there are only countably many φ–types over A with finite φ–rank, $x = x$ must have infinite φ–rank, proving the subclaim.

Fix a $\varphi \in \Delta$ with $R_\varphi(x = x) \geq \omega$. By compactness, $\Gamma(\varphi, \kappa)$ is consistent for all κ. Thus, for any infinite λ there is a set A of cardinality λ (namely the elements which interpret the constants in $\Gamma(\varphi, \lambda)$) such that $|S_\Delta(A)| = 2^\lambda$. This proves the claim.

Turning to the main body of the proof, notice that (i) $\Longrightarrow$ (ii) and (iii) $\Longrightarrow$ (iv) are trivial, while (ii) $\Longrightarrow$ (iii) is a restatement of the claim. For (iv) $\Longrightarrow$ (i), suppose (i) does not hold and A is an infinite set. Any Δ–type over A is contained in only finitely many elements of $S_\Delta(A)$ of the same Δ–rank. Furthermore, any element of $S_\Delta(A)$ contains a finite Δ–type of the same rank. Since there are $|A|$ many finite Δ–types over A, $|S_\Delta(A)| = |A|$; i.e., (iv) fails. This proves the proposition.

Corollary 5.1.11. *For T a complete theory, the following are equivalent.*

(i) *T is stable.*
(ii) *For all finite Δ and $p \in S(\mathfrak{C})$, $R_\Delta(p) < \omega$.*
(iii) *For all finite Δ and all sets A, $|S_\Delta(A)| = |A| + \aleph_0$.*
(iv) *For all sets A, $|S(A)| \leq |A|^{|T|}$.*
(v) *T is λ–stable for any λ such that $\lambda = \lambda^{|T|}$.*
(vi) *For some infinite λ, T is λ–stable.*

Proof. The equivalence of (i), (ii) and (iii) follows immediately from the proposition. Since any $p \in S(A)$ is simply the union of $p \restriction \varphi$, for φ a formula, (iii) $\Longrightarrow$ (iv). Trivially, (iv) $\Longrightarrow$ (v) $\Longrightarrow$ (vi). Finally, (ii) $\Longrightarrow$ (iii) of Proposition 5.1.1 shows that (vi) $\Longrightarrow$ (ii).

As stated earlier there are many definitions of the term "stable" in the literature. The earliest definition was: T is stable if T is λ–stable for some $\lambda \geq |T|$. Corollary 5.1.11 proves the equivalence of this definition with the one used here. Other standard definitions are:

(a) T is stable if T does not have the order property, and
(b) T is stable if for all sets A, every element of $S(A)$ is definable over A.

The reader will prove the equivalence of (b) with our definition in the exercises. We proved in Lemma 5.1.6 that when T is stable in our sense it does not have the order property. The converse, which is significantly more difficult to prove, is in [She90, §I.2]. Another equivalent involving the so-called fundamental order will be mentioned in Section 5.1.4. (See Theorem 5.1.2, specifically.)

5.1.3 Morley Sequences and Indiscernibles

As in t.t. theories indiscernible sets can be constructed by taking successive nonforking extensions of stationary types. The precise definition is

Definition 5.1.11. *Let T be stable, p a stationary type and B a set on which p is based. We call I a* Morley sequence over B in p *if I is a B–independent set of realizations of $p|B$.*

Remark 5.1.9. (i) Notice that being a Morley sequence is invariant under parallelism: if p and q are stationary and parallel then a set is a Morley sequence in p if it is a Morley sequence in q.

(ii) Let $p \in S(A)$ be a strongly minimal type. Then for B a set on which p is based, I is a Morley sequence in p if it is a set of realizations of $p|B$ which is algebraically independent over B.

(iii) More generally, any Morley sequence in a t.t. theory as defined in Definition 3.3.6 is a Morley sequence (by Corollary 5.1.4).

Lemma 5.1.14. *Let T be stable and p a stationary type based on B.*

(i) Given $n < \omega$ and $a = (a_0, \ldots, a_n)$, $b = (b_0, \ldots, b_n)$ independent sequences of realizations of $p|B$, $tp(a/B) = tp(b/B)$. Moreover, $tp(a/B)$ is stationary.

(ii) A Morley sequence over B in p is an indiscernible set over B.

Proof. Since (ii) follows immediately from (i) we only need to prove the first part, which is done by induction on n. Let $a' = (a_0, \ldots, a_{n-1})$ and $b' = (b_0, \ldots, b_{n-1})$. By induction there is an automorphism f fixing B and taking a' to b'. Since $f(tp(a_n/B \cup a')) = p|(B \cup b') = tp(b_n/B \cup b')$, $tp(a/B) = tp(b/B)$. The stationarity of $tp(a/B)$ follows from:

- the first sentence in (i) is true when B is replaced by $acl(B)$, and
- a complete type over an algebraically closed set is stationary.

Most of the properties established for indiscernibles in t.t. theories generalize directly to stable theories. In fact, combining Lemma 3.3.5 and Corollary 5.1.11 proves

Lemma 5.1.15. *Let T be stable and $(I, <)$ an infinite indiscernible sequence over A. Then*

(i) I is an indiscernible set over A.

(ii) For any formula $\varphi(x, y)$ over A there is an $n < \omega$ such that for all a,

$$|\{ b \in I : \models \varphi(a, b) \}| \le n \text{ or } |\{ b \in I : \models \neg\varphi(a, b) \}| \le n.$$

We will see momentarily that for any indiscernible set I in a stable theory all but a "small" subset J of I is a Morley sequence over J in some type. The relevant stationary type is defined here:

Definition 5.1.12. *Let T be stable, I an infinite set of indiscernibles and A a set. The* average type of I over A, *denoted $Av(I/A)$, consists of*

$$\{ \varphi(x) : \varphi \text{ is a formula over } A \text{ and } \models \varphi(a) \text{ for all but finitely many } a \in I \}.$$

Lemma 5.1.16. *Let T be stable, I and infinite set of indiscernibles and A a set. Then $Av(I/A)$ exits and is complete.*

Proof. Left to the reader in Exercise 5.1.19.

Average types are quite easy to understand when dealing with Morley sequences in a t.t. theory. Let T be t.t., p stationary, I a Morley sequence in p over A and $B \supset A$. For any $b \in B$ there is a finite $J \subset I$ such that $I \setminus J$ is a Morley sequence in p over $A \cup \{b\}$ (by Corollary 3.3.1). Thus, the average of I over any set $B \supset A$ is $p|B$.

Remark 5.1.10. Recall that $FE(A)$ denotes the set of finite equivalence relations over A (in a fixed sort determined by context). Let I be an infinite set of indiscernibles over A in a stable theory, $a \in I$ and $E \in FE(A)$. Then $E(x, a) \in Av(I/A \cup \{a\})$ since E has only finitely many classes and I is infinite and indiscernible.

Arbitrary indiscernible sets are reduced to Morley sequences with the following result.

Lemma 5.1.17. *If T is stable and I is an infinite set of indiscernibles, then $p = Av(I/\mathfrak{C})$ is based on any infinite $J \subset I$. Moreover, for any infinite $J \subset I$ $I \setminus J$ is a Morley sequence in p over J.*

Proof. The following claim (whose proof is left to the reader) indicates how to enlarge a set of indiscernibles without changing the average type.

Claim. If a realizes $p \restriction I$, then $I_0 = I \cup \{a\}$ is indiscernible and $Av(I_0/\mathfrak{C}) = p$.

By repeated applications of the claim we can assume, without loss of generality, that $|I| \geq |T|^+$. Let A be a set of cardinality $\leq |T|$ over which p does not fork. There is a set $I' \subset I$ of cardinality $\leq |T|$ such that

- A is independent from I over I' and
- for any formula φ over $A \cup I'$, if $|\{ a \in I : \models \varphi(a) \}|$ is finite then $\{ a \in I : \models \varphi(a) \} \subset I'$.

Then any $a \in I \setminus I'$ is independent from A over I' and $tp(a/A \cup I') = Av(I/A \cup I')$; i.e., $p \restriction (A \cup I') = tp(a/A \cup I')$ and this type does not fork over I'. That is to say, p does not fork over I'.

Now, let J be any infinite subset of I. For Δ a finite set of formulas there is a $\varphi(x, a_0, \dots, a_n) \in p$, where $a_0, \dots, a_n \in I'$ are distinct, such that $R_\Delta(p) = R_\Delta(\varphi(x, a_0, \dots, a_n))$. By the indiscernibility of I, for any distinct $b_0, \dots, b_n \in J$, $\varphi(x, b_0, \dots, b_n) \in p$ and $R_\Delta(p) = R_\Delta(\varphi(x, b_0, \dots, b_n))$. Thus, p does not fork over J. It remains to show that $p \restriction J$ is stationary. Fix $b \in J$ and $J' = J \setminus \{b\}$. Since J' is also infinite, p does not fork over J'. If $E \in FE(J')$, then $E(x, b) \in p$ by Remark 5.1.10. Thus, not only is $tp(b/J') = p \restriction J'$, but $tp(b/acl(J')) = p \restriction acl(J')$. By Corollary 5.1.8(i), $p \restriction J$ is stationary, proving that p is based on J. That $I \setminus J$ is a Morley sequence in p over J follows immediately.

Corollary 5.1.12. *Let T be stable, $p \in S(A)$ a stationary type and I a Morley sequence in p. Then $Av(I/\mathfrak{C})$ is parallel to p.*

Proof. See Exercise 5.1.16.

When T is t.t. this lemma can be improved to:

Corollary 5.1.13. *If T is t.t. and I is an infinite set of indiscernibles, then $p = Av(I/\mathfrak{C})$ is based on some finite $J \subset I$. Moreover, for any such J, $I \setminus J$ is a Morley sequence in p over J.*

We stated in Remark 5.1.3 that for any $p \in S(\mathfrak{C})$ (where T is stable) there is a set A of cardinality $\leq |T|$ over which p does not fork. With the above lemma this can be improved to

Corollary 5.1.14. *For T a stable theory and $p \in S(\mathfrak{C})$ there is a countable set A such that p is based on A.*

Proof. Let B be any set over which p is based and I a countably infinite Morley sequence over B in p. By Lemma 5.1.17, p is based on I.

Corollary 5.1.15. *Let T be stable, $p \in S(\mathfrak{C})$, I an infinite set of indiscernibles with $Av(I/\mathfrak{C}) = p$ and A any set. Then for all $\varphi(x,y)$, $d_p x\varphi(x,y)$ is equivalent to a formula over I. In particular, p is definable over $p \restriction A(\mathfrak{C})$.*

The final result connects average types to type diagrams.

Lemma 5.1.18. *If T is stable and I and J are infinite sets of indiscernibles with the same average type over $\mathfrak{C}$ then $D(I) = D(J)$.*

5.1.4 The Fundamental Order

Our intuition is that Morley rank in a t.t. theory and the $\Delta-$ranks in stable theories provide a measure of the complexity of types. The fundamental order is an alternative such measure.

Definition 5.1.13. *Let $\mathfrak{C}$ be the universal domain of a complete theory T.*

(i) Given $B \subset A \subset \mathfrak{C}$, $p \in S(A)$ a type in the variable v and $\varphi(v,w)$ a formula over B, we say that $\varphi(v,w)$ is represented *in p if there is an $a \in A$ such that $\varphi(v,a) \in p$. The* representation class of p over B, *$\chi_B(p)$, is* $\{\varphi(v,w) :$ φ is over B and φ is represented in $p\}$. *When $B = \emptyset$ we write $\chi(p)$ for $\chi_B(p)$.*

(ii) The fundamental order *of $\mathfrak{C}$ is*

$$\mathcal{O} = \{\chi(p) : p \in S(M) \text{ for } M \text{ a model of } T\}$$

under the partial order $\leq$ of reverse inclusion. That is, $\chi(p) \leq \chi(q)$ if and only if $\chi(q) \subset \chi(p)$.

(iii) For M, N models of T and $p \in S(M)$, $q \in S(N)$, we write $p \leq q$ if $\chi(p) \leq \chi(q)$ in the fundamental order. The relation $\leq$ on the collection of types over models is also called the fundamental order of T.

(iv) For $M \subset N$ models of T and $p \in S(M)$, $q \in S(N)$ with $p \subset q$, we call q an heir *of p if $\chi_M(q) = \chi_M(p)$.*

Remark 5.1.11. Let $\mathcal{P} = \{p : p \text{ is a complete type over a model }\}$.

(i) As usual when working with types over models we apply the definitions and results to the elements of $S(\mathfrak{C})$ as well.

(ii) The fundamental order over B is defined in the obvious way.

(iii) For $p \in S(M)$, M a model, there is a model $N \subset M$ of cardinality $|T|$ such that $\chi(p) = \chi(p \restriction N)$. Clearly, $\mathcal{O}$ has cardinality $\leq 2^{|T|}$. It does not follow (immediately) that p is an heir of $p \restriction N$ since that relation requires $\chi_N(p) = \chi_N(p \restriction N)$. However, there is a chain of models $N_0 \subset N_1 \subset \ldots$ such that for each $i < \omega$, $|N_i| = |T|$ and every formula over N_i represented in p is represented in $p \restriction N_{i+1}$. Then p is an heir of $p \restriction (\bigcup_{i<\omega} N_i)$.

(iv) $p \in \mathcal{P}$ is minimal in the fundamental order if and only if $v = w$ is represented in p; i.e., p is realized in its domain.

(v) Given $M \subset N$ models and $p \in S(M)$, $q \in S(N)$ with $p \subset q$, q is an heir of p if and only if whenever $\varphi(v, w)$ over M is represented in q there is an $a \in M$ such that $\varphi(v, a) \in q$.

(vi) For any given complete theory there is a fundamental order corresponding to each sort. For simplicity we usually assume the order is on the sort of equality.

The representation class of $p = tp(a/M)$ is one measure of the amount of information p determines about a. When $\chi(p) < \chi(q)$, then to some degree p gives more information about a realization than does q. If $q \in S(N)$ is an heir of p then all of the information contained in q (given by representation) is already contained in p; i.e., all of the information in q is inherited from p.

Example 5.1.5. (i) Let T be the theory of one equivalence relation E with infinitely many infinite classes and no finite classes. In the sort of equality the fundamental order $\mathcal{O}$ contains 3 elements. The unique minimal class is the one containing $v = w$. There is a unique maximal class which can be described as the one not containing $E(v, w)$. Strictly between these two classes in $\mathcal{O}$ is the class containing $E(v, w)$ and not containing $v = w$.

(ii) Let $(I, <)$ be a linear order and E_i a binary relation symbol for each $i \in I$. Let T be the theory expressing for i, $j \in I$:

(a) Each E_i is an equivalence relation with infinitely many infinite classes and no finite classes.
(b) If $i < j$ E_i refines E_j and each E_j−class contains infinitely many E_i−classes.

Then T is quantifier-eliminable and stable. For each cut J of I there is a $p_J \in S(\mathfrak{C})$ such that p_J represents E_j if and only if $j \notin J$. There is a type

$p_0 \in S(\mathfrak{C})$ representing $v = w$, hence every E_i. There is also a $p_1 \in S(\mathfrak{C})$ not representing $v = w$ or any E_j. Let $\mathcal{J}$ be the set of cuts of I ordered by inclusion, and let $\mathcal{J}_1$ be $\mathcal{J}$ with the addition of a minimal element. Then, the fundamental order is isomorphic to $\mathcal{J}_1$.

(iii) Let K be the universal domain of algebraically closed fields of characteristic 0. The fundamental order $\mathcal{O}$ on 2−types is described as follows. The reader can verify that every representation class in $\mathcal{O}$ contains an element of $S_2(K)$. The algebraic types in $S_2(K)$ are the minimal elements of $\mathcal{O}$ and there are $\aleph_0$ many different representation classes of algebraic types. There is a unique element of $S_2(K)$ having Morley rank 2. For strongly minimal $p, q \in S_2(K)$, $\chi(p) \subset \chi(q)$ if and only if $\chi(p) = \chi(q)$ if and only if p is conjugate to q. There are $\aleph_0$ many strongly minimal elements of $S_2(K)$ up to conjugacy. Thus, $\mathcal{O}$ has a unique maximal element, $\aleph_0$ many minimal elements and the remainder is a set of $\aleph_0$ many pairwise incomparable elements.

In each of the above (stable) examples, when M is a model and $p \in S(\mathfrak{C})$, p is an heir of $p \upharpoonright M$ if and only if p does not fork over M.

(iv) Let $\mathfrak{C}$ be the universal domain of dense linear orders without endpoints. The fundamental order $\mathcal{O}$ on 1−types has one minimal element. A nonminimal $o \in \mathcal{O}$ is determined by whether it contains $v < w$ and not $v > w$, $v > w$ and not $v < w$, or both $v < w$ and $v > w$.

Let M be a model, $p_{-\infty} = \{v < a : a \in M\}$ and $p_{+\infty} = \{v > a : a \in M\}$. The type $p_{-\infty}$ has a unique heir in $S(\mathfrak{C})$, namely $\{v < a : a \in \mathfrak{C}\}$. Similarly, $p_{+\infty}$ has a unique heir in $S(\mathfrak{C})$. Let J be a cut of M such that $M \setminus J \neq \emptyset$ and $\sup J$ does not exist in M. Let $p_0 = \{v > a : a \in J\} \cup \{v < a : a \in M \setminus J\}$. Any nonalgebraic extension of p_0 in $S(\mathfrak{C})$ is an heir of p_0. Thus, p_0 has $2^{|\mathfrak{C}|}$ many heirs in $S(\mathfrak{C})$.

The first lemma connects the fundamental order with simple inclusion.

Lemma 5.1.19. *Let $\mathcal{O}$ be the fundamental order of a complete theory and $o_1 \leq o_2$ elements of $\mathcal{O}$. Then, for $i = 1, 2$ there is p_i a complete type over a model such that $\chi(p_i) = o_i$ and $p_1 \supset p_2$.*

Proof. The proof is omitted for brevity. The reader can find it in [LP79, 2.3].

Lemma 5.1.20. *Let $\mathfrak{C}$ be the universal domain of a complete theory, M an $\aleph_0$−saturated model and $p \in S(M)$. Then there is an heir of p in $S(\mathfrak{C})$.*

Proof. It suffices to show that p has an heir in $S(N)$ for an arbitrary model $N \supset M$. Let $q(v)$ be the set of formulas over N which contains $p(v)$ and $\neg\varphi(v, a)$ for any $\varphi(v, w) \notin \chi_M(p)$, and any $a \in N$. In outline the consistency of $q(v)$ is proved as follows. Let $b \in M$, $a \in N$, $\varphi_0(v, w_0), \ldots, \varphi_k(v, w_k) \notin \chi_M(p)$ and $q_0(v) = p \upharpoonright b \cup \{\neg\varphi_0(v, a), \ldots, \neg\varphi_k(v, a)\}$. Since M is $\aleph_0$−saturated there is an elementary map f fixing b and $\varphi_0(v, w_0), \ldots, \varphi_k(v, w_k)$, and taking a to $a' \in M$. Then $f(q_0) \subset p$, so q is consistent.

Any completion of q over N is an heir of p over N, proving the lemma.

In a stable theory we get a stronger existence theorem for heirs.

Lemma 5.1.21. *Let $\mathfrak{C}$ be the universal domain of a stable theory, M a model, $p \in S(M)$ and $q \in S(\mathfrak{C})$ an extension of p. Then q is an heir of p if and only if q is a nonforking extension of p.*

Proof. First suppose q is a nonforking extension of p. Let $\varphi(v, w)$ be a formula over M represented in q. Let $\psi(w)$ be a formula over M defining $q \restriction \varphi$. Since φ is represented in q, ψ is consistent, hence ψ is satisfied by some element of M. Thus q is an heir of p.

Conversely suppose q forks over M and $\varphi(v, a) \in q$ witnesses this forking, where $\varphi(v, w)$ is over M. Let $\theta(w)$ be a formula over M which defines $p \restriction \varphi$. Then θ has the property:

$$\forall b(\, p \cup \{\varphi(v, b)\} \text{ does not fork over } M \iff \models \theta(b) \,).$$

Thus, $\varphi(v, w) \wedge \neg\theta(w)$ is represented in q but not represented in p, proving that q is not an heir of p.

The fundamental order is tied to stability with

Theorem 5.1.2. *The complete theory T is stable if and only if*

(*) *for all models M and $p \in S(M)$, p has at most one heir in $S(\mathfrak{C})$.*

Proof. If T is stable, M is a model and $p \in S(M)$, then $q \in S(M)$ is an heir of p if and only if q is a nonforking extension of p by Lemma 5.1.21.

Now suppose that (*) holds. Let M be an $\aleph_0$−saturated model and $\lambda = |M|$. Since any $p \in S(M)$ has an heir in $S(\mathfrak{C})$ (by Lemma 5.1.20), when N is a submodel of M, each $p \in S(N)$ has at most one heir in $S(M)$. For each $p \in S(M)$ there is (by Remark 5.1.11(iii)) a model $N \subset M$ of cardinality $|T|$ such that p is an heir of $p \restriction N$. By (*) each element of $S(M)$ is in one-to-one correspondence (by the heir relation) to a type over a set of cardinality $|T|$. Thus, $|S(M)| \leq \lambda^{|T|}$. For some choice of λ, $\lambda^{|T|} = \lambda$, hence T is stable (by Corollary 5.1.11).

The fundamental order can be used in conjunction with the forking relation to deepen our understanding of stable theories. See, for example, [Bue85b] and [HLP+92].

Historical Notes. Globally speaking all of these results are due to Shelah [She90]. In detail our development of stable theories follows the first section of [Hru86], which is based on notes from a course by Harrington. Lemma 5.1.11 (the Open Mapping Theorem) is due to Lascar and Poizat [LP79], however Corollary 5.1.9 is stated for superstable theories in [Las76]. The results in the second subsection are explicitly due to Shelah and can be found in [She71].

The fundamental order was developed by Lascar and Poizat in [LP79].

Exercise 5.1.1. Let T be the theory of a single equivalence relation E with infinitely many infinite classes and no finite classes. Let $\Delta = \{E(x,y), x = y\}$. Prove that for all $p \in S_1(\mathfrak{C})$, $MR(p) = R_\Delta(p)$.

Exercise 5.1.2. Show that when $\Gamma(x) \supset \Delta(x)$, $R_\Gamma(\varphi) \geq R_\Delta(\varphi)$, for all formulas φ. Also, when Γ is the set of all formulas with object variable x, $R_\Gamma(p) = MR(p)$. Thus $MR(p) \geq R_\Delta(p)$.

Exercise 5.1.3. Given a stable theory show that any $p \in S(acl(A))$ does not fork over A. (Prove this without using the existence of nonforking extensions, whose proof depends on this property.)

Exercise 5.1.4. Prove Corollary 5.1.4.

Exercise 5.1.5. Suppose that T is a complete theory with the property that for all A, every element of $S(A)$ is definable over A. Prove that T is stable. (HINT: Use Corollary 5.1.11.)

Exercise 5.1.6. Let T be stable, M a model and $p \in S(M)$. Let $\varphi(x,y)$ and $\psi(y)$ be formulas over M such that ψ defines $p \upharpoonright \varphi$. Show that ψ defines $q \upharpoonright \varphi$, where $q \in S(\mathfrak{C})$ is the nonforking extension of p.

Exercise 5.1.7. Prove: Given a stable theory, a model M and $p \in S(M)$, if p is definable over $A \subset M$, then p does not fork over A.

Exercise 5.1.8. Prove Remark 5.1.5.

Exercise 5.1.9. Let $p \in S(A)$ be a stationary type in a stable theory, $B \supset A$, and $q \in S(B)$ a forking extension of p. Show that q is also a forking extension of $p|C$ for $C \subset B$ any set on which p is based.

Exercise 5.1.10. Prove that a countable stable theory has a saturated model of cardinality κ^+, when $\kappa^+ \geq \kappa^{\aleph_0}$.

Exercise 5.1.11. Suppose that T is t.t., p is a stationary type and $C = Cb(p)$. Prove that there is a $c \in C$ such that $C = dcl(c)$.

Exercise 5.1.12. Let T be the theory in the Example 5.1.1 and $p \in S(\mathfrak{C})$ a type in x, where x has the same sort as the equivalence relations. Describe $Cb(p)$.

Exercise 5.1.13. Prove: If T is stable and I is an infinite set of indiscernibles over A, then $a, b \in I \implies stp(a/A) = stp(b/A)$. Prove, in fact, that I is indiscernible over $acl(A)$.

Exercise 5.1.14. Give a quick proof of the Open Mapping Theorem when $A = acl(A)$.

Exercise 5.1.15. Prove the claim in the proof of Lemma 5.1.17.

Exercise 5.1.16. Prove Corollary 5.1.12.

Exercise 5.1.17. Prove Corollary 5.1.15.

Exercise 5.1.18. Let T be stable and $p \in S(\mathfrak{C})$. Prove that $Cb(p) = dcl(C)$ if and only if for all $f \in \mathrm{Aut}(\mathfrak{C})$, $f(p) = p$ if and only if f fixes C pointwise.

Exercise 5.1.19. Prove Lemma 5.1.16.

Exercise 5.1.20. Prove Lemma 5.1.18.

5.2 The Stability Spectrum and $\kappa(T)$

In Corollary 5.1.11 we proved that a complete theory T is stable if and only if it is λ−stable for some infinite λ.

Definition 5.2.1. *Let T be stable.*
(i) The stability spectrum of T *is* $\{\lambda : T$ *is* λ−*stable*$\}$.
(ii) The first stability cardinal of T, $\lambda(T)$, *is the minimum infinite cardinal* λ *such that* T *is* λ−*stable.*

For a stable theory T what are the possibilities for the stability spectrum? We will see that the possibilities are controlled by $\lambda(T)$ and another important invariant of the theory, $\kappa(T)$. Many subsequent results have hypotheses involving these numbers.

The next lemma follows from Corollary 5.1.11.

Lemma 5.2.1. *If T is stable then* $|T| \leq \lambda(T) \leq 2^{|T|}$.

The following invariant helps to measure the complexity of the forking relation on a stable theory.

Definition 5.2.2. *Let T be a stable theory. The invariant $\kappa(T)$ is the least infinite cardinal κ such that whenever $\{A_i : i < \kappa\}$ is a sequence of sets with $i < j < \kappa \implies A_i \subset A_j$ and $p \in S(\bigcup_{i<\kappa} A_i)$, there is an i such that $p \restriction A_{i+1}$ does not fork over A_i.*

We let $\kappa_r(T)$ denote the least regular cardinal $\geq \kappa(T)$ (thus, $\kappa_r(T)$ is $\kappa(T)$ or $\kappa(T)^+$).

Remark 5.2.1. We leave it to the reader to see that $\kappa(T) \leq |T|^+$. Thus, for countable theories $\kappa(T)$ can only be $\aleph_0$ or $\aleph_1$. When T is t.t. $\kappa(T)$ is $\aleph_0$.

It is possible for $\kappa(T)$ to be singular when T is uncountable, creating technical difficulties which require using $\kappa_r(T)$ instead of $\kappa(T)$ in some settings.

Independence is further related to $\kappa(T)$ in the following proposition (whose proof is left to the exercises).

Proposition 5.2.1. *Let T be stable.*

(i) For all elements b and sets C there is $A \subset C$ of cardinality $< \kappa(T)$ such that $b \underset{A}{\perp} C$.

(ii) For all sets B and C there is $A \subset C$ such that $B \underset{A}{\perp} C$ and

- $|A| < \kappa(T) + |B|^+$ *if $\kappa(T)$ is regular, and*
- $|A| \leq \kappa(T) + |B|$ *otherwise.*

For λ and κ cardinals let $\lambda^{<\kappa} = \sup\{\lambda^\mu : \mu < \kappa\}$. Let ${}^{\kappa\geq}\lambda$ be the set of all functions $f : \alpha \to \lambda$, where $\alpha \leq \kappa$ (which is denoted $lh(f)$). Our eventual goal is

Theorem 5.2.1 (Stability Spectrum). *A stable theory T is λ−stable if and only if $\lambda = \lambda(T) + \lambda^{<\kappa(T)}$.*

The bulk of the proof is contained in

Lemma 5.2.2. *If T is stable and $\lambda < \lambda^{<\kappa(T)}$ then T is not λ−stable.*

Proof. Typical of such problems, we will construct many types over a set of cardinality λ by recursion. Let κ be the least cardinal such that $\lambda^\kappa > \lambda$. Since $\kappa < \kappa(T)$ there is a sequence of sets $\{A_i : i < \kappa\}$ and a $p \in S(\bigcup_{i<\kappa} A_i)$ such that

- $i < j < \kappa \implies A_i \subset A_j$, and
- $p \restriction A_{i+1}$ forks over A_i, for all $i < \kappa$.

Without loss of generality, we can require that $A_{i+1} \setminus A_i$ is a finite set a_i and $A_\delta = \bigcup_{i<\delta} A_i$, when δ is a limit ordinal. Let $A_\kappa = \bigcup_{i<\kappa} A_i$ and notice that $|A_\kappa| = \kappa \leq \lambda$. There is a tree of sets such that each branch is conjugate to $\{A_i : i \leq \kappa\}$, each node has λ many successors and these λ many successors are independent over their predecessor. It is left to the reader to see that this can be accomplished with the construction of a family of elementary maps f_ξ, $\xi \in {}^{\kappa\geq}\lambda$, such that for all $\xi, \zeta \in {}^{\kappa\geq}\lambda$,

(1) $dom(f_\xi) = A_{lh(\xi)}$,
(2) if $\zeta \subset \xi$, $f_\zeta \subset f_\xi$,
(3) if $\delta = lh(\xi)$ is a limit ordinal, $f_\xi = \bigcup_{\beta<\delta} f_{\xi\restriction\beta}$, and
(4) if $\alpha = lh(\xi)$, $B_\xi = \{f_{\xi^\frown i}(A_{\alpha+1}) : i < \lambda\}$ is independent over $f_\xi(A_\alpha)$ and B_ξ is independent from $\bigcup\{f_\eta(A_\alpha) : lh(\eta) = \alpha\}$ over $f_\xi(A_\alpha)$.

Let $\Phi = {}^\kappa\lambda$, $F = \{f_\xi : \xi \in \Phi\}$ and $A = \bigcup_{\xi\in\Phi} f_\xi(A_\kappa)$ (a set of cardinality $\lambda^{<\kappa} = \lambda$). For $\xi \in \Phi$ let $p_\xi \in S(A)$ be any nonforking extension of $f_\xi(p)$ and let $P = \{p_\xi : \xi \in \Phi\}$.

Claim. $\xi \neq \zeta \in \Phi \implies p_\xi \neq p_\zeta$.

Let b_ξ and b_ζ realize p_ξ and p_ζ, respectively. Let α be the maximal ordinal for which $\xi \upharpoonright \alpha = \zeta \upharpoonright \alpha = \eta$ and let $C = f_\eta(A_\alpha)$. By (4) and the transitivity of independence, $f_\xi(A_\alpha)$ is independent from $f_\zeta(A_\kappa)$ over C. Since b_ζ is independent from A over $f_\zeta(A_\kappa)$, the transitivity of independence again implies that b_ζ is independent from $f_\xi(A_\alpha)$ over C. However, b_ξ depends on $f_\xi(A_\alpha)$ over C (since $p_\xi = f_\xi(p)$). Thus, $p_\xi \neq p_\zeta$, as claimed.

We conclude that $|P| = \lambda^\kappa$. Since $|A| = \lambda < \lambda^\kappa$, T is not λ–stable.

Remark 5.2.2. It follows immediately from this lemma that $\lambda(T) \geq \kappa(T)$.

Definition 5.2.3. *If $p \in S(A)$ is a complete type in a stable theory we define the* multiplicity of p, Mult(p), *to be*

$$|\{\, q \in S(\mathfrak{C}) : q \supset p \text{ and } q \text{ does not fork over } A \,\}|.$$

Let $\mu(T)$ be the supremum of $\{\, \mathrm{Mult}(p) : p \text{ a complete type} \,\}$.

As stated in Remark 5.1.3, $\mathrm{Mult}(p) \leq 2^{|T|}$. A complete type is stationary if and only if it has multiplicity 1.

Lemma 5.2.3. *If T is stable, $\mu(T) + \kappa(T) \leq \lambda(T)$.*

Proof. By Remark 5.2.2, $\lambda(T) \geq \kappa(T)$. Let $p \in S(A)$ be any complete type and $B \subset A$ a set of cardinality $< \kappa(T)$ over which p does not fork. Then, $\mathrm{Mult}(p) \leq$ the multiplicity of $q = p \upharpoonright B$. Since $\kappa(T)$ and $|T|$ are both $\leq \lambda(T)$ there is a model $M \supset B$ of cardinality $\lambda(T)$. Every nonforking extension of q in $S(\mathfrak{C})$ is parallel to an element of $S(M)$, so $|S(M)| \geq \mathrm{Mult}(q)$. Since T is $\lambda(T)$–stable, $\mathrm{Mult}(p) \leq \mathrm{Mult}(q) \leq \lambda(T)$, as required.

To complete the proof of the Theorem 5.2.1 we need only prove

Lemma 5.2.4. *If T is stable and $\lambda \geq \lambda(T)$ is a cardinal such that $\lambda = \lambda^{<\kappa(T)}$, then T is λ–stable.*

Proof. Let A be a set of cardinality λ. Any $p \in S(A)$ is a nonforking extension of $p \upharpoonright B$ for some $B \subset A$ of cardinality $< \kappa(T)$. Furthermore, there are $\leq \mu(T)$ elements of $S(A)$ which are nonforking extensions of this type $p \upharpoonright B$. Thus,

$$\begin{aligned} |S(A)| \;\leq\; & (\text{the number of subsets of } A \text{ of cardinality } < \kappa(T)) \\ & \times(\text{the number of types over a given set of cardinality } < \kappa(T)) \\ & \times \mu(T) \\ \leq\; & \lambda^{<\kappa(T)} \cdot \lambda(T) \cdot \mu(T) \\ =\; & \lambda. \end{aligned}$$

This proves that T is λ–stable.

This completes the proof of Theorem 5.2.1.

Corollary 5.2.1. *If T is $\kappa(T)$−stable then $\kappa(T)$ is regular.*

Proof. Left as an exercise.

For countable theories the Stability Spectrum Theorem leads to a particularly simple partitioning of the stable theories.

Proposition 5.2.2. *For a countable complete theory T one of the following mutually exclusive conditions holds.*

(1) *T is λ−stable for all infinite λ.*
(2) *T is λ−stable if and only if $\lambda \geq 2^{\aleph_0}$.*
(3) *T is λ−stable if and only if $\lambda = \lambda^{\aleph_0}$.*
(4) *T is unstable.*

Proof. Suppose T is stable. If $\lambda(T) = \aleph_0$; i.e., T is $\aleph_0$−stable, then T is λ−stable for all infinite λ (by Proposition 3.3.1). Otherwise, there is a countable set A with $S(A)$ uncountable. Since the only uncountable possibility for $|S(A)|$ is $2^{\aleph_0}$ (see Lemma 2.2.4) $\lambda(T)$ is its maximum possible value $= 2^{\aleph_0}$. Since $\kappa(T) \leq |T|^+$, $\aleph_0$ and $\aleph_1$ are the only two possibilities for $\kappa(T)$ (when T is countable and stable). If $\kappa(T) = \aleph_0$ and $\lambda(T) = 2^{\aleph_0}$, then T is λ−stable if and only if $\lambda \geq 2^{\aleph_0}$. If $\kappa(T) = \aleph_1$, then T is λ−stable if and only if $\lambda = \lambda^{\aleph_0}$ ($\lambda(T)$ is necessarily $2^{\aleph_0}$ in this case).

Examples in earlier sections show that all of these possibilities do occur. There is no such clean division for uncountable theories, however, the exact possibilities for $\lambda(T)$ are given in [She90, III.5].

Definition 5.2.4. *A stable theory T is called* superstable *if $\kappa(T) = \aleph_0$.*

The superstable theories form a major subclass of the stable theories which will be studied extensively in Chapter 6. Notice that a stable theory T is superstable exactly when T is λ−stable for all sufficiently large λ. Proposition 5.2.2 partitions the countable stable theories into the categories: (a) the $\aleph_0$−stable theories, (b) the superstable theories which are not $\aleph_0$−stable (called the *properly superstable* theories) and (c) the stable theories which are not superstable (called the *properly stable* theories).

The following illustrates how to distinguish quickly between ω−stable and properly superstable countable theories.

Lemma 5.2.5. *If T is a countable properly superstable theory, then either T is not small or T has a complete type p over a finite set with infinite multiplicity (hence multiplicity $2^{\aleph_0}$).*

Proof. We are assuming that T is not ω−stable, hence there is a countable model M with $|S(M)| = 2^{\aleph_0}$. First suppose that every element of $S(M)$ is based on a finite subset of M. Then each element of $S(M)$ is the unique nonforking extension of a type over a finite set, hence there are $2^{\aleph_0}$ complete types over finite subsets of M proving that T is not small. On the other hand, suppose $q \in S(M)$ is not based on a finite set. Let $A \subset M$ be a finite set over which q does not fork and let $p = q \restriction A$. Then p has infinite multiplicity since otherwise there is a finite set A', $A \subset A' \subset acl(A)$, with q the unique nonforking extension of $q \restriction A'$. That a type of infinite multiplicity (in a countable stable theory) must have multiplicity $2^{\aleph_0}$ is left to the exercises.

Corollary 5.2.2. *If T is ω−categorical and superstable, then T is ω−stable.*

Proof. Suppose to the contrary that T is ω−categorical and properly superstable. Certainly, an ω−categorical theory is small, so Lemma 5.2.5 yields a complete type $p \in S(A)$, where A is finite, which has infinite multiplicity. Recall Lemma 5.1.10 linking conjugacy over $acl(A)$ with $FE(A)$ = the set of finite equivalence relations over A. There is a subset $\{ E_i(x,y) : i < \omega \}$ of $FE(A)$ such that each E_{i+1} refines E_i and $\models E_i(a,b)$, for all i, if and only if $tp(a/acl(A)) = tp(b/acl(A))$. Let a realize p. Since p has infinite multiplicity, $p \cup \{ E_i(x,a) \wedge \neg E_{i+1}(x,a) \}$ is consistent for infinitely many i. These infinitely many types over $A \cup \{a\}$ contradict the ω−categoricity of T.

Historical Notes. The original source for these results is [She71]. They are also found in [She90].

Exercise 5.2.1. Show that $\kappa(T) \leq |T|^+$, when T is stable.

Exercise 5.2.2. Prove Proposition 5.2.1.

Exercise 5.2.3. Let $\mathfrak{C}$ be the universal domain of a superstable theory and A a set. Then $Th(\mathfrak{C}_A)$ is also superstable.

Exercise 5.2.4. Give examples of countable theories in each of the classes delineated in Proposition 5.2.2.

Exercise 5.2.5. Let T be a countable stable theory and $p \in S(A)$ a type with infinite multiplicity. Show that the multiplicity of p is $2^{\aleph_0}$.

Exercise 5.2.6. Prove: If T is superstable then for every infinite set of indiscernibles I, $Av(I/\mathfrak{C})$ is based on a finite $J \subset I$.

5.3 Stable Groups and Modules

In this section we generalize the treatment of generics for ω−stable groups to the stable setting. Besides the ω−stable groups the stable groups include all modules (see Section 5.3.2). Group actions play a central role in stability theory today. Here we develop a theory of generic types for group actions specializing to a theory of generic types for groups (since a group acts on itself by translation). Examples of ω–stable groups and group actions were given previously.

In the subsection on modules we develop the most basic model-theoretic properties of these natural mathematical objects and interpret in this setting the tools we will develop to study stable groups.

The material on 1–based groups generalizes the subject matter in Section 4.3.2.

As with ω–stable groups our study of stable groups depends on the existence of the connected component of a group and stabilizers of types.

Given a definable group in a stable theory the groups of significant model theoretic interest, such as the connected component and stabilizers of types, may not be definable—they may only be $\bigwedge$–definable. Since we need a theory of generics for these groups as well we must work with $\bigwedge$–definable groups (and group actions) from the beginning. The explicit definitions are as follows. (Parts (i) and (ii) are simply restatements of Definition 3.5.11(i) and (ii).)

Definition 5.3.1. *Let T be a complete theory.*

(i) We call $(G, \cdot)$ an $\bigwedge$–definable group over *A if*

- *$(G, \cdot)$ is a group,*
- *G is a subset of $\mathfrak{C}$, $\bigwedge$–definable over A, and*
- *there is a function f, definable over A in $\mathfrak{C}$, such that $f \upharpoonright G \times G$ defines the binary operation $\cdot$ on G.*

(ii) Similarly, a group action $(G, \cdot, X, \star)$ is an $\bigwedge$–definable group action over *A if $(G, \cdot)$ is an $\bigwedge$–definable group over A in $\mathfrak{C}$, X is a subset of $\mathfrak{C}$, $\bigwedge$–definable over A, and $\star$ is the restriction to $G \times X$ of an A–definable function.*

(iii) A stable group *(*stable group action*) is an $\bigwedge$–definable group (group action) in a stable theory.*

The usual conventions about dropping the A when it is $\emptyset$ are adopted. When confusion seems unlikely the $\cdot$ and $\star$ are omitted from expressions and we simply write gh for $g \cdot h$ and gx for $g \star x$. When $G = X$ and $\star$ is multiplication on the left (right), $\star$ may be called *left* (*right*) *translation.*

If the type Φ defines an $\bigwedge$–definable group action in $\mathfrak{C}$ and N is another model of the theory, then Φ defines such a group action in N (when it contains the relevant parameters). We then call $\Phi(N)$ an *$\bigwedge$–definable group action in N.*

Let G be a stable group. If G is definable we can restrict the universe to G without altering the set of definable relations. Since restriction to an $\bigwedge$–definable set is not so well-behaved we must continue to mention the ambient theory when studying an $\bigwedge$–definable group. Instead of studying the models of $Th(G)$ where this theory is stable (as we did with ω–stable groups), we study the groups $\Phi(M)$, where $\Phi(\mathfrak{C})$ is a group and M ranges over the models of the theory.

When dealing with groups we abandon some of our notational abbreviations. By $a \in G$ we really mean that a is an element of G, not that a is a finite sequence from G. When X is the set of realizations of the type $\Phi(x)$ in some model M and A is a set, $S^X(A)$ denotes the elements of $S(A)$ which extend $\Phi(x)$. Given a set of formulas $\Delta(x)$, $S^X_\Delta(A)$ denotes $\{ p \restriction \Delta : p \in S^X(A) \}$.

As with $\omega-$stable groups, a stable group action gives rise to an action of the group on a collection of types.

Definition 5.3.2. *Let $(G, X, \star)$ be a stable group action, $\Phi(x)$ the type defining X and $p(x)$ a type over M containing Φ. Given $a \in G$, $ap = \{ \varphi(a^{-1} \star x) : \varphi \in p \}$ and is called the* left translate of p by a. *The type pa is defined similarly, and (when $X = G$) p^{-1} is obtained by replacing x by x^{-1} in p.*

This definition specializes to the earlier definition of translation in an $\omega-$stable group G by taking $X = G$ and the action to be multiplication on the left.

Let $\bar{G} = (G, X, \star)$ be a stable group action and $\Delta(x)$ a set of formulas over $\emptyset$, where x has the same sort as X. Let $\Delta^*(x) = \{ \varphi(y \star x, z) : \varphi(x, z) \in \Delta \}$, where $\varphi(y \star x, z)$ has object variable x and a new parameter variable y. We call Δ *invariant* if for any $p \in S_\Delta(\mathfrak{C})$ and $a \in G$, $ap \in S_\Delta(\mathfrak{C})$. Notice that Δ^* is invariant and every element of $S_\Delta(\mathfrak{C})$ is a Δ^*-type (since any $\varphi(x, z) \in \Delta$ is equivalent to $\varphi(1 \star x, z)$). Warning: For Δ an arbitrary finite set of formulas and $p \in S_\Delta(\mathfrak{C})$ it is possible that $ap \notin S_\Delta(\mathfrak{C})$.

Lemma 5.3.1. *Let (G, X) be a stable group action, Φ the type defining X and Δ an invariant set of formulas. Then G acts on $S^X_\Delta(\mathfrak{C})$ and for any type p containing Φ and $a \in G$, $(R, \mathrm{Mult})_\Delta(p) = (R, \mathrm{Mult})_\Delta(ap)$.*

Proof. See Exercise 5.3.2.

Thus, if H is an $\bigwedge-$definable subgroup of the stable group G, Δ is invariant with respect to the action of G on itself and $a \in G$, then H and aH have the same $\Delta-$rank and multiplicity. (Remember, the $\Delta-$rank of H is, by definition, the $\Delta-$rank of the type defining it.)

Definition 5.3.3. *Let (G, X) be a stable group action, Δ an invariant set of formulas and $p \in S^X(\mathfrak{C})$. The* $\Delta-$stabilizer of p *is $stab(p, \Delta) = \{ a \in G : a(p \restriction \Delta) = p \restriction \Delta \}$. The* stabilizer of p, *$stab(p)$, is $\{ a \in G : ap = p \}$; i.e., the $L(x)-$stabilizer of p, where $L(x)$ is the set of formulas over $\emptyset$ with object variable x. Equivalently, $stab(p) = \bigcap \{ stab(p, \Delta) : \Delta$ invariant $\}$.*

Let X be an $\bigwedge-$definable set in $\mathfrak{C}$ and $p(x, y)$ a type. A subset Y of X is called *definable-by-p* if there is a b such that $Y = p(\mathfrak{C}, b) \cap X$. Occasionally, p will have no parameter variables, in which case a definable-by-p set is just a subset of $p(\mathfrak{C})$. For $\Delta(x)$ a set of formulas as usual we call Y *definable-by-Δ* if there is a $\Delta-$type $p(x, b)$ such that $Y = p(\mathfrak{C}, b) \cap X$. Y is called *definable-by-L* if it is definable-by-ψ for some formula ψ; equivalently Y is the intersection of X and a definable set.

Lemma 5.3.2. *If G is a stable group and $\Delta(x)$ is a finite set of formulas over $\emptyset$, where x has the same sort as G, then the collection of definable-by-Δ subgroups of G has the ascending and descending chain conditions.*

Proof. Suppose, for example, that $G_0 \supset G_1 \supset G_2 \supset \ldots$ is a strictly descending chain of definable-by-Δ subgroups. Then, for each i, G_i contains two cosets of G_{i+1}, each having the same Δ^*–rank and multiplicity as G_{i+1}. Since G_{i+1} and each coset of it is defined by a Δ^*–type, $(R, \text{Mult})_{\Delta^*}(G_i) > (R, \text{Mult})_{\Delta^*}(G_{i+1})$. This contradicts the existence of Δ^*–rank.

The nonexistence of an infinite ascending chain of definable-by-Δ subgroups follows from roughly the same argument using that Δ^*–rank of any type is finite.

Definition 5.3.4. *If G is a stable group and Δ is finite (and contains $x = x$, so that G is definable-by-Δ), there is a unique minimal definable-by-Δ subgroup of G having finite index in G, which is called the* Δ–connected component of G.

Let G be a stable group. The connected component of G, *denoted G^o, is the intersection of all of the Δ–connected components, as Δ ranges over all finite sets of formulas. G is* connected *if $G^o = G$.*

If G is $\bigwedge$–definable over A, then G^o is also $\bigwedge$–definable over A. The connected component is a normal subgroup which is itself connected.

Lemma 5.3.3. *Let $(G, X, \star)$ be a stable group action, $p \in S^X(\mathfrak{C})$ and Δ a finite invariant set of formulas. Then, for any set A over which $p \upharpoonright \Delta$ is definable there is a formula $\psi(x)$ over A such that stab($p \upharpoonright \Delta$) is definable-by-ψ. Thus, the stabilizer of p is $\bigwedge$–definable over $Cb(p)$.*

Proof. Given $g \in G$, $g \in stab(p \upharpoonright \Delta)$ if and only if for all $\delta(x, y) \in \Delta$,

$$\forall y(\ \delta(x, y) \in p \upharpoonright \Delta \iff \delta(g \star x, y) \in p \upharpoonright \Delta\ )$$

Using that $p \upharpoonright \Delta$ is definable over A this equivalence yields a formula ψ over A defining $stab(p \upharpoonright \Delta)$.

We frequently need to work with types over sets rather than elements of $S^X(\mathfrak{C})$. The following is used to translate theorems about the action of G on $S^X(\mathfrak{C})$ into facts about the action on other types. (This generalizes Corollary 3.5.1.)

Lemma 5.3.4. *Let (G, X) be a stable group action, p, $q \in S^X(\mathfrak{C})$ and $a \in G$. Suppose that p and q are definable over A. Then,*

(1) *$q = ap$ if and only if*
(2) *there is a b realizing $p \upharpoonright A$ such that b is independent from a over A, ab realizes $q \upharpoonright A$ and ab is independent from a over A.*

Proof. Interpolating a few more equivalences will make the proof easy.

Claim. The following are equivalent.

(i) $q = ap$
(ii) for all sets $B \supset A \cup \{a\}$, $q \restriction B = ap \restriction B$;
(iii) there is a set $B \supset A \cup \{a\}$, $q \restriction B = ap \restriction B$.

This is proved like the claim in Corollary 3.5.1, replacing Morley rank and degree by Δ–rank and Δ–multiplicity. (In the detailed proof the reader should remember that for all finite Δ, $\mathrm{Mult}_\Delta(q) = \mathrm{Mult}_\Delta(q \restriction A) = 1$, by Remark 5.1.5.)

Turning to the lemma per se, that (1) implies (2) is just a matter of unraveling the notation. Now assume that a and b meet the conditions in (2). Let $B = A \cup \{a\}$ and let Δ be an invariant set. Since p and q are both definable over A and both b and ab are independent from a over A, b realizes $p \restriction B$ and ab realizes $q \restriction B$. Thus, $q = ap$, proving the lemma.

Since independence in a stable theory is defined with a scheme of ranks instead of a single rank a generic type cannot be defined as a type of maximal rank (as in ω–stable groups). Instead, genericity is defined in terms of forking independence (asking the reader to prove in the exercises that the two notions are equivalent for ω–stable groups).

Definition 5.3.5. *Let $(G, X, \star)$ be a stable group action, $\bigwedge$–definable over A, Φ the type defining X and $p \in S^X(\mathfrak{C})$.*

- *p is called* generic *if for all $a \in G$, ap does not fork over A.*
- *An arbitrary stationary type q is* generic *if $q|\mathfrak{C}$ is generic.*
- *An element a of X is said to be* generic over B *if $stp(a/B)$ is generic, shortening the term "generic over A" to simply "generic".*
- *If $X = G$ and $\star$ is left (right) translation we call p a left (right)* generic *of G.*

When we say "p is a generic of G", p is understood to be a left generic.

Notice that the translate of a generic is itself generic.

Lemma 5.3.5. *Let $(G, X, \star)$ be a stable group action which is $\bigwedge$–definable over A, Φ the type defining X and p a stationary type extending Φ. Then p is generic if and only if for all sets $B \supset A$ over which p is based, all a realizing $p|A$, and $g \in G$,*

$$g \underset{A}{\perp} a \Longrightarrow g \star a \underset{A}{\perp} g.$$

Proof. See Exercise 5.3.5.

Lemma 5.3.6. *Let (G, X) be a stable group action which is $\bigwedge$–definable over A and let Φ be the type defining X.*

(i) There is a generic type in $S^X(\mathfrak{C})$.

(ii) If $p \in S^X(\mathfrak{C})$ is generic, $G^o \subset stab(p)$. For any $p \in S^G(\mathfrak{C})$, $stab(p) \subset G^o$ (so when p is generic, $stab(p) = G^o$).

(iii) If $a, b \in G$ are $A-$independent generics then $b^{-1}a$ is $A-$independent from a and $A-$independent from b. Thus $b^{-1}a$ is a generic.

(iv) $S^{G^o}(\mathfrak{C})$ contains a left generic and a right generic.

Proof. To make the notation simpler we take A to be $\emptyset$.

(i) Let x be a variable in the same sort as X. A formula $\varphi(x)$ (over $\mathfrak{C}$) is called *small* if for some $a \in G$, $a\varphi(x)$ forks over $\emptyset$. Let $\Psi = \{\neg\varphi : \varphi \text{ small }\} \cup \Phi$ and suppose, towards a contradiction, that Ψ is inconsistent. Then, there are small formulas $\varphi_0, \ldots, \varphi_n$ such that every type in $S^X(\mathfrak{C})$ contains one of φ_i's. Let $\varphi_i = \varphi_i(x, a_i)$, where $\psi_i = \varphi_i(x, y_i)$ is over $\emptyset$ and let $\Delta = \{\psi_0, \ldots, \psi_n\}^*$. Let p_0 be an element of $S^X(acl(\emptyset))$ with $R_\Delta(p_0)$ maximal, and let $p \in S^X(\mathfrak{C})$ be the nonforking extension of p_0. Then one of $\varphi_0, \ldots, \varphi_n$, say $\varphi_i = \varphi$, is in p. Pick $a \in G$, so that $a\varphi$ forks over $\emptyset$. Since Δ is invariant, $R_\Delta(p) = R_\Delta(ap) = k$. By the maximality of $R_\Delta(p)$, $R_\Delta(q_0) = k$, where $q_0 = ap \restriction acl(\emptyset)$. There is only one complete $\Delta-$type over $\mathfrak{C}$ consistent with q_0 and having the same $\Delta-$rank as q_0, namely the restriction to Δ of $q_0|\mathfrak{C}$ (Lemma 5.1.8). Thus, $ap \restriction \Delta$ does not fork over $\emptyset$, contradicting that this type contains $a\varphi$. This proves the consistency of Φ. Using Corollary 5.1.8(iii) the reader can verify that any completion of Φ in $S^X(\mathfrak{C})$ is a generic, proving (i).

(ii) Let Δ be a finite invariant set of formulas and $H = stab(p, \Delta)$. There is a one-to-one correspondence between the cosets of H in G and $\{ap \restriction \Delta : a \in G\}$. Since each translate of p is a generic, and hence does not fork over $\emptyset$, there are at most $2^{|T|}$ many types in $\{ap \restriction \Delta : a \in G\}$. Since H is definable-by-ψ for some formula ψ over $acl(\emptyset)$, this bound on $[G : H]$ forces H to have finite index in G (see the exercises). Thus H contains G^o. This is true for any invariant Δ and $stap(p) = \bigcap_\Delta stab(p, \Delta^*)$, so $stab(p) \supset G^o$.

Let $p \in S^G(\mathfrak{C})$. The key observation needed to prove that $stab(p) \subset G^o$ is the following. The proof of the claim is assigned as an exercise at the end of the section.

Claim. Suppose $H \subset G$ is a definable-by-ψ subgroup of finite index, where ψ is a formula over $acl(\emptyset)$. Then for any coset B of H in G there is a formula θ over $acl(\emptyset)$ such that B is definable-by-θ. Thus,

$$\text{for any } a \text{ and } b \text{ in } G,\ stp(a) = stp(b) \implies aG^o = bG^o. \tag{5.3}$$

The proof is assigned to the reader in Exercise 5.3.6.

Let $g \in stab(p)$, B a set on which p is based, $q = p|B$ and a a realization of q independent from g over B. Since g is in the stabilizer of p, $g \cdot a$ also realizes q (Lemma 5.3.4) hence $stp(a) = stp(g \cdot a)$. By (5.3), $g = (g \cdot a) \cdot a^{-1} \in G^o$. Thus, $stab(p) \subset G^o$, completing the proof of (ii).

(iii) Since $stp(a)$ is generic and $a \perp b^{-1}$, $b^{-1}a$ is independent from b^{-1} (by Lemma 5.3.5) hence $b^{-1}a \perp b$. Similarly, the inverse of $b^{-1}a$ ($= a^{-1}b$) is independent from a^{-1}, hence $b^{-1}a \downarrow a$. By Lemma 5.3.4, $b(stp(b^{-1}a)|\mathfrak{C}) = stp(a)|\mathfrak{C}$; i.e., $stp(b^{-1}a)|\mathfrak{C}$ is a translate of a generic, hence a generic itself.

(iv) Let a and b be independent generics with respect to left translation which realize the same strong type p over $\emptyset$. By (5.3), $b^{-1}a$ is in G^o, and this is a generic element (by (iii)). The proof that there is a generic with respect to right translation in G^o is found simply by switching from left to right in the above proof.

Proposition 5.3.1. *Let (G, X) be a transitive stable group action and Φ the type defining X.*

(i) G acts transitively on the set of generics in $S^X(\mathfrak{C})$.

(ii) For any generic p and invariant set Δ, $R_\Delta(p) = R_\Delta(\Phi)$. For any finite set of formulas $\Delta(x)$ (where x is in the sort of X), $\{p \restriction \Delta : p \in S^X(\mathfrak{C})$ is generic$\}$ is finite.

Proof. For simplicity, suppose that G and X are $\bigwedge$−definable over $\emptyset$.

(i) Let p and q be generics in $S^X(\mathfrak{C})$, $p_0 = p|acl(\emptyset)$ and $q_0 = q|acl(\emptyset)$. Let a and b be independent realizations of p_0 and q_0, respectively, and $g = ba^{-1}$. Let $h \in G^o$ be a generic with respect to right translation which is independent from $\{g, a, b\}$. Since $stab(q) \supset G^o$ and $h \downarrow b$, $hga = hb$ realizes q_0. We claim that hg is independent from a. First, h is a right generic independent from g, so $hg \downarrow g$. Moreover, hg and a are independent over g (since h is independent from $\{g, a\}$ and hg is interalgebraic with h over g). By the transitivity of independence, $hg \downarrow a$, as needed. Since a is generic, $hg \downarrow hga$; i.e., $hg \downarrow hb$. By Lemma 5.3.4, $hg(p|\mathfrak{C}) = q|\mathfrak{C}$, proving (i).

(ii) Let Δ be an invaraint finite set of formulas. That there is a generic $p^* \in S^X(\mathfrak{C})$ with $R_\Delta(p^*) = R_\Delta(\Phi)$ is implicit in the proof of Lemma 5.3.6(i). All generic types $p \in S^X(\mathfrak{C})$ have the same Δ−rank (by (i)). Thus, the cardinality of $\{p \restriction \Delta : p \in S^X(\mathfrak{C})$ is generic $\}$ is bounded by (actually equal to) the Δ−multiplicity of Φ.

The following makes a good summary of what is known about generics in stable groups.

Corollary 5.3.1. *Let G be a stable group, $\bigwedge$−definable over A, and $p \in S^G(\mathfrak{C})$.*

(i) The following are equivalent.

(1) *The left stabilizer of p is G^o.*
(2) *p is a right generic.*
(3) *The right stabilizer of p is G^o.*
(4) *p is left generic.*

(ii) If a and b are generic, $G^oa = G^ob \implies stp(a/A) = stp(b/A)$.
(iii) If a, $b \in G^o$ are generic, then $tp(a) = tp(b) = q$ and q is stationary.
(iv) If p is generic then so is p^{-1}.

(v) p is generic if and only if p is a translate of the unique generic in $S^{G^o}(\mathfrak{C})$. *If p is generic and a realizes* $p|acl(\emptyset)$, *then* $a^{-1}p$ *is the generic in* $S^{G^o}(\mathfrak{C})$.

Proof. (i) (1) $\Longrightarrow$ (2) Assuming that the left stabilizer of p is G^o we need to show that each right translate of p does not fork over $\emptyset$. Since each right translate of p also has G^o as its left stabilizer it suffices to show that p does not fork over $\emptyset$. Suppose that p is based on A, $q = p|A$ and a realizes q. Let $I = \{ g_i : i < \omega \}$ be an independent set of right generics in G^o which is independent from $A \cup \{a\}$. Since $g_i \in G^o$ is independent from a over A, $g_i a$ also realizes q. In fact, the A–independence of a and I implies that $g_i a$ realizes the nonforking extension of p over $A \cup \{a\} \cup (I \setminus \{g_i\})$. Thus, $J = \{ g_i a : i < \omega \}$ is a Morley sequence in p over A, over which p is based by Lemma 5.1.17. However, since I is an independent set of right generics, J is independent over $\emptyset$ (see the exercises). Being the average type of J, p does not fork over $\emptyset$.

(2) $\Longrightarrow$ (3) and (4) $\Longrightarrow$ (1) are by Lemma 5.3.6(ii) and (3) $\Longrightarrow$ (4) is proved by switching left and right in the proof of (1) $\Longrightarrow$ (2).

(ii) Suppose that a and b are generics and $G^o a = G^o b$. Without loss of generality, $a \perp b$. For $c = ba^{-1}$, $c \perp a$ and $c \perp b$ (since b is generic). Letting $p = stp(a)|\mathfrak{C}$ and $q = stp(b)|\mathfrak{C}$, $cp = q$. Furthermore, $c \in G^o = stab(p)$ since a and b have the same coset with respect to G^o. Thus, $p = q$, proving (ii).

(iii) This follows immediately from (ii).

(iv) If p is a left generic, then p^{-1} is a right generic, hence also a left generic by (i).

(v) This is just a summary of previous results. The details are assigned as an exercise.

5.3.1 1–based Groups and Modules

Throughout Chapter 4 groups played a key role in our detailed analysis of uncountably categorical theories. Both the strongly minimal sets and the manner in which the universe is constructed from the strongly minimal sets are "simpler" when the theory is 1–based. Here, some of the results from Section 4.3.2 (principally Theorem 4.3.3) are restated in the stable context. The purpose is not to prove the more general results in detail, but to point the reader in their direction. The details can be found in [Pil]. They are not significantly different from the uncountably categorical case.

In the next subsection theories of modules are introduced as examples of 1–based theories.

Definition 5.3.6. *A stable theory is called* 1–based *if for all sets A and B, A is independent from B over* $acl(A) \cap acl(B)$.

As usual, the universal domain of a stable theory T is called 1–based if T is 1–based.

Remark 5.3.1. The following are equivalent for $\mathfrak{C}$ the universe of a stable theory.

(1) $\mathfrak{C}$ is 1−based.
(2) For all elements a and sets A, the canonical base of $tp(a/acl(A))$ is contained in $acl(a)$.

(The proof is virtually identical to the uncountably categorical case found in Remark 4.3.3.)

This equivalent definition explains the term "1−based". An uncountably categorical theory is 1−based when, given a stationary type p and q the nonforking extension of p in $S(\mathfrak{C})$, q is based on $acl(a)$ for any single a realizing p.

The concept of an abelian structure was specified in Definition 4.3.6 for *definable* groups. For $\bigwedge$−definable groups we need a slightly more complicated notion.

Definition 5.3.7. *Let G be a group, $\bigwedge$−definable over A in the universal domain of a complete theory. Let*

$$\mathcal{H} = \{ H : H \text{ is a subgroup of } G^n, \text{ for some } n, \\ \text{which is definable-by-}L \text{ over } acl(A) \}.$$

G is called an abelian structure *if for every $n < \omega$, every definable-by-L subset of G^n is equal to a boolean combination of cosets of elements of $\mathcal{H}$.*

Most results about 1−based groups depend on

Theorem 5.3.1. *Let G be a group, $\bigwedge$−definable in a 1−based (stable) theory. Then G is an abelian structure.*

The statements of the lemmas giving the proof of this theorem are virtually identical to the statements of corresponding lemmas in the proof of Theorem 4.3.3. The proofs of those earlier lemmas involved the action of a 1−based (uncountably categorical) group G on the types over G, Morley rank independence and canonical parameters. After substituting forking independence for Morley rank independence and canonical bases for canonical parameters the same proofs yield the generalized lemmas. The reader is referred to [Pil] for the details.

One preliminary lemma that deserves to be singled out is

Lemma 5.3.7. *Let G be a stable group, $\bigwedge$−definable over A. Then, G is an abelian structure if and only if*

(∗) *for any $n < \omega$ and $p \in S_n^G(G)$ there is a connected group $H \subset G^n$, definable-by-L over $acl(A)$, such that p is a left (or right) translate of the generic type of H.*

Corollary 5.3.2. *Let G be a stable group in a 1–based theory. Then, any connected $\bigwedge$–definable subgroup of G^n is $\bigwedge$–definable over $acl(\emptyset)$.*

Corollary 5.3.3. *A connected stable group in a 1–based theory is abelian.*

5.3.2 Modules

The purpose of this subsection is to introduce modules as natural examples of 1–based groups. Until stated otherwise we will work in a 1–sorted language.

Let R be a ring with identity. Here, the term R–module means *right-R–module*. The *language of R–modules* is $L_R = \{0, +, r\}_{r \in R}$, where 0 is a constant symbol, + a binary operator and each r is a unary operator. It is an elementary exercise to find a theory T_R in L_R whose models are exactly the right R–modules.

Fixing a ring R, $\varphi(\bar{v}) = \varphi(v_1, \ldots, v_n)$ is a *positive-primitive formula (*pp-formula*) over* 0 if it is equivalent in the theory T_R to one of the form

$$\exists w_1 \cdots w_l \bigwedge_{j=1}^{m} (\sum_{i=1}^{n} v_i r_{ij} + \sum_{k=1}^{l} w_k s_{kj} = 0),$$

where r_{ij}, $s_{kj} \in R$ and 0 is a tuple of 0's of the appropriate length. In matrix notation this can be written as

$$\exists w_1 \cdots w_l (v_1, \ldots, v_n, w_1, \ldots, w_l) \begin{pmatrix} r_{11} & \cdots & r_{1m} \\ \vdots & & \vdots \\ r_{n1} & \cdots & r_{nm} \\ s_{11} & \cdots & s_{1m} \\ \vdots & & \vdots \\ s_{l1} & \cdots & s_{lm} \end{pmatrix} = 0. \tag{5.4}$$

This is compressed to $\exists \bar{w}(\bar{v}\bar{w})H = 0$, where

$$H = \begin{pmatrix} \mathbf{r} \\ \mathbf{s} \end{pmatrix}$$

and $\mathbf{r}$, $\mathbf{s}$ are the matrices of r_{ij}'s and s_{kj}'s. In yet another form $\varphi(\bar{v})$ can be written as $\exists \bar{w}(\bar{v}\mathbf{r} = -\bar{w}\mathbf{s})$. As this final form suggests, a *pp*–formula can be thought of as a generalized divisibility condition. (An element a is said to be *divisible by s* if there is a b such that $a = bs$, equivalently a satisfies the formula $\exists w(v = ws)$.)

For A a subset of some R–module the term *pp–formula over A* is defined as above except that on the right hand side of (5.4) there is a $1 \times m$ row vector of elements of A. For $\varphi(\bar{v})$ a *pp*–formula over $\bar{a}$ there is $\psi(\bar{v}, \bar{w})$ a *pp*–formula over 0 such that $\varphi(\bar{v}) = \psi(\bar{v}, \bar{a})$. In general, the term *pp*–formula refers to one with nonzero parameters.

Remark 5.3.2. Let R be a ring, $\varphi(\bar{v})$ a *pp*–formula over 0 and M an R–module.

(i) In the exercises the reader is asked to verify that $\varphi(M)$ is a subgroup of M^n, where n is the length of $\bar{v}$.

(ii) Furthermore, if r is in the center of R (i.e., the subring of elements commuting with every element of R) then $\models \varphi(\bar{a}) \implies \models \varphi(\bar{a}r)$, where $(a_1, \ldots, a_n)r = \bar{a}r = (a_1 r, \ldots, a_n r)$. Thus, if R is commutative $\varphi(\bar{v})$ defines a submodule of M. If R is not commutative the subgroup defined by a *pp*–formula need not be a submodule. (Consider, for example, the ring R of 2×2 matrices over a field K with $M = R_R$. Let

$$e = \begin{pmatrix} 1 & 0 \\ 0 & 0 \end{pmatrix}$$

and $\varphi(v) = \exists w(v = we)$. Then

$$\varphi(M) = \begin{pmatrix} K & 0 \\ K & 0 \end{pmatrix},$$

is a left-ideal but not a right-ideal, hence not a submodule of M.)

(iii) More generally, if $\varphi(\bar{v}, \bar{w})$ is a *pp*–formula over 0 and M is an R–module then $\varphi(M, \bar{a})$ is empty or a coset of the subgroup $\varphi(M, 0)$ of M^l, where l is the length of $\bar{v}$ and 0 denotes a tuple of 0's of the same length as $\bar{w}$ (exercise). When $\varphi(M, \bar{a}) \neq \emptyset$, $\varphi(\bar{v}, \bar{a})$ is equivalent to $\varphi(\bar{v} - \bar{b}, 0)$, for some $\bar{b}$.

Theorem 5.3.2 (Elimination of Quantifiers). *Let R be a ring, L_R the language of R–modules and T_R the theory of right R–modules. Then for any formula $\varphi(\bar{v})$ in L_R without parameters there is $\varphi'(\bar{v})$, a boolean combination of pp–formulas, such that $T_R \models \forall \bar{v}(\varphi(\bar{v}) \longleftrightarrow \varphi'(\bar{v}))$.*

For a proof of this theorem by Baur, Garavaglia and Monk see [Zie84] or [Pre88].

In stability theory complete types are often more useful than formulas. What does the elimination of quantifiers have to say about complete types?

Definition 5.3.8. *Let M be a module over a ring R, $A \subset M$ and $p \in S_n(A)$.*

(i) The pp–part *of p, denoted p^+, is* $\{\varphi \in p : \varphi$ is a *pp*–formula over $A\}$, *while* $p^- = \{\neg\varphi \in p :\ \varphi$ is a *pp* – formula over $A\}$.

(ii) If $\bar{a}$ is a sequence from M the pp–type *of $\bar{a}$* over *A, $pp(\bar{a}/A)$ is $tp(\bar{a}/A)^+$. Some contexts involve modules of different complete theories (i.e., we are not working in a single universal domain), in which case the notation $pp^M(\bar{a}/A)$ is used.*

(iii) A pp–type *over A is a consistent set Γ of pp–formulas over A. Γ is* complete *(in M) if it is the pp–type over A of a sequence from the universal domain of $Th(M)$.*

Let M be a module over a ring R and p a complete type in $T = Th(M)$. By the elimination of quantifiers down to pp–formulas, p is equivalent to $p^+ \cup p^-$ in T and for $\bar{a}$, $\bar{b}$ sequences from M, $tp(\bar{a}/A) = tp(\bar{b}/A) \iff pp(\bar{a}/A) = pp(\bar{b}/A)$.

We now restrict our attention to 1–types, although the same facts hold for n–types since M^n is also an R–module.

For $p \in S_1(M)$ let

$$\Phi_p = \{\varphi(x) : \varphi \text{ is a } pp\text{–formula over } 0 \text{ and for some } a \in M,\ \varphi(x-a) \in p\}.$$

Equivalently, Φ_p is the set of pp–formulas $\varphi(x)$ over 0 such that for a realizing p, the coset $a + \varphi(\mathfrak{C})$ is represented in M. Notice that the difference of two realizations of p is in $\Phi_p(\mathfrak{C})$.

Given a pp–formula $\psi(x, \bar{b}) \in p$ there is a pp–formula $\varphi(x)$ over 0 and an $a \in M$ such that $\varphi(x - a)$ is equivalent to $\psi(x, \bar{b})$. Thus,

$$p^+ \text{ is equivalent to } \{\varphi(x-a) \in p : \varphi \in \Phi_p\}. \tag{5.5}$$

Working in M^{eq} let $C(p) = \{\ulcorner\varphi(x-a)\urcorner : \varphi(x-a) \in p \text{ and } \varphi \in \Phi_p\}$. Equivalently, $C(p)$ is the set of $b \in M^{eq}$ such that b is the name of a coset of a subgroup H of $\mathfrak{C}$, pp–definable over 0, and the formula expressing "$x+H = b$" is in p. Notice that $C(p) \subset dcl(c)$ for any c realizing $p \restriction C(p)$. By (5.5),

$$\text{For } p,\ q \in S_1(M),\ p = q \iff p^+ = q^+ \iff C(p) = C(q). \tag{5.6}$$

As a first consequence of this reduction:

Corollary 5.3.4. *The complete theory of an infinite module is stable.*

Proof. Let M be an R–module of cardinality κ, where $\kappa = \kappa^{|R|}$. For any $p \in S_1(M)$, $|C(p)| \leq |R|$. In M^{eq} there are $\leq \kappa^{|R|}$ sets of the form $C(p)$, as p ranges over $S_1(M)$. Thus, by (5.6), $|S_1(M)| \leq \kappa^{|R|} = \kappa$. Similarly, $|S_n(M)| \leq \kappa$ for all $n < \omega$, proving the corollary.

Let M be an infinite R–module, $p \in S_1(M)$ and identify p with its unique extension in $S_1(M^{eq})$. Since Φ_p is a pp–type over 0, $\Phi_p(\mathfrak{C})$ is a subgroup of $\mathfrak{C}$. Suppose there is an $a \in M$ realizing $p \restriction C(p)$ (as there will be if M is $|R|^+$–saturated). Thus, a and a realization b of p have the same coset with respect to $\varphi(\mathfrak{C})$ for any $\varphi \in \Phi_p$, in other words, $a + \Phi_p(\mathfrak{C}) = b + \Phi_p(\mathfrak{C})$.

It is a short trip from (5.6) to

Lemma 5.3.8. *For $p \in S_1(\mathfrak{C})$, $Cb(p) = dcl(C(p))$ in $\mathfrak{C}^{eq}$.*

Proof. It suffices to show that for $f \in \text{Aut}(\mathfrak{C})$, $f(p) = p$ if and only if f fixes $C(p)$ pointwise. Observe that for any $f \in \text{Aut}(\mathfrak{C})$, $f(C(p)) = C(f(p))$. Thus, if $f \in \text{Aut}(\mathfrak{C})$ and $f(p) = p$, f pointwise fixes $C(p)$. Conversely, if $f \in \text{Aut}(\mathfrak{C})$ is the identity on $C(p)$, then $C(p) = C(f(p))$, so $p = f(p)$ by (5.6).

Thus, $p \in S_1(\mathfrak{C})$ is also determined by $C(p)$ in the sense that it is definable over $C(p)$.

Corollary 5.3.5. *The theory of an infinite module is* $1-$*based.*

Proof. Suppose $p = tp(a/A)$ is stationary and let q be the nonforking extension of p over $\mathfrak{C}$. Without loss of generality, p is a $1-$type. Then $Cb(q) \subset dcl(A)$, so a realizes $q \upharpoonright Cb(q) \supset q \upharpoonright C(q)$. Since $C(q) \subset dcl(a)$ and $Cb(q) \subset dcl(C(q))$ (by Lemma 5.3.8), $Cb(q) \subset dcl(a)$.

One step in the proof that a $1-$based group G is an abelian structure is to show that any $p \in S^G(\mathfrak{C})$ is a translate of the generic in $stab(p)$ and $stab(p)$ is connected and definable over $acl(\emptyset)$. Below (in Proposition 5.3.2) we give an independent proof of this fact when G is a module.

The group $\Phi_p(\mathfrak{C})$ is one we are already very familiar with:

Lemma 5.3.9. *Let* $\mathfrak{C}$ *be the universal domain of a complete theory of modules, For* $p \in S_1(\mathfrak{C})$, $\Phi_p(\mathfrak{C}) = stab(p)$.

Proof. Let M be a saturated model on which p is based and $q = p|M$. Note: $q \supset p \upharpoonright C(p)$ and for any $\varphi \in \Phi_p$ there is a $b \in M$ such that $\varphi(x-b) \in q$. Let s be the type over $C(p)$ such that $stab(p) = s(\mathfrak{C})$. Since M is saturated, to prove that $s(\mathfrak{C}) = \Phi_p(\mathfrak{C})$ it suffices to show that $H = s(M) = \Phi_p(M)$. If $g \in H$ and a realizes q, then $g + a$ also realizes q. Thus, $g \in \Phi_p(M)$ since the difference of any two realizations of q is in $\Phi_p(\mathfrak{C})$; i.e., $H \subset \Phi_p(M)$.

Now suppose that $g \in \Phi_p(M)$ and let $r = tp(g + a/M)$. To prove that $r = q$ we will show that $C(q) = C(r)$ and quote (5.6). Given $\varphi \in \Phi_p$, if $\varphi(x-b) \in q$, then $\models \varphi(a-b)$ and $\varphi(g)$, so $\models \varphi(g+a-b)$. Thus, $\Phi_r \supset \Phi_p$. Conversely, given $\varphi \in \Phi_r$ with $\varphi(x-b) \in r$, $\models \varphi(a-(b-g))$, hence $\varphi \in \Phi_p$. We conclude that $\Phi_p = \Phi_r$. Since $g \in \Phi_p(M)$, a and $g + a$ have the same coset with respect to $\varphi(\mathfrak{C})$ for every $\varphi \in \Phi_p$, hence $C(q) = C(r)$. This proves (by (5.6) that $r = q$. Thus, $g \in H$; i.e., $\Phi_p(M) \subset H$.

Corollary 5.3.6. *Let* $\mathfrak{C}$ *be the universal domain of a complete theory of modules and* $p \in S_1(\mathfrak{C})$. *Then* $stab(p)$ *is* $\bigwedge-$*definable over* $\emptyset$.

Proposition 5.3.2. *Let* $\mathfrak{C}$ *be the universal domain of a complete theory of* $R-$*modules and* $p \in S_1(\mathfrak{C})$. *Then* p *is a translate of a generic type in* $stab(p)$. *Furthermore,* $stab(p)$ *is connected.*

Proof. The connectedness of the stabilizer will be proved later. The bulk of the proof lies in showing

Claim. Let a and b to be realizations of $p|C(p)$ which are independent over $C(p)$. Then $a - b$ is a generic of $stab(p)$ which is independent from a and from b.

By Lemma 5.3.9, $a - b \in G = stab(p)$. It remains to show that $a - b$ is a generic of G independent from a and independent from b. This is accomplished by finding a', b' with the desired properties which realize $tp(ab/C(p))$. We will use repeatedly throughout the proof the fact that $C(p)$ is contained in the definable closure of any realization of $p|C(p)$. As a first step suppose that a' is a realization of $p|C(p)$ and g is a generic of $G = stab(p)$ which is independent from a'. Then,

$$g + a' \text{ realizes } p|C(p). \tag{5.7}$$

For a and b as given let $g \in G$ be generic over $\{a, b\}$. Then $(g + a) - b = g+(a-b)$ is a generic of G over $\{a, b\}$. Letting h be a generic of G independent from $\{a, b, g\}$, $(g + a) - (h + b) = (g + a - b) - h$ is also generic over $\{a, b, g\}$. Let $a' = g+a$ and $b' = h+b$, both realizations of $p|C(p)$ with $a' - b'$ a generic of G. We claim that

$$a' \perp a' - b',\ b' \perp a' - b' \text{ and } a' \underset{C(p)}{\perp} b'. \tag{5.8}$$

(Since $(g + a) - (h + b)$ is independent from $\{a, b, g\}$ it is independent from $g + a$. By symmetry and transitivity of independence, g is independent from $\{a, b, h\}$, hence the same argument shows that $(g+a)-(h+b)$ is independent from $h + b$. Since a and b are independent over $C(p) \subset acl(a) \cap acl(b)$ the independence of $\{g, h, \{a, b\}\}$ implies that $\{g, a\}$ and $\{h, b\}$ are independent over $C(p)$. Hence, $a' = g + a$ and $b' = h + b$ are independent over $C(p)$.)

By (5.7) a' and b' are realizations of $p|C(p)$. Since they are independent over $C(p)$, $tp(a', b'/C(p)) = tp(a, b/C(p))$. Thus there is an automorphism of $\mathfrak{C}^{eq}$ fixing $C(p)$ and taking a' to a and b' to b. The conditions in (5.8) show that a and b meet the requirements of the claim.

Now let M be a $|R|^+$–saturated model containing $C(p)$, a a realization of $p|C(p)$ in M and b a realization of $p|M$. Let $g = b - a$ and $q = tp(g/M)$. By the claim, g is a generic of $stab(p)$ and is independent from a over $\emptyset$. Since p is based on a, b is independent from M over a, hence g is also independent from M over a. By transitivity, g is independent from M over $\emptyset$ implying that q is a generic type of $stab(p)$. Since $a + q = p|M$, p is a translate of a generic of $stab(p)$, completing the proof of the main part of the proposition.

To prove the connectedness of $G = stab(p)$ first remember that a type in $S_1(\mathfrak{C})$ and its translates have the same stabilizer. Thus, if $q \in S_1(\mathfrak{C})$ is a generic of G such that q is a translate of p, $stab(q) = G$. In fact, since the action of G on the generics is transitive, any generic of G has G as its stabilizer. Let $r \in S_1(\mathfrak{C})$ be an arbitrary generic of G. By Corollary 5.3.1(v), for a a realization of $r|acl(\emptyset)$, $-a + r$ is the generic in G^o. Since $-a$ is also in the stabilizer of r, $r = -a + r$ must be the generic in G^o. That is, G is connected.

This yields the particularly simple picture of a module $\mathfrak{C}$. Any $p \in S_1(\mathfrak{C})$ is a translate of a generic of a subgroup $\Phi(\mathfrak{C})$ ($= stab(p)$), where Φ is a pp–type over $\emptyset$.

The proposition translates many properties of types into properties of their stabilizers. For example, if $p \subset q$ are stationary types in a module, then q is a forking extension of p if and only if $stab(q|\mathfrak{C})$ has infinite index in $stab(p|\mathfrak{C})$ (left to the exercises).

Historical Notes. Historically speaking, the results of this section have the same source as those in the subsection on $\omega-$stable groups. We owe much of our knowledge about generic types outside of the $\omega-$stable setting to Poizat [Poi81], whose work was carried on by Berline and Lascar in [BL86] and [Ber86]. Lemma 5.3.2 is due to Baldwin and Saxl [BS76]. The detailed treatment here is taken from Hrushovski's dissertation [Hru86], which is also found in [Hru90b].

The logical analysis of the theory of modules begins with Szmielew's quantifier elimination theorem for abelian groups [Szm55]. The work of Eklof, Fisher, Sabbagh and Baur (see [EF72], [ES71]) took a more model-theoretic approach. The elimination of quantifiers theorem for modules is due independently to Baur [Bau76], Garavaglia and L. Monk.

Exercise 5.3.1. Suppose the types Φ and Ψ are such that $(\Phi(\mathfrak{C}), \Psi(\mathfrak{C}), \star)$ defines a group action in $\mathfrak{C}$, where Φ, Ψ and $\star$ are all over $\emptyset$. Given any model M prove that $(\Phi(M), \Psi(M), \star)$ is a group action.

Exercise 5.3.2. Prove Lemma 5.3.1.

Exercise 5.3.3. Let G be an $\omega-$stable group. Prove that $p \in S^G(G)$ is generic (as defined in this section) if and only if $MR(p) = MR(G)$.

Exercise 5.3.4. Suppose that G is a stable group in $\mathfrak{C}$ and H is a subgroup of G which is definable-by-ψ (for some formula ψ) with $[G : H] < |\mathfrak{C}|$. Prove that $[G : H]$ is finite.

Exercise 5.3.5. Prove Lemma 5.3.5.

Exercise 5.3.6. Prove the first claim in the proof of Lemma 5.3.6(ii).

Exercise 5.3.7. Suppose that $(G, X, \star)$ is a stable group action, I is an independent set of generics in X and $a \in G$ is independent from I. Show that $\{ a \star b : b \in I \}$ is an independent set of generics.

Exercise 5.3.8. Suppose that $(G, \cdot)$ and $(G, \odot)$ are both stable groups. Show that $(G, \cdot)$ is connected if and only if $(G, \odot)$ is connected.

Exercise 5.3.9. Prove Corollary 5.3.1(v).

Exercise 5.3.10. Prove that a *pp*−formula over 0 defines a subgroup of a modules. More generally, the *pp*−formula $\varphi(x, \bar{a})$ defines a coset of $\varphi(x, 0)$ (unless it is inconsistent).

Exercise 5.3.11. Prove: If $p \subset q$ are stationary types in a module, then q is a forking extension of p if and only if $stab(q|\mathfrak{C})$ has infinite index in $stab(p|\mathfrak{C})$.

5.4 Saturated Models

It follows directly from Lemma 2.2.6 that if T is κ–stable, there is a saturated model of T of cardinality κ^+. In fact, roughly the same proof gives the existence of a saturated model of cardinality κ when κ is regular and T is κ–stable. In the next proposition the restriction to regular cardinals will be removed and the assumption of κ–stability is proved to be necessary when κ is sufficiently large.

Proposition 5.4.1. *If the theory T is κ–stable, then T has a saturated model of cardinality κ. If, on the other hand, $\kappa \geq \lambda(T)$ and T is not κ–stable then T does not have a saturated model of cardinality κ.*

Proof. Let T be κ–stable. Then $cf(\kappa) \geq \kappa(T)$ since $\kappa^{cf(\kappa)} > \kappa$ and $\kappa^{<\kappa(T)} = \kappa$ (by Theorem 5.2.1).

A saturated model of cardinality κ is found using

Claim. There is a model M of T of cardinality κ such that for all $A \subset M$ of cardinality $< \kappa(T)$ and $p \in S(acl(A))$, M contains a Morley sequence over $acl(A)$ in p of cardinality κ.

The model M is constructed via an elementary chain, M_i, $i < \kappa$. Let M_0 be any model of T of cardinality κ. Given M_i, let M_{i+1} be an elementary extension of cardinality κ such that for all $p \in S(M_i)$, M_{i+1} contains a Morley sequence in p of cardinality κ. Such a model exists since $|S(M_i)| = \kappa$ and T is κ–stable. If j is a limit ordinal, let $M_j = \bigcup_{i<j} M_i$. Let $M = \bigcup_{i<\kappa} M_i$. If A is a subset of M of cardinality $< \kappa(T)$, then there is an $i < \kappa$ such that $A \subset M_i$ (since $cf(\kappa) \geq \kappa(T)$). By construction, any strong type over A, M contains a Morley sequence of cardinality κ, proving the claim.

To prove that M is saturated let $A \subset M$ have cardinality $\lambda < \kappa$ and let $p \in S(A)$. Let $B \subset A$ be a set of cardinality $< \kappa(T)$ over which p does not fork and let $q \in S(acl(B))$ be such that $q|acl(A) \supset p$. It suffices to find an a realizing q which is independent from A over B. By the claim there is a set $I \subset M$ of cardinality κ which is a Morley sequence over $acl(B)$ in q. By Proposition 5.2.1 there is a set $J \subset I$ of cardinality $< \kappa(T) + |A|^+$ (if $\kappa(T)$ is regular) and $\leq \kappa(T) + |A|$ (otherwise) such that A is independent from I over $B \cup J$. In any case, $|J| < \kappa$ (since $\kappa \geq \kappa(T)$ and $\kappa = \kappa(T)$ can only occur when $\kappa(T)$ is regular by Corollary 5.2.1). Then $a \in I \setminus J$ is a realization of q which is independent from A over $J \cup B$, hence independent from A over B (by the transitivity of independence). Thus, a realizes p, proving the first part of the proposition.

Now suppose that $\kappa \geq \lambda(T)$ and T is not κ–stable. By The Stability Spectrum Theorem (Theorem 5.2.1), $\kappa < \kappa^{<\kappa(T)}$ and $\kappa \geq \kappa(T)$ (since $\lambda(T) \geq \kappa(T)$). The proof of the nonexistence of a saturated model of cardinality κ is split into two cases, the first reducing largely to cardinal arithmetic.

Case 1. $cf(\kappa) \geq \kappa(T)$.

Choose $\mu < \kappa(T)$ such that $\kappa^\mu > \kappa$. Express κ as $\sup_{i<cf(\kappa)}\kappa_i$, where $\kappa_i < \kappa$. Every subset of κ of cardinality $< \kappa(T)$ is contained in one of the κ_i's by the assumption of this case, hence $\kappa^\mu = \sup_{i<cf(\kappa)}\kappa_i^\mu$. Thus, there is $\kappa_i < \kappa$ with $\kappa_i^\mu > \kappa$. Since $\mu < \kappa(T)$ the proof of Lemma 5.2.2 yields a set A of cardinality κ_i such that $|S(A)| = \kappa_i^\mu$. Thus, there cannot be a saturated model of cardinality κ.

Case 2. $cf(\kappa) < \kappa(T)$.

Let $\lambda = cf(\kappa)$ and $\kappa = \sup_{i<\lambda}\kappa_i$, where $\kappa_i < \kappa$. Suppose, towards a contradiction, that M is a saturated model of cardinality κ. Write M as $\bigcup_{i<\lambda} M_i$, where M_i is a submodel of M of cardinality κ_i and when $\delta < \lambda$ is a limit ordinal, $M_\delta = \bigcup_{i<\delta} M_i$. Since $\lambda < \kappa(T)$ there is a chain of sets A_i, $i \leq \lambda$, such that

- $A_{i+1} \setminus A_i$ is finite,
- $A_\delta = \bigcup_{i<\delta} A_i$, if δ is a limit ordinal, and
- there is a $p \in S(A_\lambda)$ such that $p \restriction A_{i+1}$ forks over A_i, for all $i < \lambda$.

Since M is saturated and $|M_i| < \kappa$ we can choose the A_i's and M_i's so that $A_i \subset M_i$, for all $i < \lambda$, and A_{i+1} is independent from M_i over A_i. Since $|A_\lambda| = \lambda < \kappa(T) \leq \kappa$ and M is saturated there is an $a \in M$ realizing p. Find $i < \lambda$ such that $a \in M_i$. This contradicts that $p \restriction A_{i+1}$ forks over A_i and A_{i+1} is independent from M_i over A_i, proving the proposition.

In the exercises the reader is asked to investigate when the union of an elementary chain of saturated models of a stable theory is also saturated.

5.4.1 a−models

One useful property of t.t. theories is that every type over a model is based on a finite subset of the model. The same may not be true in a properly superstable theory, even though every type does not fork over a finite set. Consider, for example, the theory T in the language $L = \{E_i(x, y) : i < \omega\}$ which says that each E_i is an equivalence relation with only infinite classes, $E_0(x, y)$ is $x = x$ and each E_i−class is refined into two E_{i+1}−classes. Let M be any countable model. There is a one-to-one correspondence between stationary types over $A \subset M$ and elements of $S(M)$ which are definable over A. Since T is small there can only be countably many types over M definable over a finite subset of M, while there are $2^{\aleph_0}$ types over M. If, however, we chose M to be $\aleph_1$−saturated instead of countable, $tp(a/M)$ would be based on any $b \in M$ realizing $tp(a/acl(\emptyset))$. The following property of a model M is potentially weaker than κ^+−saturation, but is enough to guarantee that every complete type over M is based on a subset of M of cardinality $< \kappa^+$.

Definition 5.4.1. *A model M is* almost κ−saturated, *(a, κ)−saturated, for short, if for all sets $A \subset M$ of cardinality $< \kappa$, every strong type over A is realized in M. When T is stable the terms almost $\kappa(T)$−saturated model,* almost saturated model, a–saturated model, *and* a–model *are used interchangeably.*

Restating the definition, M is (a, κ)–saturated if for all $A \subset M$ of cardinality $< \kappa$, every element of $S(acl(A))$ is realized in M. This form of the definition makes it clear that a model which is $(\kappa + |T|^+)$–saturated is (a, κ)–saturated and an (a, κ)–saturated model is κ–saturated. In any context using almost κ–saturated models for some $\kappa > \kappa(T)$ we can usually reach the same goal using a–models. The term almost κ–saturated is a little misleading because an (a, κ)–saturated model is certainly κ–saturated. The "almost" comes from the fact that an (a, κ)–saturated model must realize every type "almost over a subset A"; i.e., a type consisting of formulas almost over A.

These models do have the desired property:

Lemma 5.4.1. *Let T be stable and M an a–model. For any $p \in S(M)$ there is a subset of M of cardinality $< \kappa(T)$ on which p is based.*

Proof. Let $A \subset M$ be a set of cardinality $< \kappa(T)$ such that p does not fork over A. Since M is an a–model there is $a \in M$ realizing $p \upharpoonright acl(A)$. By Corollary 5.1.8(i) p is based on $A \cup \{a\}$, a set of cardinality $< \kappa(T)$.

Almost saturated models will be used in Section 5.6.2 to develop a good theory of domination. The "dimension theory" developed in Section 5.6 is easiest to apply in the context of a–models. Indeed, many of the theorems in [She90] related to Morley's Conjecture apply to the class of a–models of a superstable theory.

Types over a–models are determined by the elements realizing a subtype (compare this with Lemma 5.1.9):

Lemma 5.4.2. *Let M be an a–model, A a subset of M of cardinality $< \kappa(T)$, $q \in S(A)$ and $b \subset q(\mathfrak{C})$. Then, $tp(b/A \cup q(M)) \models tp(b/M)$.*

Proof. Let e be an element of M and b' an arbitrary realization of $tp(b/A \cup q(M))$. We need to show that b and b' have the same type over e. Let E be a subset of $q(M)$ of cardinality $< \kappa(T)$ such that e is independent from $q(M) \cup A$ over $E \cup A$. In fact, e is independent from $q(\mathfrak{C}) \cup A$ over $E \cup A$. (Suppose that e depends on the finite $c \subset q(\mathfrak{C})$ over $E \cup A$. Since M is $\kappa(T)$–saturated there is a $c' \subset q(\mathfrak{C})$ realizing $tp(c/E \cup A \cup \{e\})$; a contradiction.) Let $b_0 \in M$ realize $stp(b/E \cup A)$. Since this strong type is based on $\{b_0\} \cup E \cup A \subset q(M) \cup A$, $tp(b/q(M) \cup A)$ is stationary. Thus, b and b' have the same type over $q(M) \cup A \cup \{e\}$, proving the lemma.

This alternative kind of saturation is further connected to ordinary saturation in

Lemma 5.4.3. *(i) If T is t.t. then a model M is $(a, \kappa)-$saturated if and only if M is $\kappa-$saturated. Thus, the $a-$models of a t.t. theory are exactly the $\aleph_0-$saturated models.*

(ii) If T is a countable stable theory then an $\aleph_1-$saturated model is $a-$saturated. Thus, if T is properly stable a model is $\aleph_1-$saturated if and only if it is $a-$saturated.

Proof. (i) Suppose that M is $\kappa-$saturated, $A \subset M$ has cardinality $< \kappa$ and $p \in S(acl(A))$. Since T is t.t. a complete type has only finitely many nonforking extensions. Thus, there is a set B, $A \subset B \subset acl(A)$, of cardinality $< \kappa$ such that p is implied by its restriction to B. Since M is $\kappa-$saturated there is an $a \in M$ realizing $p \upharpoonright B$, hence also p.

(ii) If T is countable, $\kappa(T)$ is $\leq |T|^+ = \aleph_1$. Thus, an $\aleph_1-$saturated model is $(\kappa(T) + |T|^+)-$saturated, hence $a-$saturated. If T is properly stable $\kappa(T)$ is $\aleph_1$, hence an $a-$model must be $\aleph_1-$saturated.

Corollary 5.4.1. *Let T be a countable stable theory and M a model of T. If T is properly stable then M is an $a-$model if and only if M is $\aleph_1-$saturated. If T is $\aleph_0-$stable then M is an $a-$model if and only if M is $\aleph_0-$saturated.*

Remark 5.4.1. If T is a countable properly superstable theory and M is $\aleph_0-$saturated, M may or may not be an $a-$model, depending on detailed properties of T.

Historical Notes. Proposition 5.4.1 was proved for $\omega-$stable theories by Harnik in [Har73] and generalized to stable theories by Shelah in [She90, III, 3.10 and 3.12]. The notion of an $a-$model is due to Shelah [She90].

Exercise 5.4.1. If T is a countable properly superstable theory what is the least cardinal in which T has an $a-$model?

Exercise 5.4.2. Let T be stable, M an $a-$model and λ an infinite cardinal. Show that M is $\lambda-$saturated if and only if for every infinite set of indiscernibles $I \subset M$ there is a set of indiscernibles J, $I \subset J \subset M$, of cardinality λ.

Exercise 5.4.3. Let T be a countable stable theory and M_i, $i < \aleph_1$, an elementary chain of $\lambda-$saturated models (for some λ). Prove that $M = \bigcup_{i<\aleph_1} M_i$ is also $\lambda-$saturated.

Exercise 5.4.4. Let T be a stable theory and $\{M_i : i < \delta\}$ a chain of $\lambda-$saturated models, where $\kappa(T) \leq cf(\delta)$. Show that $M = \bigcup_{i<\delta} M_i$ is $\lambda-$saturated.

5.5 Prime Models

In most applications of prime models an essential ingredient is their uniqueness. For prime models in countable theories this was proved in Section 2.1, where we also remarked that an uncountable theory may have prime models which are not isomorphic or a nonatomic prime model. We proved in Lemma 3.1.5 the existence of prime models over arbitrary sets in an $\aleph_0$−stable theory without addressing the uniqueness issue. In the proofs of certain results, such as Corollary 3.1.4, the nonexistence of Vaughtian pairs was used in place of the uniqueness of prime models to see that prime models over conjugate sets are isomorphic. Outside of the context of uncountably categorical theories, no such replacement is possible. The uniqueness of prime models (over sets) in any t.t. theory is proved in this section.

Prime models over sets may not exist in theories which are not totally transcendental. However, we will see that in stable theories there are prime models over sets relative to the class of a−models. (The $\aleph_0$−prime models of Section 3.1 are an example of such models.) We will also prove the uniqueness of these so-called a−prime models.

5.5.1 Prime Models in a t.t. Theory

The proof of the existence of prime models over sets in an $\aleph_0$−stable theory did use the countability of the theory, so another proof which handles all t.t. theories is needed. The models we find are not only prime, but are constructed as such in the following sense.

Notation. If $\{ a_\beta : \beta < \alpha \}$ is a set of elements indexed by an ordinal α and $\beta < \alpha$, $A_\beta = \{ a_\gamma : \gamma < \beta \}$.

Definition 5.5.1. *(i) A set* $\{ (a_\beta, \varphi_\beta) : \beta < \alpha \}$ *is called a* t−construction over A *if, for all* $\beta < \alpha$, φ_β *is a formula over* $A \cup A_\beta$ *which isolates* $tp(a_\beta / A \cup A_\beta)$. *When the particular isolating formulas are irrelevant we may simply write* $\{ a_\beta : \beta < \alpha \}$ *for the* t−*construction. A set is* t−constructible over A *if some enumeration of it is a* t−*construction over* A.

(ii) A model $M \supset A$ *(of the theory in question) is called* strictly prime over A *if it is* t−*constructible.*

Remark 5.5.1. In Shelah's terminology isolated types are in the class $\mathbf{F}^{\mathbf{t}}_{\aleph_0}$, meaning that the type is isolated by a subtype of cardinality $< \aleph_0$. Moreover, a t−construction is called an $\mathbf{F}^{\mathbf{t}}_{\aleph_0}$−construction in [She90].

Remark 5.5.2. If $\mathfrak{C}$ is t.t. and A is a set, the theory of $\mathfrak{C}$ with constants for the elements of A is also t.t. Thus, a theorem stated to hold for the prime models of any t.t. theory actually holds for the prime models over sets in any t.t. theory. In practice, though, it is common to state a major result about the prime models over a set A in a t.t. theory, and say "Without loss of generality, $A = \emptyset$" when beginning the proof. In lemmas we may omit the set in the statement.

The results of the section are summarized in

Theorem 5.5.1. *Suppose that T is a t.t. theory and A is a set.*
(i) There is a strictly prime model over A.
(ii) Any two strictly prime models over A are isomorphic over A.
(iii) The following are equivalent:

(1) *M is strictly prime over A.*
(2) *M is prime over A.*
(3) *M is atomic over A and does not contain an uncountable set of indiscernibles over A.*

Corollary 5.5.1. *If T is t.t. and M, N are prime models of T, then $M \cong N$.*

The first part of the theorem is handled rather easily:

Lemma 5.5.1. *If T is a t.t. theory then there is a strictly prime model M. In addition, M is prime and atomic.*

Proof. The main point is contained in

Claim. For all sets B the isolated points are dense in $S(B)$.

Let φ be a formula over B. Let $p \in S(B)$ be an element containing φ which has minimal Morley rank α. Let $\psi \in p$ be such that p is the only element of $S(B)$ containing ψ and having Morley rank α. Since we can assume that ψ implies φ the minimal rank assumption shows that ψ isolates p, to prove the claim.

The construction of a strictly prime model and the verification that it is prime and atomic is carried out exactly as in Lemma 3.1.5.

Corollary 5.5.2. *A $t-$constructible set in a t.t. theory is atomic.*

The following basic fact about $t-$constructible sets arises repeatedly in the proof.

Lemma 5.5.2. *If $B = \{ b_\beta : \beta < \alpha \}$ is a $t-$construction over A and $B' \subset B$ is finite, then $\{ b_\beta : \beta < \alpha \}$ is a $t-$construction over $A \cup B'$.*

Proof. For each $\beta < \alpha$, $\{ b_\beta : \beta < \alpha \}$ is a $t-$construction over $A \cup B_\beta$, hence B is atomic over $A \cup B_\beta$. In particular, $tp(b_\beta \bar{b}'/A \cup B_\beta)$ is isolated, where $\bar{b}'$ is an enumeration of B'. By the transitivity of isolation, $tp(b_\beta/A \cup B_\beta \cup B')$ is isolated, as desired.

Part (ii) of the theorem is proved in the following lengthy result.

Proposition 5.5.1. *If T is t.t. and A is a set, then any two strictly prime models over A are isomorphic over A.*

Proof. Adding constants to the language for the elements of A results in another t.t. theory, so we may as well assume that $A = \emptyset$. Let $\{ (a_\beta, \varphi_\beta) : \beta < \alpha \}$ be a $t-$construction of a strictly prime model M over $\emptyset$.

A set $B \subset M$ is *closed* if for all $a_\beta \in B$ the parameters of φ_β are in B. Note: any union of closed sets is closed. In the proof we need the following basic properties of closed sets.

Claim. (i) For every $\beta < \alpha$ there is a finite closed set C containing a_β.
(ii) If B is closed M is atomic over B.

(i) This is proved by induction on β. Let $\{c_0, \dots, c_n\} \subset A_\beta$ be the set of parameters appearing in φ_β. Each c_i is contained in a finite closed set C_i, hence $C_0 \cup \ldots \cup C_n \cup \{a_\beta\}$ is a closed set containing a_β.

(ii) Since a strictly prime model is atomic it suffices to show that for all $\beta < \alpha$, $tp(a_\beta/B \cup A_\beta)$ is isolated. This is accomplished by showing that for all $\beta < \alpha$, $a_\beta \in B$ or φ_β isolates $tp(a_\beta/B \cup A_\beta)$. Fix $\beta < \alpha$ and suppose $a_\beta \notin B$. To show that φ_β isolates $tp(a_\beta/B \cup A_\beta)$ we prove (inductively), for $\delta < \alpha$ and $B^\delta = B \cap A_\delta$, φ_β isolates $tp(a_\beta/B^\delta \cup A_\beta)$. If $\delta < \beta$, $B^\delta \subset A_\beta$, so we can assume that $\delta > \beta$ and $\delta = \gamma + 1$ (since the case for limit ordinals is easy). By the inductive hypothesis, φ_β isolates $tp(a_\beta/B^\gamma \cup A_\beta)$. Since B is closed φ_γ is over B^γ, hence it isolates $tp(a_\gamma/B^\gamma \cup A_\beta \cup \{a_\beta\})$. Thus, $\varphi_\beta(x) \wedge \varphi_\gamma(y)$ isolates $tp(a_\beta, a_\gamma/B^\gamma \cup A_\beta)$. This, in turn, implies that φ_β isolates $tp(a_\beta/B^\gamma \cup \{a_\gamma\} \cup A_\beta) = tp(a_\beta/B^\delta \cup A_\beta)$, proving the claim.

With this claim in hand we can proceed with the main body of the proof. Let $\{ (b_\beta, \psi_\beta) : \beta < \alpha' \}$ be a $t-$construction over $\emptyset$ of a model M'. Without loss of generality, $\alpha \leq \alpha'$. We define by recursion on $\beta < \alpha$ closed sets $C_\beta \subset M$ and $C'_\beta \subset M'$ and elementary maps f_β from C_β onto C'_β such that:

(1) If $\gamma < \beta$ then $C_\gamma \subset C_\beta$, $C'_\gamma \subset C'_\beta$ and $f_\gamma \subset f_\beta$.
(2) If β is a limit ordinal then $C_\beta = \bigcup_{\gamma<\beta} C_\gamma$, $C'_\beta = \bigcup_{\gamma<\beta} C'_\gamma$ and $f_\beta = \bigcup_{\gamma<\beta} f_\gamma$.
(3) If η is a limit ordinal and $\beta = \eta + (2n+2)$, then $a_{\eta+n} \in C_\beta$; if $\beta = \eta + (2n+1)$ then $b_{\eta+n} \in C'_\beta$.

Consider a typical case in the recursion such as $\beta = \gamma + 1 = \eta + (2n+2)$, where η is a limit ordinal. There is a finite closed set $B_0 \subset M$ containing $a_{\eta+n}$ (by (i) of the claim). Since M is atomic over C_γ (by (ii) of the claim) there is an elementary map g_0 extending f_γ and taking B_0 onto some $B'_0 \subset M'$. Now find a finite B'_1 with $B'_0 \subset B'_1 \subset M'$ such that B'_1 is closed and note that $C'_\gamma \cup B'_1$ is closed. Since M' is atomic over C'_γ and B'_0 is finite, M' is atomic over $C'_\gamma \cup B'_0$. Hence, there is an elementary map g_1 extending g_0 and taking $C_\gamma \cup B_1$ (for some $B_1 \subset M$) onto $C'_\gamma \cup B'_1$. Continuing in this manner produces chains of finite sets $B_i \subset M$ and $B'_i \subset M'$ and elementary maps g_i from $C_\gamma \cup B_i$ onto $C'_\gamma \cup B'_i$ such that B_i is closed if i is even and B'_i is closed if i is odd. Finally, the sets $C_\beta = C_\gamma \cup \bigcup_{i<\omega} B_i$, $C'_\beta = C'_\gamma \cup \bigcup_{i<\omega} B'_i$ and $f_\beta = \bigcup_{i<\omega} g_i$ satisfy (1)–(3).

In the end the construction yields an isomorphism $f = f_\alpha$ from M onto the closed subset B'_α of M'. Since M' is an atomic model over B'_α, M' must equal B'_α. Thus, M and M' are isomorphic, proving the proposition.

The equivalents in part (iii) of the theorem are handled in the next two lemmas. In Lemma 5.5.1 (1) $\Longrightarrow$ (2) was proved. The next lemma proves (2) $\Longrightarrow$ (3).

Lemma 5.5.3. *If T is t.t. and M is prime over A, then M is atomic over A and there is no uncountable subset of M which is indiscernible over A.*

Proof. It follows from Lemma 5.5.1 that any prime model over A is atomic. (A prime model is elementarily embeddable into a strictly prime model N over A, and any subset of N is atomic over A.) The model M is also prime over $acl(A)$. An infinite set of indiscernibles over A is indiscernible over $acl(A)$ (by Exercise 5.1.13), so we may as well assume that $A = acl(A)$. By working over A we may take A to be $\emptyset$. Suppose, towards a contradiction, that M contains an uncountable indiscernible set I. Without loss of generality, M is strictly prime (since a prime model is contained in a strictly prime model). By Corollary 5.1.13 there is a finite set $J \subset I$ such that $I \setminus J$ is a Morley sequence over J. Since J is finite M is strictly prime over J. So, we may as well assume that $J = \emptyset$ and I is a Morley sequence over $\emptyset$ in a stationary type $p \in S(\emptyset)$.

Let I_0 be a countable infinite subset of I and M' a prime (hence atomic) model over I_0 which is contained in M. Since M is prime over $\emptyset$ it can be elementarily embedded into M', hence M' contains an uncountable Morley sequence J in p. We prove that $a \in J \implies a \not\perp I_0$ as follows. There is a formula φ over a finite $I' \subset I_0$ which isolates $tp(a/I_0)$. If $a \perp I'$, then a and any $b \in I_0 \setminus I'$ have the same type over I' (since p is stationary and I_0 is independent over $\emptyset$). Thus, any $b \in I_0 \setminus I'$ also satisfies φ contradicting that φ isolates a type over I_0. Thus, $a \not\perp I_0$.

Since I_0 is countable there is a countable $J_0 \subset J$ such that I_0 and J are independent over J_0. A fortiori, any $b \in J \setminus J_0$ (which exists since J is uncountable) is independent from I_0 over J_0. However, b is independent from J_0, hence independent from I_0 by transitivity. This contradicts what was proved in the previous paragraph, to establish the lemma.

The proof of (3) $\Longrightarrow$ (1) of Theorem 5.5.1(iii) is

Lemma 5.5.4. *Suppose that T is t.t., M is atomic over A and does not contain an uncountable set of indiscernibles over A. Then M is strictly prime over A.*

Proof. The structure of the proof is set with the following claim.

Claim. It suffices to show that for any set $A' \subset M$ such that M is atomic over A' and M does not contain an uncountable indiscernible set over A' there is a set B such that $(*)$ $A' \subset B \subset M$, such that M is atomic over B and $B \neq A'$ is $t-$constructible over A'. In fact, it suffices to find a set B satisfying $(*)$ with $A' = A$.

Notice that M does not contain an uncountable set of indiscernibles over any subset containing A. Thus, assuming $(*)$, we can apply it again with A replaced by B to obtain a set $B_1 \neq B$, $B \subset B_1 \subset M$, which is $t-$constructible over B and over which M is atomic. The constructions of B over A and B_1 over B can be pieced together to give a $t-$construction of B_1 over A. If B_i, $i < \delta$, is a chain where each B_i is a $t-$construction over which M is atomic and the $t-$construction of B_{i+1} extends the $t-$construction of B_i, then $\bigcup_{i<\delta} B_i$ is also a $t-$construction over which M is atomic. From these facts we obtain a chain of sets $A \subset B \subset B_1 \subset B_2 \subset \ldots \subset M$, each of which is $t-$constructible over A, and whose union is all of M. This proves that M is strictly prime over A as desired.

A set D, $C \subset D \subset M$, is said to be *full over* C if whenever a and b in M have the same type over C, $a \in D \iff b \in D$.

Claim. Let $A' \subset M$ be any set such that M is atomic over A' and M does not contain an uncountable indiscernible set over A'. If $D \subset M$ is full over A', then M is atomic over D.

Let a be an arbitrary element of M and B a finite subset of D such that $tp(a/D) = q$ does not split over B. Since B is finite, M is atomic over $A' \cup B$, yielding a formula $\varphi(x)$ which isolates $tp(a/A' \cup B)$. Assume, towards a contradiction, that φ does not isolate $tp(a/D)$ and $\psi(x, b) \in q$ is such that $\models \exists x(\varphi(x) \wedge \neg\psi(x, b))$. Thus, for θ a formula isolating $tp(b/A' \cup B)$ there is a $c \in M$ such that $\models \theta(c)$ and $\neg\psi(a, c)$. Then $c \in D$, because D is full over A. Since $\models \psi(a, b)$ and $\models \neg\psi(a, c)$ we contradict that q does not split over B, proving the claim.

Claim. Let $A' \subset M$ be any set such that M is atomic over A' and M does not contain an uncountable indiscernible set over A'. If $D \subset M$ is full over A', then D is $t-$constructible over A'.

This is proved for all D and A' by induction on the least ordinal α such that $a \in D \implies MR(a/A') < \alpha$. If $\alpha = 1$ any enumeration of D is a $t-$construction. Suppose α is a limit ordinal and for $\beta < \alpha$ let $B_\beta = \{a \in D : MR(a/A) < \beta\}$. Since each B_β is full over A', M is atomic over B_β. In fact, $B_{\beta+1}$ is full over B_β, hence is $t-$constructible over B_β, by induction. Piecing these constructions together end to end yields a $t-$construction of $D = \bigcup_{\beta<\alpha} B_\beta$ over A'.

Now suppose that $\alpha = \beta + 1$ and $\{p_i : i < \lambda\}$ is an enumeration of the complete types over A' realized in D. We will find by recursion on $i < \lambda$ a

t–construction of $B_i = p_i(D)$ over $C_i = A \cup \bigcup_{j<i} B_j$. A t–construction of D over A' is again obtained by piecing together these constructions end to end.

At stage i in the recursion let B' be a maximal A'–independent subset of B_i which is also independent from C_i over A'. Observe that B' must be countable. (There are only finitely many strong types over A' extending p_i. Assuming B' to be uncountable there is an uncountable $B'' \subset B'$ such that the elements of B'' have the same strong type over A'. Then B'' would be an uncountable Morley sequence, hence indiscernible set over A', in contradiction to the hypotheses.) Let $\{ b_m : m < \omega \}$ be an enumeration of B' (with repetitions if B' is finite). For $m < \omega$ let $D_m = \{ b \in B_i : MR(b/C_i \cup \{b_0, \ldots, b_m\}) < \alpha \}$ and set $D_{-1} = \emptyset$. We will show that each D_m is t–constructible over $C_i \cup D_{m-1}$, and to aid in the recursion, that M is atomic over $C_i \cup D_m$. Assuming that M is atomic over $C_i \cup D_{m-1}$ it is also atomic over $C_i \cup D_{m-1} \cup \{b_m\}$. Examining the definition shows that D_m is full over $C_i \cup D_{m-1} \cup \{b_m\}$, hence M is atomic over $C_i \cup D_m$. Furthermore, for any $b \in D_m$, $MR(b/C_i \cup D_{m-1} \cup \{b_m\}) < \alpha$, so D_m is t–constructible over $C_i \cup D_{m-1} \cup \{b_m\}$ (by induction). In fact, D_m is t–constructible over $C_i \cup D_{m-1}$ (since the type of b_m over $C_i \cup D_{m-1}$ is isolated). Finally, observe that these constructions can be concatenated to produce a t–construction of $B_i = \bigcup_{m<\omega} D_m$ over C_i, as desired. This proves the claim.

The reduction obtained in the first claim will be used to complete the proof. Suppose $a \in M \setminus A$. Let $D \supset A$ be a subset of M, full over A, which contains a. By the second claim M is atomic over D, and by the third claim D is t–constructible over A. Thus, D satisfies (*). By the first claim the lemma is proved.

Following is an example of a superstable theory with a prime model but no atomic model. Of course, the language is necessarily uncountable. This shows, in particular, that the restriction to t.t. theories in Theorem 5.5.1 is necessary.

Example 5.5.1. Let M_0 be the direct product of $\aleph_0$ copies of the two element group $(\mathbb{Z}_2, +, 0)$, (with universe ${}^{\omega}2$) and $M_1 = M_0 \times M_0$. Let H be the subgroup $M_0 \times \{0\}$. Let π be the composition of the projection of M_1 onto $\{0\} \times M_0$ followed by the natural identification of this group with H. For $i < \omega$ let $U_i = \{ (f, g) : f, g \in {}^{\omega}2 \text{ and } f(j) = g(j) = 0 \text{ for } j \leq i \}$, a subgroup of M_1 of index 2^{i+1}. Let $M_2 = (M_1, H, U_i, +, 0, \pi)_{i<\omega}$, a structure in a language $L = \{H, U_i, +, 0, \pi\}_{i<\omega}$. Then, $T_0 = Th(M_2)$ is quantifier-eliminable, countable and superstable (of ∞–rank 2, see Section 6.1). The connected component of $\mathfrak{C}$ is $\bigcap_{i<\omega} U_i(\mathfrak{C})$, so $\mathfrak{C}$ has $2^{\aleph_0}$ many generics. Also $H(\mathfrak{C})$ has $2^{\aleph_0}$ many generic types since $H(\mathfrak{C}) \cap \mathfrak{C}^o$ has $2^{\aleph_0}$ many cosets in $H(\mathfrak{C})$. Note that any nonzero element of $H(\mathfrak{C})$ is a generic element. Similarly, for any $a \in H(\mathfrak{C})$ the set $\pi^{-1}(a) = \{ b : \pi(b) = a \}$ is partitioned into continuum many cosets of $\mathfrak{C}^o$. Moreover, any two elements of $\pi^{-1}(a)$ with the same coset of $\mathfrak{C}^o$ have the same type over $acl(a)$. Furthermore, π induces an isomorphism π^* between the groups $H(\mathfrak{C})$ and $\mathfrak{C}/H(\mathfrak{C})$ which preserves the U_i's.

Let c^* be a nonzero element of $H(\mathfrak{C})$. Let $C = \{ c_\eta : \eta < 2^{\aleph_0} \}$ be a subset of $\pi^{-1}(c^*)$ such that $C + \mathfrak{C}^o \supset \pi^{-1}(c^*)$ and $\eta \neq \eta' \implies c_\eta + \mathfrak{C}^o \neq c_{\eta'} + \mathfrak{C}^o$. Finally, let T be the theory of N with a constant symbol for each c_η. It is easy to verify that for any $M' \models T$ and $a \in H(M')$ every strong type over a extending $\pi(v) = a$ is realized in M'.

Elimination of quantifiers gives the existence of a model M with the property that each element of $H(M)$ is the difference of two c_η's (and each element of $M/H(M)$ is π^* of an element of $H(M)$). We claim that M is a prime model which is not atomic. Let M' be any model of the theory. Certainly there is an elementary mapping F of $H(M) \cup (\pi^{-1}(c^*) \cap M)$ into M', so it remains to map the sets $\pi^{-1}(b) \cap M$ into $\pi^{-1}(Fb) \cap M'$, for $b \in H(M)$. Given $b \in H(M) \setminus \{0, c^*\} = X$ let $\hat{b}$ be an element of $\pi^{-1}(b) \cap M$. Then, $\pi^{-1}(b) \cap M$ is simply $\hat{b} + H(M)$. Elimination of quantifiers guarantees that $\{ \hat{b} : b \in X \}$ is independent over $X \cup (\pi^{-1}(c^*) \cap M)$. As we argued above, $F(stp(\hat{b}/b))$ is realized in M' by some element $\hat{b}_F$. Since $\{ \hat{b}_F : b \in X \}$ must be independent over $F(X \cup (\pi^{-1}(c^*) \cap M))$, F extends to an elementary map f taking $\hat{b}$ to $\hat{b}_F$. Since $\pi^{-1}(b) \cap M = \hat{b} + H(M)$, f extends to an isomorphism of M into M'. Hence M is prime. Furthermore, for any choice of $\hat{b}$, $tp(\hat{b}/\{b\})$ is nonisolated (left as an exercise). Thus, M is not atomic.

5.5.2 a-prime Models

In the proof of Morley's Categoricity Theorem we dealt with $\aleph_0$−prime models; i.e., prime models over sets relative to the class of $\aleph_0$−saturated models. In general,

Definition 5.5.2. *If $\mathcal{K}$ is a class of models and A is a set we call $M \supset A$ a* prime model over A relative to $\mathcal{K}$ *if $M \in \mathcal{K}$ and for any $N \in \mathcal{K}$ such that $N \supset A$, M is elementarily embeddable over A into N.*

We will see that by taking $\mathcal{K}$ to be the class of a−models we obtain relatively prime models in properly superstable or properly stable theories which act somewhat like prime models in totally transcendental theories. The relevant notions of isolation are defined summarily.

Definition 5.5.3. *Let T be stable and κ an infinite cardinal, $\mathcal{K}$ the class of κ−saturated models of T and $\mathcal{K}_a$ the class of (a, κ)−saturated models of T.*

(i) A model M is κ−prime over A *if M is a prime model over A relative to $\mathcal{K}$. We call M (a, κ)−prime over A if M is prime over A relative to $\mathcal{K}_a$. When $\kappa = \kappa(T)$ we abbreviate (a, κ)−prime to* a−prime.

(ii) Let $B \subset A$ have cardinality $< \kappa$. A type $p \in S(A)$ is κ−isolated *over B if $p \restriction B \models p$. A strong type p over A is* (a, κ)−isolated over B *if $p \restriction acl(B) \models p$; equivalently, for all a realizing p, $stp(a/B) \models p$. We call*

p κ–isolated (respectively, (a, κ)–isolated) if it is κ–isolated (respectively, (a, κ)–isolated) over some subset of A. When $\kappa = \kappa(T)$ we say a–isolated *for (a, κ)–isolated.*

Only types over algebraically closed sets are relevant in (a, κ)–isolation, so it requires a little more care in handling than κ–isolation. However, most proofs for (a, κ)–isolation and κ–isolation are very similar. The next lemma is stated for (a, κ)–isolation, but the proofs are much the same (or easier) for κ–isolation.

Lemma 5.5.5. *Let T be stable, κ an infinite cardinal, p a strong type over A and $B \subset A$.*

(i) If p is (a, κ)–isolated over B, then p does not fork over B.

(ii) p is (a, κ)–isolated over B if and only if it is $(\kappa + |T|^+)$–isolated over acl(B).

(iii) (Existence) *If $\kappa \geq \kappa(T)$ then for any A, $B \subset A$ of cardinality $< \kappa$ and strong type q over B there is an (a, κ)–isolated strong type r over A extending q.*

Proof. Part (i) is clear since p is implied by a strong type over B. Part (ii) is left as an exercise to the reader.

(iii) Without loss of generality, $A = acl(A)$. By the definition of $\kappa(T)$, there is a set C, $B \subset C \subset A$, with $|C \setminus B| < \kappa(T)$ and a type $r \in S(acl(C))$ extending p such that no extension of r over A forks over C. Then $|C| < \kappa$ and the unique nonforking extension of r over $acl(A)$ is an (a, κ)–isolated extension of p.

When $\kappa < \kappa(T)$, (a, κ)–isolated types may not exist as in (iii) of the previous lemma. There are few uses for (a, κ)–isolation when $\kappa > \kappa(T)$ and as (ii) indicates the notion collapses to κ–isolation when κ is sufficiently large. To focus on the main concept, we will only deal with $(a, \kappa(T))$–isolation; i.e., a–isolation, in the remainder of the section.

Our treatment is further specialized by considering only theories in which $\kappa(T)$ is regular (as it is when T is countable). Indeed, many of the results do not hold without this restriction. This property is used to obtain the following, which is part of Proposition 5.2.1.

> **(*)** Suppose that $\kappa(T)$ is regular, $\kappa \geq \kappa(T)$ and $|B| < \kappa$. Then for all sets A, there is a set $C \subset A$ of cardinality $< \kappa$ such that B is independent from A over C.

Definition 5.5.4. *Let T be stable.*

*(i) A set $B = \{ b_\beta : \beta < \alpha \}$ is called an a–*construction over *A if for all $\beta < \alpha$, stp($b_\beta / A \cup B_\beta$) is a–isolated. A set C is a–*constructible over *A if some enumeration of C is an a–construction over A.*

(ii) A model $M \supset A$ is strictly a–prime over *A if M is an a–model and is a–constructible over A.*

*(iii) A set B is $a-$*atomic over *A if for all finite sequences b from B, $stp(b/A)$ is $a-$isolated.*

(iv) Let $B = \{ b_\beta : \beta < \alpha \}$ be an $a-$construction over some set A and, for $\beta < \alpha$, $C_\beta \subset B_\beta \cup A$ a set of cardinality $< \kappa(T)$ over which $stp(b_\beta/B_\beta \cup A)$ is $a-$isolated. Then, a set $D \subset B$ is called closed *if whenever $b_\beta \in D$, $C_\beta \subset D \cup A$.*

In analogy to Theorem 5.5.1 we will prove below

Theorem 5.5.2. *Suppose that T is stable, $\kappa(T)$ is regular and A is any set.*

(i) There is a strictly $a-$prime model over A and every such model is $a-$prime over A.

(ii) Any two strictly $a-$prime models over A are isomorphic over A.

(iii) A strictly $a-$prime model over A is $a-$atomic over A and does not contain a set of cardinality $> \kappa(T)$ which is indiscernible over A.

(iv) An $a-$prime model over A is strictly $a-$prime over A.

Unlike the ordinary prime case we will not prove the converse of part (iii): If M is $a-$atomic over A and does not contain a set over indiscernibles over A of cardinality $> \kappa(T)$, then M is strictly $a-$prime over A. The proof of this harder result can be found in [Her92], which improves a slightly weaker result proved in [She90, IV, 4.14].

Lemma 5.5.6. *Suppose that T is stable and $\kappa(T)$ is regular.*

(i) If B is $a-$constructible over A and $C \subset B$ there is a closed set $C' \subset B$ containing C of cardinality $< |C|^+ + \kappa(T)$.

(ii) If the strong type p over A is $a-$isolated and p does not fork over $B \subset A$, then $q = p \restriction acl(B)$ is $a-$isolated. In fact, p is $a-$isolated over a subset of B.

(iii) If B is $a-$constructible over A, then B is $a-$atomic over A.

(iv) If B is $a-$atomic over A and $|B| < \kappa(T)$, then any enumeration of B is an $a-$construction over A. If B is $a-$atomic over A and $|B| \leq \kappa(T)$, then B is $a-$constructible over A.

(v) (Transitivity) *If $|B| < \kappa(T)$ and B is $a-$atomic over A, then $stp(b/A \cup B)$ is $a-$isolated if and only if $B \cup \{b\}$ is $a-$atomic over A.*

(vi) If B is $a-$atomic over A and $C \subset B$ has cardinality $< \kappa(T)$, then B is $a-$atomic over $A \cup C$.

(vii) If B is $a-$constructible over A and $C \subset B$ is closed, then C is $a-$constructible over A and B is $a-$constructible over $A \cup C$.

Proof. (i) Let $B = \{ b_\beta : \beta < \alpha \}$ be an $a-$construction of B over A. For $x \in B$ and $b_\beta = x$ let D_x be a subset of B_β of cardinality $< \kappa(T)$ such that $stp(b_\beta/B_\beta \cup A)$ is $a-$isolated over $D_x \cup A$. For $X \subset B$ let $X' = \bigcup \{ D_x : x \in X \} \cup X$. Then the set $D = \bigcup_{n<\omega} C^{(n)}$, where $C^{(0)} = C$ and $C^{(n+1)} = (C^{(n)})'$ is $a-$constructible. (Enumerate C as $\{ c_\beta : \beta < \alpha' \}$ where $\gamma < \beta < \alpha'$, $c_\gamma = b_\delta$ and $c_\beta = b_\epsilon$ implies that $\delta < \epsilon$. The reader can check that this enumeration gives an a-construction of C over A.) To check the cardinality of D we consider only when $|C| < \kappa(T)$, leaving the general case to the reader. If $\kappa(T) = \aleph_0$ then each $C^{(n)}$ is finite and for some n, $C^{(n+1)} = C^{(n)}$, hence D is finite. If $\kappa(T)$ is uncountable, then the regularity of the cardinal implies that $|C^{(n)}| < \kappa(T)$ and $\sup_{n<\omega}|C^{(n)}|$ is also $< \kappa(T)$, completing the proof of (i).

(ii) Without loss of generality, $A = acl(A)$ and $B = acl(B)$. Let $r \subset p$ be a strong type over a set A_0 of cardinality $< \kappa(T)$ which is equivalent to p. Since $\kappa(T)$ is regular there is a set $C \subset B$, the algebraic closure of a set of cardinality $< \kappa(T)$, such that A_0 is independent from B over C and p is based on C. We will show that $s = q \restriction C = p \restriction C$ is equivalent to p, which is sufficient to prove this part. Suppose, to the contrary, that a realizes s, but a does not realize q. Without loss of generality, a is independent from A_0 over B, hence a is independent from A_0 over C (by transitivity). Since the nonforking extension of s over A is p, a realizes $p \restriction A_0 = r$. Hence, a realizes $p \supset q$, a contradiction which proves (ii).

(iii) Without loss of generality, $A = acl(A)$. It suffices to show that when $B = \{ b_\gamma : \gamma \leq \beta \}$ is an a-construction over A and B_β is a-atomic over A, B is also a-atomic over A. Suppose this to be false. Let $a = b_\beta$ and let $\bar{b}$ be a finite sequence from B_β such that $stp(a\bar{b}/A)$ is not a-isolated. Let $C_0 \supset \bar{b}$ be a subset of $B_\beta \cup A$ of cardinality $< \kappa(T)$ such that $stp(a/B_\beta \cup A)$ is a-isolated over C_0. Then C_0 is a-atomic over A, in fact, $C = acl(C_0)$ is also a-atomic over A (see the exercises). Let $A_0 \subset A$ be the algebraic closure of a set of cardinality $< \kappa(T)$ such that $C \cup \{a\}$ is independent from A over A_0. Since $stp(a\bar{b}/A)$ is not a-isolated there is a sequence $a'\bar{b}'$ realizing $stp(a\bar{b}/A_0) = tp(a\bar{b}/A_0)$ which does not realize $tp(a\bar{b}/A)$. Let f be an automorphism of the universe fixing A_0 and mapping $a\bar{b}$ to $a'\bar{b}'$; let $D = f(C)$. Since C is a-atomic over A and independent from A over A_0, (ii) implies that $tp(\bar{c}/A_0) \models tp(\bar{c}/A)$ for any finite sequence $\bar{c}$ from C. This implies that for all $\bar{c}$ from C, $tp(\bar{c}/A) = tp(f(\bar{c})/A)$, hence there is an automorphism g of $\mathfrak{C}$ which fixes A and maps $f(c)$ to c, for any $c \in C$. Since $f(a) = a'$, $g(tp(a'/D \cup A_0)) = tp(a/C \cup A_0)$. Again using (ii), $tp(a/C \cup A_0) \models tp(a/C \cup A)$, hence $g(a')$ realizes $tp(a/C \cup A)$. This contradicts that $g(\bar{b}') = \bar{b}$, g fixes A and $a'\bar{b}'$ does not realize $tp(a\bar{b}/A)$, completing the proof of (iii).

(iv) The case for $|B| < \kappa(T)$ is proved by induction on $|B|$. For finite B this follows immediately from

Claim. If $stp(ab/A)$ is a-isolated over B then $stp(a/A \cup \{b\})$ is a-isolated over $B \cup \{b\}$ and $stp(b/A)$ is a-isolated over B.

Since $stp(ab/B) \models stp(ab/A)$ it is clear that $stp(b/B) \models stp(b/A)$. Suppose that a' realizes $stp(a/B \cup \{b\})$. Then, $a'b$ realizes $stp(ab/B)$, hence $a'b$ is independent from A over B. This implies that a' (like a) is independent from $A \cup \{b\}$ over $B \cup \{b\}$, thus a' realizes $stp(a/A \cup \{b\})$, proving the claim.

Now suppose that B is infinite and let $\{ b_\beta : \beta < \alpha \}$ be any enumeration of B. Let $\beta < \alpha$, $\kappa = |B_\beta|$ and write B_β as $\bigcup_{i<\kappa} C_i$, where $|C_i| < \kappa$ and $i < j \implies C_i \subset C_j$. Since each set $C_i \cup \{b_\beta\}$ is a-atomic over A the inductive hypothesis implies that any enumeration of the set is an a-construction. Thus, $stp(b_\beta/C_i \cup A)$ is a-isolated. This allows us to write $stp(b_\beta/B_\beta \cup A)$ as $\bigcup_{i<\kappa} p_i$, where the p_i's form a chain and each p_i is a-isolated. We leave it as an exercise to the reader to show that since $\kappa < \kappa(T)$ this union is a-isolated. This proves that $\{ b_\beta : \beta < \alpha \}$ is an a-construction of B.

If $|B| = \kappa(T)$, then the first part of (iv) says that an enumeration $\{ b_\alpha : \alpha < \kappa(T) \}$ of B is an a-construction over A.

(v) Both directions of the biconditional follow from (iii) and (iv) of this lemma.

(vi) The proof of this part is left for the reader in the exercises.

(vii) Let $\{ b_i : i < \alpha \}$ be an a-construction of B and let $<$ denote the induced well-ordering of B. For $i < \alpha$ let $E_i \subset B_i \cup A$ be the distinguished set of cardinality $< \kappa(T)$ such that $stp(b_i/B_i \cup A)$ is a-isolated over E_i. Let $D = B \setminus C$ and $C = \{ c_i : i < \gamma \}$, $D = \{ d_i : i < \delta \}$ enumerations of these sets which respect the enumeration of B (i.e., if $c_i = b_\beta$ and $c_j = b_{\beta'}$ then $i < j \Longleftrightarrow \beta < \beta'$). Let $f : \delta \longrightarrow \alpha$ and $g : \gamma \longrightarrow \alpha$ be defined by: $f(i) = j$ if $d_i = b_j$ and $g(i) = j$ if $c_i = b_j$.

Since C is closed, $i < \gamma$ implies that $E_{g(i)} \subset C_i \cup A$, hence the given enumeration is an a-construction of C over A. To prove that D is a-constructible over $A \cup C$ it suffices to show that for $i < \delta$, $stp(d_i/D_i \cup A \cup C)$ is a-isolated over $E_{f(i)}$. To prove this we fix i and show by induction on $k < \gamma$ that $stp(d_i/D_i \cup A \cup C_k \cup E_{f(i)})$ is a-isolated over $E_{f(i)}$. (This is sufficient because a quick check of the definitions shows that $E_{f(i)}$ is contained in $D_i \cup A \cup C$.) Since $C_0 = \emptyset$ and $D_i \subset B_{f(i)}$ there is nothing to prove when $k = 0$. The condition is also preserved at limit ordinals, so suppose that $k = l + 1$. If c_l precedes d_i in the ordering of B then $D_i \cup C_l \subset B_{f(i)}$ and again there is nothing to prove. In the remaining case, when $f(i) < g(l)$, we see that $E_{f(i)} \subset A \cup D_i \cup C_l$, $A \cup D_{i+1} \cup C_l \subset A \cup B_{g(l)}$ and $stp(c_l/A \cup D_{i+1} \cup C_l)$ is a-isolated over a subset of $A \cup C_l$ (since $E_{g(l)} \subset A \cup C_l$). A fortiori, $stp(c_l/A \cup C_l \cup D_i) \models stp(c_l/A \cup C_l \cup D_{i+1})$. Switching the roles of c_l and d_i in this equation proves that $stp(d_i/A \cup C_l \cup D_i) \models stp(d_i/A \cup C_{l+1} \cup D_i)$. Since $stp(d_i/A \cup E_{f(i)}) \models stp(d_i/A \cup C_l \cup D_i)$ we have shown that $stp(d_i/A \cup C_{l+1} \cup D_i)$ is a-isolated over $E_{f(i)}$, as required to complete the proof.

Parts of the following proposition are proved like the corresponding results for ordinary isolation in t.t. theories. There are differences in that finite is replaced by "of cardinality $< \kappa(T)$", but the properties proved above fill the gaps. With this proposition we prove parts (i)–(iii) of Theorem 5.5.2.

Proposition 5.5.2. *Suppose that T is stable, $\kappa(T)$ is regular and A is any set.*

(i) There is a strictly a-prime model M over A. The model M is a-prime over A and its cardinality is $\leq$ the first cardinal $\geq |A|$ in which T is stable.

(ii) Any two strictly a-prime models over A are isomorphic over A.

(iii) A strictly a−prime model over A is a−atomic over A and does not contain a set of cardinality $> \kappa(T)$ which is indiscernible over A.

It remains to prove Theorem 5.5.2(iv), which is done in

Proposition 5.5.3. *Suppose that T is stable and $\kappa(T)$ is regular. Then for all sets A, if M is a−prime over A, M is strictly a−prime over A. In fact, if $M \supset B \supset A$, B is a−constructible.*

Proof. We know that there is a strictly a−prime model over A and M can be embedded into this model over A. Thus, it suffices to show that when M' is a strictly a−prime model over A, any set B, $A \subset B \subset M'$, is a−constructible over A. Let $\{a_\alpha : \alpha < \nu\}$ be an a−construction of M' over A. We will find, by recursion on $\alpha < \mu$ (for some $\mu < |M'|^+$) sets $D_\alpha \supset A$ such that

(1) Each D_α is closed.
(2) If $\alpha < \beta$, $D_\alpha \subset D_\beta$ and $D_\alpha = \bigcup_{\beta<\alpha} D_\beta$ when α is a limit ordinal.
(3) $|D_{\alpha+1} \setminus D_\alpha| \leq \kappa(T)$.
(4) $M' = \bigcup_{\alpha<\mu} D_\alpha$.
(5) D_α and B are independent over $B_\alpha = D_\alpha \cap B$.

That B is a−constructible will follow rather quickly from this.

The set $D_{\alpha+1}$ will be the union of D_α and the sets $D^n_{\alpha+1}$, $n < \omega$, which are defined as follows.

(i) $D^0_{\alpha+1} = \{a_\alpha\}$.
(ii) If $n > 0$ is even, $D^n_{\alpha+1}$ is a closed set of cardinality $< \kappa(T)$ containing $D^{n-1}_{\alpha+1}$.
(iii) If n is odd we first let $C \subset B$ be a set of cardinality $< \kappa(T)$ such that $D^{n-1}_{\alpha+1}$ is independent from $B \cup D_\alpha$ over $C \cup D_\alpha$, and then set $D^n_{\alpha+1} = D^{n-1}_{\alpha+1} \cup C$.

Finally, let $D_{\alpha+1} = D_\alpha \cup \bigcup_{n<\omega} D^n_{\alpha+1}$. A union of closed sets is closed, so (1) holds. Since $D_{\alpha+1} \setminus D_\alpha$ is the union of countably many sets of cardinality $< \kappa(T)$ this difference has cardinality $\leq \kappa(T)$. To verify the fifth condition suppose that d is a finite subset of $D_{\alpha+1}$ and n is minimal with $d \subset D^n_{\alpha+1}$. By construction there is a set $C \subset B \cap D_{\alpha+1}$ such that d is independent from B over $D_\alpha \cup C$. Since B and D_α are independent over $B \cap D_\alpha$, d is independent from B over $B \cap D_{\alpha+1}$, so (5) holds. The other conditions hold trivially.

We claim that $B_{\alpha+1}$ is a−atomic over B_α. Since D_α is closed, Lemma 5.5.6 says that M' is a−constructible over D_α, hence a−atomic over D_α. That is, for any finite set b from $B_{\alpha+1}$, $stp(b/D_\alpha)$ is a−isolated. Since b is independent from D_α over B_α, Lemma 5.5.6(ii) implies that $stp(b/B_\alpha)$ is a−isolated, as desired. Then $B_{\alpha+1}$ is a−constructible over B_α (by Lemma 5.5.6(iv)). Piecing together all of these constructions gives an a−construction of B, completing the proof of the proposition.

Corollary 5.5.3. *Suppose that T is stable and $\kappa(T)$ is regular. If M is a−prime over A and $M \supset B \supset A$, then M is a−prime over B.*

Historical Notes. Shelah first proved the uniqueness of prime models for ω-stable theories by induction on rank in [She72]. The proof given here was inspired by an unpublished proof by Ressayre. All of the other results are by Shelah and can be found in [She90, IV].

Exercise 5.5.1. Prove the property (*) preceding the definition of an a-construction (Definition 5.5.4).

Exercise 5.5.2. Prove that the union of a chain of $< \kappa(T)$ many a-isolated types is a-isolated.

Exercise 5.5.3. Prove that the algebraic closure of a set C is a-atomic over A whenever C is a-atomic over A.

Exercise 5.5.4. Prove (vi) of Lemma 5.5.6.

Exercise 5.5.5. Prove Corollary 5.5.3.

5.6 Orthogonality, Domination and Weight

We saw in the proofs of the Morley Categoricity Theorem and the Baldwin-Lachlan Theorem how the dimension theory on strongly minimal sets can be used to determine when models are isomorphic. This section is one facet in the development of a dimension theory for arbitrary stationary types which is based on the forking dependence relation.

Definition 5.6.1. *Let T be stable. Given a stationary type p over A and a set B, I is a* basis for p in B *if it is a maximal Morley sequence over A in p which is contained in B.*

Some key features of the dimension theory for strongly minimal sets in uncountably categorical theories are

1. If M is a model, φ is a strongly minimal formula over $A \subset M$ and p is the unique nonalgebraic completion of φ over A, then all bases for p in M have the same cardinality (which is called the *dimension* of $p(M)$ or $\varphi(M)$).
2. If ψ is another strongly minimal formula over M and M is uncountable, then $\varphi(M)$ and $\psi(M)$ have the same dimension.

Using also the fact that M is prime over any strongly minimal set in M we obtained the Morley Categoricity Theorem. To prove the Baldwin-Lachlan Theorem (about countable models) we need a finer result on dimension: If, in addition, φ' is a strongly minimal formula conjugate to φ, then $\varphi(M)$ and $\varphi'(M)$ have the same dimension.

When working in an uncountably categorical theory strongly minimal sets have the same dimension in any uncountable model. Other t.t. theories may contain strongly minimal sets whose dimensions in models can differ widely. Here is a simple example.

Example 5.6.1. Let T be the theory of one equivalence relation E with infinitely many infinite classes and no finite classes. For any a, $E(\mathfrak{C}, a)$ is a strongly minimal set. For any pair of infinite cardinals κ, λ there is a model M containing a and b such that $E(M, a)$ has dimension κ and $E(M, b)$ has dimension λ.

This freedom to find models with varying dimensions of strongly minimal sets can be used to show there are nonisomorphic models in each uncountable cardinal. In detail, let κ be uncountable and Λ the set of cardinals $\leq \kappa$ and Λ^+ the set of infinite cardinals $\leq \kappa$. Let $\Phi = \{f :$ f is a function from Λ^+ into Λ such that $f(\kappa) \neq 0$ or $\kappa \in$ range $(f)\}$. For any model M of T of cardinality κ let F_M be the element of Φ such that for any $\lambda \in \Lambda^+$, $F_M(\lambda)$ is the number of E–classes in M of cardinality λ. Then,

- any element of Φ is F_M for some model M of T, and
- for models M, N of T of cardinality κ, $M \cong N$ if and only if $F_M = F_N$.

Thus, the number of models of T of cardinality κ, up to isomorphism, is $|\Phi|$.

In generalizing the observed behavior of strongly minimal sets to a collection P of stationary types in a stable theory the most basic questions are:

1. Given stationary types p, $q \in P$ (over, say, $\emptyset$) let $(*)$ denote the condition: for all sufficiently large cardinals κ, λ there is a model M containing a basis for p of cardinality κ and a basis for q of cardinality λ. What basic properties of p and q cause $(*)$ to hold?
2. When $(*)$ fails for a pair p, q how widely can the cardinality of bases of p and q vary as we range over models of the theory.
3. Can we isolate a broad class of types Q such that dimension is well-defined for any $p \in Q$? (That is, for any $p \in Q$ and model M containing the domain of p, all bases of p in M have the same cardinality.)

These items are the subjects (in order) of the three subsections: orthogonality, domination and weight.

Before getting to the main topics we make the conceptual jump of introducing types and strong types in infinitely many variables.

Notation. If A and B are sets, *tp*(A/B) is the set of formulas in a potentially infinite sequence of variables obtained in the expected way. (In the background we have fixed an arbitrary enumeration of A.) Such types, called $*-$*types*, are only used to conveniently speak of the class of all sets conjugate to A over B: A' realizes *tp*(A/B) if there is an automorphism of the universe fixing B and taking A' to A. Extending also the notation for strong types, $stp(A/B) = tp(A/acl(B))$.

Most terms associated to the forking relation on types can be generalized to $*$–types using the obvious definitions. For example, $tp(A/B)$ is stationary if it has a unique nonforking extension over $\mathfrak{C}$, equivalently, $tp(A/B) \models stp(A/B)$.

If p and q are stationary types over A then all pairs ab, where a realizes p, b realizes q and a is independent from b over A, realize the same stationary type over A. This type is denoted $p \otimes q$. More generally,

Notation. If $p_i = stp(B_i/A)$, $i \in I$, is a family of stationary $*$–types over A and $\{ B_i : i \in I \}$ is A–independent then $stp(\bigcup_{i \in I} B_i/A)$ is denoted $\bigotimes_{i \in I} p_i$. If p is a strong type over A and λ is a cardinal, $p^{(\lambda)}$ is the strong type over A of a Morley sequence over A in p of cardinality λ (equivalently, the $\otimes$–product of λ copies of p).

Throughout the entire section we assume the underlying theory to be stable. The stability hypothesis may be restated for emphasis in key definitions and results.

5.6.1 Orthogonality

Intuitively, stationary types are orthogonal when there are models in which bases for the types have widely varying cardinalities. The actual definition (given subsequently) specifies a property which guarantees this behavior.

Definition 5.6.2. *(i) The $*$–types p and q over A are said to be* almost orthogonal, *written $p \overset{a}{\perp} q$, if for all B realizing p and C realizing q, B and C are independent over A. The negation of the relation is denoted $p \overset{a}{\not\perp} q$.*

(ii) The stationary $*$–types p and q are called orthogonal, *written $p \perp q$, if for all sets A on which both p and q are based, $p|A \overset{a}{\perp} q|A$.*

(iii) The $*$–types p and q are orthogonal *if $p' \perp q'$ whenever $p', q' \in S(\mathfrak{C})$ are nonforking extensions of p and q, respectively.*

Example 5.6.2. As trivial examples of orthogonal and nonorthogonal types consider a single unary predicate U, the theory T_1 saying that U and $\neg U$ define infinite sets, and the theory $T_2 \supset T_1$ saying that an additional function symbol F defines a bijection between the sets defined by U and $\neg U$.

Then, for p and q the unique nonalgebraic types containing U and $\neg U$, respectively, $p \perp q$ in T_1 and $p \not\perp q$ in T_2.

Let T be the theory in Example 5.6.1. Let $a \neq b$ in $\mathfrak{C}$ and p_a, p_b the strongly minimal types extending $E(x, a)$, $E(x, b)$, respectively. Then p_a is orthogonal to p_b.

The finite character of forking allows us to prove most facts about orthogonality by considering only types (instead of $*$−types). (Use the fact that $stp(B/A)$ is almost orthogonal to $stp(C/A)$ if and only if $stp(b/A) \overset{a}{\perp} stp(c/A)$, for all finite subsets b of B and c of C.) In the sequel we will normally state results only for types, leaving the extension to $*$−types to the reader.

Lemma 5.6.1. *Let T be stable.*

(i) For stationary types p and q, the following are equivalent.

(1) $p \perp q$.

(2) *For some set A on which p and q are both based and all sets $B \supset A$, $p|B \overset{a}{\perp} q|B$.*

(3) *For some a−model M on which p and q are both based, $p|M \overset{a}{\perp} q|M$.*

(ii) If p and q_i, $i \in I$, are stationary types, $p \perp \bigotimes_{i \in I} q_i$ if and only if $p \perp q_i$, for all $i \in I$.

Proof. The proof of (ii) is left to the exercises, while (1) $\Longrightarrow$ (2) $\Longrightarrow$ (3) of (i) are vacuously true. To prove (3) $\Longrightarrow$ (1) let p be nonorthogonal to q and M an a−model on which both p and q are based. Suppose A is a set on which both p and q are based such that $p|A \overset{a}{\not\perp} q|A$. For any $B \supset A$, $p|B \overset{a}{\not\perp} q|B$, so there is a $B \supset M$ such that $p|B \overset{a}{\not\perp} q|B$. Let a be a realization of $p|B$, b a realization of $q|B$ and c a finite subset of B such that a depends on b over $M \cup \{c\}$. Let $C \subset M$ be of cardinality $< \kappa(T)$ over which $stp(abc/M)$ is based. Since M is an a−model there is $c' \in M$ realizing $stp(c/C)$. Now choose elements a' and b' such that $stp(a'b'c'/C) = stp(abc/C)$ and $a'b'$ is independent from M over $C \cup \{c'\}$. These conditions imply that a' is not only independent from M over $C \cup \{c'\}$, but also over C; i.e., a' realizes $p|M$, and similarly, b' realizes $q|M$. Since a' and b' are dependent over $C \cup \{c'\}$, $p|M \overset{a}{\not\perp} q|M$ as required.

Remark 5.6.1. As a simple application of the lemma:

> If T is uncountably categorical and φ is a strongly minimal formula, then every nonalgebraic stationary type p is nonorthogonal to φ.

(A type is nonorthogonal to φ if it is nonorthogonal to the unique nonalgebraic $q \in S(\mathfrak{C})$ containing φ. Let M be an a−model containing the parameters in φ and over which p is based. Let a realize $p|M$ and N be the prime model over $M \cup \{a\}$. Since T has no Vaughtian pair there is a $b \in N \setminus M$ satisfying φ, hence realizing $q|M$. By Corollary 3.3.4, a and b are dependent over M, witnessing the nonorthogonality of p and q (by the lemma).)

The argument used to verify this remark can be generalized. When p and q are orthogonal stationary types over an a−model M there is a larger a−model realizing p and omitting q. (This is proved using Corollary 5.6.1 below.) We can iterate this process to find a "long" Morley sequence in p in a a−model $N \supset M$ which omits q. Hence the cardinalities of bases for p can q can vary widely as we range over a−models containing M. (See Corollary 5.6.2 below.)

Given nonorthogonal stationary types p and q the first question which comes to mind is: What must a set A contain to ensure that $p|A$ and $q|A$ are not almost orthogonal? We saw in Lemma 5.1.17 that p is based on an indiscernible set whose average type is parallel to p. The next result shows that such indiscernibles also control the manner in which the stationary type interacts with other types.

Proposition 5.6.1. *For T stable and p, $q \in S(\mathfrak{C})$ the following are equivalent:*

(1) $p \perp q$.
(2) *For some infinite sets of indiscernibles I, J such that $Av(I/\mathfrak{C}) = p$ and $Av(J/\mathfrak{C}) = q$, $p|(I \cup J) \overset{a}{\perp} q|(I \cup J)$.*
(3) *For some set A on which p and q are both based and for all n, $(p|A)^{(n)}$ is almost orthogonal to $(q|A)^{(n)}$.*

Proof. With regard to (2), recall from Lemma 5.1.17 that p and q are based on I and J, respectively, hence $p|(I \cup J)$ and $q|(I \cup J)$ exist. That (1) $\Longrightarrow$ (2) is trivial. For (2) $\Longrightarrow$ (3) suppose that there are such sets of indiscernibles I and J. Let I' and J' be infinite and coinfinite subsets of I and J, respectively. By Lemma 5.1.17 both p and q are based on $A = I' \cup J'$. Assuming that for some n $(p|A)^{(n)} \overset{a}{\not\perp} (q|A)^{(n)}$ yields minimal sets $I_0 \subset I$ and $J_0 \subset J$ such that $stp(I_0/A) \overset{a}{\not\perp} stp(J_0/A)$. Let $a \in I_0$, $b \in J_0$, $I_1 = I_0 \setminus \{a\}$ and $J_1 = J_0 \setminus \{b\}$. By the minimality assumption on I_0 and J_0,

$$I_0 \underset{A}{\perp} J_1 \text{ and } I_1 \underset{A}{\perp} J_0.$$

Combining this with the initial assumption about I_0 and J_0 shows that for $B = A \cup I_1 \cup J_1$, $stp(a/B) = p|B$ is not almost orthogonal to $stp(b/B) = q|B$. Since $B \subset I \cup J$ this contradicts that $p|(I \cup J) \overset{a}{\perp} q|(I \cup J)$, proving that (2) $\Longrightarrow$ (3).

To prove (3) $\Longrightarrow$ (1) suppose that p is nonorthogonal to q and A is any set on which p and q are based. By Lemma 5.6.1(i) there is a set $B \supset A$, a realizing $p|B$ and b realizing $q|B$ such that a and b are dependent over B. Let $I = \{ a_i b_i : i < \omega \}$ be an infinite Morley sequence in $stp(ab/B)$. Then $\{ a_i : i < \omega \}$ and $\{ b_i : i < \omega \}$ are Morley sequences over B (and A) in p. Since ab realizes the average type of I over B and this average is based on I there must be some n and m such that a_n and b_m are dependent over $A \cup \{ a_i : i < n \} \cup \{ b_i : i < m \}$. The sequences $(a_0, \ldots, a_n)$ and $(b_0, \ldots, b_m)$ witness that $(p|A)^{(n+1)} \overset{a}{\not\perp} (q|A)^{(m+1)}$, hence if $n \geq m$, $(p|A)^{(n+1)} \overset{a}{\not\perp} (q|A)^{(n+1)}$, completing the proof of the lemma.

The next example exhibits a simple situation in which nonorthogonal types p and q are based on a set A, but are almost orthogonal over this set.

Example 5.6.3. Let M be an infinite direct sum of copies of the group $\mathbb{Z}_2$ (in the language $\{+, 0\}$) and let N be a subgroup of M of index 2. Add to the language a unary predicate P and let M' be the expansion of M interpreting P by N. Then, $P(x)$ and $\neg P(x)$ are strongly minimal formulas. Let p and q be the unique nonalgebraic completions of $P(x)$, $\neg P(x)$, respectively, over $\emptyset$. For any realizations a and a' of q in M' there is an automorphism of the model fixing N pointwise and taking a to a'. As the reader can verify, this implies that $p \stackrel{a}{\perp} q$. However, any element of $M' \setminus N$ is interalgebraic with an element of N over a, hence $q|a \stackrel{a}{\not\perp} p|a$.

Remark 5.6.2. Let p, $q \in S(A)$ be nonorthogonal regular types in a stable theory. By the previous proposition there are m and n such that $p^{(n)}$ and $q^{(m)}$ are not almost orthogonal. We can ask: What are the minimal such m and n. For example, when p and q are modular strongly minimal types (in an uncountably categorical theory) we can take m and n to be 1 by Corollary 4.3.5. However, when p and q are both locally modular and nonmodular the minimal pair may be $m = n = 2$. In general, the answer to this question is a deep result in geometrical stability theory worked out in [Hru89].

Definition 5.6.3. *The complete type p is orthogonal to the set A, written $p \perp A$, if $p \perp q$ for all $q \in S(A)$.*

The results in [She90] and [SHM84] show how families of nonorthogonal types can lead to many nonisomorphic models in a fixed cardinality. The presence of a type orthogonal to a set leads to arbitrarily large families of pairwise orthogonal types.

Example 5.6.4. Consider first the $\omega-$stable, $\omega-$categorical theory T_1 of a single equivalence relation E with infinitely many infinite classes and no finite classes. Let M be a model (which is an $a-$model as it is $\aleph_0-$saturated). For some $a \in M$ let $p \in S_1(M)$ be the unique nonalgebraic type containing $E(x, a)$. It is left to the reader to show (using the elimination of quantifiers) that $tp(\bar{b}/M)$ is nonorthogonal to p only if some element of the sequence $\bar{b}$ is $E-$equivalent to a, hence forks over $\emptyset$. If $q \in S(\mathfrak{C})$ does not fork over $\emptyset$ then q is based on M. We have shown that $q|M$, hence q, is orthogonal to p. Thus, p is orthogonal to $\emptyset$.

Now consider the theory T_2 of two equivalence relations E and E' such that

- E and E' have infinitely many infinite classes,
- E and E' have no finite classes, and
- for all a and b there are infinitely many elements which are $E-$equivalent to a and $E'-$equivalent to b.

(T_2 is known as the theory of two cross-cutting equivalence relations.) The reader can check that for any a and b, $E(x, a)$ and $E'(x, b)$ both have Morley rank 2, $E(x, a) \wedge E'(x, b)$ has Morley rank 1 and each of these formulas has degree 1. Let M be a model, $a \in M$, p the unique element of $S_1(M)$ of Morley rank 2 containing $E(x, a)$ and q the unique element of $S_1(M)$ containing

$\neg E(x,b) \wedge \neg E'(x,b)$ for all $b \in M$. Let c be a realization of q and d an element satisfying $E(x,a) \wedge E'(x,c)$. The specified equivalences imply that d realizes p and depends on c over M. Thus, p is nonorthogonal to q. Since q is based on $\emptyset$ this type witnesses that p is nonorthogonal to $\emptyset$. This gives a type which forks over $\emptyset$ but is nonorthogonal to $\emptyset$.

Unraveling the definitions shows that $p \perp A$ if and only if $p \perp acl(A)$. The next proposition is the key to understanding this relation.

Proposition 5.6.2. *For T a stable theory, p a stationary type and A a set, the following are equivalent.*

(1) $p \perp A$.
(2) *If p is based on B and A' is independent from B over A, then p is orthogonal to A'.*
(3) *For any set B on which p is based, if f is an automorphism fixing $acl(A)$ with $f(B) \underset{A}{\perp\!\!\!\perp} B$, then $p|B \perp f(p|B)$.*

Proof. We can assume without loss of generality that $A = acl(A)$.

(1) $\Longrightarrow$ (2). Suppose p is based on B, $p \perp A$, $A' = acl(A')$ is independent from B over A and, to the contrary, that $q \in S(A')$ is nonorthogonal to p. By Proposition 5.6.1 there are a and b which are finite Morley sequences over $A' \cup B$ in p and q, respectively, and are dependent over $A' \cup B$. Since a is independent from A' over B, a must depend on $A' \cup \{b\}$ over B. Since B is independent from $A' \cup \{b\}$ over A this set witnesses that $stp(a/B)$ is nonorthogonal to A. For some n, $stp(a/B) = (p|B)^{(n)}$, so Lemma 5.6.1(ii) implies that p is nonorthogonal to A, a contradiction which proves the implication.

(2) $\Longrightarrow$ (3). This part holds trivially.

(3) $\Longrightarrow$ (1). Suppose $p \in S(B)$, $p \not\perp q$ for some $q \in S(A)$ and, towards a contradiction, that p is orthogonal to $p' = f(p)$, where f is as in (3). Let $\{B_i : i \in I\}$ be an A-independent family of realizations of $stp(B/A)$, where $|I| = \kappa(T)$, and let p_i be a conjugate of p over B_i. Since all independent pairs of realizations of $stp(B/A)$ have the same type over A the pair (p_i, p_j) is conjugate to (p, p'), hence $p_i \perp p_j$, for $i \neq j \in I$. Let M be an a-model containing $A \cup \bigcup_{i \in I} B_i$, $q' = q|M$ and $p'_i = p_i|M$ (for $i \in I$). Since q is a stationary type over A and each p_i is conjugate over A to p, $q \not\perp p_i$, for each $i \in I$. By Lemma 5.6.1(i) there are b realizing q' and a_i realizing p'_i such that b depends on a_i over M, for each $i \in I$. Let $I' \subset I$ be a set of cardinality $< \kappa(T)$ such that b is independent from $\{a_i : i \in I\}$ over $M \cup \{a_i : i \in I'\}$. Then, for a fixed $j \in I \setminus I'$, the dependence over b and a_j over M forces a_j to be dependent on $\{a_i : i \in I'\}$ over M. However, Lemma 5.6.1(ii) implies that p_j is orthogonal to $\bigotimes_{i \in I'} p_i$. This contradiction completes the proof of the proposition.

5.6.2 Domination

The domination relation on stationary types addresses the second motivating question at the beginning of the section. Namely: When p and q are nonorthogonal stationary types, must bases of p and q (in a given model) have the same cardinality? This question is approached by defining the domination relation on types, which forces bases of p and q to have the same cardinality, then discussing (under the subsection on weight) when domination agrees with nonorthogonality. In this subsection we also point out some useful connections between a–isolation and dependence.

The domination relation on triples of sets was introduced in Definition 3.4.2 in the context of totally transcendental theories. Before discussing domination between types we study the extension of this notion to triples of sets in stable theories.

Definition 5.6.4. *For sets A, B and C we say that A is* dominated by B over C, *and write $A \lhd B\ (C)$, if for all sets D,*

$$D \underset{C}{\perp} B \implies D \underset{C}{\perp} A.$$

The sets A and B are said to be domination equivalent over C, *written $A \sqsupseteq\!\sqsubseteq B\ (C)$, if $A \lhd B\ (C)$ and $B \lhd A\ (C)$.*

The basic properties of the relation, especially as they relate to independence are found in

Lemma 5.6.2. *(i) Suppose that $A_0 \subset A$ and C is independent from A over A_0. Then*

$$B \lhd C\ (A_0) \implies B \lhd C\ (A).$$

(ii) If $A_0 \subset A$ and $B \cup C$ is independent from A over A_0, then

$$B \lhd C\ (A) \implies B \lhd C\ (A_0).$$

(iii) If $B_0 \subset B$, $B \lhd B_0 \cup C\ (A)$ and $C \underset{A}{\perp} B$, then $B \lhd B_0\ (A)$.

(iv) If $\{ A_i : i \in I \}$ is a family of sets which is independent over C and B_i is dominated by A_i over C for all $i \in I$, then $\{ B_i : i \in I \}$ is independent over C.

(v) (Transitivity) *If $A \supset B \supset C \supset D$, $A \lhd B\ (C)$ and $B \lhd C\ (D)$ then $A \lhd C\ (D)$.*

Proof. (i) Let D be independent from C over A. Then $D \cup A$ is independent from C over A_0 (by the transitivity of independence). Since $B \lhd C\ (A_0)$, $D \cup A$ is independent from B over A_0. We conclude that D is independent from B over A.

(ii) Suppose that D is independent from C over A_0. Without loss of generality, D is independent from A over $A_0 \cup C \cup B$. (If there is such a

D which is dependent on B over A_0, there is one satisfying this condition.) Since A is independent from $B \cup C$ over A_0, $D \cup B \cup C$ is independent from A over A_0. Thus, D is independent from $A \cup C$ over $A_0 \cup C$ which, combined with the original assumption about D, shows that D is independent from $A \cup C$ over A_0. Thus, D is independent from C over A and D is independent from B over A (since $B \lhd C\ (A)$). From the independence of D from A over A_0 we conclude that D and B are independent over A_0, as desired.

(iii) The proof of this, which is similar to (ii), is assigned in the exercises.

(iv) To simplify the notation let $C = \emptyset$. To prove the independence of the family of B_i's we prove by induction on $|X|$ that whenever X is a finite subset of I, $\{ A_i : i \in I \setminus X \} \cup \{ B_i : i \in X \}$ is independent. When $X = \emptyset$ it is true by hypothesis. For the inductive step let $j \in X$ and $Y = X \setminus \{j\}$. Since $\{ A_i : i \in I \setminus X \} \cup \{A_j\} \cup \{ B_i : i \in Y \}$ is independent and B_j is dominated by A_j, $\{ A_i : i \in I \setminus X \} \cup \{B_j\} \cup \{ B_i : i \in Y \}$ is independent, as required.

(v) The straightforward proof is left to the reader.

Definition 5.6.5. *For stationary $*-$types p and q we say that p* is dominated *by q and write $p \lhd q$ if there is a set A on which both p and q are based and realizations C and D of $p|A$ and $q|A$, respectively, such that C is dominated by D over A.*

If $p \lhd q$ and $q \lhd p$ we call p and q domination equivalent *and write $p \sqsupseteq q$.*

Remark 5.6.3. Notice that $\lhd$ is transitive on stationary types and $x \sqsupseteq y$ is an equivalence relation (exercise). Also, for p, q and r stationary, $p \lhd q \implies p \otimes r \lhd q \otimes r$. Thus, if p_i, q_i are stationary and $p_i \sqsupseteq q_i$, for $1 \le i \le n$, then $p_1 \otimes \ldots \otimes p_n$ is domination equivalent to $q_1 \otimes \ldots \otimes q_n$.

Again, we only prove properties for types, leaving the extension to $*-$types to the reader. Domination links two types within $a-$models through the following connections to $a-$isolation. The first result is little more than a rewording of the definitions. The subsequent proposition is a direct generalization of Lemma 3.4.7. Both proofs are left to the reader in the exercises.

Lemma 5.6.3. *If $B \supset C$ are sets and a is an element, then $\{a\} \cup B \lhd B\ (C)$ if and only if for all sets D independent from B over C, $stp(a/B) \models stp(a/D \cup B)$.*

Proposition 5.6.3. *Let M be an $a-$model and $B \supset A$. If B is $a-$atomic over $M \cup A$ then B is dominated by A over M. If A is finite (or of cardinality $< \kappa(T)$ when $\kappa(T)$ is regular) and B is dominated by A over M, then B is $a-$atomic over $M \cup A$.*

As an immediate consequence of the proposition we obtain:

Corollary 5.6.1. *If M is an $a-$model and B is $a-$atomic over $A \cup M$, then for any $b \in B \setminus M$, $b \not\perp_M A$.*

Corollary 5.6.2. *Suppose M is an a–model and p, $q \in S(M)$ are orthogonal. Then, for all cardinals κ, λ there is an a–model $N \supset M$ and sets I, J such that I is a basis for p in N, J is a basis for q in N, $|I| = \kappa$ and $|J| = \lambda$.*

Proof. Assigned as Exercise 5.6.9.

Proposition 5.6.4. *For p and q stationary types and M an a–model on which p and q are based, the following are equivalent:*

(1) $p \lhd q$.
(2) *There is a b realizing $q|M$ and an a realizing $p|M$ such that $a \lhd b\,(M)$.*
(3) *For any b realizing $q|M$ and any a–model $N \supset M \cup \{b\}$, $p|M$ is realized in N.*

Proof. Most of the work goes into proving (1) $\Longrightarrow$ (2). Suppose that A is a set on which p and q are based and there are c, d realizing $p|A$, $q|A$, respectively, with $c \lhd d\,(A)$. Without loss of generality, cd is independent from M over A. Let B be a subset of M of cardinality $< \kappa(T)$ over which both p and q are based. Choosing $A_0 \subset A$ a set of cardinality $< \kappa(T)$ over which $tp(cd/A)$ does not fork, Lemma 5.6.2 implies that $c \lhd d\,(A_0)$. Since M is an a–model there is a set $A_1 \subset M$ realizing $tp(A_0/B)$, and there are ab such that $tp(abA_1/B) = tp(cdA_0/B)$ and ab is independent from M over $A_1 \cup B$. The types of c and d over $M \cup A_0$ are $p|(M \cup A_0)$ and $q|(M \cup A_0)$, respectively, and these types are based on B, so a realizes $p|M$ and b realizes $q|M$. Furthermore, by $a \lhd b\,(A_1)$, $tp(c/M \cup A_0) = p|(M \cup A_0)$, the independence of ab from M over A_1, and Lemma 5.6.2(i), $a \lhd b\,(M)$. This completes the proof of (1) $\Longrightarrow$ (2).

(2) $\Longrightarrow$ (3) Letting a, b and M be as in (2), the proof of this part follows immediately from

Claim. There is an element a' realizing $p|M$ such that $stp(a'/M \cup \{b\})$ is a–isolated.

Let $A_0 \subset M$ be a set of cardinality $< \kappa(T)$ over which $tp(ab/M)$ does not fork. Let N be an a–prime model over $M \cup \{b\}$ and a' a realization of $stp(a/A_0 \cup \{b\})$ in N. By Lemma 5.6.2(ii) a' is dominated by b over A_0. Thus, the A_0–independence of b and M yields the A_0–independence of a' and M. We conclude that a' realizes $p|M$ and $stp(a'/M \cup \{b\})$ is a–isolated (simply because $a' \in N$), proving the claim.

That (3) $\Longrightarrow$ (1) follows from Proposition 5.6.3.

By the previous proposition and Proposition 5.6.3, given an a–model M and p, $q \in S(M)$ such that $p \lhd q$, there are a realizing p and b realizing q such that $ab \lhd b\,(M)$.

Corollary 5.6.3. *Given an a–model M, p, $q \in S(M)$ domination equivalent and an a–model $N \supset M$, there is a base for p in N of cardinality λ if and only if there is a base for q in N of cardinality λ.*

Proof. Suppose I is a basis for p in N of cardinality λ. For $a \in I$ there is a $b_a \in N$ realizing q such that $\{b_a, a\}$ is dominated by a over M. Let $J = \{ b_a : a \in I \}$. By Lemma 5.6.2(iv), J is a Morley sequence in q of cardinality λ.

5.6.3 Weight

The most basic problem in the development of a dimension theory is: Given a set A, do all maximal independent subsets of A have the same cardinality? If the answer is negative is there at least a measure of how widely the cardinalities of maximal independent sets can differ? Weight addresses these issues.

Definition 5.6.6. *Let T be stable and p a complete $*$−type over A.*

(i) Let $PWT(p)$ be the set of all λ such that given B realizing p there is an A−independent set C such that $c \not\downarrow_A B$, for all $c \in C$ and $|C| = \lambda$.

(ii) The pre-weight *of p ($pwt(p)$) is* sup $PWT(p)$.

(iii) Suppose that p is stationary. The weight *of p ($wt(p)$) is the supremum of $\{ pwt(p|C) : p$ is based on $C \}$.*

For any set B the pre-weight of B over A *is $pwt(tp(B/A))$, which we denote by $pwt(B/A)$. The* weight of B over A *($wt(B/A)$) is $wt(stp(B/A))$. If $A = \emptyset$ we omit it as usual.*

Remark 5.6.4. In Definition 3.4.1 pre-weight was defined for complete types in a t.t. theory. The reader can verify that the two notions agree in t.t. theories.

Pre-weight and weight are invariant under conjugacy. Thus, $pwt(B/A) = pwt(stp(B/A))$, and $wt(B/A) = wt(B'/A)$ for all sets A, B and B' with $tp(B'/A) = tp(B/A)$. Also, given $B' \subset B$, $pwt(B'/A) \leq pwt(B/A)$ and $wt(B'/A) \leq wt(B/A)$.

Remark 5.6.5. Given a superstable theory and a complete type p in finitely many variables, $PWT(p) \subset \omega$. (Suppose $p \in S(A)$, a realizes p and I is an A−independent set such that each $b \in I$ depends on a over A. There is a finite $J \subset I$ such that a is independent from $I \cup A$ over $J \cup A$. Any $b \in I \setminus J$ is independent from $A \cup \{a\}$ over $J \cup A$, hence independent from a over A. Thus, $I = J$, proving that I is finite.)

Remark 5.6.6. A complete strongly minimal type has weight 1. (See Exercise 5.6.10.)

If λ is any cardinal, λ^- is κ if $\lambda = \kappa^+$ and $\lambda^- = \lambda$ if λ is a limit cardinal. For countable stable theories, $\kappa(T)^-$ is always $\aleph_0$.

Lemma 5.6.4. *Let T be stable.*

(i) $wt(B/A) \leq \kappa(T)^- + |B|$.

(ii) If p is a stationary type and M is an a–model over which p is based, $wt(p) = pwt(p|M)$.

(iii) If p and q are stationary types and $p \lhd q$, then $wt(p) \leq wt(q)$.

(iv) Domination equivalent stationary types have the same weight.

Proof. (i) Since A is arbitrary here it suffices to prove the inequality for pre-weight instead of weight. Let $C = \{ c_i : i \in I \}$ be an A–independent set with c_i dependent on B over A, for all $i \in I$. By Proposition 5.2.1 there is a set $J \subset I$ of cardinality $< \kappa(T) + |B|^+$ (if $\kappa(T)$ is regular) and $\leq \kappa(T) + |B|$ (otherwise) such that B is independent from $C \cup A$ over $A \cup \{ c_i : i \in J \}$. The stated conditions on C force J to equal I, hence $|I|$ satisfies the restrictions placed on $|J|$. Whether $\kappa(T)$ is regular or singular, $|I| \leq \kappa(T)^- + |B|$. Since $pwt(B/A)$ is the supremum of the cardinalities of such sets I this proves (i).

(ii) It suffices to show that whenever

- A is a set on which p is based,
- a is a realization of $p|A$, and
- $C = \{ c_i : i \in I \}$ an A–independent set such that c_i depends on a over A, for all $i \in I$,

there are

- b realizing $p|M$ and
- an M–independent set $\{ c'_i : i \in I \}$ such that c'_i depends on b over M for all $i \in I$.

Let A be a set on which p is based, a a realization of $p|A$ and $C = \{ c_i : i \in I \}$ an A–independent set such that c_i depends on a over A, for all $i \in I$. We can assume that $A \supset M$ (an exercise left to the reader). The argument in (i) shows that $|I| < \kappa(T)$, so there is an $A' \subset A$ of cardinality $\leq \kappa(T)$ such that $\{a\} \cup C$ is independent from A over A' (see Proposition 5.2.1). Let $B \subset M$ be a set of cardinality $< \kappa(T)$ over which p is based. Since M is $\kappa(T)$–saturated there is a set $A'' \subset M$ realizing $tp(A'/B)$. In fact, A'' realizes $tp(A'/B \cup \{a\})$ since p is based on B and a is independent from both A' and M over B. Let $C' = \{ c'_i : i \in I \}$ be a family of sets such that $C' \cup A''$ is conjugate to $C \cup A$ over $B \cup \{a\}$ and C' is independent from M over $A'' \cup \{a\}$. These conditions imply that C' is not only A''–independent, but M–independent and a depends on c'_i over M for all $i \in I$. Thus, $|I| \leq pwt(p|M)$, completing the proof of (ii).

(iii) Let M be an a–model on which both p and q are based. By Proposition 5.6.4 there are a realizing p and b realizing q such that a is dominated by b over M. If $\{ c_i : i \in I \}$ is an M–independent set such that c_i depends on a over M for all $i \in I$, then the domination hypothesis guarantees that c_i depends on b over M for all $i \in I$. Thus, $pwt(p) \leq pwt(q)$. By (ii) and the fact that M is an a–model, $wt(p) \leq wt(q)$, completing the proof.

The proof of (i) leads us to ask if the weight of any stationary type is actually $< \kappa(T)$. However, even when T is superstable it is not clear that $stp(a/A)$ has finite weight. It is at least conceivable that there are A−independent sets of size n for arbitrarily large n witnessing that $wt(a/A) \geq n$. This will be shown to be impossible in the next subsection in a lengthy argument.

The next result links the weight of a union of two sets to the weights of the component sets.

Proposition 5.6.5 (Additivity). *Let T be stable and $B = \bigcup\{ B_i : i < \alpha \}$ a family of sets. Then*
(i) $wt(B/A) \leq \sum wt(B_i/A \cup B_{<i})$, where $B_{<i} = \bigcup\{ B_j : j < i \}$, and
(ii) If $\{ B_i : i < \alpha \}$ is A−independent then $wt(B/A) = \sum wt(B_i/A)$.

Proof. (i) Replacing A by a larger set A' independent from B over A such that $pwt(B/A') = wt(B/A)$, it suffices to show $pwt(B/A) \leq \sum wt(B_i/A \cup B_{<i})$. Without loss of generality, $A = \emptyset$. Let $\{ c_i : i \in I \}$ be an independent family witnessing that $pwt(B) \geq |I|$. We will write I as a union of a disjoint family of sets J_i, $i < \alpha$, such that $\{ c_j : j \in J_i \}$ witnesses that $wt(B_i/B_{<i}) \geq |J_i|$. Since $|I| = \sum_{i<\alpha} |J_i|$ this proves the necessary inequality.

Preliminarily, we choose by recursion, $I_i \subset I$, for $i < \alpha$, a maximal subset of $I_i^{\cap} = \bigcap_{j<i} I_j$ such that $\{ c_j : j \in I_i \} \cup \{B_{<i+1}\}$ is independent. The sets I_i, $i < \alpha$, form a descending chain of subsets of I whose intersection is empty since every c_i depends on $B = B_\alpha$. Let $J_i = I_i^{\cap} \setminus I_i$. (Intuitively, J_i is what is lost from the chain at stage i). Since $\bigcap_{i<\alpha} I_i = \emptyset$, $\bigcup_{i<\alpha} J_i = I$ and the J_i's certainly form a pairwise disjoint family. The definition of the chain of sets directly implies that $C_{I_j} = \{ c_k : k \in I_j \}$ is independent from $B_{<i+1}$, and c_j depends on B_i over $B_{<i} \cup C_{I_j}$ for every $j \in J_i$. Thus, the pre-weight of B_i over $B_{<i} \cup C_{I_j}$ is $\geq |J_i|$. By the independence of C_{I_j} from $B_{<i+1}$ we conclude that $wt(B_i/B_{<i}) \geq |J_i|$. This proves (i).

(ii) Replacing A by a larger set if necessary and then letting $A = \emptyset$ it suffices to show (by (i)) that $pwt(B) \geq \sum_{i<\alpha} pwt(B_i)$. Let $C_i = \{ c_j^i : j \in J_i \}$ be an independent family such that c_j^i depends on B_i for all j. Without loss of generality, C_i is independent from $\{ B_j \cup C_j : j \neq i \}$ over B_i. By the transitivity of independence, $\{ B_i \cup C_i : i < \alpha \}$ is independent. Thus, $C = \bigcup_{i<\alpha} C_i$ is independent and each element of C depends on B. Since $|C| = \sum_{i<\alpha} |C_i|$, $pwt(B) \geq \sum_{i<\alpha} pwt(B_i)$.

As the following corollary suggests we can expect a reasonable dimension theory on the realizations of a weight 1 type.

Corollary 5.6.4. *Let T be stable and $\mathcal{B}$ a collection of sets such that $wt(B/A) = 1$, for all $B \in \mathcal{B}$. Let $\bar{B} = \{ B_i : i \in I \}$ and $\bar{C} = \{ C_j : j \in J \}$ be maximal A−independent subsets of $\mathcal{B}$. Then, $|I| = |J|$.*

Proof. By Proposition 5.6.5(ii), $wt(\bigcup \bar{B}/A) = |I|$. Since $\bar{B}$ is a maximal independent subset of $\mathcal{B}$, C_j depends on $\bar{B}$ over A, for all $j \in J$, hence $|J| \leq wt(\bar{B}/A) = |I|$. For the same reasons, $|I| \leq |J|$.

Definition 5.6.7. *If $p \in S(A)$ is stationary and has weight* 1, *then all bases for p in C (where C is some set) have the same cardinality by the corollary. We call this cardinality the* dimension of p in C, *denoted* $\dim(p, C)$.

In the next subsection we show that in a stable theory in which $PWT(p) \subset \omega$, for all stationary p, a stationary type p is domination equivalent to a finite product of weight 1 types. This goes a long way towards reducing all problems about types vis-a-vis orthogonality and domination to the class of weight 1 types, as well as giving a good dimension theory on $\mathfrak{C}$.

Remark 5.6.7. While weight 1 types do have dimension, they have a weakness in one area. A frequently used feature of dimension on strongly minimal sets is its additivity: If φ is a strongly minimal formula defined over A and $N \supset M$ are models with $A \subset M$, $\dim(\varphi(N)/A) = \dim(\varphi(N)/M) + \dim(\varphi(M)/A)$. There are weight 1 types on which a corresponding additivity result fails (even for a−models). We can eliminate this pathology by working with a special class of weight 1 types called regular types. We will prove that every weight 1 stationary type in a superstable theory is domination equivalent to a regular type, giving us this more robust dimension theory in superstable theories.

We end with two corollaries which address the issue of linking the dimensions of different weight 1 types.

Corollary 5.6.5. *If T is stable and p, q are stationary types with p of weight* 1, *then $p \not\perp q \iff p \lhd q$. Thus, if q also has weight* 1, $p \not\perp q \iff p \sqsupseteq\!\sqsubseteq q$.

Proof. It follows directly from the definitions that $p \lhd q \implies p \not\perp q$ for any two stationary types. Supposing p and q to be nonorthogonal let M be an a−model on which both p and q are based and a, b realizations of p, q, respectively, which are dependent over M (see Lemma 5.6.1(i)). Suppose, towards a contradiction, that a is not dominated by b over M. Then there is a c independent from b over M which depends on a over M. This is impossible since $wt(a/M) = 1$. Thus, $p \lhd q$.

Corollary 5.6.6. *Let T be stable, M an a−model and p, $q \in S(M)$ weight 1 types.*

(i) If $p \perp q$, then for all $\kappa \geq |M|$ there is an a−model $N \supset M$ such that $dim(p, N) = \kappa$ and $dim(q, N) = 0$.

(ii) If $p \not\perp q$, then for all a−models $N \supset M$, $dim(p, N) = dim(q, N)$.

Proof. The reader is asked to combine the relevant results in the exercises.

Thus, all of the questions on page 274 can be answered quite satisfactorily for dimension on weight 1 types over a−models.

5.6.4 Finite Weight

For T a stable theory let

$$PWT(T) = \bigcup \{ PWT(p) : p \text{ a stationary type in } T \}.$$

(In this definition we allow only types, not $*$−types.) In this subsection we study weight and domination in a stable theory in which $PWT(T) \subset \omega$. This class of theories includes the superstable theories (see Remark 5.6.5). The key result, which follows, goes a long way towards reducing any type to weight 1 types (at least as far as orthogonality, domination and dimension go).

Theorem 5.6.1. *Let T be a stable theory with $\kappa(T)$ regular and $PWT(T) \subset \omega$.*

(i) Then every stationary type in T has finite weight.

(ii) Moreover, given a stationary type p there are weight 1 types $q_1, \ldots, q_n$ (where $n = wt(p)$) such that $p \sqsupseteq q_1 \otimes \ldots \otimes q_n$.

Corollary 5.6.7. *Every stationary type in a superstable theory has finite weight.*

All of the work in the proof goes into showing (ii); (i) will follow quickly using some previously established facts about weight and domination.

Remark 5.6.8. Suppose T is stable, $wt(a/A) = 1$ and a depends on b over A. Then a is dominated by b over A. (Suppose, to the contrary, that there is a c independent from b over A such that c depends on a over A. Then the pair b, c witnesses that $wt(a/A) \geq 2$; contradiction.)

The following proposition, due to Tapani Hyttinen [Hyt95], is the key. The proof given here is largely due to Pillay.

Proposition 5.6.6. *Let T be stable and p a stationary type. If there is no weight 1 stationary type q dominated by p then for some nonforking extension p' of p, $\aleph_0 \in PWT(p')$.*

Proof. Let A_0 be a set over which p is based. Certainly p has weight > 1, so there is a set A_1 such that $pwt(p|A_1) > 1$. Choose b realizing $p|A_1$ and elements a_1, c_1 such that

(∗) $\{a_1, c_1\}$ is A_1−independent, b depends on a_1 over A_1 and b depends on c_1 over A_1.

Suppose there is a set $A_{11} \supset A_1$ such that

$$b \underset{A_0}{\perp} A_{11}, \quad c_1 \underset{A_1}{\not\perp} A_{11} \quad \text{and} \quad c_1 A_{11} \underset{A_1}{\perp} a_1. \tag{5.9}$$

Then, $\{a_1, c_1\}$ is A_{11}−independent, b depends on a_1 over A_{11} and b depends on c_1 over A_{11}. Iterating this process, let $A_1 \subset A_{11} \subset A_{1\alpha} \subset \ldots$, $\alpha < \lambda$ be a chain of sets such that for $A_1' = \bigcup_{\alpha<\lambda} A_{1\alpha}$ and all $\alpha < \beta < \lambda$,

$$b \underset{A_0}{\perp} A_1', \quad c_1 \underset{A_\alpha}{\not\perp} A_\beta \ \text{ and } \ c_1 A_1' \underset{A_1}{\perp} a_1.$$

Thus, $\lambda < \kappa(T)$. By choosing the $A_{1\alpha}$'s to be a maximal chain and replacing A_1 by A_1' we can require (in addition to $(*)$) that

$(**)$ there is no set $B \supset A_1$ such that

$$b \underset{A_0}{\perp} B, \quad c_1 \underset{A_1}{\not\perp} B \ \text{ and } \ c_1 B \underset{A_1}{\perp} a_1. \tag{5.10}$$

Without loss of generality, A_1 is an a–model.

Claim. $wt(c_1/A_1) \geq 2$.

Suppose $wt(c_1/A_1) = 1$. Since b depends on c_1 over A_1, c_1 is dominated by b over A_1. This contradicts that p does not dominate a weight 1 stationary type, to prove the claim.

We chose A_1 to be an a–model, so $pwt(c_1/A_1) \geq 2$ (Lemma 5.6.4(ii)). Let $\{a_2, c_2\}$ be an A_1–independent set such that c_1 depends on a_2 over A_1 and c_1 depends on c_2 over A_1. Choose a_2 and c_2 so that $c_2 a_2$ is independent from $b c_1 a_1$ over $A c_1$. Thus, $\{a_2, c_2, a_1\}$ is A_1–independent. Let $B = A_1 \cup \{c_2\}$. Then c_1 depends on B over A_1 and a_1 is independent from $c_1 B$ over A_1. By $(**)$, b depends on B over A_1; i.e., b depends on c_2 over A_1. As above there is a set $A_2 \supset A_1$ such that

- b is independent from A_2 over A_0,
- $\{a_1, a_2, c_2\}$ is A_2–independent,
- b depends on any element of $\{a_1, a_2, c_2\}$ over A_2 and
- there is no set $B \supset A_2$ such that

$$b \underset{A_0}{\perp} B, \quad c_2 \underset{A_2}{\not\perp} B \ \text{ and } \ c_2 B \underset{A_2}{\perp} a_1 a_2.$$

Continuing in this manner yields elements $a_1, a_2, a_3, \ldots$ and sets $A_1 \subset A_2 \subset A_3 \subset \ldots$ such that, letting $\bar{A} = \bigcup_{j<\omega} A_j$, for each $i < \omega$, b is independent from $\bar{A}$ over A_0, $\{a_1, \ldots, a_i\}$ is A_i–independent and independent from $\bar{A}$ over A_i, and b depends on a_i over each A_j, $j \geq i$. Thus, $\{a_i : i < \omega\}$ is $\bar{A}$–independent and b depends on each a_i over $\bar{A}$. Since $tp(b/\bar{A})$ is a nonforking extension of p the proposition is proved.

Corollary 5.6.8. *Let T be a stable theory such that $PWT(T) \subset \omega$. Then for any a–models $M \subset N$, $M \neq N$, there is a $a \in N \setminus M$ such that $tp(a/M)$ has weight* 1.

Proof. See Exercise 5.6.14.

Lemma 5.6.5. *Suppose that T is stable with $\kappa(T)$ regular, M is an a–model, $B \supset M$ and $p \in S(B)$ is a weight* 1 *stationary type nonorthogonal to M. Then there is a weight* 1 *type $q \in S(M)$ domination equivalent to p.*

Proof. Let $B_0 \subset B$ be a set of cardinality $< \kappa(T)$ such that p is based on $acl(B_0)$. Let $A \subset M$ be a set of cardinality $< \kappa(T)$ such that B_0 and M are independent over A (which exists since $\kappa(T)$ is regular) and $p \not\perp A$. Since M is an a-model there is a $B_1 \subset M$ realizing $stp(B_0/A)$. Since B_0 and B_1 are A-independent, Proposition 5.6.2 indicates that $p|acl(B_0)$ is nonorthogonal to its conjugate p' over $acl(B_1)$. Weight is preserved under conjugacy, so p' also has weight 1. Thus, $q = p'|M$ is a weight 1 type nonorthogonal to p. On weight 1 types nonorthogonality is the same as domination equivalence (Corollary 5.6.5). This proves the lemma.

Proof of Theorem 5.6.1. We prove part (ii) first. Without loss of generality, $p \in S(M)$ for an a-model M.

Claim. Let $N \supset M$ be an a-model and $C \subset N$ a maximal M-independent set of realizations of weight 1 types in $S(M)$. Then N is dominated by C over M.

Let M' be a maximal subset of N which is dominated by C over M. Notice that M' is an a-model. (Let $M'' \subset N$ be the a-prime model over M'. By Proposition 5.6.3, M'' is dominated by M' over M, hence by C over M. The maximality of M' forces M'' to equal M'.) Suppose, towards a contradiction, that $M' \neq N$. By Corollary 5.6.8 there is an $a \in N \setminus M'$ such that $wt(a/M') = 1$. If $tp(a/M')$ is orthogonal to M, $M' \cup \{a\}$ is dominated by M' over M, contradicting the maximality of M'. Thus, $tp(a/M')$ is nonorthogonal to M, yielding a $q \in S(M)$ of weight 1 domination equivalent to $tp(a/M')$ (by Lemma 5.6.5). Proposition 5.6.4 then gives a $b \in N$ such that $tp(b/M')$ is a nonforking extension of q. Then $C \cup \{b\}$ is an M-independent set of realizations of weight 1 types, contradicting the maximality of C to prove the claim.

Let a be a realization of p, N the a-prime model over $M \cup \{a\}$ and $C \subset N$ a maximal M-independent set of realizations of weight 1 types over M. Since every element of C depends on a over M (Corollary 5.6.1) and $pwt(p)$ is finite, C is finite. Let $C = \{c_1, \ldots, c_n\}$ and $q_i = tp(c_i/M)$, for $1 \leq i \leq n$. Since C is finite Proposition 5.6.3 implies that $a \sqsupseteq\!\!\sqsubseteq C\ (M)$. The type of C over M is $r = q_1 \otimes \ldots \otimes q_n$, so $p \sqsupseteq\!\!\sqsubseteq r$, proving (ii).

(i) Let p be a stationary type and $q_1, \ldots, q_n$ weight 1 types such that $p \sqsupseteq\!\!\sqsubseteq q_1 \otimes \ldots \otimes q_n$. By Proposition 5.6.5(ii), $wt(q_1 \otimes \ldots \otimes q_n) = n$. Part (i) now follows from (ii) and Lemma 5.6.4(iv) (which says that domination equivalent types have the same weight). This proves the theorem.

As stated in Remark 5.6.7 a full-featured dimension theory requires an additivity condition which may fail for weight 1 types. Simply knowing that every type in a superstable theory has finite weight does, however, have its applications. A good example is the following theorem by Lachlan, whose original proof (before weight was developed) was much harder.

Theorem 5.6.2 (Lachlan). *A countable superstable theory has* 1 *or infinitely many countable models.*

Proof. Assume, to the contrary, that T is a countable superstable theory which is not $\aleph_0$-categorical, but has finitely many countable models. By Lemma 2.3.1, T has a countable model M which realizes every complete type over $\emptyset$ and is prime over a finite set a. Let $n = wt(a)$. Since T is not $\aleph_0$-categorical there is a nonisolated type $p \in S(\emptyset)$. Let $b = \{b_0, \ldots, b_n\}$ be an independent set of realizations of p. Since M realizes p we may as well assume that $b \in M$. Since $tp(b_i)$ is nonisolated and $tp(b_i/a)$ is isolated Corollary 5.1.9 indicates that $a \not\perp b_i$, for all $i \leq n$. The independence of $\{b_0, \ldots, b_n\}$ now contradicts that $wt(a) = n$ to prove the theorem.

Remark 5.6.9. The alert reader will notice that the Baldwin-Lachlan Theorem is a special case of this theorem. Indeed, parts of the proof of the Baldwin-Lachlan Theorem given earlier are restricted versions of the proof of Lachlan's result.

Historical Notes. The concepts of orthogonality and weight are due to Shelah and found in [She90]. The domination relation on sets and types was developed by Lascar in [Las82]. Our exposition owes a great debt to Makkai [Mak84]. Theorem 5.6.1 is found for regular types (instead of weight 1 types) in [She90]. The generalization to stable theories with $PWT(T) \subset \omega$ was done by Pillay, with the key step due to by Hyttinen. Theorem 5.6.2 was proved by Lachlan in [Lac73] with an alternative proof found in [Las76].

Exercise 5.6.1. Prove Lemma 5.6.1(ii).

Exercise 5.6.2. Let p be a stationary type based on a set A, r a stationary type nonorthogonal to p and r' a conjugate of r over A. Show that r' is also nonorthogonal to A.

Exercise 5.6.3. Prove Lemma 5.6.2(iii).

Exercise 5.6.4. Prove Lemma 5.6.3.

Exercise 5.6.5. Prove Proposition 5.6.3.

Exercise 5.6.6. Prove that $x \lhd y$ is transitive on stationary types and $x \sqsubseteq y$ defines an equivalence relation.

Exercise 5.6.7. Suppose that p and q are strongly minimal types in a stable theory. Prove that $p \not\perp q$ if and only if $p \sqsubseteq q$ (without using Corollary 5.6.5).

Exercise 5.6.8. Suppose that $M \supset N$ are a-models, p, $q \in S(N)$ are domination equivalent and I is a basis for p in M. Show that there is a basis J for q in M such that $|J| = |I|$.

Exercise 5.6.9. Prove Corollary 5.6.2.

Exercise 5.6.10. Prove Remark 5.6.6.

Exercise 5.6.11. Let p be a weight 1 type nonorthogonal to $\emptyset$ and p' a conjugate of p over $acl(\emptyset)$. Prove that $p \not\perp p'$.

Exercise 5.6.12. Alter the vector space example (Example 5.6.3) slightly to produce two strongly minimal types p and q such that $p^{(2)} \overset{a}{\perp} q$, $p \overset{a}{\perp} q^{(2)}$, but $p^{(2)} \overset{a}{\not\perp} q^{(2)}$.

Exercise 5.6.13. Prove Corollary 5.6.6.

Exercise 5.6.14. Prove Corollary 5.6.8.

6. Superstable Theories

In Section 5.6 we defined the notion of a basis of a type p (relative to a set) and posed several questions on the behavior of the corresponding dimension function (including its well-definedness). We proved in Section 5.6.3 that on the class of weight 1 types dimension is well-defined, and nonorthogonality is the same as domination equivalence. In Section 5.6.4 we showed that in a superstable theory every type is domination equivalent to a finite product of weight 1 types. For a full-featured dimension theory, though, we need an additional property (additivity) which may fail for a weight 1 type (see Remark 5.6.7). In the second section of this chapter we develop the theory of a class of weight 1 types called the regular types in an arbitrary superstable theory. A regular type in a superstable theory satisfies the additivity property missing for weight 1 types (Proposition 6.3.2) and every weight 1 type is domination equivalent to a regular type.

This well-behaved dimension theory is at the heart of the solution of such problems as Morley's Conjecture for countable first-order theories (mentioned in the Preface). Regular types will be used to characterize the models of a "bounded" t.t. theory in Section 7.1.1.

Before turning to regular types we develop two notions of rank which are used in virtually every study of superstable theories.

6.1 More Ranks

Many of the properties proved for t.t. theories relied heavily on the existence of Morley rank; i.e., the fact that every type has Morley rank $< \infty$. The family of Δ–ranks served to define the forking dependence relation but, because it is a family of ranks instead of a single rank, it is missing many of properties of an ordinal-valued rank. Here we define two ranks which exist in superstable theories and provide a sharper measure of the complexity of formulas and types with respect to the forking relation.

Throughout the section we assume any mentioned theory to be stable.

Definition 6.1.1. *(i) In a stable theory we define the rank $U(-)$ on complete types by the following recursion. For p a complete type and α an ordinal,*

*$U(p) \geq \alpha$ if for all $\beta < \alpha$ there is a forking extension q of p such that $U(q) \geq \beta$. We write $U(p) = \alpha$ and say the $U-$*rank of *p* is *α if $U(p) \geq \alpha$ and $U(p) \not\geq \alpha + 1$. If $U(p) \geq \alpha$ for all α we write $U(p) = \infty$ and say the $U-$rank of p does not exist.*

(ii) For consistent formulas $\varphi(x)$ and α an ordinal or -1 the relation $R^\infty(\varphi)$ is defined by the following recursion.

(1) $R^\infty(\varphi) = -1$ *if φ is inconsistent;*
(2) $R^\infty(\varphi) = \alpha$ *if*

$$\{p \in S_n(\mathfrak{C}) : \varphi \in p \text{ and } \neg\psi \in p \text{ for all formulas } \psi \text{ with } R^\infty(\psi) < \alpha\}$$

is nonempty and has cardinality $< |\mathfrak{C}|$.

The relation $R^\infty(\varphi) = \alpha$ is read the $\infty-$rank of p is α. *When φ is consistent and $R^\infty(\varphi) \neq \alpha$ for all ordinals α we write $R^\infty(\varphi) = \infty$ and say that the $\infty-$rank of p does not exist. For p an arbitrary type, $R^\infty(p) =$* $\inf\{R^\infty(\varphi) : \varphi \text{ is implied by } p\}$.

When the $U-$rank of every complete type exists it gives a direct measure of forking: When $p \subset q$ are complete types, $U(p) = U(q)$ if and only if q is a nonforking extension of p (see the exercises). Thus, when the $U-$rank of every complete type exists it is natural to define $U(p)$, for $p \in S(\mathfrak{C})$, to be $U(p \restriction A)$, where A is any set over which p does not fork (the rank is the same over any such A). Be aware that $U-$rank is only defined on *complete* types (a point we will expand on later).

As the reader can see, the definition of $\infty-$rank is very similar to the definition of Morley rank, except that here the appropriate set of types is only required to have cardinality $< |\mathfrak{C}|$ rather than $< \aleph_0$. Many of the most basic properties of Morley rank extend to $\infty-$rank with the same proofs:

Lemma 6.1.1. *Let T be a complete theory, p an $n-$type and α an ordinal.*

(i) If $p \in S(\mathfrak{C})$ then $R^\infty(p) = 0$ if and only if p is algebraic.

(ii) $R^\infty(p) = \alpha$ if and only if $\{q \in S_n(\mathfrak{C}) : p \subset q$ and *$R^\infty(q) = \alpha\}$ is nonempty and has cardinality $< |\mathfrak{C}|$.*

(iii) If $R^\infty(p) = \alpha$ there is a $q \in S_n(\mathfrak{C})$ such that $q \supset p$ and $R^\infty(q) = \alpha$.

(iv) $R^\infty(p) \geq \alpha$ if and only if, for all $\beta < \alpha$ and all φ implied by p, $\{q \in S_n(\mathfrak{C}) : \varphi \in q$ and *$R^\infty(q) \geq \beta\}$ and has cardinality $< |\mathfrak{C}|$.*

(v)

$$R^\infty(\varphi) \text{ is the least ordinal } \alpha \text{ such that } \{p \in S_n(\mathfrak{C}) : \varphi \in p \text{ and } R^\infty(p) \geq \alpha\} \text{ has cardinality } < |\mathfrak{C}|. \tag{6.1}$$

In Shelah's terminology, $R^\infty(-) = R(-, L, \infty)$.

The basic existence properties are proved in

Lemma 6.1.2. *Let T be a stable theory.*

(i) Then T is superstable if and only if for every formula φ, $R^\infty(\varphi) < \infty$.

(ii) For every complete type p, $U(p) \le R^\infty(p)$.

Proof. The proofs of both parts rely on

Claim. For $p \in S(A)$, $q \supset p$ and α an ordinal, if $R^\infty(p) = \alpha$, then

$$q \text{ does not fork over } A \iff R^\infty(q) = \alpha.$$

Without loss of generality, $q \in S(\mathfrak{C})$. Suppose $R^\infty(q) = \alpha$ and let $Q \subset S(\mathfrak{C})$ be the set of extensions on p of ∞−rank α. Then Q has cardinality $< |\mathfrak{C}|$. Moreover, every conjugate over A of q is in Q, so q does not fork over A by Lemma 5.1.13. Now suppose that q does not fork over A. There is an extension r of p in $S(\mathfrak{C})$ having ∞−rank α, which we just proved does not fork over A. All nonforking extensions of p in $S(\mathfrak{C})$ are conjugate over A (by Corollary 5.1.8(ii)), hence r and q are conjugate. Therefore, $R^\infty(q) = \alpha$, proving the claim.

(i) First suppose each formula has ∞−rank. By the claim, each element of $S(\mathfrak{C})$ does not fork over a finite set, hence T is superstable.

Now suppose there is a formula φ which does not have ∞−rank. Then there is a complete type p over a set A such that $R^\infty(p) = \infty$. The nonsuperstability of T will follow from

Claim. If $p \in S(A)$ and $R^\infty(p) = \infty$ there is a forking extension q of p with $R^\infty(q) = \infty$.

Suppose, towards a contradiction, that there is no such q. Let $\alpha = \sup\{ R^\infty(q) : q \text{ is a forking extension of } p\}$. Consider $Q = \{q \in S(\mathfrak{C}) : q \supset p \text{ and } R^\infty(q) \ge \alpha + 1\}$. Then Q is nonempty and contains only nonforking extensions of p. Thus, $|Q| < |\mathfrak{C}|$. By Lemma 6.1.1(v), $R^\infty(p) < \infty$, a contradiction which proves the claim.

Proceeding with the main body of the proof, iterated use of the preceding claim generates an infinite chain $p_0 \subset p_1 \subset \ldots \subset p_i \subset \ldots$ of complete types such that p_{i+1} is a forking extension of p_i, for all i. This proves that T is not superstable, completing the proof of (i).

(ii) We need to show that $U(p) \ge \alpha \implies R^\infty(p) \ge \alpha$, for all complete types p and ordinals α. Using the first claim this becomes an easy induction which is relegated to the exercises.

It follows quickly from the first claim in the proof of the lemma that in a superstable theory any formula φ has $\le 2^{|T|}$ extensions in $S(\mathfrak{C})$ of the same ∞−rank; i.e., the bound in the definition of ∞−rank can be taken to be $2^{|T|}$ instead of an arbitrary cardinal $< |\mathfrak{C}|$.

Remark 6.1.1. It is actually the case that a stable theory T is superstable if and only if $R^\infty(x = x) < |T|^+$. This, however, is significantly harder to prove than (i) of the lemma (see [She90, II]. Since this refined bound has few uses we will not prove it here.

Example 6.1.1. (A formula of $\infty-$rank 1 which does not have Morley rank) Let T be the theory of infinitely many refining equivalence relations with only infinite classes such that E_0 has one class and E_{i+1} splits each E_i class into two classes. Then, the formula $x = x$ has $2^{\aleph_0}$ many nonalgebraic completions in $S(\mathfrak{C})$, hence $R^\infty(x = x) = 1$. Every nonalgebraic consistent formula in x has continuum many extensions in $S(\mathfrak{C})$, hence there is no formula of Morley rank > 0.

Definition 6.1.2. *(i) A formula of* $\infty-$*rank* 1 *is also known as a* weakly minimal formula.

(ii) A set defined by a weakly minimal formula in some model is called a weakly minimal set.

(iii) A type having a unique nonalgebraic completion over $\mathfrak{C}$ *is called a* minimal type.

(iv) The set of realizations of a minimal type in some model is called a minimal set.

Remark 6.1.2. A weakly minimal formula has $\leq 2^{|T|}$ many nonalgebraic completions over $\mathfrak{C}$. A complete type is minimal if and only if it is stationary and has $U-$rank 1. A formula φ over A is weakly minimal if and only if every nonalgebraic completion of φ over $acl(A)$ is minimal.

A strongly minimal formula is both a weakly minimal formula and a minimal type.

In some superstable theories $U-$rank and $\infty-$rank agree on the complete types, however there are relatively simple examples where $U-$rank and $\infty-$rank differ.

Example 6.1.2. (Where $U-$rank and $\infty-$rank agree) Consider the theory T_0 of a single equivalence relation E with infinitely many infinite classes and no finite classes. Up to conjugacy, there are three types in $S_1(\mathfrak{C})$: those containing $x = a$, for some a; the nonalgebraic types containing $E(x, a)$, for some a; and those containing $\neg E(x, a)$, for all a. These types have $U-$rank, $\infty-$rank and Morley rank 0, 1, and 2, respectively. Similarly, these ranks agree on $S_n(\mathfrak{C})$.

(Where $U-$rank and $\infty-$rank differ) Now we will define a theory consisting of "infinitely many disjoint copies of T_0". The language contains unary predicates U_i and binary predicates E_i, for $i < \omega$. The axioms for T say that the U_i's are pairwise disjoint and E_i defines a copy of T_0 on the elements satisfying U_i. Certainly, T is superstable (in fact, $\omega-$stable) and quantifier eliminable. Let $q \in S(\emptyset)$ be the unique 1−type containing $\neg U_i(x)$, for all $i < \omega$. Then $U(q) = 1$ and $R^\infty(q) = 2$. (Any formula in q is consistent with some U_i, hence has $\infty-$rank 2.)

There are identifiable properties of a theory which guarantee that $U-$rank and $\infty-$rank agree on all complete types. Isolating fairly broad classes of theories in which this is true is a difficult matter which is relegated to Section 7.2.

For now we simply remark that Morley rank and ∞−rank agree in uncountably categorical and $\aleph_0$−categorical, $\aleph_0$−stable theories, and Morley rank and ∞−rank agree with U−rank on the complete types in such theories.

Definition 6.1.3. *A map R which associates an ordinal with a complete type p is called a* notion of rank *if it satisfies:*

(1) *$R(p) = R(fp)$, for any $f \in \mathrm{Aut}(\mathfrak{C})$.*
(2) *If $q \supset p$ is complete, $R(q) \leq R(p)$.*
(3) *For each A containing the domain of p there is a $q \in S(A)$ containing p with $R(q) = R(p)$.*
(4) *If $p \in S(B)$ there is a finite $B' \subset B$ such that $R(p) = R(p \restriction B')$.*
(5) *There is a cardinal λ such that any type has at most λ extensions of the same rank over $\mathfrak{C}$.*

Let R be a notion of rank.

- *R is called* connected *if $R(p) = \alpha$ and $\beta \leq \alpha$ implies the existence of a complete type $q \supset p$ with $R(q) = \beta$.*
- *R is called* continuous *if whenever $p \in S(A)$ and $R(p) = \alpha$ there is a $\varphi \in p$ such that $R(q) \leq \alpha$ for all $q \in S(A)$ containing φ.*

Properties (1)–(5) are exactly what is needed of an ordinal-valued function on types to induce a freeness relation on the universe (see Definition 3.3.1). Morley rank, ∞−rank and the Δ−ranks are all continuous. A unique property of U−rank is that there are superstable theories in which it is *not* continuous. It follows that there is no notion of rank defined via formulas (as was ∞−rank) which agrees with U−rank on complete types in every superstable theory. (The example given above where U−rank and ∞−rank differ is also a counter-example to the continuity of U−rank.) On the other hand, U−rank is connected (see the exercises) while ∞−rank is not. (The type q in Example 6.1.2 has ∞−rank 2 and every forking extension has ∞−rank 0.)

It is tempting to extend U−rank to the incomplete types in a superstable theory T with the rule:

$$U(p) = \sup\{\, U(q) : \ q \supset p, \ q \in S(\mathfrak{C}) \,\},$$

where p is an arbitrary type in T. But there are choices for T and p for which this supremum is not attained; i.e., there is no $q \in S(\mathfrak{C})$, $q \supset p$, such that $U(q) = \sup\{\, U(q) : \ q \supset p, \ q \in S(\mathfrak{C}) \,\}$. This seriously limits the usefulness of U−rank on incomplete types. There is one general setting, however, in which it is appropriate to speak of the U−rank of an incomplete type. Suppose that G is a superstable group (i.e., a stable group $\bigwedge$−definable in a superstable theory). Proposition 5.3.1 says that the action of G on the generics in $S^G(G)$ is transitive, hence all generics have the same U−rank α (see the exercises). For p the type defining G we then define $U(p)$ to be α (which is the supremum of $\{\, U(q) : \ q \supset p$ and q complete$\}$).

That U–rank fails to be continuous is offset by the fact that it satisfies some additivity properties reminiscent of dimension in a strongly minimal set. If D is strongly minimal and $\bar{a}$, $\bar{b}$ are tuples from D, then $\dim(\bar{a}\bar{b}) = \dim(\bar{a}/\bar{b}) + \dim(\bar{b})$. We will see momentarily that U–rank satisfies a corresponding identity when all the relevant ranks are finite. When some of the types have infinite U–rank the irregularities of ordinal addition create problems. For example, if the sequences $\bar{a}$ and $\bar{b}$ from D are independent, then $\dim(\bar{a}/\bar{b}) = \dim(\bar{a})$ and $\dim(\bar{b}/\bar{a}) = \dim(\bar{b})$, so $\dim(\bar{a}\bar{b}) = \dim(\bar{a}) + \dim(\bar{b})$ and this is also $\dim(\bar{b}/\bar{a}) + \dim(\bar{a}) = \dim(\bar{b}) + \dim(\bar{a})$. Rewriting this with U–rank replacing dimension results in the identity: $U(\bar{a}) + U(\bar{b}) = U(\bar{b}) + U(\bar{a})$ for any two independent tuples in a superstable theory. Since addition of infinite ordinals is noncommutative this may not hold. We can, however, obtain useful inequalities if we replace ordinal addition by the *natural sum* (or *Hessenberg sum*) on ordinals (which is commutative). The *natural sum* of two ordinals α and β, denoted $\alpha \oplus \beta$, is defined recursively by the clause:

$$\alpha \oplus \beta = \sup(\{\, \alpha \oplus \beta' + 1 : \beta' < \beta \,\} \cup \{\, \alpha' \oplus \beta + 1 : \alpha' < \alpha \,\}).$$

The important features of $\oplus$ are:

(1) $\oplus$ agrees with $+$ when the ordinals are finite.
(2) $\oplus$ is commutative.
(3) If $\beta' < \beta$, $\alpha \oplus \beta' < \alpha \oplus \beta$.

We prove

Proposition 6.1.1 (Additivity). *If T is superstable, then for all a, b and A,*

$$U(a/A \cup \{b\}) + U(b/A) \leq U(ab/A) \leq U(a/A \cup \{b\}) \oplus U(b/A).$$

Proof. Without loss of generality, $A = \emptyset$. We prove by induction that for $\alpha = U(a/b) + U(b)$, $U(ab) \geq \alpha$. When α is a limit ordinal this follows quickly by induction. Suppose $\alpha = \beta + 1$. Then $U(b)$ must be a successor, say $\gamma + 1$. By the connectedness of U–rank there is a set B such that $U(b/B) = \gamma$. Without loss of generality, a is independent from B over b. Since $\delta = U(a/b)$ is an ordinal such that $\delta + \gamma + 1 = \beta + 1$ and $\delta = U(a/B \cup \{b\})$, $\delta + \gamma = U(a/B \cup \{b\}) + U(b/B) = \beta$. By induction, $U(ab/B) \geq \beta$. Since ab depends on B over $\emptyset$, $U(ab) \geq \beta + 1$, completing the proof of this part of the proposition.

Next, we prove by induction that $U(ab) \geq \alpha \implies U(a/b) \oplus U(b) \geq \alpha$. Again, it suffices to consider the case $\alpha = \beta + 1$. Thus, there is a set B on which ab depends such that $U(ab/B) = \beta$. Then b depends on B or a depends on B over b and $U(a/B \cup \{b\}) \oplus U(b/B) \geq \beta$ (by induction). These dependencies imply that $U(a/B \cup \{b\}) < U(a/b)$ or $U(b/B) < U(b)$, respectively. In either case, $U(a/B \cup \{b\}) \oplus U(b/B) < U(a/b) \oplus U(b)$. Thus, $U(a/b) \oplus U(b) \geq \beta + 1$, completing the proof.

Simply because $+$ and $\oplus$ agree on finite ordinals,

Corollary 6.1.1. *Suppose that T is superstable and $U(b/A)$ and $U(a/A \cup \{b\})$ are finite. Then*
(i) $U(ab/A) = U(a/A \cup \{b\}) + U(b/A)$ and
(ii) $U(a/A) - U(a/A \cup \{b\}) = U(b/A) - U(b/A \cup \{a\})$.

The second part of the corollary can be viewed as a strong form of the Symmetry Lemma. It says that a depends on b the same amount that b depends on a. In the following discussion we generalize this corollary to types of infinite U−rank.

While the above definition of $\oplus$ makes it easy to prove properties of the operation, it makes it difficult to compute particular values. The following equivalent definition shows how to perform the operation. Given an ordinal α there are ordinals $\beta_1 > \ldots > \beta_k$ and natural numbers $n_1, \ldots, n_k$ such that $\alpha = \omega^{\beta_1} \cdot n_1 + \omega^{\beta_2} \cdot n_2 + \ldots + \omega^{\beta_k} \cdot n_k$. This expression is unique (when each n_i is nonzero) and is called the *Cantor normal form* of α. Given two ordinals α and α', expand them as $\alpha = \omega^{\beta_1} \cdot n_1 + \ldots + \omega^{\beta_k} \cdot n_k$ and $\alpha' = \omega^{\beta_1} \cdot n_1' + \ldots + \omega^{\beta_k} \cdot n_k'$, where $\beta_1 > \ldots > \beta_k$ and n_i, $n_i' < \omega$. Then

$$\alpha \oplus \alpha' = \omega^{\beta_1} \cdot (n_1 + n_1') + \ldots + \omega^{\beta_k} \cdot (n_k + n_k').$$

(This is the definition given in [Hau62]. The equivalence of the two definitions is left to the reader.)

Ordinals of the form ω^β act as "limit points" of the operation $\oplus$ in the sense that $\alpha, \alpha' < \omega^\beta \implies \alpha \oplus \alpha' < \omega^\beta$. Generalizing, we write $\alpha \ll \omega^{\alpha_1} \cdot m_1 + \ldots + \omega^{\alpha_l} \cdot m_l$ (where $\alpha_1 > \ldots > \alpha_l$) if $\alpha < \omega^{\alpha_l}$. Then, $\alpha, \alpha' \ll \beta \implies \alpha \oplus \alpha' \ll \beta$. As the notation suggests $\alpha \ll \beta$ is one way of formally saying that α is much less than β.

Lemma 6.1.3. *Suppose that T is superstable and p, q are complete types such that $U(q) \ll U(p)$. Then p is orthogonal to q.*

Proof. Without loss of generality, p and q are stationary. Suppose, to the contrary, that p is nonorthogonal to q. Let A be a set on which both p and q are based such that there are a realizing $p|A$ and b realizing $q|A$ which are dependent over A. Let $U(p) = \beta + \omega^\alpha \cdot n$, where $\beta = 0$ or $\omega^\alpha \ll \beta$, and $U(q) = \gamma \ll U(p)$ (equivalently $\gamma < \omega^\alpha$). By Proposition 6.1.1, $\beta + \omega^\alpha \cdot n = U(a/A) \le U(ab/A) \le U(a/A \cup \{b\}) \oplus U(b/A)$. Since $U(a/A \cup \{b\}) < \beta + \omega^\alpha \cdot n$ and $\gamma < \omega^\alpha$, $U(a/A \cup \{b\}) \oplus U(b/A) = \beta + \omega^\alpha \cdot (n-1) + \gamma'$ for some $\gamma' < \omega^\alpha$. This contradicts the inequality and the stated value of $U(a/A)$.

Some applications of the additivity of U−rank are most easily stated through the following equivalence relation. For ordinals α, β and γ we write $\beta \approx_\alpha \gamma$ if there are β', $\gamma' < \omega^\alpha$ such that $\beta + \beta' = \gamma + \gamma'$. Intuitively, ordinals are $\approx_\alpha$ if they are "= modulo ω^α". The key ingredient in the proof of the following corollary is:

$$\beta,\ \gamma < \omega^{\alpha+1} \implies \beta + \gamma \approx_\alpha \beta \oplus \gamma. \tag{6.2}$$

(This follows quickly from the fact that an ordinal $< \omega^{\alpha+1}$ can be written as $\omega^\alpha \cdot n + \epsilon$, where $\epsilon < \omega^\alpha$.) Furthermore, we can do basic arithmetic "modulo ω^α" on ordinals. Given β, γ, δ, $\epsilon < \omega^{\alpha+1}$,

$$\beta \approx_\alpha \gamma \text{ and } \delta \approx_\alpha \epsilon \implies \beta + \delta \approx_\alpha \gamma + \epsilon,$$

and

$$\beta + \gamma \approx_\alpha \delta + \gamma \implies \beta \approx_\alpha \delta.$$

Corollary 6.1.2. *Suppose that T is superstable and both $U(a)$ and $U(b)$ are $< \omega^{\alpha+1}$. Then*
(i) $U(ab) \approx_\alpha U(a/b) + U(b)$ and
(ii) $U(a) \approx_\alpha U(a/b) + \omega^\alpha \cdot m \iff U(b) \approx_\alpha U(b/a) + \omega^\alpha \cdot m$.

Proof. (i) This follows immediately from Proposition 6.1.1 and (6.2).

(ii) It suffices to show the $\implies$ direction. Suppose that $U(a) \approx_\alpha U(a/b) + \omega^\alpha \cdot m$. Two applications of (i) and (6.2) show that $U(a/b) + U(b) \approx_\alpha U(b/a) + U(a)$. Substituting yields,

$$U(a/b) + U(b) \approx_\alpha U(b/a) + U(a/b) + \omega^\alpha \cdot m$$

and cancelation gives the desired equation $U(b) \approx_\alpha U(b/a) + \omega^\alpha \cdot m$.

Historical Notes. The rank $R^\infty(-)$ is defined in [She90, II] as $R(-, L, \infty)$ and Lemma 6.1.2(i) is Theorem 3.14 of that chapter. Lascar's U−rank is defined and developed in [Las76], where Proposition 6.1.1 is proved.

Exercise 6.1.1. Let T be superstable and $p \subset q$ complete types. Show that $U(p) = U(q)$ if and only if q is a nonforking extension of p.

Exercise 6.1.2. Prove Lemma 6.1.2(ii).

Exercise 6.1.3. Suppose that T is superstable and $a \in acl(A \cup \{b\})$. Then $R^\infty(a/A) \le R^\infty(ab/A) = R^\infty(b/A)$ and $U(a/A) \le U(ab/A) = U(b/A)$.

Exercise 6.1.4. Suppose that T is superstable, φ is formula over A and $R^\infty(\varphi) \ge k$, where $k < \omega$. Prove that there is a $p \in S(A)$ containing φ with $U(p) \ge k$. (HINT: Use induction.)

Exercise 6.1.5. Prove the connectedness of U−rank.

Exercise 6.1.6. Show that when T is stable and a and b are interalgebraic over A, $U(a/A) = U(b/A)$.

Exercise 6.1.7. Show that in a superstable group G all generic types have the same U−rank.

Exercise 6.1.8. Let T be a superstable theory and E a definable equivalence relation in T. Let a be an element and b the name for the $E-$class of a. Supposing that $U(a) = 5$ and $U(a/b) = 2$, compute $U(b)$.

Exercise 6.1.9. Give an example of a superstable theory containing a complete type of $U-$rank ω.

6.2 Geometrical Matters: A Dichotomy Theorem

Throughout Chapter 4 we saw how detailed information about the pregeometries in an uncountably categorical theory effected the overall structure of the theory. For example, when an uncountably categorical theory contains a locally modular strongly minimal set, the theory is 1−based. Here we prove (Theorem 6.2.1), saying that many superstable theories contain minimal sets which are locally modular. Basic consequences of this theorem will be stated without proof.

Let D be a minimal set, $\bigwedge-$definable over a set A in a stable universal domain $\mathfrak{C}$. Let $c\ell(-)$ be $acl(- \cup A) \cap D$. As with strongly minimal sets, $(D, c\ell)$ is a homogeneous pregeometry. As usual, when P is a property of a pregeometry and p is a minimal type we say that p has property P when $p(\mathfrak{C})$ has property P.

Theorem 6.2.1 (Dichotomy Theorem). *Let T be superstable and D a minimal set which is not locally modular. Then D is strongly minimal.*

This will be proved with several component results. First we reduce the problem to considering only minimal sets which are weakly minimal; i.e., minimal sets of $\infty-$rank 1.

Lemma 6.2.1. *Let T be superstable and D a minimal set which is nontrivial. Then D is weakly minimal.*

Proof. Without loss of generality, $D = p(\mathfrak{C})$ for a stationary type $p \in S(\emptyset)$. Since D is nontrivial we can also assume that there is $\{a, b, c\} \subset D$ which is pairwise independent but dependent. Let $\varphi(x, y, z) \in tp(abc)$ be such that, for all a', b', c',

$$\models \varphi(a', b', c') \implies a' \in acl(b', c') \text{ and } b' \in acl(a', c').$$

Let $\alpha = R^{\infty}(p)$ and $\theta \in p$ a formula of $\infty-$rank α. Let $\psi(x)$ be the formula $d_p y(\exists z(\varphi(x, y, z)))$ (see the notation on page 224). Then $\psi(x) \in p$ (since $\models \psi(a)$).

Claim. ψ is weakly minimal.

It suffices to show that for any $a' \in \psi(\mathfrak{C})$, $U(a') \leq 1$. (If $R^\infty(\psi) \geq 2$, there is a $q \in S(\emptyset)$ containing ψ such that $U(q) \geq 2$ by Exercise 6.1.4.) Choose a' satisfying ψ. By the definition of ψ there are b' and c' such that $b' \in D$, $b' \perp a'$, and $\models \varphi(a', b', c') \wedge \theta(c')$. Since $b' \in acl(a', c')$, $\alpha = R^\infty(b'/a') \leq R^\infty(c'/a') \leq R^\infty(c') \leq \alpha$. Thus, c' is independent from a' over $\emptyset$. Since $a' \in acl(b', c')$, $U(a'/c') \leq U(b'/c') \leq 1$. Thus, $U(a') \leq 1$, proving the claim and the lemma.

Definition 6.2.1. *Let D be a minimal set, $\bigwedge$−definable over $\emptyset$ in a superstable theory. A* plane curve *in D is a minimal subset of D^2.*

We call D linear *if whenever $p(\mathfrak{C})$ is a plane curve in D and c is interalgebraic with $Cb(p)$, $U(c) \leq 1$.*

Remark 6.2.1. Let D be a minimal set in a superstable theory, $\bigwedge$−definable over $\emptyset$, and D' a strongly minimal set over $\emptyset$ in a t.t. theory. Plane curves in D behave very much like plane curves in D'. The principle difference is that the canonical parameter of a plane curve in D' is an element of $(D')^{eq}$ and the canonical base of plane curve in D is an infinite subset of D^{eq}. However, if $X = p(\mathfrak{C})$ is a plane curve in D and $C = Cb(p)$, there is a $c \in Cb(p)$ such that c is interalgebraic with C. The resemblance between the two types of plane curves is even stronger. There is a $c \in Cb(p)$ such that, letting $q = tp(a/c)$ for $a \in X$, any $b \in D^2$ realizing q is in X. (This is left for the reader to prove.) It follows that $Cb(p) = dcl(c)$. Thus, we will call c a canonical base of X and identify X with the element c.

Lemma 6.2.2. *Let D be a minimal set in a superstable theory, $\bigwedge$−definable over $\emptyset$. The following are equivalent.*

(1) *D is locally modular,*
(2) *D/A is locally modular, for some set A,*
(3) *D is linear.*

Proof. This is proved just like the corresponding result for strongly minimal sets, Lemma 4.2.4.

Proof of Theorem 6.2.1. Let D be a minimal set, $\bigwedge$−definable over $\emptyset$, which is not locally modular. By the preceding lemma there is a plane curve $X \subset D^2$ such that for c a canonical base of X, $U(c) > 1$. By the connectedness of U−rank there is a set A such that $U(c/A) = 2$. Working over A, X is a plane curve in D/A with canonical base c, so we may as well assume that $U(c/\emptyset) = 2$.

An element of X has the form $a = (a_0, a_1) \in D^2$, where $a_0 \in acl(c, a_1)$ and $a_1 \perp c$. We will prove that $r = tp(a/c)$ has Morley rank 1. It follows that $tp(a_1/c)$ has Morley rank 1, hence D has Morley rank 1 (since $a_1 \perp c$), proving the theorem. The formula in $tp(a/c)$ of Morley rank 1 is found with a compactness argument applied to the following theory. Below a formula

$\psi(x, y)$ is called "provably algebraic in y" if for any b, $\psi(x, b)$ is algebraic; i.e., $\models \forall y \exists^{<n} x \psi(x, y)$, for some n. Given a realization d of $tp(c)$, let r_d denote the conjugate of r over d.

The lemmas proved earlier about families of plane curves in strongly minimal sets are true for minimal sets after $\dim(-)$ is replaced by $U(-)$. (This uses that $U-$rank satisfies the identities in Corollary 6.1.1.) Since r defines a plane curve whose canonical base is c, for any $d \neq c$ realizing $tp(c)$, $r(x) \cup r_d(x)$ is algebraic. By compactness, there are formulas $\sigma(x, y)$ $\psi(x, yz)$ such that

- $\psi(x, yz)$ is provably algebraic in yz,
- for any d realizing $tp(c)$, if $d \neq c$, $\sigma(x, c) \wedge \sigma(x, d)$ implies $\psi(x, cd)$.

Since D is weakly minimal, X is weakly minimal and we can choose σ so that

$$\begin{aligned} &\sigma(x, c) \text{ is weakly minimal and for all} \\ &a' \in \sigma(\mathfrak{C}, c),\ a' \not\perp c \text{ and } R^{\infty}(a') \leq 2. \end{aligned} \tag{6.3}$$

Combining this with the generalization of Lemma 4.2.6 yields

$$\begin{aligned} \text{for any } a' \in \sigma(\mathfrak{C}, c) \setminus acl(c),\ U(a') = R^{\infty}(a') = 2 \text{ and} \\ U(c/a') = U(c) - 1 = 1. \end{aligned} \tag{6.4}$$

With these pieces in place we can prove

Claim. $\sigma(x, c)$ has Morley rank 1.

We need to show that the formula $\sigma(x, c)$ has finitely many nonalgebraic completions in $S(acl(c))$. Let d be a realization of $stp(c)$ which is independent from c over $\emptyset$. Since $\sigma(x, c) \wedge \sigma(x, d)$ implies $\psi(x, cd)$ and $\psi(x, cd)$ is algebraic it suffices to show

(*) any nonalgebraic element of $S(acl(c))$ containing $\sigma(x, c)$ is realized in $\sigma(\mathfrak{C}, c) \cap \sigma(\mathfrak{C}, d)$.

Let $a' \in \sigma(\mathfrak{C}, c) \setminus acl(c)$. By (6.4), $U(a') = 2$ and $U(c/a') = 1$. Choose d' realizing $stp(c/a')$ which is independent from c over a'. We show that d' is independent from c with the following $U-$rank calculation. $U(d'ca') = U(d'c/a') + U(a') = U(d'/ca') + U(c/a') + U(a') = 1 + 1 + 2 = 4$. Also, $U(d'ca') = U(a'/d'c) + U(d'c) = U(a'/d'c) + U(d'/c) + U(c) = 0 + U(d'/c) + 2$. We conclude that $U(d'/c) = 2$, hence d' is independent from c. Since d' and d have the same strong type over $\emptyset$ and both are independent from c, $stp(d'/c) = stp(d/c)$. Let f be an automorphism of $\mathfrak{C}$ which fixes $acl(c)$ and maps d' to d. Then $f(a')$ is a realization of $stp(a'/c)$ in $\sigma(\mathfrak{C}, c) \cap \sigma(\mathfrak{C}, d)$. This proves (*), hence the claim and the theorem.

In Theorem 4.3.1 we showed that an uncountably categorical theory T has a locally modular strongly minimal set if and only if T is 1−based. This extends the simplicity of the pregeometry on a strongly minimal set to the simplicity of the entire universal domain with respect to forking dependence.

The following generalizes Theorem 4.3.1.

Theorem 6.2.2. *Let T be a superstable theory such that for each complete type p*

- *$U(p) < \omega$, and*
- *if $U(p) = 1$, p is locally modular.*

Then T is $1-$based.

We will not prove this theorem here. All of the key ideas are found in the proof of Theorem 4.3.1. The proof used the fact that the universal domain of an uncountably categorical theory is asm-constructible. We simply need a concept corresponding to "almost strongly minimal set" which is appropriate for minimal sets. The "semi-minimal sets" fill the gap. (See [Bue86] or [Pil]. There is also an exposition of this material in [Bue93].)

The theme underlying Theorem 6.2.1 is that geometrical and topological complexity cannot coexist. (A minimal set over $acl(A)$ which is not strongly minimal is a relatively complicated point in the Stone space topology on $S(acl(A))$.) The application which best displays Theorem 6.2.1 and Theorem 6.2.2 is the following corollary. We need to borrow a definition and results from later sections. A stable theory T is *unidimensional* if all nonalgebraic stationary types are nonorthogonal (Definition 7.1.1). In a unidimensional theory T every complete type has finite $U-$rank (by Corollary 7.2.1) and there is no strongly minimal set unless T is uncountably categorical. (See Examples 7.1.1 and 7.1.2.) Combining these facts with the Theorems 6.2.1 and 6.2.2 proves

Corollary 6.2.1. *Let T be a superstable unidimensional theory which is not uncountably categorical. Then T is $1-$based.*

Many of the results in Chapter 4 have faithful generalizations to superstable theories in which each type has finite $U-$rank. There is also a "geometrical theory" surrounding regular types (instead of minimal types). This material is expounded in [Pil].

The proof of Vaught's conjecture for superstable theories of finite rank depended heavily on the results in this section.

Historical Notes. Theorems 6.2.1 and 6.2.2 are due to Buechler, see [Bue85a] and [Bue86], respectively.

6.3 Regular Types

In this section we prove the facts about regular types outlined in the introduction of the chapter.

We assume throughout the section that any theory mentioned is stable.

Definition 6.3.1. *The nonalgebraic stationary type p is* regular *if for any set A over which p is based, any extension of* $p|A$ *which forks over A is orthogonal to p.*

It is clear that any strongly minimal type, in fact minimal type is regular. As (iii) of the next lemma says, one of the most basic properties of a regular type is that forking dependence is transitive on its set of realizations.

Lemma 6.3.1. *(i) For p a nonalgebraic stationary type the following are equivalent:*

(1) *p is regular;*
(2) *for some set A on which p is based, any extension of* $p|A$ *which forks over A is orthogonal to p;*
(3) *for some set A on which p is based,*

$$\text{when } a, b \text{ realize } p|A,\ b \underset{A}{\perp} C \text{ and } a \underset{A\cup C}{\not\perp} b,\quad a \underset{A}{\perp} C.$$

(ii) If $p \in S(A)$ *is a nonalgebraic stationary type, where* $|A| < \kappa(T)$*, and* $M \supset A$ *is an a−model, then p is regular if p is orthogonal to every forking extension of p in* $S(M)$.

(iii) (Transitivity) *Let* $p \in S(A)$ *be a regular type,* $\{a\} \cup \{b_i : i \in I\} \subset p(\mathfrak{C})$ *and C a set.*

$$\text{If } a \underset{A}{\not\perp} \{b_i : i \in I\} \text{ and } b_i \underset{A}{\not\perp} C \text{ (for } i \in I) \text{, then } a \underset{A}{\not\perp} C.$$

Proof. (i) The equivalence of (2) and (3) is simply a matter of rewording the relevant definitions, while (1) $\Longrightarrow$ (2) is trivial.

To prove (2) $\Longrightarrow$ (1) suppose (2) holds with A is in the statement, and B is a set on which p is based. Let $q = tp(c/C)$ be a stationary type which is a forking extension of $p|B$ and assume to the contrary that $q \not\perp p$. Since p is based on B any conjugate of q over B also satisfies these conditions. Thus, we can assume $C \cup \{c\}$ to be independent from A over B. In particular, c realizes the nonforking extension of $p|B$ over $A \cup B$; i.e., $p|(A \cup B)$. Thus, $q|(A \cup C)$ is a forking extension of the types: $p|B$, $p|(A \cup B)$ and $p|A$. By (2) q is orthogonal to $p|A$, contradicting the assumption that $q \not\perp p$.

(ii) This is left to the reader in the exercises.

(iii) Assume to the contrary that a is independent from C over A and, without loss of generality, $C \supset A$. By the transitivity of independence, a depends on $\{b_i : i \in I\}$ over C. Thus, there are $J \subset I$ and $j \in I \setminus J$ such that a is independent from $D = C \cup \{b_i : i \in J\}$ over C and a depends on b_j over D. However, the strong type of b_j over D is a forking extension of $p|A$, hence is orthogonal to p. Since $stp(a/D)$ is parallel to p this contradicts the dependence of a and b_j over D to prove (iii).

As promised, dimension is well-defined on regular types:

Proposition 6.3.1. *Every regular type has weight* 1.

Proof. We must show, given a regular type p based on A, $pwt(p|A) = 1$. For notational simplicity, take $A = \emptyset$. Suppose, towards a contradiction, that a is a realization of p and there are sets B and C such that $B \perp C$, $a \not\perp B$ and $a \not\perp C$. Without loss of generality, B is the universe of an a-model. Let $I \subset B$ be an infinite indiscernible set whose average type over B is $tp(a/B)$. Since $a \not\perp B$, $Av(I/B) \neq p|B$, hence $\{a\} \cup I$ must be dependent over $\emptyset$. Let J be a minimal subset of I such that $\{a\} \cup J$ is dependent. Since $\{a\} \cup I$ is indiscernible (see Lemma 5.1.17), any proper subset of $\{a\} \cup J$ is independent, hence a Morley sequence in p. Let $b \in J$ and $J' = J \setminus \{b\}$. By Lemma 6.3.1(iii),

$$b \not\perp_{J'} a \text{ and } a \not\perp C \implies b \not\perp_{J'} C.$$

This contradicts the independence of B and C to prove the proposition.

Following is the canonical example of a weight 1 type which is not regular.

Example 6.3.1. Let M be the direct sum of $\aleph_0$ copies of the group $(\mathbb{Z}_4, +)$. (This is also Example 3.5.1(iii).) We will show that the generic type of M has weight 1, but is not regular. Since M is simply a module over $\mathbb{Z}$ its quantifier eliminability down to the positive primitive formulas implies (after a little work):

- $T = Th(M)$ is categorical in every infinite cardinal and has Morley rank 2.
- M contains a unique strongly minimal subset definable over 0, namely $2M$ = the elements of order 2 in M.
- M and $2M$ are connected.

Let $p \in S(M)$ be the unique generic type and $q \in S(M)$ the generic type of $2M$. Since q is strongly minimal it is regular. To prove that p has weight 1 it suffices (by Lemma 5.6.4(iii) and Remark 5.6.6) to show that $p \triangleleft q$. Since every model of this theory is an a-model we need only show that when b is a realization of q and N is the prime model over $M \cup \{b\}$, p is realized in N (see Proposition 5.6.4). Let a be any element of N such that $2a = b$. An analysis of the possibilities for $tp(a/M)$ using the pp-elimination of quantifiers leads to the conclusion that a realizes p as desired.

Now let a and b be independent realizations of p and $a' = a + 2b$. Then, $a \not\perp a'$, $b \not\perp \{a, a'\}$ and $a \perp b$. Combining this with Lemma 6.3.1(ii), shows that p is not regular.

By Corollary 5.6.4, dimension is well-defined on a collection of realizations of a regular type. Furthermore, nonorthogonality is the same as domination equivalence on regular types (see Corollary 5.6.5). Combining previous facts concerning weight 1 types and the transitivity of dependence on the realizations of a regular type results in the following critical additivity property of dimension.

Proposition 6.3.2 (Additivity of Dimension). *Let T be stable, $M \subset N$ a–models, $A \subset M$ a set of cardinality $< \kappa(T)$ and $p \in S(A)$ a regular type. Then,*

$$\dim(p, N) = \dim(p, M) + \dim(p|M, N).$$

Proof. Let I be a basis for p in M and J a basis for $p|M$ in N. It suffices to show that $I \cup J$ is a basis for p in N. If a is an arbitrary element of $p(N)$, $\{a\} \cup J$ depends on M over A. By Lemma 5.4.2, $tp(ab/p(M) \cup A) \models tp(ab/M)$ for any finite $b \subset J$. Thus, $\{a\} \cup J$ depends on $p(M)$ over A, in fact $\{a\} \cup J$ depends on I over A by Lemma 6.3.1(ii). This proves that $I \cup J$ is a basis for p in N, as required.

Corollary 6.3.1. *If $p \in S(A)$ is regular and $C \supset B$ are subsets of $p(\mathfrak{C})$, then* $\dim(C/A) = \dim(C/A \cup B) + \dim(B/A)$.

Proof. Left to the exercises.

The proof of the proposition contains a proof of

Corollary 6.3.2. *Let M be an a–model, $A \subset M$ of cardinality $< \kappa(T)$, $p \in S(A)$ a regular type and I a basis for p in M. Then, given a Morley sequence J in p which depends on M over A, J depends on I over A.*

The collection of regular types provide us with a class on which dimension in a–models is particularly well-behaved:

Proposition 6.3.3. *Let T be superstable and $p \in S(A)$ a regular type, where A is finite.*

(i) If $M \supset A$ is a–prime over a finite set, then p has dimension $\aleph_0$ in M.

(ii) If I is an infinite Morley sequence in p and M is a–prime over $A \cup I$, then I is a basis for p in M.

(iii) For any a–model $M \supset A$, if $B \subset M$ is finite and $q \in S(B)$ is a regular type nonorthogonal to p, then $\dim(p, M) = \dim(q, M)$.

(iv) For any $\kappa \geq \lambda(T)$ there is an a–model $M \supset A$ of cardinality κ such that $\dim(p, M) = \kappa$ *and* $\dim(q, M) = \aleph_0$ *for any regular type q over a finite subset which is orthogonal to p.*

(v) Let M_0 be an a–model of cardinality $\lambda \geq \lambda(T)$ and $X \subset S(M_0)$ a set of regular types over M_0 such that there is a regular type $q \in S(M_0)$ orthogonal to every element of X. Then, there is an a–model M of cardinality λ containing M_0 such that $\dim(r, M) = 0$, *for all $r \in X$, and* $\dim(q, M) = \lambda$ *for any regular type q over a finite subset of M such that $q \perp r$ for all $r \in X$.*

Proof. (i) Suppose $M \supset A$ is a–prime over the finite set B. Recall from Theorem 5.5.2(iii) that when M is a–prime over a set it does not contain an uncountable set of indiscernibles over that set. Let I be a basis for p in M. There is a finite $J \subset I$ such that I is independent from B over $J \cup A$. Then

$I \setminus J$ is Morley sequence in $p|(A \cup B)$, which hence has cardinality $\aleph_0$ since M is a-prime over $A \cup B$.

(ii) This follows from the proof of Theorem 5.5.2(iii), which is relatively easy.

(iii) Let $N \subset M$ be an a-prime model over $A \cup B$. Since p and q are nonorthogonal regular types, they are domination equivalent. By Corollary 5.6.3, $\dim(p|N, M) = \dim(q|N, M)$. From (i) we conclude that $\dim(p, N) = \dim(q, N) = \aleph_0$, so (ii) follows from the additivity of dimension (Proposition 6.3.2).

(iv) Let I be a Morley sequence in p of cardinality κ and M an a-prime model over $A \cup I$. Let q be a regular type over a finite subset B of M. Let $J \subset I$ be a countably infinite set with $stp(B/A \cup J)$ a-isolated. Thus, there is $N \subset M$ an a-prime model over $A \cup J$ containing B. By Corollary 5.5.3, M is a-prime over $N \cup I$. By (i), J is a basis for p in N and by Corollary 6.3.2, $I \setminus J = I'$ is a Morley sequence in $p|N$. Since M is a-prime over $N \cup I'$, every element of $M \setminus N$ depends on I' over N (by Corollary 5.6.1). Thus, every type over N realized in $M \setminus N$ is nonorthogonal to p. In particular, $q|N$ is omitted in M, so the additivity of dimension implies that $\dim(q, M) = \dim(q, N) = \aleph_0$.

(v) We define an elementary chain of a-models, $M_0 \subset M_1 \subset \ldots \subset M_n \subset \ldots$, each of cardinality λ so that the union M has the desired properties. Suppose M_n has been defined and let $Q_n = \{q \in S(M_n) :$ q is a regular type orthogonal to every $r \in X\}$. Then Q_n is nonempty and $|Q_n| \leq \lambda$ (since T is λ-stable). Let I be an M_n-independent set of cardinality λ so that for each $q \in Q_n$, I contains a Morley sequence in q of cardinality λ and $tp(I/M_n) \perp r$, for all $r \in X$. Let M_{n+1} be an a-prime model over $M \cup I$. All of the desired properties are easily verified.

Remark 6.3.1. The relationships between the dimensions of nonorthogonal regular types when these dimensions are finite is a complicated matter. Consider, for example, a model M of an uncountably categorical theory and complete strongly minimal (hence regular) types p, q over a finite set A. If p and q are modular, then $\dim(p, M) = \dim(q, M)$. (See Corollary 4.3.5.) Using deeper results from geometrical stability theory it is possible to show that $\dim(p, M) = \dim(q, M)$ when both p and q are locally modular and nonmodular. (This was proved by Hrushovski and Laskowski in [Las88].) The relationship between $\dim(p, M)$ and $\dim(q, M)$ when p and q are nonorthogonal regular types over a finite set $A \subset M$ and M is a model is related to how the nonorthogonality of p and q is witnessed. As stated in Remark 5.6.2 this is a deep problem in geometrical stability theory.

The usefulness of regular types in superstable theories depends heavily on the following existence result.

Lemma 6.3.2 (Existence). *Suppose that T is superstable, $M \subset N$ are a-models, $A \subset M$ is finite and $q \in S(A)$ is realized in $N \setminus M$. Then, there is a regular type $p \in S(M)$ containing q which is realized in N.*

Proof. Let $p \in S(M)$ be an extension of q of least U-rank which is realized in $N \setminus M$. We will show that p is regular. Let $B \subset M$ be a finite set containing A on which p is based. Assuming, to the contrary, that p is not regular there is a finite set C, $B \subset C \subset M$ and an $r \in S(C)$ which is a forking extension of $p|B$ nonorthogonal to p (see Lemma 6.3.1(ii)). We contradict the minimality assumption on $U(p)$ as follows. Enlarging C if necessary we can assume by the nonorthogonality of the relevant types that there are a realizing $p|C$ and b realizing r which are dependent over C. Since N is an a-model realizing p we can require that $a \in N$ realizes p and b is in N. Since a depends on b over C and a is independent from M over C, b must be in $N \setminus M$. Since $U(r) < U(p)$ and $r \supset q$ this contradicts the minimality assumption on $U(p)$. We conclude that p is regular.

In the proof of the lemma we also show

Corollary 6.3.3. *For T a superstable theory, M an a-model and $p \in S(M)$, the following are equivalent.*

(1) *p is regular.*
(2) *There is an a-model $N \supset M$ realizing p and a finite set $A \subset M$ such that whenever $b \in N \setminus M$ realizes $p \restriction A$, b realizes p.*

This corollary can be useful in verifying that certain types are regular. Take, for example, the theory of an equivalence relation E with infinitely many infinite classes and no finite classes. Then, T is ω-stable and ω-categorical, so every model is an a-model. To verify that the unique $q \in S_1(\emptyset)$ is regular let M be any model, a a realization of $q|M$ and N the prime model over $M \cup \{a\}$. Our knowledge of the models of this theory tells us that every element of $N \setminus M$ is E-equivalent to a, hence not E-equivalent to any element of M. That is, every element of $N \setminus M$ realizes $q|M$. We conclude that q is regular by the corollary.

The proof of the lemma also shows that for T a 1-sorted superstable theory, every regular type in T^{eq} is nonorthogonal to a regular 1-type in the sort of T (see the exercises).

Concerning possible extensions of Lemma 6.3.2, it is true that for any two distinct *models* $M \subset N$ of a superstable theory there is a regular type over M realized in N. However, the proof of this more general result, found in [SB89], is considerably more difficult than the one above and requires more sophisticated machinery. We will not reproduce the proof here since the above result together with the existence lemma for strongly regular types in t.t. theories proved later, are sufficient to prove a high percentage of the existing results. The hypothesis that T is superstable in the lemma is necessary, though. There is an example of a countable stable theory having distinct $\aleph_1$-saturated models $M \subset N$ with no regular type over M realized in N.

Corollary 6.3.4. *Given a weight 1 type p in a superstable theory there is a regular type q domination equivalent to p.*

Proof. Without loss of generality, $p \in S(M)$, for M some a–model. Let a be a realization of p and N an a–prime model over $M \cup \{a\}$. By Lemma 6.3.2 there is a regular type $q \in S(M)$ realized in N. This forces p and q to be nonorthogonal, hence $p \sqsupseteq\!\!\!\!\sqsubseteq q$, as needed.

Corollary 6.3.5. *Let T be superstable, $M \subset N$ a–models and $C \subset N$ a maximal M–independent set such that, for each $c \in C$, $tp(c/M)$ is regular. Then N is dominated by C over M.*

Proof. Combine the claim in the proof of Theorem 5.6.1 with the preceding corollary.

The following decomposition theorem allows us to deal only with regular types in many settings. Recall from Corollary 5.6.7 that every stationary type in a superstable theory has finite weight. This can be extended to:

Corollary 6.3.6 (Decomposition Theorem). *If T is superstable and p is a stationary type there are regular types $q_1, \ldots, q_n$, for $n = wt(p)$, such that $p \sqsupseteq\!\!\!\!\sqsubseteq q_1 \otimes \ldots \otimes q_n$. In fact, if M is an a–model on which p is based, a realizes $p|M$ and N is a–prime over $M \cup \{a\}$, N is also a–prime over $M \cup \{b\}$, where $b = \{b_0, \ldots, b_n\}$ is a maximal M–independent sequence of elements realizing regular types over M.*

Proof. Given a stationary type p, by Theorem 5.6.1 there are weight 1 types $r_1 \ldots, r_n$ such that $p \sqsupseteq\!\!\!\!\sqsubseteq r_1 \otimes \ldots \otimes r_n$. From Corollary 6.3.4 we get, for each $1 \leq i \leq n$, a regular type q_i domination equivalent to r_i. Using Remark 5.6.3, p, $r_1 \otimes \ldots \otimes r_n$ and $q_1 \otimes \ldots \otimes q_n$ are domination equivalent.

The second part of the corollary follows from the claim in the proof of Theorem 5.6.1 and Corollary 6.3.4.

Any sequence $\{q_0, \ldots, q_{n-1}\}$ of regular types such that $p \sqsupseteq\!\!\!\!\sqsubseteq \bigotimes_{i<n} q_i$ is called a *regular decomposition of p*. Part (i) of the following corollary addresses the uniqueness of a regular decomposition. The other parts of the corollary indicate the degree to which this decomposition theorem reduces orthogonality and domination on all stationary types to considering only regular types.

Corollary 6.3.7. *Let T be superstable.*

(i) If $\{q_0, \ldots, q_{n-1}\}$ and $\{q'_0, \ldots, q'_{n'-1}\}$ are regular decompositions of p and p', respectively, then $p' \lhd p$ if and only if there is a one-to-one map of indices $j < n'$ into indices $i_j < n$ such that $q'_j \sqsupseteq\!\!\!\!\sqsubseteq q_{i_j}$ for all $j < n'$. Thus, $p \sqsupseteq\!\!\!\!\sqsubseteq p'$ if and only if there is a bijection $j \mapsto i_j$ such that $q'_j \sqsupseteq\!\!\!\!\sqsubseteq q_{i_j}$ for all $j < n'$.

(ii) Given stationary types p, p', r and r', if $r \sqsupseteq\!\!\!\!\sqsubseteq r'$ then $p' \otimes r' \lhd p \otimes r$ implies $p' \lhd p$.

(iii) For any stationary p and p', $p \not\perp p'$ if and only if there is a regular q nonorthogonal to both p and p'.

Proof. (i) Suppose that $M \subset N$ are a-models and C is an M-independent set of realizations of regular types over M such that N is dominated by C over M. Let X be an equivalence class of the regular types over M with respect to domination equivalence and C_X the elements c of C with $tp(c/M) \in X$.

Claim. C_X is a basis for $D = \{ d \in N : tp(d/M) \in X \}$.

Since C is M-independent and every element of $C \setminus C_X$ realizes a type over M orthogonal to the elements of X, $tp(C/M \cup C_X)$ is orthogonal to each element of X. For $d \in D$,

$$\text{since } d \not\downarrow_M C \text{ and } d \downarrow_M C_X,\ d \not\downarrow_{M \cup C_X} C,$$

a contradiction from which we conclude that d depends on C_X over M, as claimed.

Now suppose that p, p', the q_i's and q_i''s are as hypothesized, all are based on the a-model M, $p' \lhd p$, a realizes $p|M$ and N is the a-prime model over $M \cup \{a\}$. Let $c = (c_0, \ldots, c_{n-1})$ and $c' = (c_0', \ldots, c_{n'-1}')$ be realizations of $\bigotimes_{i<n} q_i|M$ and $\bigotimes_{j<n'} q_j'|M$, respectively, in N. Since N is a-prime over $M \cup \{c\}$, c is a basis for the realizations of the regular types over M in N. The argument in the first paragraph shows that we can partition the c_i's and c_i''s into equivalence classes with respect to the nonorthogonality of their types and the dimension of a class of c_i''s is $\le$ the dimension of the corresponding class of c_i's. This indicates how to define the one-to-one map required in the statement.

The other direction of the biconditional is clear.

Parts (ii) and (iii) follow quickly from (i).

Applications of this decomposition in terms of regular types will be seen in Section 7.1.

6.3.1 Rank Considerations

We assume every theory mentioned in this subsection to be superstable. In Lemma 6.1.3 we showed that widely different U-ranks imply the orthogonality of the two types. It is a short step from there to

Lemma 6.3.3. *If p is a stationary complete type of U-rank ω^α for some α then p is regular.*

Proof. Since $\beta < \omega^\alpha \implies \beta \ll \omega^\alpha$ any forking extension of p is orthogonal to p by Lemma 6.1.3.

We prove in the next proposition that the types of U-rank ω^α are the canonical regular types. Properties of dependence on these special regular types are occasionally easier to prove.

Proposition 6.3.4. *If p is a stationary complete type of weight 1 and $U(p) = \gamma + \omega^\alpha \cdot n$ where $\omega^\alpha \ll \gamma$ or $\gamma = 0$, then*

- *there is a set A on which p is based,*
- *an a realizing $p|A$, and*
- *a $c \in acl(A \cup \{a\})$ with $U(c/A) = \omega^\alpha$.*

A fortiori, p is nonorthogonal to a type of U−rank ω^α.

The proposition says, for instance, that when (in addition) $U(p)$ is finite p is nonorthogonal to a type of U−rank 1. Dependence on the set of realizations of a U−rank 1 type is algebraic dependence which is particularly easy to study. This reduction to types of U−rank 1 is often helpful when working in the finite rank context. Before turning to the main body of the proof we prove a general result allowing us to consider only the case $\gamma = 0$.

Lemma 6.3.4. *If $U(c) = \omega^{\alpha_1} \cdot n_1 + \ldots + \omega^{\alpha_k} \cdot n_k$ where $\alpha_1 > \ldots > \alpha_k$ and $1 \leq l \leq k$, there is a $c' \in acl(c)$ such that $U(c/c') = \omega^{\alpha_1} \cdot n_1 + \ldots + \omega^{\alpha_l} \cdot n_l$ and $U(c') = \omega^{\alpha_{l+1}} \cdot n_{l+1} + \ldots + \omega^{\alpha_k} \cdot n_k$.*

Proof. First consider the case $l = 1$ and simplify the notation by writing $U(c) = \omega^\alpha \cdot n + \beta$, where $\beta < \omega^\alpha$.

By the connectivity of U−rank there is a d such that $U(c/d) = \omega^\alpha \cdot n$. Without loss of generality, $d \in Cb(c/d)$.

Claim. $U(d) < \omega^\alpha$.

Choosing $\{c_0, \ldots, c_m\}$ to be a Morley sequence in $stp(c/d)$ on which this strong type is based, $d \in acl(c_0, \ldots, c_m)$. By repeated application of Corollary 6.1.2, $U(c_0 \cdots c_m d) = U(c_0 \cdots c_m) \approx_\alpha \omega^\alpha \cdot n(m+1) \approx_\alpha U(c_0 \cdots c_m/d)$. Since $d \in acl(c_0, \ldots, c_m)$, $U(d)$ is certainly $< \omega^{\alpha+1}$. Thus, Corollary 6.1.2 says that $U(d) \approx_\alpha 0$; i.e., $U(d) < \omega^\alpha$.

Let c' be an element of $C = Cb(d/c)$ such that d is independent from c over c'. Then $c' \in acl(c)$ and $C \subset acl(c')$. Since c is independent from d over c', $U(c/c') = U(c/c'd) \leq U(c/d) = \omega^\alpha \cdot n$. As in the proof of the claim, c' is algebraic in a Morley sequence in $stp(d/c)$ and $U(c') < \omega^\alpha$. It is easy to verify that $\delta' \ll \delta \implies \delta + \delta' = \delta \oplus \delta'$. Thus, the additivity of U−rank implies $U(c/c') + U(c') = U(cc') = U(c)$; i.e., $\omega^\alpha \cdot n + U(c') = \omega^\alpha \cdot n + \beta$. Since $\beta < \omega^\alpha$ we conclude that $U(c') = \beta$, as desired.

Applying the $l = 1$ case repeatedly yields elements $c_1, \ldots, c_{k-1}$ such that, letting $c_0 = c$, $c_{i+1} \in acl(c_i)$, $U(c_i/c_{i+1}) = \omega^{\alpha_{i+1}} \cdot n_{i+1}$ and $U(c_{i+1}) = \omega^{\alpha_{i+2}} \cdot n_{i+2} + \ldots + \omega^{\alpha_k} \cdot n_k$. The argument used at the end of the previous paragraph shows that $U(c/c_i) = \omega^{\alpha_1} \cdot n_1 + \ldots + \omega^{\alpha_i} \cdot n_i$ whenever $1 \leq i < k$. Thus, the conditions of the lemma are satisfied by taking $c' = c_l$.

Proof of Proposition 6.3.4. By Lemma 6.3.4 there is a set B on which p is based, a realizing $p|B$ and $a' \in acl(A \cup \{a\})$ with $U(a'/A) = \omega^\alpha \cdot n$. Since

$wt(a'/A) = 1$ it suffices to prove the proposition for $stp(a'/A)$ instead of p, so without loss of generality, $U(p) = \omega^\alpha \cdot n$.

Choose a set A and element b such that for some a realizing $p|A$, a depends on b over A, $U(a/A \cup \{b\}) \geq \omega^\alpha \cdot (n-1)$ and $U(b/A)$ is minimal with respect to these restrictions. Without losing this minimality assumption we can assume that $b \in Cb(a/A \cup \{b\})$ (since the canonical base is algebraic in $A \cup \{b\}$). Without loss of generality, $A = \emptyset$.

Claim. $U(b) = \omega^\alpha$.

We first show that $U(b) \approx_\alpha \omega^\alpha$. Let $a_0, \ldots, a_k$ be a Morley sequence in $stp(a/b)$ on which this strong type is based. Then $U(a_i/b) \approx_\alpha \omega^\alpha \cdot (n-1)$ for each i. If $U(a_i/a_0) \approx_\alpha \omega^\alpha \cdot n$, then $a_i \perp a_0$. The data: a_i realizes $p|a_0$, $U(b/a_0) < U(b)$ and $U(a_i/a_0 b) \approx_\alpha \omega^\alpha \cdot (n-1)$ contradict the above minimality assumption on $U(b/A)$. From Corollary 6.1.2 and $U(a_i/a_0) \approx_\alpha \omega^\alpha \cdot (n-1)$ (for $1 \leq i \leq k$) we get $U(a_0 \cdots a_k b) = U(a_0 \cdots a_k) \approx_\alpha \omega^\alpha \cdot n + \omega^\alpha \cdot (n-1)k$. Since $U(a_0 \cdots a_k/b) \approx_\alpha \omega^\alpha \cdot (n-1)(k+1)$, the relation $U(b) \approx_\alpha \omega^\alpha$ follows from that same corollary.

Now suppose, towards a contradiction, that $U(b) = \omega^\alpha + \beta$ for some $\beta > 0$. By Lemma 6.3.4 there is an element $b' \in acl(b)$ with $U(b/b') = \omega^\alpha$ and $U(b') = \beta$. Since $\beta \ll U(a)$, $a \perp b'$. The properties: a realizes $p|b'$, a depends on b over b', and $U(a/A \cup \{b, b'\}) \geq \omega^\alpha \cdot (n-1)$, contradict the minimality assumption on $U(b/A)$, completing the proof of the claim.

Now we reverse the roles of a and b to find c. First notice that $U(b/a) \approx_\alpha 0$. Choose c interalgebraic with $Cb(b/a)$ and $\{b_0, \ldots, b_m\}$ a Morley sequence in $stp(b/a)$ in which c is algebraic. Since $wt(a) = 1$ and a depends on each b_i, $b_i \not\perp b_0$ for all i, hence $U(b_i/b_0) < \omega^\alpha$. Since $U(b_i/c) \approx_\alpha 0$ repeating the argument in the previous paragraph shows that $U(c) \approx_\alpha \omega^\alpha$. Reasoning as at the end of the proof of the claim yields a $d \in acl(c)$ such that a realizes $p|d$, $c \in acl(ad)$ and $U(c/d) = \omega^\alpha$. This proves the proposition.

In the exercises the reader is asked to show that we can take A to be any a-model M on which p is based. It is interesting to note that we use the weight 1 hypothesis only near the end where we prove that $\{b_0, \ldots, b_m\}$ cannot be pairwise independent.

Corollary 6.3.8. *If p is a nonalgebraic complete type of finite U-rank, then p is nonorthogonal to a minimal type.*

Historical Notes. Regular types were developed by Shelah in [She90, V]. Our treatment of their properties follows Makkai [Mak84] to some degree. The results on U-rank in the subsection are by Lascar [Las84].

Exercise 6.3.1. Prove: For T a 1-sorted superstable theory, every regular type in T^{eq} is nonorthogonal to a regular 1-type in the sort of T.

Exercise 6.3.2. Give an example of a regular type of Morley rank ω (in some t.t. theory).

Exercise 6.3.3. Prove Lemma 6.3.1(ii).

Exercise 6.3.4. Prove Corollary 6.3.1.

Exercise 6.3.5. Write out a direct proof of Corollary 6.3.3.

Exercise 6.3.6. For p a type in a superstable theory let q be a stationary type of least ∞−rank nonorthogonal to p. Show that q is regular.

Exercise 6.3.7. Show that we can take A to be any a−model on which p is based in Proposition 6.3.4.

6.4 Strongly Regular Types

The dimension theory for regular types and the class of a−models in a superstable theory (see Proposition 6.3.3) does not generalize completely to the class of all models. In this section we restrict to a t.t. theory and develop a similar dimension theory for a subclass of regular types (the strongly regular types) and the class of all models of the theory. While all of the results here hold in an arbitrary t.t. theory we will simplify the notation by restricting to countable theories; i.e., **we assume throughout the section that every theory is ω−stable**. The relevant types are the following:

Definition 6.4.1. *Let $p \in S(A)$ be a stationary nonalgebraic type and $\varphi \in p$. The pair (p, φ) is called* strongly regular *if for all sets $B \supset A$ and $q \in S(B)$, $\varphi \in q$ and $q \not\perp p$ implies that $q = p|B$. p is* strongly regular *if (p, ψ) is strongly regular for some $\psi \in q$.*

It is common to write SR instead of strongly regular. It is easy to see that a SR type is regular. (Let $q \in S(B)$ be a forking extension of $p \in S(A)$. Then $\varphi \in q$ and $q \neq p|B$, hence $q \perp p$.) If (p, φ) is SR then (p, ψ) is also SR for any formula $\psi \in p$ which implies φ. Thus, when p is SR there is a $\varphi \in p$ such that (p, φ) is SR, $MR(\varphi) = MR(p)$ and $\deg(\varphi) = 1$. For $p \in S(A)$ stationary and $\varphi \in p$ let A_0 be a finite set such that p is based on A_0 and φ is over A_0. Then, (p, φ) is SR if and only if $(p \restriction A_0, \varphi)$ is SR. Thus, when checking to see if a pair is strongly regular we can always assume the type to be over a finite set.

Compare the following equivalents with Corollary 6.3.3.

Lemma 6.4.1. *For $p \in S(A)$ a stationary type and $\varphi \in p$ the following are equivalent.*

(1) *(p, φ) is SR.*

(2) *For any model $M \supset A$ there is a model $N \supset M$ realizing $p|M$ such that any $a \in \varphi(N) \setminus M$ realizes $p|M$.*

(3) *There is a model $M \supset A$ and a model $N \supset M$ such that $(*)$ $p|M$ is realized in N and any $a \in \varphi(N) \setminus M$ realizes $p|M$.*

Proof. (2) $\Longrightarrow$ (3) holds trivially.

(1) $\Longrightarrow$ (2). Let b realize $p|M$ and N be the prime model over $M \cup \{b\}$. By Corollary 3.3.4, any $a \in N \setminus M$ depends on b over M. Thus, $a \in \varphi(N) \setminus M \implies tp(a/M) \not\perp p|M \implies tp(a/M) = p|M$ (by the strong regularity of (p, φ)).

Before turning to the remaining nontrivial implication, (3) $\Longrightarrow$ (1), we prove

Claim. If (p, φ) is not SR then for any a–model $M \supset A$ there is $q \in S(M)$ containing φ such that $q \not\perp p$ and $q \neq p|M$.

Let r be a stationary type witnessing that (p, φ) is not SR. Let A_0 be a finite subset of A containing the parameters in φ and on which p is based. We can assume r to be over a finite set $B \supset A_0$. Let $B' \subset M$ realize $tp(B/A_0)$ and r' be a conjugate of r over B'. Since p is based on A_0, the nonorthogonality of r and p implies the nonorthogonality of r' and p (see Exercise 5.6.2). Again, using that p is based on A_0, $p|\mathfrak{C}$ does not split over A_0, hence the specified automorphism taking B to B' over A_0 maps $p|B$ to $p|B'$. Thus, r' is not $p|B'$. Since r' contains φ it witnesses that (p, φ) is not strongly regular.

(3) $\Longrightarrow$ (1) As we said above, we can assume A to be a finite set, which we take to be $\emptyset$ (without loss of generality). Let $M_0 \subset N_0$ be models satisfying $(*)$ in (3) and $a \in N$ a realization of $p|M_0$. We will use a theory in an expanded language to show

Claim. There are a–models M and N satisfying $(*)$.

Let L be the language of the relevant theory T, P a new unary predicate and $L' = L \cup \{P\} \cup \{a\}$. Let Ψ be the collection of formulas $\psi(x, y)$ over $\emptyset$ such that $\models \psi(a, b) \implies a \not\perp b$. Let $T' \supset T$ be the theory in L' expressing the following properties of a model M' (the detailed formulation is left to the reader):

- $P(M')$ is an elementary submodel of M' with respect to the language L;
- a realizes $p \restriction \emptyset$ and $a \notin P(M')$;
- for all $b \in \varphi(M') \setminus P(M')$, b realizes $p \restriction \emptyset$;
- for all $b \in \varphi(M') \setminus P(M')$ and for all formulas $\psi(x, y) \in \Psi$, $\models \neg\psi(b, c)$ for all $c \in P(M')$.

The model obtained by interpreting P by M on N gives a model of T', proving its consistency. By a now standard elementary chain argument there is a model M' of T' such that M' and $P(M')$ are $\aleph_0$–saturated (i.e., a–saturated) as models of T. This proves the claim.

Assuming M and N to be a–models it suffices to show (by the first claim) that $p|M$ is the only type in $S(M)$ which contains φ and is nonorthogonal to p. Suppose, to the contrary, that $q \neq p|M$ in $S(M)$ contains φ and is nonorthogonal to p. Let $B \subset M$ be a finite set on which q is based such that $p|B \overset{a}{\not\perp} q|B$. Since N is an $a-$model there is a $b \in N$ realizing $q|B$ depending on a over B. Since $a \perp M$, b is not in M. Since b satisfies φ the hypothesis implies that b realizes $p|M$. This contradicts that b realizes $q|B \neq p|B$.

Lemma 6.4.2 (Existence). *For any distinct models $M \subset N$ there is an $a \in N$ such that $tp(a/M)$ is strongly regular. We may choose a to satisfy any formula over M which is satisfied in $N \setminus M$. Moreover, if $q \in S(M)$ is an SR type nonorthogonal to $tp(a/M)$ then q is realized in N.*

Proof. Given a formula θ over M satisfied in $N \setminus M$, let $a \in N \setminus M$ be such that $MR(a/M)$ is minimal in $\{ MR(b/M) : b \in N \setminus M \text{ and } \models \theta(b) \}$. Choose $\varphi \in p = tp(a/M)$ such that $MR(\varphi) = MR(p)$ and $\deg(\varphi) = 1$. Without loss of generality, φ implies θ. The minimal rank condition guarantees that every element of $N \setminus M$ satisfying φ has the same Morley rank as p. Since φ has degree 1 every such element realizes $p|M$. By Lemma 6.4.1, (p, φ) is SR.

Turning to the moreover part of the lemma suppose that $q \in S(M)$ is strongly regular and nonorthogonal to p. Choose $\psi \in q$ such that (q, ψ) is a strongly regular pair. Without loss of generality, N is prime over $M \cup \{a\}$, hence every type over M realized in N is nonorthogonal to p. Let b be an element independent from a over M such that there is c realizing $q|(M \cup \{b\})$ with a and c dependent over $M \cup \{b\}$. Let d be an element of M over which $tp(abc/M)$ is based. Then a depends on cb over d and we can assume ψ to be a formula over d. Let $\theta(x, y, a)$ be a formula in $tp(cb/ad)$ with the properties:

(1) $\exists y \theta(x, y, a)$ implies ψ, and
(2) $\models \theta(c', b', a)$ implies $a \underset{d}{\not\perp} c'b'$.

Since b is independent from a over M and $tp(a/M)$ is definable over d there is a b' in M such that $\models \exists x \theta(x, b', a)$. Let $c' \in N$ satisfy $\theta(x, b', a)$. Item (1) and the dependence guaranteed by (2) force c' to be in $\psi(N) \setminus M$. Thus, $tp(c'/M)$ is nonorthogonal to $tp(a/M)$, hence nonorthogonal to q (since these are regular types). Since (q, ψ) is SR $tp(c'/M)$ must be q as required.

Corollary 6.4.1. *Every regular type is nonorthogonal to a strongly regular type. In fact, a stationary type of least Morley rank nonorthogonal to a given regular type is strongly regular.*

Proof. See Exercise 6.4.2.

As with Lemma 6.4.1 the scheme throughout the section is to generalize properties about regular types and $a-$models in superstable theories to strongly regular types and models in t.t. theories. The difficult part of this extension is

Proposition 6.4.1. *If a type p is nonorthogonal to a model M, then there is a strongly regular type over M nonorthogonal to p.*

Proof. Again the idea is to use a theory in an expanded language to reduce our attention to a–models, then use properties proved earlier for regular types.

Without loss of generality p is a regular type over a $|M|^+$–saturated model $N \supset M$. Let a realize p and N' be an a–prime model over $N \cup \{a\}$. Let $q \in S(N)$ be a nonalgebraic type realized in N' such that $MR(q \restriction M)$ is minimal among all such. Require, furthermore, that $MR(q)$ is minimal in $\{ MR(r) : r \in S(N)$ and $MR(r \restriction M) = MR(q \restriction M) \}$. It follows that q is strongly regular. (Take $\psi \in q$ the conjunction of two formulas, one which determines $MR(q \restriction M)$ and another which determines $MR(q)$. By Corollary 6.4.2 there is a SR type over N containing ψ and realized in N'. The minimal rank assumptions on q guarantee that this SR type is q.) The rest of the proof is needed to show that q does not fork over M.

Fix a formula $\psi \in q$ so that (q, ψ) is SR, $MR(\psi) = MR(q)$ and $\deg(\psi) = 1$. Let $d \in N$ contain the parameters in ψ (in which case q is based on d). Since p is nonorthogonal to M and $p \sqsupseteq q$ (these are nonorthogonal regular types) q is nonorthogonal to M. Since N is $|M|^+$–saturated there is a $d' \in N$ realizing $tp(d/M)$ and independent from d over M. By Proposition 5.6.2 the type q' over d' conjugate to $q|d$ over M is nonorthogonal to q, hence domination equivalent. This yields (using Proposition 5.6.1):

- finite Morley sequences $I \subset N$ in $q|d$ and $I' \subset N'$ in q', and
- elements b realizing q and b' realizing $q'|N$ in N' such that b depends on b' over $I \cup I' \cup \{d, d'\}$.

Since $I \cup \{d\}$ is M–independent from $I' \cup \{d'\}$ and conjugate over M we may as well absorb these Morley sequences into the original domains and assume that b and b' are dependent over dd'. Let $e \in M$ be an element on which $tp(dd'/M)$ is based. Without loss of generality there is a formula $\psi_0 \in q_0 = q \restriction e$ with Morley rank $MR(q \restriction M)$. Note: the original assumptions about q imply that any element of $N' \setminus N$ satisfying ψ_0 also satisfies $q_0|M = q \restriction M$.

To use previously proved facts about regular types we need to work within the class of a–models.

Claim. There are a–models M_0, N_0 and N_0' such that

(1) $M_0 \subset N_0 \subset N_0'$;
(2) $e \in M_0$, $d, d' \in N_0$ and $b, b' \in N_0'$;
(3) $d \underset{M_0}{\perp} d'$;
(4) b realizes $q|N_0$ and b' realizes $q'|N_0$;
(5) for any $b^* \in N_0' \setminus N_0$ satisfying ψ_0, $tp(b^*/M_0) = q_0|M_0$.

As in the proof of the second claim in Lemma 6.4.1 we proceed by expressing the desired properties with a first-order theory in an expanded language with predicates P and Q representing the two models M_0 and N_0. In that earlier proof we showed how to express, e.g., b realizes $q|Q(N^*)$ (where N^*

is an arbitrary model of the expanded theory). Notice that the condition "d and d' are $P(N^*)$–independent" can be obtained by requiring dd' to realize $stp(dd'/e)|P(N^*)$. The reader is asked to fill in the details from these hints in the exercises.

Now fix a–models $M_0 \subset N_0 \subset N'_0$ as in the claim. The elements b and b' witness the nonorthogonality of $q|N_0$ and $q'|N_0$. These types are based on d and d', respectively, and $\{d, d'\}$ is independent over e. Since M_0 is an a–model and $tp(d/M_0)$ is based on e, an element $d'' \in M_0$ realizing $tp(d'/e)$ also realizes $stp(d'/ed)$. Thus, the conjugate q'' of $q|d$ over d'' is a strongly regular type nonorthogonal to q. Since q and q'' are domination equivalent, there is $b'' \in N'_0$ realizing $q''|N'_0$. Certainly, b'' satisfies ψ_0. We assumed that q forks over M; i.e., has Morley rank $< MR(\psi_0)$. Thus, $MR(b''/M_0) \leq MR(b''/d''e) < MR(\psi_0)$, contradicting the last condition listed in the claim. This contradiction proves that q does not fork over M, and finally completes the proof of the proposition.

Most matters involving orthogonality of types in superstable theories reduce to properties of regular types through:

> for all distinct a–models $M \subset N$ with $wt(N/M)$ finite there is I, an M–independent set of realizations of regular types over M such that N is a–prime over $M \cup I$.

The literal generalization of this result to models and SR types is:

> for all distinct models $M \subset N$ with $wt(N/M)$ finite there is I, an M–independent set of realizations of strongly regular types over M such that N is prime over $M \cup I$.

However, the result in the a–model context was proved by first showing N is dominated by I over M, then noting that domination and a–atomicity (with respect to a finite set) are equivalent over a–models. The next lemma states the only general connection between domination and ordinary atomicity, so we must be content with the subsequent proposition.

Lemma 6.4.3. *If A is atomic over $B \cup M$, where M is a model, then $A \cup B$ is dominated by B over M.*

Proof. This is a mild generalization of Lemma 3.4.7. Without loss of generality, $A = a$ and $B = b$ are finite. Suppose, to the contrary, there is a c independent from b over M which depends on ab over M; i.e., c depends on a over $M \cup \{b\}$. Let $d \in M$ be an element over which $tp(abc/M)$ is based. Let $\psi(x, y)$ be a formula in $tp(ac/bd)$ such that

- $\exists y \psi(x, y)$ isolates $tp(a/M \cup \{b\})$ and
- whenever a' realizes $tp(a/bd)$ and $\models \psi(a', c')$, a' and c' are dependent over bd.

Since b and c are $M-$independent and $tp(b/M)$ is based on d there is a $c' \in M$ such that $\exists x\psi(x, c')$. Let a' satisfy $\psi(x, c')$. Then a' realizes $tp(a/M \cup \{b\})$, hence, $tp(a'/M \cup \{b\})$ does not fork over bd, contradicting the dependence of a' and c' over bd.

Proposition 6.4.2. *For models $M \subset N$, $M \neq N$, there is I, an $M-$independent set of realizations of strongly regular types over M such that N is dominated by $M \cup I$ over M.*

Proof. Let I be a maximal $M-$independent subset of N consisting of realizations of SR types over M. Let $M' \subset N$ be a maximal set dominated by I over M. Note that M' is a model. (The prime model $M'' \subset N$ over M' is dominated by M' over M by Lemma 6.4.3. By the transitivity of domination, M'' is dominated by I over M, hence $M'' = M'$.) Supposing, towards a contradiction, that $M' \neq N$ let $a \in N$ realize a strongly regular type over M'. Assuming first that $tp(a/M')$ is orthogonal to M, $M' \cup \{a\}$ is dominated by M' over M, contradicting the maximality of M'. Thus, $tp(a/M')$ is nonorthogonal to M. By Proposition 6.4.1 there is a strongly regular $p \in S(M)$ nonorthogonal to $tp(a/M')$. By Corollary 6.4.2 there is a $b \in N$ realizing $p|M'$. This element b contradicts the maximality of I, completing the proof.

Throughout Section 5.6 we studied orthogonality and domination relative to the class of $a-$models. Using the proven facts about SR types we can extend some of these results to the class of all models of a t.t. theory. As a first installment:

Lemma 6.4.4. *If M is a model and p, $q \in S(M)$ are nonorthogonal types, then $p \overset{a}{\not\perp} q$.*

Proof. Let a realize p and N be a prime model over $M \cup \{a\}$. By Proposition 6.4.2 there is an $M-$independent sequence $(b_0, \ldots, b_n) = b$ in N such that $p_i = tp(b_i/M)$ is SR and N is dominated by b over M. Let c be a realization of q, N' a prime model over $M \cup \{c\}$ and $(d_0, \ldots, d_m) = d \in N'$ an $M-$independent sequence dominating N' over M with $q_i = tp(d_i/M)$ strongly regular. Since p and q are nonorthogonal some p_i must be nonorthogonal to some q_j, say $p_0 \not\perp q_0$. By Corollary 6.4.2 q_0 is realized by some d_0' in N. Replacing N' by some conjugate over M we can assume that $d_0 \in N$. Since N is dominated by a over M and N' is dominated by c over M, a and c are dependent over M. We conclude that $p \overset{a}{\not\perp} q$ as desired.

The domination relation on types was motivated by the question:

> For M an $a-$model, when does realizing $p \in S(M)$ in some $a-$model $N \subset M$ force $q \in S(M)$ to be realized in N.

In this generalization from $a-$models to models the natural definition is:

Definition 6.4.2. *For M a model and p, $q \in S(M)$ we write $p <_{RK} q$ if p is realized in any model $N \supset M$ which realizes q. If $p <_{RK} q$ and $q <_{RK} p$ we write $p \sqsubseteq_{RK} q$ and says that p and q are RK-*equivalent.

The relation $<_{RK}$ abbreviates the term *Rudin-Keisler ordering* which is an ordering on ultraproducts. (We will not go into the rationale behind this definition here. See [Las86].) Parsing the definition, in a t.t. theory

$$
\begin{aligned}
p <_{RK} q \quad &\iff \quad p \text{ is realized in the prime model over } M \cup \{b\}, \\
&\qquad\quad \text{where } b \text{ realizes } q \\
&\iff \quad \text{there are } a \text{ realizing } p \text{ and } b \text{ realizing } q \\
&\qquad\quad \text{with } tp(a/M \cup \{b\}) \text{ isolated.}
\end{aligned}
$$

While we did not include the model M in the notation, $<_{RK}$ is only defined on $S(M)$ for a fixed model M. For instance, $<_{RK}$ is not invariant under parallelism. (When p, $q \in S(M)$ and $N \supset M$ it is possible that p is not $<_{RK} q$ while $p|N <_{RK} q|N$.) The reason $\triangleleft$ is better behaved than $<_{RK}$ is because domination is equivalent to $a-$isolation over an $a-$model, but not equivalent to isolation (in general).

Certainly, $p <_{RK} q$ implies $p \not\perp q$. In fact

Remark 6.4.1. Given a model M and p, $q \in S(M)$, $p <_{RK} q$ implies $p \triangleleft q$. (See Exercise 6.4.5.)

For example, $<_{RK}$ is not a parallelism invariant; the relation is firmly attached to the model M. An exception is found when dealing with SR types.

Lemma 6.4.5. *Let M be a model.*

(i) If $p \in S(M)$ is an SR type nonorthogonal to $q \in S(M)$, then $p <_{RK} q$.

(ii) Nonorthogonal SR types are RK-equivalent.

(iii) An element of $S(M)$ is minimal in the $<_{RK}$ order if and only if it is RK-equivalent to a strongly regular type.

Proof. To prove the first part of the lemma choose b realizing q and N a prime model over $M \cup \{b\}$. The nonorthogonality of p and q forces p to be nonorthogonal to some regular type r nonorthogonal to q. In fact, we can take r to be an SR type over M realized in N. By Corollary 6.4.2, p is realized in N.

(ii) This follows immediately from (i).

Part (iii) of the lemma follows quickly from

Claim. An SR type $p \in S(M)$ is $<_{RK}-$minimal.

Let a realize p (an arbitrary SR type) and N be prime over $M \cup \{a\}$. Let $r \in S(M)$ be a $<_{RK}-$minimal type realized in N. Then, r is nonorthogonal to p, hence $p <_{RK} r$ (by the first part of the lemma). By the minimality of r, p is $RK-$equivalent to r, so p is also $<_{RK}-$minimal.

This leads to an interesting minimality condition, whose proof is left to Exercise 6.4.4:

Remark 6.4.2. Suppose that M is a model, $tp(a/M)$ is SR and N is prime over $M \cup \{a\}$. Then any model $N' \neq M$, $M \subset N' \subset N$, is isomorphic to N over M.

As the following example shows RK–equivalence and domination equivalence can differ even in rather simple ω–stable theories.

Example 6.4.1. The language consists of a binary relation symbol E and a unary function symbol S. The axioms for the theory say that E is an equivalence relation with infinitely many infinite classes and no finite classes, S defines a bijection on the universe with no cycles and $\forall v E(v, s(v))$. This theory is complete, quantifier-eliminable and ω–stable. Let M be any model, a an element not E–equivalent to any element of M and b such that for all n, $b \neq S^n(a)$ and $a \neq S^n(b)$. Let $p = tp(a/M)$ and $q = tp(ab/M)$. It is easy to see that $p \sqsupseteq q$ and p and q are not RK–equivalent.

Turning our attention to the dimensions of SR types in models, the following are proved like the corresponding results about regular types and a–models. The details are left to the reader.

Proposition 6.4.3 (Additivity of Dimension). *Let $M \subset N$ be models, $A \subset M$ a finite set and $p \in S(A)$ a strongly regular type. Then,*

$$\dim(p, N) = \dim(p, M) + \dim(p|M, N).$$

(The property of SR types which corresponds to Lemma 5.4.2 is Lemma 5.1.9. When (p, φ) is SR (as in the statement of the proposition) and I is a basis for p in M the strong regularity of the pair implies that $tp(\varphi(M)/A \cup I) \perp p$. The proposition is proved by inserting these changes in the earlier proof.)

Corollary 6.4.2. *Let M be a model, $A \subset M$ finite, $p \in S(A)$ a strongly regular type and I a basis for p in M. Given J a Morley sequence in p depending on M over A, J depends on I over A.*

Combining these results, the proof of Proposition 6.3.3 and the fact that nonorthogonal SR types are RK–equivalent yields

Proposition 6.4.4. *Let $p \in S(A)$ be a strongly regular type where A is finite.*

(i) If $M \supset A$ is prime over a finite set, p has dimension $\leq \aleph_0$ in M. Furthermore, if I is an infinite Morley sequence in p and M is prime over $A \cup I$, I is a basis for p in M.

(ii) For any model $M \supset A$, if $B \subset M$ is finite and $q \in S(B)$ is a SR type nonorthogonal to p, then $\dim(p, M) + \aleph_0 = \dim(q, M) + \aleph_0$.

(iii) For any $\kappa \geq |T|$ there is a model $M \supset A$ of cardinality κ such that $\dim(p, M) = \kappa$ *and for any strongly regular type q over a finite subset of M,* $q \perp p \implies \dim(q, M) \leq \aleph_0$.

(iv) Let M_0 be a model of cardinality λ and X a set of strongly regular types over M_0. Then, there is a model M of cardinality λ containing M_0 such that $\dim(q, M) = 0$, *for all $q \in X$, and* $\dim(r, M) = \lambda$ *for all strongly regular types r over a finite subset of M such that $r \perp q$ for all $q \in X$.*

(The first part of (i) is obvious since the prime model over a finite set in an ω-stable theory is countable. The result is, however, true more generally in any t.t. theory.)

We close with another lemma which expresses the intuition that the SR types in an ω-stable theory forms a basis for all of the complete types. The proof is assigned as Exercise 6.4.7.

Lemma 6.4.6. *Let T be ω-stable and M a countable model of T. Suppose that for any SR type p over a finite subset of M,* $\dim(p, M)$ *is infinite. Prove that M is saturated.*

Historical Notes. Strongly regular types were defined by Shelah in Definition 3.6 of [She90, V], although he does not require the type to be stationary. Lemma 6.4.1 is found in Theorem 3.18 of that chapter, as is Corollary 6.4.2. The results 6.4.1 through 6.4.5 are by Lascar [Las82], as is the Rudin-Keisler order. Proposition 6.4.3 is stated by Bouscaren and Lascar explicitly in [BL83, 4.2]. Analogues of Proposition 6.4.4 can be found in [Mak84], where Makkai attributes them to handwritten notes by Shelah on his proof of Vaught's conjecture for ω-stable theories.

Exercise 6.4.1. Prove that a stationary type of least Morley rank which is nonorthogonal to a given type p is strongly regular.

Exercise 6.4.2. Prove Corollary 6.4.1.

Exercise 6.4.3. Prove the claim in the proof of Proposition 6.4.1.

Exercise 6.4.4. Prove: Suppose that M is a model, $tp(a/M)$ is SR and N is prime over $M \cup \{a\}$. Then any model $N' \neq M$, $M \subset N' \subset N$, is isomorphic to N over M.

Exercise 6.4.5. Prove Remark 6.4.1.

Exercise 6.4.6. Suppose that $p = tp(a/M)$ and N is a prime model over $M \cup \{a\}$. Let $C = \{c_0, \ldots, c_n\} \subset N$ be a maximal M-independent set of realizations of SR types over M and let $q_i = tp(c_i/M)$, for $i \leq n$. Show that $p \sqsubseteq q_0 \otimes \ldots \otimes q_n$. (Hence, the q_i's are a regular decomposition of p.)

Exercise 6.4.7. Prove Lemma 6.4.6.

7. Selected Topics

As the title suggests this chapter is a collection of various more advanced topics. The first section, on bounded and unbounded theories, both contains useful facts about a natural class of theories and illustrates how the regular type machinery can be used to classify the models of a theory with relatively simple invariants. The second section delves more deeply into the properties of our notions of rank in some very special theories such as the uncountably categorical ones.

7.1 Bounded and Unbounded Theories

We work in a stable theory throughout the section.

Definition 7.1.1. *(i) The theory T is called* bounded *if there are $< |\mathfrak{C}|$ domination equivalence classes of nonalgebraic stationary types; T is* unbounded *if it is not bounded.*

(ii) The theory T is unidimensional *if any two nonalgebraic types are nonorthogonal.*

Shelah (and many others) call unbounded theories *multidimensional* and bounded theories *nonmultidimensional* or *nmd*, for short. There are many examples of such theories:

Lemma 7.1.1. *The theory of any infinite module is bounded.*

Proof. Let $\mathfrak{C}$ be the universal domain of the relevant theory. By Proposition 5.3.2 and Lemma 5.3.9 an element p of $S_1(\mathfrak{C})$ is the translate of the generic type in *stab*(p), a group $\bigwedge-$definable over $\emptyset$. Certainly p is domination equivalent to this generic type. Since there are $\leq 2^{|T|}$ many such groups this is a bound on the number of domination equivalence classes. The same argument establishes this bound for types in other sorts (i.e., $n-$types in the $1-$sorted theory of the module), proving the lemma.

We will say little here about bounded theories which are properly stable. The superstable ones become easier to handle using

Lemma 7.1.2. *The following are equivalent for a superstable theory T:*

(1) *Every nonalgebraic type is nonorthogonal to* $\emptyset$.
(2) *Every regular type is nonorthogonal to* $\emptyset$.
(3) *For every stationary type* p *and* $f \in \mathrm{Aut}(\mathfrak{C})$ *fixing* $acl(\emptyset)$ *pointwise,* $p \sqsupset\!\sqsubset f(p)$.
(4) T *is bounded.*

Proof. (1) $\Longrightarrow$ (2) holds trivially.

(2) $\Longrightarrow$ (3) Suppose (2) holds. Let $p \in S(A)$ be stationary and $f \in \mathrm{Aut}(\mathfrak{C})$ which fixes $acl(\emptyset)$ pointwise.

Claim. To prove (3) it suffices to consider the case when A is independent from $f(A)$.

Suppose the property in the claim to be true. For the given stationary $p \in S(A)$ and f there is a $g \in \mathrm{Aut}(\mathfrak{C})$ fixing $acl(\emptyset)$ such that $g(A) \perp A$ and $g(A) \perp f(A)$, hence p, $g(p)$ and $f(p)$ are all domination equivalent, proving the claim.

Assuming now that $A \perp f(A)$ let $q_0 \otimes \ldots \otimes q_n$ be a product of regular types domination equivalent to p. We can take A large enough so that $q_i \in S(A)$, for $i \leq n$, while still assuming that $A \perp f(A)$. Since each q_i is nonorthogonal to $\emptyset$, Proposition 5.6.2 says that q_i is nonorthogonal to $f(q_i)$, hence $q_i \sqsupset\!\sqsubset f(q_i)$ (since they have weight 1). Domination equivalence is preserved under $\otimes$ so p, $q_0 \otimes \ldots \otimes q_n$, $f(q_0) \otimes \ldots \otimes f(q_n)$ and $f(p)$ are all domination equivalent, as desired.

(3) $\Longrightarrow$ (4) This is clear since there are $\leq 2^{|T|}$ many nonalgebraic stationary types up parallelism and conjugacy over $acl(\emptyset)$.

(4) $\Longrightarrow$ (1) This is left as an exercise.

Definition 7.1.2. *For* T *a bounded superstable theory let* $DIM(T)$*, called the* set of dimensions of T *denote the set of equivalence classes of regular types with respect to nonorthogonality. The cardinality of* $DIM(T)$ *is called the* width of T *(or the* number of dimensions of T*) and denoted* $ND(T)$.

Corollary 7.1.1. *Given a bounded superstable theory* T*, the width of* T *is* $\leq 2^{|T|}$ *and* $\leq |T|$ *when* T *is totally transcendental.*

Proof. This follows immediately from Lemma 7.1.2.

These definitions explain the term unidimensional theory; it is a bounded theory with 1 dimension (when the theory is superstable). As the following shows we have already encountered numerous examples of unidimensional theories (for countable theories the right hand side is simply the definition of "uncountably categorical").

Proposition 7.1.1. *A t.t. theory* T *is* $\lambda-$*categorical for all* $\lambda > |T|$ *if and only if* T *is unidimensional.*

Proof. ($\Longrightarrow$) The categoricity assumption and Proposition 6.4.4(iii) implies that all SR types are nonorthogonal, hence all nonalgebraic stationary types are nonorthogonal.

($\Longleftarrow$) Let $N \supsetneq M$ be models of T and φ is a nonalgebraic formula over M. Corollary 6.4.2 and the unidimensionality of T directly yield an $a \in N \setminus M$ satisfying φ (with $tp(a/M)$ strongly regular). Thus, T does not have a Vaughtian pair. For countable theories Theorem 3.1.2 immediately implies the uncountable categoricity of T. For arbitrary t.t. theories we simply repeat the proof of that earlier theorem using deeper results about t.t. theories (such as the existence of prime models over sets) when necessary.

There are, however, rather simple countable unidimensional theories which are not uncountably categorical:

Example 7.1.1. (A weakly minimal unidimensional theory) We begin by letting G be the direct product of $\aleph_0$ many copies of the group $\mathbb{Z}_2$. Let H_i, for $1 \le i < \omega$, be the subgroup consisting of the elements whose first i coordinates are 0. Let $M = (G, +, 0, H_i)_{i<\omega}$ in the language consisting of +, 0 and predicate symbols P_i interpreted by the H_i's, and let $T = Th(M)$ (as a 1−sorted theory). It is easy to show that this theory is quantifier eliminable. Thus, for $\mathfrak{C}$ the universe, $\bigcap_{i<\omega} P_i(\mathfrak{C}) = \mathfrak{C}^o$ is a vector space over $\mathbb{Z}_2$ and there is no other structure on this group induced by the formulas (every vector space automorphism of $\mathfrak{C}^o$ extends to an automorphism of $\mathfrak{C}$). Thus the type $\bigcap_i P_i(x)$ is minimal. For an arbitrary element a, the set of realizations of $stp(a)$ is simply $a + \mathfrak{C}^o$. It follows immediately that $\mathfrak{C}$ is a weakly minimal set and any nonalgebraic element of $S_1(\mathfrak{C})$ is a translate of the generic in $\mathfrak{C}^o$. Thus, all nonalgebraic stationary types are nonorthogonal to the generic type of $\mathfrak{C}^o$. Since the generic of $\mathfrak{C}^o$ is minimal, all nonalgebraic stationary types are nonorthogonal.

Example 7.1.2. The theory of the group of integers, $(\mathbb{Z}, +)$ is also weakly minimal, unidimensional and not t.t. (See the analysis in [BBGK73].)

Remark 7.1.1. Hrushovski showed in [Hru90b] that every stable unidimensional theory is superstable.

For bounded theories the Decomposition Theorem (Theorem 6.3.6) can be strengthened by limiting the collection of needed regular types.

Lemma 7.1.3. *If T is a bounded superstable theory and M is any a−model, then for all stationary types p there are regular types $q_0, \ldots, q_n \in S(M)$ such that $p \sqsupseteq q_0 \otimes \ldots \otimes q_n$.*

Proof. In a bounded superstable theory every regular type is nonorthogonal to $\emptyset$. By Lemma 5.6.5 every regular type is nonorthogonal to one in $S(M)$, from which the lemma follows.

This leads us to what can be viewed as a basis for an arbitrary $a-$model (or model in the t.t. case) with respect to a fixed set of regular types.

Proposition 7.1.2. *(i) Let M be an $a-$model of a bounded superstable theory T and $N \subset M$ an $a-$prime model over $\emptyset$. If $C \subset M$ is a maximal $N-$independent set of realizations of regular types over N then M is $a-$prime over $N \cup C$ (in fact, there is no $a-$model M', $N \cup C \subset M' \subsetneq M$).*

(ii) Let M be a model of a bounded t.t. theory and $N \subset M$ a prime model over $\emptyset$. If $C \subset M$ is a maximal $N-$independent set of realizations of strongly regular types over N then M is prime over $N \cup C$ and minimal over $N \cup C$.

Proof. (i) Suppose, to the contrary, that $M' \subsetneq M$ is an $a-$model containing $N \cup C$. There is an $a \in M \setminus M'$ with $p = tp(a/M')$ regular. Since T is bounded we can take p to be based on N, hence a is independent from $M' \supset C$ over N. This contradicts the maximality of C to prove that $M' = M$. Since M contains a model $a-$prime over $N \cup C$ we also conclude that M is $a-$prime over $N \cup C$.

(ii) This is proved exactly like (i), using Proposition 6.4.1 and Corollary 6.4.2 when necessary.

Part (ii) of the proposition gives a decomposition theorem for bounded t.t. theories which generalizes Theorem 6.3.6 more directly than Proposition 6.4.2.

Corollary 7.1.2. *In a bounded t.t. theory a complete nonalgebraic stationary type over a model is RK-equivalent to a finite $\otimes-$product of SR types.*

A *representation theorem* for the models in a class $\mathcal{K}$ is a result yielding a function $\mathcal{I}(-)$ from $\mathcal{K}$ into a collection of sets such that for all $M, N \in \mathcal{K}$, $M \cong N \iff \mathcal{I}(M) = \mathcal{I}(N)$. The set $\mathcal{I}(M)$ is called an *isomorphism invariant of M in $\mathcal{K}$*. A representation theorem is a *structure theorem* when the isomorphism invariant is set-theoretically "simple". The reader is referred to [She85] for a discussion of the complicated matter of making the term "simple" more precise. Here we accept as simple invariants: cardinal numbers, sequences of cardinals of length $\leq 2^{|T|}$ and quotients of such objects by equivalence relations. Following are some examples of structure theorems.

1. A vector space is determined up to isomorphism by its dimension.
2. A divisible abelian group can be written as a direct sum of copies of $\mathbb{Q}$ and $\mathbb{Z}_{p^\infty}$ (for various primes p). The isomorphism type of the group is determined by the number of copies of each of these groups in any such decomposition.
3. Fixing a strongly minimal formula φ over the prime model the isomorphism type of a model M of an uncountably categorical theory T is determined by the dimension in M of a conjugate of φ. (This is Morley's Categoricity Theorem when M is uncountable, and the Baldwin-Lachlan Theorem in general.)

Proposition 7.1.2(ii) is a good first approximation of a structure theorem for the models of a bounded t.t. theory T. Let $M \models T$ and fix a copy of the prime model M_0 in M. For $i \in DIM(T)$ let $d_i(M)$ be $\dim(p, M)$, where p is any SR type in $S(M_0)$ whose nonorthogonality class is i (the choice of p does not effect this dimension). Let $\mathcal{I}_{M_0}(M)$ be $\langle d_i(M) : i \in DIM(T) \rangle$. Now let N, N' be two models containing M_0 such that $\mathcal{I}_{M_0}(N) = \mathcal{I}_{M_0}(N')$. Then there are M_0−independent sets $C \subset N$ and $C' \subset N'$ such that $C = \bigcup_{i \in DIM(T)} C_i$, where C_i is a basis for a regular type over M_0 whose nonorthogonality class is i, and similarly for C'. For each class $i \in DIM(T)$ we can choose C_i and C'_i to be sets of realizations of the same type over M_0. Thus, there is an elementary map f fixing M_0 and taking C onto C'. By Proposition 7.1.2, f extends to an isomorphism of N onto N'. Summarizing these statements, when N, N' are models containing M_0 with $\mathcal{I}_{M_0}(N) = \mathcal{I}_{M_0}(N')$, N is isomorphic to N' over M_0. However, these invariants only characterize the models up to isomorphism over M_0 rather than over $\emptyset$. We might tag a model M with the set of all $\mathcal{I}_{M_0}(M)$, as M_0 ranges over all copies of the prime model in M, however there may be $|M|^{|T|}$ many such models leading to a set-theoretically complicated (hence not very useful) "invariant". The situation is analogous to our picture of the models of an uncountably categorical theory prior to the Baldwin-Lachlan Theorem. For M a countable model of such a theory there is a realizing an isolated type and strongly minimal formula φ over a such that $\dim(\varphi(M)) = \dim(\varphi(N)) \implies M \cong N$ for any N containing a. It is conceivable, though, that there is a countable model N containing a with $\dim(\varphi(M)) \neq \dim(\varphi(N))$ which is still isomorphic to M (by a map not fixing a). This will happen exactly when there is an a' in M realizing $tp(a)$ such that the dimension in M of the conjugate of φ over a' is different from the dimension of φ in M. This is shown to be impossible in the proof of the Baldwin-Lachlan Theorem. Proposition 7.1.2(ii) generalizes Morley's Categoricity Theorem, while the following refined study of dimensions generalizes the Baldwin-Lachlan Theorem (which actually follows from the next lemma).

7.1.1 Bounded ω−stable Theories

In this subsection we restrict our attention to the models of a bounded t.t. theory. Similar results can be proved for the class of a−models in a superstable theory (left to the reader).

Lemma 7.1.4. *Let T be a t.t. theory and $p \in S(a)$ an SR type nonorthogonal to $\emptyset$ with $tp(a)$ isolated. Then for any model M of T containing a and conjugate $p' \in S(a')$ of p over M which is nonorthogonal to p, $\dim(p, M) = \dim(p', M)$. In particular, when $stp(a) = stp(a')$, p and p' have the same dimension in M.*

Proof. We adopt the notation p_b for the conjugate of p over b when b realizes $tp(a)$. Notice that every conjugate of p is nonorthogonal to $\emptyset$. By Lemma 7.1.2, all conjugates of p over $acl(\emptyset)$ are nonorthogonal, hence the last sentence of the lemma follows from the first part.

Let M be a model containing a and choose a' realizing $tp(a)$.

Claim. If a' realizes $stp(a)$, $\dim(p, M) = \dim(p_{a'}, M)$.

First consider the case when $a \perp a'$. Let $N \subset M$ be a prime model over $\{a, a'\}$. Since a and a' are independent realizations of the same strong type the pairs aa' and $a'a$ have the same type over $\emptyset$. Any elementary map on $\{a, a'\}$ extends to an automorphism of N. Thus, $\dim(p, N) = \dim(p_{a'}, N)$. Since p and $p_{a'}$ are nonorthogonal SR types, $\dim(p|N, M) = dim(p_{a'}|N, M)$. By the additivity of dimension for SR types (Lemma 6.4.3), $\dim(p, M) = \dim(p_{a'}, M)$, as required.

Turn to the general case of an arbitrary realization a' of $stp(a)$. Let b be a realization of $stp(a)$ which is independent from M and N a prime model over $M \cup \{b\}$. By the previous paragraph, $\dim(p, N) = \dim(p_b, N) = \dim(p_{a'}, N)$. Since any SR type in $S(M)$ has finite dimension in N (bounded by $wt(b)$) we derive the equality of $\dim(p, M)$ and $\dim(p_{a'}, M)$ from the equations $\dim(p|M, N) = \dim(p_{a'}|M, N)$ and $\dim(p|M, N) + \dim(p, M) = \dim(p_{a'}|M, N) + \dim(p_{a'}, M)$.

Now take a' to be an arbitrary realization of $tp(a)$ in M such that $p \not\perp p_{a'}$. Since a and a' realize an isolated type over $acl(\emptyset)$, as well as over $\emptyset$, there is a prime model $N \subset M$ containing elements b and b' realizing $stp(a)$ and $stp(a')$, respectively. Since there is an automorphism of N taking b to b', $\dim(p_b, N) = \dim(p_{b'}, N)$. Since p and $p_{a'}$ are nonorthogonal and both are nonorthogonal to $\emptyset$, the conjugates over b and b' are nonorthogonal. Thus, the additivity of dimension again gives the equality of $\dim(p_b, M)$ and $\dim(p_{b'}, M)$. By the first part of the proof, $\dim(p, M) = \dim(p_{a'}, M)$.

For the remainder of the subsection, T is a bounded ω–stable theory.

For $i \in DIM(T)$ we want to give meaning to the term "the dimension of i in a model M". To do this we make a selection of representative types q_i satisfying:

1. q_i is SR of nonorthogonality class i and $dom(q_i)$ realizes an isolated type;
2. if i, $j \in DIM(T)$ and there is a conjugate of q_i nonorthogonal to q_j, then q_i is conjugate to q_j.

Also fix $\{q_i : i \in DIM(T)\}$ for the remainder of the subsection. For $i \in DIM(T)$ and M a model we let $\dim(i, M)$ denote $\dim(r_i, M)$ where r_i is a conjugate over $acl(\emptyset)$ of q_i which is over M. By Lemma 7.1.4 this dimension is independent of the choice of r_i (any r, r' conjugate over $acl(\emptyset)$ to q_i have the same dimension in any model containing $dom(r) \cup dom(r')$). If N is a submodel of M we define $\dim(i|N, M)$ to be $\dim(r_i|N, M)$, where r_i is a conjugate over $acl(\emptyset)$ to q_i which is over N. As a preliminary step to finding invariants for models with respect to isomorphism (over $\emptyset$) we define $\mathcal{I}_1(M)$, called the *pre-invariant of* M, to be the function f from $DIM(T)$ into the class of cardinals such that $f(i) = dim(i, M)$, for $i \in DIM(T)$. It is immediate from the definition that $\mathcal{I}_1(M) = \mathcal{I}_1(N)$ when M and N are models isomorphic over $acl(\emptyset)$, but not necessarily when M and N are isomorphic over $\emptyset$. We will deal with this deficiency later, first taking care of

Lemma 7.1.5. *Given models M, N of T, if $\mathcal{I}_1(M) = \mathcal{I}_1(N)$, then $M \cong N$, in fact, M and N are isomorphic over $acl(\emptyset)$.*

Proof. The basic approach is fairly clear. We choose prime models $M_0 \subset M$ and $N_0 \subset N$ and bases J_i, K_i for the distinguished representative of $i \in DIM(T)$ over these prime models in M, N. Since the dimensions of the regular types match up in M and N we should be able to lift an isomorphism between M_0 and N_0 to one taking J_i onto K_i. However, it's possible for $\dim(i, M) = dim(i, N)$ to be $\aleph_0$, while $\dim(i|M_0, M) = \aleph_0$ and $\dim(i|N_0, N)$ is finite. By choosing the prime models carefully (using the following claim) we will eliminate this irregularity.

Claim. For M_1 a prime model there is a prime model $M_1' \supset M_1$ such that for all $i \in DIM(T)$ with $\dim(i, M_1)$ infinite, $\dim(i|M_1, M_1')$ is also infinite.

Let $i \in DIM(T)$ be such that $\dim(r_i, M_1)$ is infinite, where r_i is conjugate to q_i over $acl(\emptyset)$. Let a realize $r_i|M_1$, N a prime model over $M_1 \cup \{a\}$ and I a basis for r_i in M_1. For any $b \in M_1$ there is an $a' \in I$ realizing $tp(a/b)$ (since I is infinite) hence $tp(ab)$ is isolated. Thus, N is atomic (hence prime) over $M_1 \cup \{a\}$. Iterating this process infinitely many times for each element of $DIM(T)$ results in M_1'.

Given the original choice of the prime model $M_0 \subset M$ let $M_0' \supset M_0$ be a prime model as in the claim. Let f be an isomorphism of M_0' onto M_0 and $M_0'' = f(M_0)$. Thus, replacing M_0 by M_0'' if necessary we can assume that whenever $\dim(i, M_0)$ is infinite, $\dim(i|M_0, M)$ is infinite. Choose a prime model $N_0 \subset N$ with the same property. Now let $i \in DIM(T)$. Since $\mathcal{I}_1(M) = \mathcal{I}_1(N)$ the additivity of dimensions for SR types implies that $\dim(i, M_0) + \dim(i|M_0, M) = \dim(i, N_0) + \dim(i|N_0, N)$. If $\dim(i, M_0) = \dim(i, N_0)$ is finite we conclude automatically that $\dim(i|M_0, M) = \dim(i|N_0, N)$. If the dimension of i is infinite in a prime model, then the choice of M_0 and N_0 forces $\dim(i|M_0, M) = \dim(i|N_0, N)$.

The models M_0 and N_0 are isomorphic over $acl(\emptyset)$ via some map f. For $i \in DIM(T)$, let J_i be a basis in M for $r_i|M_0$, where r_i is as usual. Let K_i be a basis in N for $f(r_i)|N_0$. Since $r_i \perp r_j$ when $i \neq j \in DIM(T)$, $J = \bigcup_{i \in DIM(T)} J_i$ is M_0-independent. Similarly, $K = \bigcup_i K_i$ is M_0-independent. Thus, f extends to an elementary map g which takes J onto K. Because M is prime over $M_0 \cup J$, and N is prime over $N_0 \cup K$, g extends to an isomorphism of M onto N.

Thus, the pre-invariants characterize the models up to isomorphism over $acl(\emptyset)$. However, the pre-invariants may not be be preserved under arbitrary isomorphisms. For example, given $M \supset dom(q_i)$, $f : M \cong N$ may map q_i to an SR type orthogonal to q_i. This behavior is found in the following class of examples.

Example 7.1.3. Fix $n < \omega$. Let T_0 be the theory of an equivalence relation E with exactly n classes, each infinite, and let X be the set of classes of E. We define T by adding structure to X. Let G be an arbitrary group of permutations of X and add relations on X so that in the resulting universe $\mathfrak{C}$, $\{f \restriction X : f \in \mathrm{Aut}(\mathfrak{C})\} = G$. For $x \in X$ the formula $v/E = x$ defines a strongly minimal set and distinct elements of X give rise to orthogonal sets. Thus, there is a one-to-one correspondence between $DIM(T)$ and X. Let G act on $DIM(T)$ through this correspondence. Pre-invariants correspond to isomorphic models if and only if they are conjugate with respect to G.

Motivated by this example we define an action of $\mathrm{Aut}(\mathfrak{C})$ on the pre-invariants so that models are isomorphic when they have conjugate pre-invariants. We will then take as the invariant of M the conjugacy class of a pre-invariant of M. The details follow.

Any conjugate of one of the q_i's is SR, hence lies in one of the nonorthogonality classes that make up $DIM(T)$. Define an action of $\mathrm{Aut}(\mathfrak{C})$ on $DIM(T)$ by: for $f \in \mathrm{Aut}(\mathfrak{C})$ and $i \in DIM(T)$, $f(i)$ is the unique j such that $f(q_i) \not\perp q_j$. Set-theoretically a pre-invariant is a function from $DIM(T)$ into the class of cardinals. The action of $\mathrm{Aut}(\mathfrak{C})$ on $DIM(T)$ can be extended to the class of pre-invariants by:

Given pre-invariants ϕ, ϕ' and $f \in \mathrm{Aut}(\mathfrak{C})$, $f(\phi) = \phi'$ if $\phi = \phi' \circ f$.

Lemma 7.1.6. *If f is an isomorphism from the model M onto the model N, then for $i \in DIM(T)$, $\mathcal{I}_1(M)(i) = \mathcal{I}_1(N)(f(i))$; i.e., $f(\mathcal{I}_1(M)) = \mathcal{I}_1(N)$.*

Proof. By Lemma 7.1.4, when r_i is over M and conjugate to q_i over $acl(\emptyset)$, $\dim(f(r_i), N) = \dim(f(i), N)$. Thus, $\dim(i, M) = \dim(f(i), N)$, proving that $f(\mathcal{I}_1(M)) = \mathcal{I}_1(N)$.

Define an equivalence relation $\sim$ by: given pre-invariants ϕ, ϕ', $\phi \sim \phi'$ if there is an $f \in \mathrm{Aut}(\mathfrak{C})$ such that $f(\phi) = \phi'$ that is, the $\sim$ –classes are the orbits under the action of $\mathrm{Aut}(\mathfrak{C})$ on the pre-invariants. Finally,

Definition 7.1.3. *Define the* invariant of M *to be* $\mathcal{I}(M) = \mathcal{I}_1(M)/\sim$.

Remember that to qualify as a structure theorem the assigned invariants must be set-theoretically simple in some intuitive way. Let G be the group of permutations of $DIM(T)$ (and the class of pre-invariants) induced by $\mathrm{Aut}(\mathfrak{C})$ as above. Since any element of $\mathrm{Aut}(\mathfrak{C})$ which is the identity on $acl(\emptyset)$ is the identity on $DIM(T)$, G can be identified with a quotient group of the automorphism group of $acl(\emptyset)$. Thus, $|G| \leq 2^{\aleph_0}$ and for any pre-invariant ϕ, the corresponding invariant is $G\phi$. We accept this as sufficient evidence of the set-theoretic simplicity of the invariants for the models of T.

Theorem 7.1.1 (Structure Theorem). *Models M and N are isomorphic if and only if $\mathcal{I}(M) = \mathcal{I}(N)$.*

Proof. See Lemma 7.1.6 for a proof that isomorphic models have conjugate pre-invariants; i.e., the same invariant. On the other hand, if $\mathcal{I}(M) = \mathcal{I}(N)$, then there is a model N' isomorphic to N such that $\mathcal{I}_1(N') = \mathcal{I}_1(M)$. By Lemma 7.1.5, N' and M are isomorphic.

Since invariants are set-theoretically simple we can call the result a "Structure Theorem".

This completes the assignment of invariants to a bounded t.t. theory. Recall that the spectrum function of T is the function $I(-,T)$ assigning to an infinite cardinal λ the number of models of T of cardinality λ. The assignment of invariants lets us compute exactly the spectrum function for T. This is an example of how the Structure Theorem leads to additional information about the models of the theory. The number $I(\lambda, T)$ may depend on the group G (defined above) and properties of $i \in DIM(T)$ such as "there is a model M and $r \in i$ such that $\dim(r, M)$ is finite". Note: we have not yet discussed the possible values of the pre-invariant functions.

Definition 7.1.4. *Let p be a stationary type over a finite set A in a stable theory. Then p is called* eventually nonisolated *(e.n.i.) if there is a finite $B \supset A$ such that $p|B$ is nonisolated. Otherwise, p is n.e.n.i.*

If T is a bounded $\omega-$stable theory as above we call $i \in DIM(T)$ eventually nonisolated *(e.n.i.) if there is an $r \in i$ which is e.n.i.*

Observe that being e.n.i. is preserved under conjugacy of types.

Lemma 7.1.7. *Let T be a bounded $\omega-$stable theory.*

(i) $i \in DIM(T)$ is e.n.i. if and only if for any $r \in i$ and model $M \supset dom(r)$, $\dim(r, M)$ is finite.

(ii) If i is e.n.i. and M_0 is a prime model then, for some $k < \omega$, $\dim(i, M_0) = k$ and for any cardinal $\kappa \geq k$ there is a model M with $\dim(i, M) = \kappa$.

Proof. Both (i) and (ii) follow easily from two claims which are at the heart of the connection between dimension and being e.n.i.

Claim. Let r be an SR type over a finite set A, $M \supset A$ a model and q an SR type over a finite set $A' \subset M$ which is nonorthogonal to r. Then $\dim(r, M)$ is infinite if and only if $\dim(q, M)$ is infinite.

Without loss of generality, $A' = A$ and M is the prime model over A. There are a, b realizing $r|M$, $q|M$, respectively, such that $tp(a/M \cup \{b\})$ and

$tp(b/M \cup \{a\})$ are both isolated. Let $B \subset M$, $A \subset B$, be a finite set on which $tp(ab/M)$ is based (in which case $tp(b/B \cup \{a\})$ is isolated). Let I be a basis for r in M, J a basis for q in M and suppose I is infinite. For any finite $C \subset M$, $C \supset B$, there is an $a_0 \in I$ realizing $tp(a/C)$. For any such a_0 there is a $b_0 \in M$ such that $tp(a_0b_0/C) = tp(ab/C)$, in particular, b_0 realizes $q|C$. It follows that J is infinite, proving the claim.

Claim. Let r be an SR type over a finite set A. Then r is e.n.i. if and only if $\dim(r, M)$ is infinite in any model $M \supset A$.

First suppose that r is e.n.i. and $B \supset A$ is a finite set such that $r|B$ is nonisolated. Let M be a prime model over B and I a basis for r in M. If I is infinite there is an $a \in I$ realizing $r|B$. This contradicts that $r|B$ is nonisolated and M is prime over B, so I is finite.

Now suppose that r is n.e.n.i. and $M \supset A$ is a model. If $J \subset M$ is a finite Morley sequence in r over A, then $r|(I \cup A)$ is isolated, hence realized in M. Thus, $\dim(r, M)$ is infinite, proving the claim.

(i) follows immediately from the two claims.

The proof of (ii) is left as Exercise 7.1.4.

Let Δ_1 be the e.n.i. dimensions in $DIM(T)$, Δ_2 the n.e.n.i. dimensions and $\delta_i = |\Delta_i|$, for $i = 1, 2$. For simplicity we will compute the function $I^*(-, T)$, where $I^*(\lambda, T)$ is the number of models of cardinality $\leq \lambda$. The "subtraction" needed to compute $I(-, T)$ is left to the reader.

Lemma 7.1.8. *Let Φ_α be the set of pre-invariants of models of T of cardinality $\leq \aleph_\alpha$ and Φ_α/G the orbits of pre-invariants under G; i.e., the invariants of models of T. Then, $I^*(\aleph_\alpha, T) = |\Phi_\alpha/G|$ and $|\Phi_\alpha| = |\alpha + \omega|^{\delta_1} \times |\alpha + 1|^{\delta_2}$. When $|\alpha|$ is regular and uncountable, $I^*(\aleph_\alpha, T) = |\alpha|$.*

(The proof is left to the reader.) The actual value of $|\Phi_\alpha/G|$ depends on detailed information about G — there is no uniform formula for computing this cardinal in terms of $|G|$ and $|\Phi_\alpha|$ — however, for any particular theory it is easy to determine the value. We can, though, give some rough limits for some $\aleph_\alpha$. The group G is infinite only if $DIM(T)$ is infinite, in which case there are infinitely many conjugacy classes of dimensions. Thus, $I^*(\aleph_\alpha, T) \geq |\alpha + 1|^{\aleph_0}$, in fact $I^*(\aleph_\alpha, T) \geq |\alpha + \omega|^{\aleph_0}$ if δ_1 is infinite. We already know that $I^*(\aleph_\alpha, T) \leq |\alpha + \omega|^{\delta_1} \times |\alpha + 1|^{\delta_2}$, so if δ_1 is infinite, $I^*(\aleph_\alpha, T) = |\alpha + \omega|^{\aleph_0}$. If, say, δ_1 is finite, but nonzero, and δ_2 is infinite, then $I^*(\aleph_\alpha, T) = |\alpha + 1|^{\aleph_0} + \aleph_0$. It is easy to construct examples of theories satisfying each of these conditions. When $DIM(T)$ is finite the group G comes into play. It is natural to ask when T can have finitely many models in some uncountable cardinal. Of course, this is possible when T is uncountably categorical, but the above argument says it's also true exactly when $\delta_1 = 0$ and δ_2 is finite. For example, take the theory of an infinite and coinfinite predicate symbol P. This theory is ω–stable, bounded, has no e.n.i. dimensions, two n.e.n.i. dimensions and the group G consists of the identity. By the above formula, $I^*(\aleph_\alpha, T) = |\alpha + 1|^2$, which is finite exactly when α is finite. The reader who constructs other examples will probably conjecture the following result.

Lemma 7.1.9 (Lachlan). *Let T be a bounded ω–stable theory such that in some uncountable cardinal κ, T has more than one but finitely many models of cardinality κ. Then T is ω–categorical.*

Proof. It suffices to assume that T is not ω–categorical and prove it has an e.n.i. dimension. Let $\{ q_i : i \in DIM(T) \}$ be a family of SR types satisfying the conditions on page 328. Remember, $i \in DIM(T)$ is e.n.i. if and only if q_i is e.n.i. Since T is not ω–categorical there is a nonisolated complete type p_0 over $\emptyset$. Let M be a prime model and a a realization of p with $a \mathrel{\perp} M$. By Proposition 7.1.2(ii) there is a finite M–independent set C of realizations of SR types over M such that $tp(C/M \cup \{a\})$ and $tp(a/M \cup C)$ are isolated. Without loss of generality, for each $c \in C$ there is an $i \in DIM(T)$ such that c realizes $q_i|M$. Let $A \subset M$ be finite such that $tp(aC/M)$ is based on A and $tp(a/M \cup C)$ is isolated over $A \cup C$. By the usual corollary to the Open Mapping Theorem $tp(a/A)$ is nonisolated. Thus, $tp(C/A)$ is nonisolated. Let $C' \subset C$ and $c \in C$ be such that $tp(C'/A)$ is isolated and $p = tp(c/A \cup C')$ is nonisolated. Then p is parallel to some q_i, which is hence e.n.i. This proves the lemma.

This lemma about countable theories with finitely many but more than one model in some uncountable cardinality can be improved significantly. Shelah showed in [She90, VIII,1.7] that any countable theory which is not ω–stable has $\geq \min\{2^{2^{\aleph_0}}, 2^\lambda\}$ models in each uncountable cardinality λ. Moreover, Lachlan proved in [Lac75] that if T is an ω–stable theory with finitely many models in some uncountable cardinality, then T is ω–categorical and bounded (see Corollary 7.1.3). The best possible result is

> A countable theory T has finitely many but more than one model in some uncountable power if and only if T is ω–stable, ω–categorical and bounded.

The left-to-right direction uses results by Shelah (to reduce our attention to ω–stable theories) and a theorem by Lachlan. (Later we will complete Lachlan's contribution by reproducing the proof that such theories are bounded.) The right-to-left direction reduces to showing that a bounded ω–stable, ω–categorical theory has finitely many dimensions. This requires the deep "geometrical" results found in [CHL85].

We leave it to the reader to investigate other properties of the spectrum functions of bounded ω–stable theories on his one.

Besides calculating the possible spectrum functions this analysis of SR types in a bounded ω–stable theory leads to Proposition 7.1.3. Some background is needed to understand this result.

As mentioned earlier this study of bounded t.t. theories can be viewed as a generalization of the Baldwin-Lachlan Theorem. One part of their analysis was to prove the homogeneity of all countable models of an uncountably categorical theory. The following trivial example shows that not every bounded ω−stable theory has this property.

Example 7.1.4. (A bounded ω−stable theory with a nonhomogeneous countable model) Let $L = \{ P_i : i < \omega \} \cup \{ E \}$, where E is a binary relation symbol and P_i is unary. Let M be a structure for L in which $\{ P_i(M) : i < \omega \}$ form a pairwise disjoint family of infinite sets, and E defines an equivalence relation on M with two classes, each class containing infinitely many elements of $P_i(M)$ for all $i < \omega$. Then $T = Th(M)$ is a bounded ω−stable theory. A countable model N in which one E−class is contained in $\bigcup_{i<\omega} P_i(N)$ and the other contains an element not in any $P_i(N)$, is not homogeneous. (For a fixed $i \in \omega$, let $a, b \in P_i(N)$ such that $\models \neg E(a, b)$. Then $tp(a) = tp(b)$ but there is no automorphism of N mapping a to b.)

Of course, if we added constants to the language for the E−classes in this example every model would be $\aleph_0$−homogeneous, in other words, every model of the theory is $\aleph_0$−homogeneous over $acl(\emptyset)$. We'll see shortly that this is always true in a bounded ω−stable theory.

Definition 7.1.5. *A model M is* almost κ−homogeneous *(where $\kappa \geq |T|$) if M is κ−homogeneous over $acl(\emptyset)$; i.e., M is κ−homogeneous in the language with constants for $acl(\emptyset)$.*

The usual conventions for κ−homogeneous models are adopted for almost κ−homogeneous models, for example, M is almost homogeneous if it is almost $|M|$−homogeneous.

Proposition 7.1.3. *If T is a bounded t.t. theory, then every model of T is almost $\aleph_0$−homogeneous.*

Proof. Given a, b and c in M with $stp(a) = stp(b)$ we must find a $d \in M$ such that $stp(bd) = stp(ac)$. We can enlarge a and b by adjoining elements realizing isolated types over $\{a\} \cup acl(\emptyset)$ and $\{b\} \cup acl(\emptyset)$ to require that M contains $e = \{e_0, \ldots, e_n\}$ such that e is a−independent, $q_i = tp(e_i/a)$ is SR, both $tp(c/ae)$ and $tp(e/ac)$ are isolated, and $tp(c/a) \models stp(c/a)$. Let f be an automorphism fixing $acl(\emptyset)$ and mapping a to b. Then, q_i is conjugate to $r_i = f(q_i)$ over $acl(\emptyset)$ so they have the same dimension in M (by Lemma 7.1.4). Thus, there is $e' = \{e'_0, \ldots, e'_n\}$ such that $stp(ae) = stp(be')$. Since $tp(c/ae)$ is isolated there is a $d \in M$ with $tp(aec) = tp(be'd)$. Since $tp(c/a) \models stp(c/a)$, $tp(d/b) \models stp(d/b)$. Thus, $stp(bd) = stp(ac)$, as required.

Almost $\aleph_0$−homogeneous models do have some of the same relative uniqueness and universality conditions as $\aleph_0$−homogeneous models:

Lemma 7.1.10. *Let T be an ω−stable theory.*

(i) If M is an almost $\aleph_0$−homogeneous model and N is a countable model such that every type over $\emptyset$ realized in N is realized in M, then N can be elementarily embedded into M.

(ii) If M and N are countable almost homogeneous models realizing the same types over $\emptyset$, then $M \cong N$.

Proof. Both parts of the lemma follow quickly from

Claim. Let N be a countable model and M a model such that every element of $S(\emptyset)$ realized in N is realized in M. Then there is a model $N' \cong N$ such that every element of $S(acl(\emptyset))$ realized in N' is realized in M.

Let Q be the set of elements of $S(acl(\emptyset))$ realized in M. Let $N = \{ a_i : i < \omega \}$, $b_i = (a_0, \ldots, a_i)$ and $p_i = tp(b_i)$, a type in a sequence v_i of $i+1$ variables. Since T is ω−stable each p_i has finitely many extensions over $acl(\emptyset)$. Since each p_i is realized in M there is $q_i \in Q$, an extension of p_i, such that for infinitely many (hence all) $j \geq i$ there is an element of Q extending p_j whose restriction to v_i is q_i. In fact, (by König's Lemma) we can choose the q_i's so that q_i is the restriction to v_i of q_{i+1}. Thus, there is a set $\{ c_i : i < \omega \} \subset M$ such that $(c_0, \ldots, c_i)$ realizes q_i. The model $N' = \{ c_i : i < \omega \}$ is the desired isomorphic copy of N, proving the claim.

Since an almost $\aleph_0$−homogeneous model is $\aleph_0$−homogeneous over $acl(\emptyset)$ both (i) and (ii) follow from Corollaries 2.2.2 and 2.2.4.

7.1.2 Unbounded Theories

This subsection is a continuation of our study of how the isomorphism type of a model M of an ω−theory is tied to the dimensions of the regular types over M. For $p \in S(A)$ and B a set conjugate to A over $\emptyset$, p_B denotes a type over B conjugate to A. In Lemma 7.1.8 we calculated the spectrum function for a bounded ω−stable theory. We showed, for example, that when $|\alpha|$ is regular and uncountable, $I^*(\aleph_\alpha, T) = |\alpha|$, a number significantly smaller than the maximum possible value, $2^{\aleph_\alpha}$ (in general). In other words, having a bounded number of SR types, up to nonorthogonality, leads to relatively few models. In the next proposition we give a comparatively large lower bound to the spectrum function of an *un*bounded ω−stable theory.

Proposition 7.1.4. *If T is an unbounded ω−stable theory, then for all $\alpha > 0$, $I(\aleph_\alpha, T) \geq |\alpha + \omega|^{|\alpha+1|}$.*

Proof. Let $\kappa_\alpha = |\alpha + \omega|^{|\alpha+1|}$. By Lemma 7.1.2 and Proposition 6.4.1 there is an SR type $p \in S(a)$ which is orthogonal to $\emptyset$. Let $q = stp(a)$. Let Λ be the collection of all functions f from $\{ \beta : \beta \leq \alpha \}$ into $\{ \lambda : \lambda$ is a cardinal $\leq \aleph_\alpha \}$ such that $f(\alpha) = \aleph_\alpha$. Note, $|\Lambda| = \kappa_\alpha$. For each $f \in \Lambda$ we construct as follows a model M_f of cardinality $\aleph_\alpha$ so that for $f \neq g \in \Lambda$, $M_f \ncong M_g$.

Fix $f \in \Lambda$. For $\beta \leq \alpha$ let J_β be a Morley sequence in q of cardinality $f(\beta)$, chosen so that $J = \bigcup_{\beta \leq \alpha} J_\beta$ is also independent. First let M_0 be a prime model over J (which has cardinality $\aleph_\alpha$). Given $b \in J$ and I a basis for p_b in M_0, I is independent from J over b since p_b is orthogonal to q. Thus, I is indiscernible over J, which is hence countable by Theorem 5.5.1. Thus, $\dim(p_b, M_0)$ is countable for all $b \in J$. By iterated use of Proposition 6.4.4(iii) there is a model $M_1 \supset M_0$ of cardinality $\aleph_\alpha$ such that for each $\beta < \alpha$ and $b \in J_\beta$, $\dim(p_b, M_1) = \aleph_\beta$. Let $X = \{p_b : b \in J\}$. Finally, by (iv) of Proposition 6.4.4 there is a model $M_f \supset M_1$ of cardinality $\aleph_\alpha$ such that $\dim(p'|M_1, M_f) = 0$ for all $p' \in X$ and $\dim(r, M_f) = \aleph_\alpha$, for any SR type r over a finite subset of M_f which is orthogonal to each element of X.

We now verify that distinct elements of Λ give rise to nonisomorphic models. Suppose that $F : M_g \cong M_f$, where $f, g \in \Lambda$, and $J \subset M_f$ is a Morley sequence in q as in the above construction. It suffices to show that $g(\beta) \leq f(\beta)$ for all $\beta \leq \alpha$. Simply by the definition of Λ, $f(\alpha) = g(\alpha)$. Let $\beta < \alpha$ and c a realization of q in M_g with $\dim(p_c, M_g) = \aleph_\beta$. By construction, any conjugate of p over M_f which is orthogonal to each type in $\{p_b : b \in J\}$ has dimension $\aleph_\alpha$. Thus, by Proposition 6.4.4(ii), $F(p_c)$ is nonorthogonal to one of these p_b's, in fact, $F(p_c) \not\perp p_b$, for some $b \in J_\beta$. Since p is orthogonal to $\emptyset$, when $d, d' \in q(\mathfrak{C})$ are independent, $p_d \perp p_{d'}$. Thus, if J' is a Morley sequence in q such that $\dim(p_c, M_g) = \aleph_\beta$ for all $c \in J'$, F induces an injection $\phi : J' \longrightarrow J_\beta$ defined by: for $c \in J'$, $\phi(c)$ is the element b of J_β such that $F(c) \not\perp b$. We conclude that $g(\beta) \leq f(\beta)$, as required to prove the proposition.

Corollary 7.1.3. *If T is an ω−stable theory with finitely many models in some uncountable cardinal, then T is ω−categorical and bounded.*

(See Lemma 7.1.9 and subsequent remarks concerning this corollary.)

Historical Notes. Unbounded theories are defined as multi-dimensional theories in [She90, V.5.2], although unidimensional theories are defined earlier in [She90, V.2.2]. Lemma 7.1.2 is stated explicitly as [Las86, 9.7]. Proposition 7.1.1 is [She90, IX.1.8]. The main idea in Proposition 7.1.2 is found in Section 4 of [BL83] (Lemma 4.5, in particular) and is stated more explicitly as [Las86, 9.13]. Most of the results in the subsection on bounded ω−stable theories are found in [She90, IX,2.3], [BL83] and [Las86]. Proposition 7.1.3 is Corollary 5.3 of [BL83]. Lemma 7.1.10(ii) is due to Pillay [Pil82]. Lachlan's Lemma 7.1.9 is found in [Lac75]. Proposition 7.1.4 is implicit in Section 5 of [She90, V].

Exercise 7.1.1. Prove (4) $\Longrightarrow$ (1) in Lemma 7.1.2.

Exercise 7.1.2. Prove the following fact using the same ideas used to prove Proposition 7.1.2.

> If T is a bounded t.t. theory, M is an a–model and C is an M–independent set of realizations of SR types over M, then a prime model over $M \cup C$ is an a–model.

(This is due to Pillay. HINT: Take a prime model N over $M \cup C$, an a–prime model N' over $M \cup C$ containing N, and show that N' must equal N.)

Exercise 7.1.3. Following the methods used to analyze bounded ω–stable theories develop a theory of invariants for the class of a–models in a bounded countable superstable theory and prove the resulting structure theorem.

Exercise 7.1.4. Prove (ii) in Lemma 7.1.7.

Exercise 7.1.5. State and prove an analogue of Proposition 7.1.4 which holds relative to the class of a–models in a bounded countable superstable theory.

7.2 More on Ranks

This section is devoted to refining our knowledge of Morley rank, ∞–rank and U–rank in some special superstable theories. In particular, we will prove the "definability of Morley rank" and the equivalence of Morley rank and U–rank in uncountably categorical theories. We will also prove corresponding results about ∞–rank in unidimensional superstable theories.

See Definition 6.1.3 for the definition of a "notion of rank". Given a complete theory T, a map R which takes a formula of T to an ordinal is a *notion of rank on formulas* if the map R' which takes a complete type p to $\inf\{ R(\varphi) : \varphi \in p\}$ is a notion of rank.

Definition 7.2.1. *Given a complete theory T, a notion of rank R on formulas in is said to be* definable *if for all formulas $\varphi(x,a)$,*

$$\text{there is a } \theta \in tp(a) \text{ such that } \models \theta(b) \implies R(\varphi(x,a)) = R(\varphi(x,b)). \quad (7.1)$$

When θ satisfies (7.1) and $R(\varphi(x,a)) = \alpha$ we say θ proves that $R(\varphi(x,a)) = \alpha$.

One instance of the definability of Morley rank in uncountably categorical theories played an important role in our proofs of Morley's Categoricity Theorem and the Baldwin-Lachlan Theorem. Specifically, for T such a theory and $\varphi(x,y)$ a formula there is a formula $\theta(y)$ such that $MR(\varphi(x,a)) = 0 \iff \models \theta(a)$ (see Lemma 3.1.12). This fact was of critical importance in showing that there is a strongly minimal formula over the prime model. The definability of Morley rank has numerous applications in the study of uncountably categorical theories. Of equal importance is the equality of Morley rank and U–rank and the fact that these ranks are always finite (proven below). This implies that Morley rank has the same additivity properties as U–rank (Corollary 6.1.1); i.e., $MR(ab) = MR(a/b) + MR(b)$, for all a and b.

Uncountable categoricity is an important hypothesis in obtaining the definability of Morley rank. Example 3.1.3 produces a simple ω–stable theory in which this definability fails. (There is an a such that $E(x,a)$ is nonalgebraic, but for any $\theta \in tp(a)$ there is a b satisfying θ such that $E(x,b)$ is algebraic.)

Our first major goal is

Theorem 7.2.1. *If T is a unidimensional theory, then*
(i) ∞–rank is definable in T and
(ii) for all complete types p, $R^\infty(p) = U(p) < \omega$.

Part (i) is certainly the hardest. This will follow from the slightly more general

Proposition 7.2.1. *Suppose that T is a superstable theory in which every nonalgebraic type of finite ∞–rank is nonorthogonal to $\emptyset$. Then, for all formulas $\varphi(x,a)$,*

(∗) *if $R^\infty(\varphi(x,a)) = n < \omega$, there is a formula $\theta \in tp(a)$ such that*
$\models \theta(b) \implies R^\infty(\varphi(x,b)) = n$.

Fix T satisfying the hypotheses of the proposition until the completion of the proof. In the proof we will use the following which was assigned in Section 6.1 as Exercise 6.1.4 (and proved in [She90, V,7.12(5)]).

Lemma 7.2.1. *If $R^\infty(\varphi(x,a)) \geq n$ (where $n < \omega$), then there is a $p \in S(a)$ containing $\varphi(x,a)$ such that $U(p) \geq n$.*

(The proof is a relatively easy induction on $R^\infty(\varphi(x,a))$.)

The proposition is proved by induction on rank. Assume that (∗) holds for all formulas of ∞–rank $< n$ and $n = R^\infty(\varphi(x,a))$. The proof of (∗) for $\varphi(x,a)$ is divided into two similar but distinct parts. In the proofs we will use the fact that if $a' \supset a$ and there is a formula $\theta' \in tp(a')$ such that $\models \theta'(b) \implies R^\infty(\varphi(x,b)) = R^\infty(\varphi(x,a))$, then there is such a formula over a. (Simply quantify existentially over the variables satisfied by the elements of $a' \setminus a$.) This permits us to expand a to a set having additional properties.

Lemma 7.2.2. *There is a formula $\theta \in tp(a)$ such that*

$$\models \theta(b) \implies R^\infty(\varphi(x,b)) \geq n.$$

Proof. Assume the lemma fails. Then, for each $\theta \in tp(a)$ there is an a' satisfying θ such that $R^\infty(\varphi(x,a')) \leq n-1$. By Lemma 7.2.1 there is a type q over $acl(a)$ containing φ and having U–rank $\geq n$. Since U–rank is $\leq \infty$–rank for any complete type, $U(q) = n$. Since q is nonorthogonal to $\emptyset$, we can pick a sufficiently large so that if c realizes q there is a pair $c'a'$ realizing $stp(ca)$ such that

$$a' \perp ca,\ c'a' \perp a \text{ and } c' \not\perp_{aa'} c$$

(see Propositions 5.6.1 and 5.6.2). Let $k = R^\infty(c/aa'c')$ and $k' = R^\infty(c'/aa'c)$, both of which are $< n$. Without loss of generality, $k \leq k'$. Let $\psi(x, y, x', y') \in tp(cac'a')$ be such that $\models \psi(d, b, d', b')$ implies

(2) $\models \varphi(d, b) \wedge \varphi(d', b')$,
(3) $R^\infty(\psi(x, b, d', b')) \leq k$, and
(4) $R^\infty(\psi(d, b, x', b')) = k'$.

(We can require (3) and (4) by the inductive hypothesis. First pick a formula $\psi_0(x, y, x', y')$ such that $\exists x \psi_0(x, y, x', y')$ proves that $k = R^\infty(\psi_0(x, ac'a'))$. Choose $\psi(x, y, x', y')$ to be a formula implying ψ_0 such that $\exists x' \psi(x, y, x', y')$ proves that $k' = R^\infty(\psi(ac, x', a'))$. While $\exists x \psi(x, y, x', y')$ may not prove that $k = R^\infty(\psi(x, ac'a'))$, the inequality in (3) does hold.) We will obtain a contradiction by showing that $k' + n \leq k + (n - 1)$.

Since a' is independent from ca we can apply the Open Mapping Theorem to find a $\sigma \in tp(a')$ such that every element of $S(\emptyset)$ containing σ has a nonforking extension containing $\exists x' \psi(c, a, x', y)$. Since we assumed that the lemma fails there is a b, $b \perp ca$ such that $\exists x' \psi(c, a, x', b)$ and $R^\infty(\varphi(x, b)) \leq n - 1$. From (4) we know that $R^\infty(\psi(c, a, x', b)) = k'$, so by Lemma 7.2.1 there is a d satisfying $\psi(c, a, x', b)$ with $U(d/cab) = k'$. Now compute $U(dc/ab)$ using the U–rank identities of Corollary 6.1.1: $U(dc/ab) = U(d/cab) + U(c/ab) = k' + n$. We also know that $U(c/dab) \leq k$ by (3), so $U(dc/ab) = U(c/dab) + U(d/ab) \leq k + (n-1)$. This is the contradiction which proves the lemma.

To complete the proof of Proposition 7.2.1 we prove

Lemma 7.2.3. *There is a formula $\theta \in tp(a)$ such that*

$$\models \theta(b) \implies R^\infty(\varphi(x, b)) \leq n.$$

Proof. The bulk of the proof is contained in

Claim. For any $p = tp(ca)$, where $\models \varphi(c, a)$, there is an $\hat{a} \supset a$, $c \perp_a \hat{a}$, and a $\psi(x, \hat{y}) \in tp(c\hat{a})$ such that for all b, $R^\infty(\psi(x, b)) \leq R^\infty(c/a)$.

If $R^\infty(c/a) < n$ the desired formula is obtained by induction, so we can assume that $R^\infty(c/a) = n$. Assume, to the contrary, that there are no such $\hat{a}$ and ψ. Since $stp(c/a)$ is nonorthogonal to $\emptyset$, there are:

– $\hat{a} \supset a$, $c \perp_a \hat{a}$, and
– $c'\hat{a}'$ realizing $stp(c\hat{a})$

such that

$$c\hat{a} \perp \hat{a}', \; \hat{a} \perp c'\hat{a}' \text{ and } c \not\perp_{\hat{a}\hat{a}'} c'.$$

Let $R^\infty(c/c'\hat{a}\hat{a}') = k$ and $R^\infty(c'/c\hat{a}\hat{a}') = k'$, where, without loss of generality, $k' \leq k$. By induction, there is a formula $\psi(x, \hat{y}, x', \hat{y}') \in tp(c\hat{a}c'\hat{a}')$ such that $\models \psi(d, \hat{b}, d', \hat{b}')$ implies

(5) $\models \varphi(d, \hat{b}) \wedge \varphi(d', \hat{b}')$.
(6) $R^\infty(\psi(x, \hat{b}, d', \hat{b}')) = k$, and
(7) $R^\infty(\psi(d, \hat{b}, x', \hat{b}')) \leq k'$.

(More properly, in (5) we mean that $\varphi(d, b)$ holds for an appropriate subset b of $\hat{b}$.) Since $c'\hat{a}' \perp \hat{a}$ the open mapping theorem yields a formula $\theta(x', \hat{y}') \in tp(c'\hat{a}') = tp(c\hat{a})$ such that $r \in S(\emptyset)$ has a nonforking extension containing $\exists x\psi(x, \hat{a}, x', \hat{y}')$ if and only if $\theta(x', \hat{y}') \in r$. Since we have assumed the claim to fail for p there are $\hat{b}$ and d such that $\models \exists x\psi(x, \hat{a}, d, \hat{b})$, $\hat{a} \perp d\hat{b}$ and $R^\infty(d/\hat{b})$ is $\geq n+1$. Furthermore, by Lemma 7.2.1, we can assume that $U(d/\hat{b}) \geq n+1$. By (6) there is a c_0 satisfying $\psi(x, \hat{a}, d, \hat{b})$ such that $R^\infty(c_0/d\hat{a}\hat{b}) = U(c_0/d\hat{a}\hat{b}) = k$. Now compute $U(c_0d/\hat{a}\hat{b})$ in two ways. If $U(d/\hat{b})$ is infinite, then $U(c_0d/\hat{b})$ is infinite, hence $\geq k + (n+1)$. If $U(d/\hat{b}) < \omega$, $U(c_0d/\hat{a}\hat{b}) = U(c_0/d\hat{a}\hat{b}) + U(d/\hat{a}\hat{b}) \geq k + (n+1)$ (by Corollary 6.1.1). On the other hand, $U(c_0d/\hat{a}\hat{b}) = U(d/c_0\hat{a}\hat{b}) + U(c_0/\hat{a}\hat{b}) \leq k' + n$. This contradicts that $k' \leq k$, to prove the claim.

We now continue the proof to find the desired formula in $tp(a)$. Let M be a saturated model containing a. Let $r \in S(M)$ be an arbitrary type containing $\varphi(x, a)$. By the claim and the saturatedness of M there is an $\hat{a} \in M$, $\hat{a} \supset a$, and a formula $\psi(x, \hat{a}) \in r$ such that

(8) $R^\infty(\psi(x, \hat{b})) \leq R^\infty(r) \leq n$, for all $\hat{b}$.

By compactness, there is a $k < \omega$ and formulas $\psi_i(x, \hat{a}_i)$, $i \leq k$, each satisfying (8), such that any $r \in S(M)$ containing $\varphi(x, a)$ also contains one of these $\psi_i(x, \hat{a}_i)$'s. Let $\delta(x, \hat{c}a) = \bigvee_{i \leq k} \psi_i(x, \hat{a}_i)$, where $\hat{c} = \cup \hat{a}_i$, and observe that for any $\hat{d}b$, $R^\infty(\delta(x, \hat{d}b)) \leq n$. Let $\sigma(z)$ be the formula

$$\exists \hat{y}(\forall x(\varphi(x, z) \rightarrow \delta(x, \hat{y}, z)).$$

Then $\sigma \in tp(a)$ and for all b, $\models \sigma(b) \implies R^\infty(\varphi(x, b)) \leq n$. This proves the lemma.

This completes the proof of Proposition 7.2.1.

Recall that the major weakness of $U-$rank is that it is not continuous. In the class of theories presently under consideration this limitation is removed in (i) of

Lemma 7.2.4. *Let T be a superstable theory in which each type of finite U–rank is nonorthogonal to $\emptyset$. Further suppose that each type of U–rank 1 is nonorthogonal to a formula of ∞–rank 1. Then for all complete types p with $U(p) = n < \omega$*

(i) there is a $\varphi \in p$ such that for any complete type q, with $\varphi \in q$, $U(q) \leq n$, and

(ii) $U(p) = R^\infty(p)$.

Proof. We prove (i) and (ii) simultaneously by induction on rank. Both (i) and (ii) are clear when $U(p) = 0$. To make the induction work we need to handle the rank 1 case separately.

Claim. If p is a complete type of U–rank 1, then $R^\infty(p) = 1$.

Without loss of generality $dom(p) = M$ is an a–model and there are a realizing p and b such that $R^\infty(b/M) = 1$ and a depends on b over M, hence $a \in acl(M \cup \{b\})$. Thus, $R^\infty(a/M) \leq R^\infty(ab/M) = R^\infty(b/M) = 1$, implying that $R^\infty(a/M) = 1$ (since p is nonalgebraic) to prove the claim.

Turning to (i), if p is a complete type of U–rank 1 and $\varphi \in p$ has ∞–rank 1 then each complete type containing φ has U–rank ≤ 1, hence U–rank is continuous on the types of U–rank 1.

Now suppose that $U(p) = n + 1$ and both (i) and (ii) hold for complete types of U–rank $\leq n$. We first prove (i) for p. By the Open Mapping Theorem it suffices to find an appropriate formula in some nonforking extension of p (see Exercise 7.2.3), thus we can assume $dom(p) = M$ to be an a–model. By Proposition 6.3.4 (and our freedom to choose M sufficiently large) there are a realizing p and b dependent on a over M such that $U(b/M) = 1$. By the U–rank identity (Corollary 6.1.1), $U(a/M \cup \{b\}) = n$. By induction, $R^\infty(a/M \cup \{b\})$ is also n. Furthermore, (i) holds for $tp(b/M)$. Combining these facts with Proposition 7.2.1 produces a formula $\psi(x, y) \in tp(ab/M)$ such that

$$\text{whenever} \quad \models \exists x \psi(x, b'),\ R^\infty(\psi(x, b')) = n \text{ and } U(b'/M) \leq 1.$$

Suppose that $\models \exists y \psi(a', y)$. To complete the proof of (i) it suffices to show that $U(a'/M) \leq n + 1$. Let b' satisfy $\psi(a', y)$. Then $R^\infty(a'/M \cup \{b'\}) \leq n$, so $U(a'/M \cup \{b'\}) \leq n$ by Lemma 6.1.2(ii). Since $U(b'/M) \leq 1$, the U–rank identities imply that $U(a'/M) \leq U(a'b'/M) \leq U(a'/M \cup \{b'\}) + U(b'/M) \leq n + 1$, as required.

Turning to (ii), assume $U(p) = n+1$ and $R^\infty(q) = U(q)$ whenever $U(q) \leq n$. We must now show that $R^\infty(p) = n+1$. Let $\varphi \in p$ be such that $U(q) \leq n+1$ for any complete type q containing φ. Let $Q = \{ q \in S(\mathfrak{C}) : \varphi \in q \text{ and } \neg\psi \in q$ for all ψ with ∞–rank $\leq n \}$, which is nonempty since $R^\infty(p) \geq n + 1$ (U–rank is always $\leq \infty$–rank). Suppose φ is over A. If $q \in Q$, then $U(q) > n$ since (ii) holds for types of U–rank $\leq n$. Furthermore, $U(q \restriction A) \leq n + 1$ by the choice of φ, so each element of Q does not fork over A. Thus, $|Q| < |\mathfrak{C}|$, from which we conclude that $R^\infty(\varphi) = n + 1$ to complete the proof of the lemma.

Corollary 7.2.1. *If T is unidimensional, then $R^\infty(x=x) < \omega$.*

Proof. Let Φ be the set of all formulas of finite ∞−rank (in the same sort as x). By the previous lemma each type of finite U−rank has finite ∞−rank. Thus, assuming $\{\neg\varphi : \varphi \in \Phi\}$ to be consistent results in an element of $S(\mathfrak{C})$ of infinite U−rank, hence a q of U−rank ω. By Lemma 6.1.3 q is orthogonal to any type of finite U−rank. This contradiction to the unidimensionality of the theory proves the corollary.

This completes the proof of Theorem 7.2.1.

When T is a unidimensional t.t. theory we can add Morley rank to the picture. Sticking to the most interesting case we state the relevant result for countable theories.

Proposition 7.2.2. *If T is an uncountably categorical theory, then for all complete types p, $U(p) = R^\infty(p) = MR(p) < \omega$. Furthermore, Morley rank is definable in T.*

Proof. We have already shown that U−rank and ∞−rank are equal and finite in such theories. If we prove that Morley rank is equal to U−rank in T the definability of Morley rank will follow from Theorem 7.2.1. Our proof will closely parallel the proof of Proposition 7.2.1.

We proved above the continuity of U−rank in T. This is strengthened by showing

Claim. For all complete p there is a $\varphi \in p$ such that $\{q \in S(\mathfrak{C}) : \varphi \in q$ and $U(q) \geq U(p)\}$ is finite.

This is proved by induction on $U(p) = n+1$. (The result is obviously true when p is algebraic.) Again, we can assume that $dom(p) = M$ is a saturated model. There are a realizing p and b dependent on a over M such that $tp(b/M)$ is strongly minimal. By a now standard argument, $U(a/M \cup \{b\}) = R^\infty(a/M \cup \{b\}) = n$. By the definability of ∞−rank and its equality with U−rank there is a formula $\psi(x,y) \in tp(ab/M)$ such that

(9) $\exists x\psi(x,y)$ is strongly minimal,
(10) $\{q \in S(M \cup \{b\}) : \psi(x,b) \in q$ and $U(q) \geq n\}$ is finite, and
(11) for all b' satisfying $\exists x\psi(x,y)$, $R^\infty(\psi(x,b')) \leq n$ and whenever $\models \psi(a',b')$, $b' \in acl(M \cup \{a'\})$.

Suppose that $\models \exists y\psi(a',y)$ and b' satisfies $\psi(a',y)$. By (9) and (11) $U(a'/M \cup \{b'\}) \leq n$ and $U(a'/M) \leq n+1$. If $U(b'/M) = 0$, then $U(a'/M) \leq n$. Otherwise, $tp(b'/M) = tp(b/M)$ (by (9)) and a' depends on b' over M (by (11)). Thus, $U(a'/M) = n+1$, implying that $U(a'/M \cup \{b'\}) = n$ and $b' \notin M$ (by the usual U−rank computations). Thus, $tp(a'/M \cup \{b'\})$ is conjugate over M to one of the types over $M \cup \{b\}$ defined in (10). We conclude that there are finitely many types in $S(M)$ containing $\theta(x) = \exists y\psi(x,y)$ with U−rank

$n + 1$. Since each such type is stationary there are finitely many elements of $S(\mathfrak{C})$ containing θ and having U–rank $n + 1$, completing the proof of the claim.

Using this claim we can prove that $U(p) = MR(p)$ for any complete type p by an argument similar to that used to obtain (ii) of Lemma 7.2.4. The details are left to the reader in Exercise 7.2.1.

Contained within the proof of the previous proposition is a proof of

Corollary 7.2.2. *If T is a unidimensional theory containing a formula of Morley rank* 1, *then T is totally transcendental (hence uncountably categorical).*

Historical Notes. The finiteness and definability of Morley rank in an uncountably categorical theory is due to Baldwin [Bal73] and independently Zil'ber [Zil74]. Shelah extended these results to superstable unidimensional theories in [She90, IX.1.11]. There is a good exposition of these results by Saffe in [Saf84], although the proof of Theorem 4.5 of that paper does not work as written. After much prodding by Ambar Chowdhury I wrote down the proof that appears here. Bradd Hart and, independently, Predrag Tanovic have also written proofs.

Exercise 7.2.1. Finish the proof of Proposition 7.2.2.

Exercise 7.2.2. Prove Corollary 7.2.2.

Exercise 7.2.3. Let T be superstable, $p \in S(A)$, $A' \supset A$ and $p' \in S(A')$ a nonforking extension of p containing a formula φ' such that $\varphi' \in q \implies U(q) \leq U(p')$, for all $q \in S(A')$. Show that there is a $\varphi \in p$ such that $\varphi \in q \implies U(q) \leq U(p)$, for all $q \in S(A)$.

References

[Bal73] John T. Baldwin. α_T is finite for $\aleph_1$-categorical T. *Transactions of the Amer. Math. Soc.*, 181:37–51, 1973.

[Bal88] John T. Baldwin. *Fundamentals of Stability Theory.* Springer-Verlag, Berlin/ Heidelberg / New York, 1988.

[BBGK73] J. Baldwin, A. Blass, A. Glass, and D. Kueker. A 'natural' theory without a prime model. *Algebra Universalis*, 3(2):152–155, 1973.

[BL71] John T. Baldwin and Alistair H. Lachlan. On strongly minimal sets. *J. of Symbolic Logic*, 36:79–96, 1971.

[BS76] J. Baldwin and J. Saxl. Logical stability in group theory. *J. Austral. Math. Soc.*, 21:267–76, 1976.

[Bau76] W. Baur. Elimination of quantifiers for modules. *Israel J. of Math.*, 25:64–70, 1976.

[Ber86] Ch. Berline. Superstable groups; a partial answer to conjectures of Cherlin and Zil'ber. *Ann. of Pure and Applied Logic*, 30:45–61, 1986.

[BL86] Ch. Berline and D. Lascar. Superstable groups. *Ann. of Pure and Applied Logic*, 30:1–43, 1986.

[BM67] G. Birkoff and S. MacLane. *Algebra.* Macmillan, 1967.

[Bou83] E. Bouscaren. Countable models of multidimensional $\aleph_0$-stable theories. *J. of Symbolic Logic*, 48:377–83, 1983.

[BH] E. Bouscaren and E. Hrushovski. Interpreting groups in stable one-based theories. preprint, 1992.

[BL83] E. Bouscaren and D. Lascar. Countable models of non-multidimensional $\aleph_0$-stable theories. *J. of Symbolic Logic*, 48:197–205, 1983.

[Bue85a] Steven Buechler. The geometry of weakly minimal types. *J. of Symbolic Logic*, 50(4):1044–1053, 1985.

[Bue85b] Steven Buechler. Maximal chains in the fundamental order. *J. of Symbolic Logic*, 51(2), 1985.

[Bue86] Steven Buechler. Locally modular theories of finite rank. *Ann. of Pure and Appl. Logic*, 30:83–94, 1986.

[Bue87] Steven Buechler. Classification of small weakly minimal sets I. In J.T. Baldwin, editor, *Classification Theory*, Lecture Notes in Mathematics 1292, pages 32–71. Springer-Verlag, Berlin/ Heidelberg/ New York, 1987.

[Bue91] Steven Buechler. Pseudoprojective strongly minimal sets are locally projective. *J. of Symbolic Logic*, 56(4):1184–1194, 1991.

[Bue93] Steven Buechler. Vaught's conjecture for superstable theories of finite rank. to appear in Ann. of Pure and Appl. Logic, 1993.

[Can95] G. Cantor. Zur Begründung der transfiniten Mengenlehre I. *Mathematische Annalen*, 46:504–06, 1895.

[CK73] C. C. Chang and H. J. Keisler. *Model Theory.* North-Holland, Amsterdam and New York, 1973.

[Che79] G. Cherlin. Groups of small Morley rank. *Ann. Math. Logic*, 17:1–28, 1979.

[CHL85] G. Cherlin, L. Harrington, and A.H. Lachlan. $\aleph_0-$categorical, $\aleph_0-$stable structures. *Annals of Pure and Applied Logic*, 29:103–135, 1985.

[CS80] G. Cherlin and S. Shelah. Superstable fields and groups. *Ann. Math. Logic*, 18:227–70, 1980.

[EM56] A. Ehrenfeucht and A. Mostowski. Models of axiomatic theories admitting automorphisms. *Fund. Math.*, 43:50–68, 1956.

[EF72] P. Eklof and E. Fisher. The elementary theory of abelian groups. *Ann. Math. Logic*, 4:115–71, 1972.

[ES71] P. Eklof and G. Sabbagh. Model-completions and modules. *Ann. Math. Logic*, 2:251–95, 1971.

[End72] Herbert Enderton. *A Mathematical Introduction to Logic.* Academic Press, New York, 1972.

[Eng59] E. Engeler. A characterization of theories with isomorphic denumerable models. *Notices of the Am. Math. Soc.*, 6:161, 1959.

[Eri75] M. M. Erimbetov. Complete theories with 1−cardinal formulas. *Algebra i Logika*, 14:245–57, 1975. In Russian.

[Go30] K. Gödel. Die Vollständigkeit der Axiome des logischen Funktionenkalküls. *Monatshefte für Math. u. Phys.*, 37:349–60, 1930.

[Go31] K. Gödel. Eine Eigenschaft der Realisierungen des Aussagenkalküls. *Ergebnisse Math. Kolloq.*, 3:20–21, 1931.

[GMRN61] A. Grzegorczyk, A. Mostowski, and C. Ryll-Nardzewski. Definability of sets in models of axiomatic theories. *Bull. Acad. Polon. Sci. Sér. Sci. Math. Astronom. Phys.*, 9:163–167, 1961.

[Hal59] M. Hall. *The Theory of Groups.* Macmillan, New York, 1959.

[Har73] Victor Harnik. On the existence of saturated models of stable theories. *Proceedings of the Amer. Math. Soc.*, 52:361–367, 1973.

[Har93] Bradd Hart. Review of "Classification Theory, Revised Edition" by S. Shelah. *J. of Symbolic Logic*, 58:1071–1074, 1993.

[Har80] Robin Hartshorne. *Algebraic Geometry.* Springer-Verlag, 1980.

[Hau62] F. Hausdorff. *Set Theory.* Chelsea, New York, 1962.

[Her92] Alejandro Hernandez. *ω_1-saturated models of stable theories.* PhD thesis, University of California-Berkeley, 1992.

[HLP+92] B. Herwig, J. Loveys, A. Pillay, P. Tanovic, and F. Wagner. Stable theories with no dense forking chains. *Archive for Math. Logic*, 31:297–304, 1992.

[Hod87] Wilfrid Hodges. What is a structure theory? *Bull. London Math. Soc.*, 19:209–37, 1987.

[Hod93] Wilfrid Hodges. *Model Theory*, volume 42 of *Encyclopedia of Mathematics and its Applications.* Cambridge University Press, Camridge, 1993.

[Hru86] E. Hrushovski. *Contributions to stable model theory.* PhD thesis, Univ. of California-Berkeley, 1986.

[Hru87] E. Hrushovski. Locally modular regular types. In J.T. Baldwin, editor, *Classification Theory*, Lecture Notes in Mathematics 1292, pages 132–164. Springer-Verlag, Berlin/ Heidelberg/ New York, 1987.

[Hru89] E. Hrushovski. Almost orthogonal regular types. *Ann. Pure Appl. Logic*, 45:139–55, 1989.

[Hru90a] E. Hrushovski. A new strongly minimal set. *Annals of Pure and Applied Logic*, 50:117–138, 1990.

[Hru90b] E. Hrushovski. Unidimensional theories are superstable. *Annals of Pure and Applied Logic*, 50:117–138, 1990.

[HP87] E. Hrushovski and A. Pillay. Weakly normal groups. In *Logic Colloquium '85*, pages 233–244. North-Holland, Amsterdam and New York, 1987.

[Hun04] E. V. Huntington. The continuum as a type of order: an exposition of the modern theory. *Annals of Mathematics*, 6:178–9, 1904.

[Hyt95] T. Hyttinen. Remarks on structure theorems for ω_1−saturated models. *Notre J. of Formal Logic*, 1995. Spring.

[KM67] H.J. Keisler and M. Morley. On the number of homogeneous models of a given power. *Israel J. of Math.*, 5:73–78, 1967.

[Kni78] Julia F. Knight. Prime and atomic models. *J. of Symbolic Logic*, 43(3):385–393, 1978.

[Lac73] A.H. Lachlan. The number of countable models of a countable superstable theory. In *Proc. of the International Congress on Logic, Methodology and Philosophy of Science, Romania, 1971*, pages 45–56. North-Holland, Amsterdam and New York, 1973.

[Lac75] A. H. Lachlan. Theories with a finite number of models in an uncountable power are categorical. *Pacific J. of Math.*, 61:465–481, 1975.

[Lac80] A. H. Lachlan. Singular properties of Morley rank. *Fundamenta Mathematica*, 108:145–57, 1980.

[Las76] Daniel Lascar. Rank and definability in superstable theories. *Israel J. of Math.*, 23:53–87, 1976.

[Las82] Daniel Lascar. Ordre de Rudin-Keisler et poids dans les théories ω−stable. *Zeit. Math. Logik Grund. Math.*, 28(5):411–430, 1982.

[Las84] Daniel Lascar. Relation entre le rang U et le poids. *Fund. Math.*, 121(2):117–123, 1984.

[Las85] Daniel Lascar. Why some people are excited by Vaught's conjecture? *J. of Symbolic Logic*, 50:973–982, 1985.

[Las86] Daniel Lascar. *Stability in Model Theory.* Longman Scientific and Technical, Essex, 1986. Translated from the French.

[LP79] D. Lascar and B. Poizat. An introduction of forking. *J. of Symbolic Logic*, 44:330–350, 1979.

[Las88] M. C. Laskowski. Uncountable theories that are categorical in a higher power. *J. of Symbolic Logic*, 53(2):512–530, 1988.

[Lo15] L. Löwenheim. Über Möglichkeiten im Relativkalkül. *Math. Ann.*, 76:447–470, 1915.

[Mac71a] A. Macintyre. On ω_1−categorical theories of abelian groups. *Fund. Math.*, 70(3):253–270, 1971.

[Mac71b] A. Macintyre. On ω_1−categorical theories of fields. *Fund. Math.*, 71:1–25, 1971.

[Mak84] M. Makkai. A survey of basic stability theory. *Israel J. of Math.*, 49(1-3):181–238, 1984.

[Mal36] A. I. Mal'tsev. Untersuchungen aus dem Gebiete der mathematischen Logik. *Mat. Sbornik*, 1(43):323–36, 1936.

[Mar66] W. E. Marsh. *On ω_1−categorical and not ω−categorical theories.* PhD thesis, Dartmouth College, 1966.

[Mor65] Michael D. Morley. Categoricity in power. *Transactions of the A.M.S.*, 114:514–538, 1965.

[Mor70] Michael D. Morley. The number of countable models. *J. of Symbolic Logic*, 35:14–18, 1970.

[MV62] M. Morley and R. Vaught. Homogeneous universal models. *Math. Scand.*, 11:37–57, 1962.

[NP89] Ali Nesin and Anand Pillay. *The model theory of groups*, volume 11 of *Notre Dame Mathematical Lectures.* Notre Dame Press, 1989.

[New90] L. Newelski. A proof of Saffe's conjecture. *Fundamenta Mathematica*, 134:143–155, 1990.

[Pil] A. Pillay. *Geometrical Stability Theory*. Oxford University Press, to appear.

[Pil82] A. Pillay. Weakly homogeneous models. *Proc. of the Amer. Math. Soc.*, 86:126–32, 1982.

[Poi81] Bruno Poizat. Soub-groupes definissable d'un groupe stable. *J. of Symbolic Logic*, 46(1):137–46, 1981.

[Poi83a] Bruno Poizat. Groupes stables, avec types génériques réguliers. *J. of Symbolic Logic*, 48(2):339–55, 1983.

[Poi83b] Bruno Poizat. Une théorie de Galois imaginaire. *J. of Symbolic Logic*, 48:1151–1170, 1983.

[Poi87] Bruno Poizat. *Groupes stables*. Nur al-Mantiq wal-Ma'rifah, Villeurbanne, 1987.

[Pre88] M. Prest. *Model Theory and Modules*. London Math. Soc. Lecture Note Series 130. Cambridge University Press, 1988.

[Ram30] F.P. Ramsey. On a problem of formal logic. *Proc. London Math. Soc. Ser. 2*, 30:264–286, 1930.

[RN59] C. Ryll-Nardzewski. On categoricity in power $\leq \aleph_0$. *Bull. Acad. Polon. Sci. Math. Astron. Phys.*, 7:545–548, 1959.

[Row64] F. Rowbottom. The Łoś conjecture for uncountable theories. *Not. Amer. Math. Soc.*, 11(248):248, 1964.

[Saf84] J. Saffe. Categoricity and ranks. *J. of Symbolic Logic*, 49(4):1379–92, 1984.

[She69] Saharon Shelah. Stable theories. *Israel J. of Math.*, 7:187–202, 1969.

[She71] Saharon Shelah. Stability, the f.c.p. and superstability; model theoretic properties of formulas in the first order theory. *Annals of Math. Logic*, 3:271–362, 1971.

[She72] Saharon Shelah. Uniqueness and characterization of prime models over sets for totally transcendental first order theories. *J. of Symbolic Logic*, 37:107–113, 1972.

[She85] Saharon Shelah. Classification of first order theories which have a structure theory. *Bulletin of the American Mathematical Society*, 12:227–232, 1985.

[She90] Saharon Shelah. *Classification Theory and the number of nonisomorphic models*. North-Holland, Amsterdam and New York, 1990. Revised Edition.

[SB89] S. Shelah and S. Buechler. On the existence of regular types. *Annals of Pure and Applied Logic*, 45:277–308, 1989.

[SHM84] S. Shelah, L. Harrington, and M. Makkai. A proof of Vaught's conjecture for ω−stable theories. *Israel J. of Math.*, 49:259–286, 1984.

[Sho67] Joseph Shoenfield. *Mathematical Logic*. Addison-Wesley, 1967.

[Sko20] T. Skolem. Logisch-kombinatorische untersuchungen über die erfül-lbarkeit oder beweisbarkeit mathematischer sätze nebst einem theorem über dichte mengen. *Skrifter utgitt av Videnskapsselskapet i Kristiana, I, Math. Naturv. Kl.*, 4, 1920.

[Sve59] L. Svenonius. $\aleph_0$−categoricity in first-order predicate calculus. *Theoria (Lund)*, 25:82–94, 1959.

[Szm55] W. Szmielew. Elementary properties of abelian groups. *Fund. Math.*, 41:203–71, 1955.

[TV57] A. Tarski and R. L. Vaught. Arithmetical extensions of relational systems. *Comp. Math.*, 13:81–102, 1957.

[Vau61] R. Vaught. Denumerable models of complete theories. In *Infinitistic Methods*, pages 303–321. Pregamon, London, 1961.

[VdW49] B. L. Van der Waerden. *Modern Algebra*. Frederick Ungar Publishing Co., New York, 1949.

[Zie84] M. Ziegler. Model theory of modules. *Ann. of Pure and Appl. Logic*, 26:149–213, 1984.

[Zil74] B. I. Zil'ber. On the transcendence rank of formulas of an $\aleph_1$−categorical theory. *Mat. Zametki*, 15(2):321–9, 1974.

[Zil77a] B. I. Zil'ber. The construction of models of categorical theories and the problem of finite axiomatizability. Kemerovo State University, 1977.

[Zil77b] B. I. Zil'ber. Groups and rings whose theory is categorical. *Fund. Math.*, 95:173–88, 1977. in Russian.

[Zil80] B. I. Zil'ber. Strongly minimal countably categorical theories. *Sibirsk. Mat. Zh.*, 21:98–112, 1980.

[Zil84a] B. I. Zil'ber. Strongly minimal countably categorical theories. II. *Sibirsk. Mat. Zh.*, 25(3):71–88, 1984.

[Zil84b] B. I. Zil'ber. Strongly minimal countably categorical theories. III. *Sibirsk. Mat. Zh.*, 25(4):63–77, 1984.

[Zil84c] B. I. Zil'ber. The structure of models of uncountably categorical theories. In *Proc. Internat. Congress of Math., 1983*, pages 359–68. Warsaw: Panstwowe Wydawnictwo Naukowe, 1984.

[Zil91] B. I. Zil'ber. Groups and rings whose theory is categorical. *A.M.S. Translations*, 149:1–16, 1991.

[Zil93] B. I. Zil'ber. *Uncountably categorical theories*, volume 117 of *Translations of Mathematical Monographs*. American Mathematical Society, 1993.

Index

www.ingramcontent.com/pod-product-compliance
Lightning Source LLC
LaVergne TN
LVHW102009210926
848483LV00010B/235